Molecular Mechanisms of Synaptic Plasticity: Dynamic Changes in Neurons Functions

Molecular Mechanisms of Synaptic Plasticity: Dynamic Changes in Neurons Functions

Editor

Giuseppina Martella

MDPI

Basel • Beijing • Wuhan • Barcelona • Belgrade • Novi Sad • Cluj • Manchester

Editor
Giuseppina Martella
Department of Human Sciences
Pegaso Telematic University
Rome
Italy

Editorial Office
MDPI
St. Alban-Anlage 66
4052 Basel, Switzerland

This is a reprint of articles from the Special Issue published online in the open access journal *International Journal of Molecular Sciences* (ISSN 1422-0067) (available at: www.mdpi.com/journal/ijms/special_issues/mechanisms_synaptic_plasticity).

For citation purposes, cite each article independently as indicated on the article page online and as indicated below:

Lastname, A.A.; Lastname, B.B. Article Title. *Journal Name* **Year**, *Volume Number*, Page Range.

ISBN 978-3-0365-8715-8 (Hbk)
ISBN 978-3-0365-8714-1 (PDF)
doi.org/10.3390/books978-3-0365-8714-1

Contents

About the Editor

Giuseppina Martella

Giuseppina Martella currently works as an associate professor of Physiology at the Pegaso Telematic University, Department of Human Sciences. She also continues her work at the Laboratory of Physiology and Plasticity, Fondazione Santa Lucia, IRCCS, Rome. She received her Master's Degree in Biology with a focus on Pathophysiology from the University of Rome Tor Vergata. She continued her training in the Neurophysiology laboratories of Professor Giorgio Bernardi.

She has been interested in neurological pathologies since 2000. She then obtained her specialization in Clinical Pathology in 2007, continuing to work on movement disorders under the guidance of Professors Stefani, Calabresi, and Pisani. In 2012, she received her doctorate in Neuroscience. She has been the recipient of numerous fellowships and research grants.

Since 2014, she has been devoted to the study of Synaptic Plasticity phenomena and has focused her attention on Parkinson's Disease and Genetic Dystonias.

She is the author of more than 70 papers in international journals and is the recipient of several Research Awards and Grants.

Her major goal is the knowledge of physiological mechanisms and their alterations in Neurological disorders; to achieve this goal, she is currently collaborating with other researchers on translational projects that start from the molecular and continue in the patient.

Preface

Synaptic plasticity is a complex and crucial neuronal mechanism linked to principal memory and motor functions. During the developmental period into old age, the neural frame is subject to structural and functional modifications in response to external stimuli. This essential skill of neuronal cells underpins the ability to learn about mammalian organisms (Glanzman et al., 2010).

Synaptic plasticity phenomena include microscopic changes such as spine pruning and macroscopic changes such as the cortical remapping response to injury (Citri and Malenka, 2008; Hofer et al., 2009). The increase in neurological and neuropsychiatric disorders in the current century—although not occurring among the aging population—has resulted in a greater urgency to understand the aberrant processes connected to these diseases (Martella et al., 2016; 2018; Bonsi et al., 2018).

Giuseppina Martella
Editor

International Journal of *Molecular Sciences*

MDPI

Editorial

Molecular Mechanisms of Synaptic Plasticity: Dynamic Changes in Neuron Functions

Giuseppina Martella [1,2]

[1] Laboratory of Neurophysiology and Plasticity, IRCCS Fondazione Santa Lucia, 00143 Rome, Italy; g.martella@hsantalucia.it

[2] Department of Human Sciences, Faculty of Umanities Educations and Sports, Pegaso University, 80143 Naples, Italy

The human brain has hundreds of billions of neurons and at least 7 million dendrites have been hypothesized to exist for each neuron, with over 100 trillion neuron–neuron, neuron–muscle, and neuron–endocrine cell synapses [1,2].

Our body continually receives stimuli from the outer environment, and our brain's ability to respond to these stimuli is ensured through synaptic processes, motivating the foundations of this Special Issue.

This issue aims to underline the role of synaptic plasticity phenomena in our body, and clarify the mechanism operated by neurons to guarantee these phenomena. The collection of the issue comprises 14 papers, including 8 reviews and 6 original works, of which is a protocol for differentiating neurons from human stem cells and 5 are preclinical works.

Of these preclinical manuscripts, one is focalized on the glucocorticoids that may alter the gene and protein expression in catecholamine neurons. Using organotypic cultures incubated with the neurosteroid corticosterone, the authors of the paper demonstrated significant increases in tyrosine hydroxylase and dopamine transporter; conversely, modifications were recorded in phenylethanolamine N-methyltransferase. This demonstrated that dopamine signals may carry out regulations via internal glucocorticoid secretion [3].

In a mouse model of Parkinson's disease, Imbriani and coworkers showed that caspase-3, considered an effector caspase involved in neuronal apoptosis, has a hormesis-based double role in PINK1-knockout (KO) mice. In other words, when this caspase has a lower level of activation, it modulates physiological phenomena, as well as corticostriatal LTD, and only at higher activation levels is it conducive to the apoptotic process. In PINK1-KO mice, lower caspase-3 activation is capable of rescuing a defective LTD by promoting dopamine release [4].

Professor Petrosini's group demonstrates that the pathway of the hippocampus, dorsal striatum, and amygdala shows a high expression of receptors for cannabinoid type 1 (CB1). The presence of these receptors in areas deputed to spatial learning depicts the involvement of cannabinoid in navigational strategies. The authors demonstrated that treatment with the CB1 antagonist, AM251, impaired spatial learning and modified the pattern of performed navigational strategies in a mouse model of Gaucher disease. The strength of the synaptic plasticity on the behavioral output is modulated by CB1 receptors, which leads to spatial navigational strategies [5].

Dystonia is a movement disorder characterized by the co-contraction of agonist and antagonistic muscles that cause body twisting and related pain. The pathophysiology of the disease was partly related to the downward regulation and dysfunction of the dopamine D2 receptor in the striatum. Using a mouse model of DYT1 dystonia, D'Angelo and collaborators found that DYT1 mutant mice present a reduced expression of D2 receptors associated with a marked reduction in the number and size of D2-positive synapses at the level of striatal area. This rearrangement causes a longer duration and greater sphere of

Citation: Martella, G. Molecular Mechanisms of Synaptic Plasticity: Dynamic Changes in Neuron Functions. *Int. J. Mol. Sci.* **2023**, *24*, 12567. https://doi.org/10.3390/ijms241612567

Received: 2 August 2023
Accepted: 7 August 2023
Published: 8 August 2023

influence of dopamine transmission, explaining the electrophysiological characteristics of this pathology [6].

SNAP-25 is a component of the SNARE complex of proteins that have the function of coupling the synaptic vesicle and plasma membranes together, in order to favor the neurotransmitter release in the nervous system. Moreover, SNAP-25 influences synaptic plasticity in the first stages of life.

In mice of 30 days, transgenic for an unmatured isoform of SNAP-25b, Schaffer collateral CA1 synapses had a much faster kinetic release with a consequent decrement in LTP and an enhancement in LTD amplitude. Conversely, mice four-month-old mice showed that the learning process avoiding LTD upregulation rescued physiological plasticity. These data explain the correlation between long-term synaptic plasticity and cognitive learning ability [7].

Starting from two neural stem cell lines from healthy patients, Capetian et al. have identified a cultural technique that integrates and contrasts the best neuronal elements in their protocol paper. Differentiated neuronal stem cells were incorporated into a highly concentrated Matrigel (neurosphere) droplet. After inclusion, neurospheres were treated with a combination of glutaraldehyde and paraformaldehyde using various membrane combinations to optimize imaging via transmission electron microscopy. This new method represents the best way to analyze synaptic structures [8].

The reviews in this Special Issue aim to highlight the synaptic capacities of our brain in relation to their molecular mechanisms, emphasizing structural and functional changes due to plasticity.

Inhibitory networks involved in synaptic plasticity were shown to be regulated by a vast number of pathways in their synaptic connections. Because inhibitory plasticity was characterized later than excitatory synaptic plasticity, this literature revision proposed by Mapelli represents a new and interesting point of view to understand synaptic mechanisms. The authors proposed an overview of inhibitory plasticity in different brain areas [9].

Through an in-depth study of synapses, it was found that the neurocentric model, encompassing the neuronal pre- and post-synaptic terminals and the synaptic cleft, is not able to explain all the fine-tuned plastic modifications visualized in pathological and physiological events. For this reason, the tripartite synapse model including oligodendrocytes, astrocytes, and microglia was proposed. Moreover, physiological synaptic plasticity and maladaptive plasticity commonly showed a deep connection with other molecular elements of the extracellular matrix.

Alongside that, a new model of the tetrapartite synapse, including the neurovascular unit and the immune system, was recently proposed. The latter appears more congruent with the different mechanisms of physiologic adaptive and maladaptive plasticity. However, a better interpretation may be granted by the construction of predictive molecular models [10].

N-methyl-d-aspartate glutamate receptors (NMDARs) are tetramers composed of different homologous subunits of GluN1-, GluN2-, or GluN3-type, able to generate a high variety of receptor subtypes with various pharmacological and signaling properties. Their subunit composition is regulated during the development of activity-dependent synaptic plasticity. Through their literature revision, Franchini and coworkers showed the role of these subunits in both physiology and pathological synaptic plasticity [11].

Recently, 5-HT7R serotonin receptors have been shown to be involved in reshaping neuronal cytoarchitecture in many neurodevelopmental disorders. Over the past few years, the synaptic plasticity of this receptor has been demonstrated in the ILD and LTP forms. Furthermore, 5-HT7R contributes to inflammatory bowel disease. Perrone-Capano's group offers a review in which the new aspects highlighted in the digestive tract and immune system have been clarified [12].

Neuregulins (NRGs), a family of proteins acting on tyrosine kinase receptors of the ErbB family play an essential role in the development of the nervous system. They also contribute to the functioning of the adult brain through synaptic plasticity. Ledonne and

Mercuri presented a literature review that emphasizes the role of GRN signaling in modulating synaptic plasticity. Furthermore, they suggest that NRG-dependent dysregulation of synaptic plasticity could be implicated in numerous pathophysiological diseases [13].

Anti-homeostatic synaptic plasticity, first LTP, represents one of the main physiological mechanisms used by our brain in response to brain damage. Both anti-homeostatic and homeostatic synaptic mechanisms contribute to shaping brain networks. In multiple sclerosis, inflammatory mediators induce altered synaptic functioning that corresponds to the addition of demyelination and gray matter atrophy. In the revision proposed by Stampanoni Bassi and colleagues, an altered LTP expression was shown to contribute to disrupting the brain network [14].

Different forms of synaptic plasticity have been established: anti-homeostatic (i.e., Hebbian) and homeostatic plasticity (i.e., synaptic scaling). The balance between these forms is necessary to correspond to the architecture of the brain system.

In the review made by Stampanoni Bassi and co-workers, the key features of the brain network architecture were introduced, which resulted from fine-tuning between two different forms of synaptic plasticity [15].

Inflammatory/immunity mediators are closely related to diseases of the central nervous system. Moreover, these mediators are involved in synaptic plasticity pathways and synaptic plasticity is hardly affected by the immune system. Additionally, the immune system is linked to synaptic processes.

The two main custom clearance systems are autophagy and ubiquitin–protease (UPS). In their review, Fornai et al. discuss the role of autophagy and UPS in connecting immunity with synaptic plasticity in health and disease [16].

At present, the increase in neurological and neuropsychiatric diseases urges understanding the function of the brain network and the fine mechanism underlying brain dysfunction, which this Special Issue aims to satisfy.

Conflicts of Interest: The author declares no conflict of interest.

References

1. Herculano-Houzel, S. The Human Brain in Numbers: A Linearly Scaled-up Primate Brain. *Front. Hum. Neurosci.* **2009**, *3*, 31. [CrossRef] [PubMed]
2. Colbran, R.J. Thematic Minireview Series: Molecular Mechanisms of Synaptic Plasticity. *J. Biol. Chem.* **2015**, *290*, 28594–28595. [CrossRef] [PubMed]
3. Busceti, C.L.; Ferese, R.; Bucci, D.; Ryskalin, L.; Gambardella, S.; Madonna, M.; Nicoletti, F.; Fornai, F. Corticosterone Upregulates Gene and Protein Expression of Catecholamine Markers in Organotypic Brainstem Cultures. *Int. J. Mol. Sci.* **2019**, *20*, 2901. [CrossRef] [PubMed]
4. Imbriani, P.; Tassone, A.; Meringolo, M.; Ponterio, G.; Madeo, G.; Pisani, A.; Bonsi, P.; Martella, G. Loss of Non-Apoptotic Role of Caspase-3 in the PINK1 Mouse Model of Parkinson's Disease. *Int. J. Mol. Sci.* **2019**, *20*, 3407. [CrossRef] [PubMed]
5. Laricchiuta, D.; Balsamo, F.; Fabrizio, C.; Panuccio, A.; Termine, A.; Petrosini, L. CB1 Activity Drives the Selection of Navigational Strategies: A Behavioral and c-Fos Immunoreactivity Study. *Int. J. Mol. Sci.* **2020**, *21*, 1072. [CrossRef] [PubMed]
6. D'angelo, V.; Giorgi, M.; Paldino, E.; Cardarelli, S.; Fusco, F.R.; Saverioni, I.; Sorge, R.; Martella, G.; Biagioni, S.; Mercuri, N.B.; et al. A2a Receptor Dysregulation in Dystonia Dyt1 Knock-out Mice. *Int. J. Mol. Sci.* **2021**, *22*, 2691. [CrossRef] [PubMed]
7. Gopaul, K.R.; Irfan, M.; Miry, O.; Vose, L.R.; Moghadam, A.; Subah, G.; Hökfelt, T.; Bark, C.; Stanton, P.K. Developmental Time Course of SNAP-25 Isoforms Regulate Hippocampal Long-Term Synaptic Plasticity and Hippocampus-Dependent Learning. *Int. J. Mol. Sci.* **2020**, *21*, 1448. [CrossRef] [PubMed]
8. Capetian, P.; Müller, L.; Volkmann, J.; Heckmann, M.; Ergün, S.; Wagner, N. Visualizing the Synaptic and Cellular Ultrastructure in Neurons Differentiated from Human Induced Neural Stem Cells—An Optimized Protocol. *Int. J. Mol. Sci.* **2020**, *21*, 1708. [CrossRef] [PubMed]
9. Gandolfi, D.; Bigiani, A.; Porro, C.A.; Mapelli, J. Inhibitory Plasticity: From Molecules to Computation and Beyond. *Int. J. Mol. Sci.* **2020**, *21*, 1805. [CrossRef] [PubMed]
10. De Luca, C.; Colangelo, A.M.; Virtuoso, A.; Alberghina, L.; Papa, M. Neurons, Glia, Extracellular Matrix and Neurovascular Unit: A Systems Biology Approach to the Complexity of Synaptic Plasticity in Health and Disease. *Int. J. Mol. Sci.* **2020**, *21*, 1539. [CrossRef] [PubMed]
11. Franchini, L.; Carrano, N.; Di Luca, M.; Gardoni, F. Synaptic GluN2A-Containing NMDA Receptors: From Physiology to Pathological Synaptic Plasticity. *Int. J. Mol. Sci.* **2020**, *21*, 1538. [CrossRef] [PubMed]

12. Crispino, M.; Volpicelli, F.; Perrone-Capano, C. Role of the Serotonin Receptor 7 in Brain Plasticity: From Development to Disease. *Int. J. Mol. Sci.* **2020**, *21*, 505. [CrossRef] [PubMed]
13. Ledonne, A.; Mercuri, N.B. On the Modulatory Roles of Neuregulins/ErbB Signaling on Synaptic Plasticity. *Int. J. Mol. Sci.* **2019**, *21*, 275. [CrossRef] [PubMed]
14. Stampanoni Bassi, M.; Iezzi, E.; Gilio, L.; Centonze, D.; Buttari, F. Synaptic Plasticity Shapes Brain Connectivity: Implications for Network Topology. *Int. J. Mol. Sci.* **2019**, *20*, 6193. [CrossRef] [PubMed]
15. Stampanoni Bassi, M.; Iezzi, E.; Pavone, L.; Mandolesi, G.; Musella, A.; Gentile, A.; Gilio, L.; Centonze, D.; Buttari, F. Modeling Resilience to Damage in Multiple Sclerosis: Plasticity Meets Connectivity. *Int. J. Mol. Sci.* **2019**, *21*, 143. [CrossRef] [PubMed]
16. Limanaqi, F.; Biagioni, F.; Busceti, C.L.; Ryskalin, L.; Soldani, P.; Frati, A.; Fornai, F. Cell Clearing Systems Bridging Neuro-Immunity and Synaptic Plasticity. *Int. J. Mol. Sci.* **2019**, *20*, 2197. [CrossRef] [PubMed]

International Journal of *Molecular Sciences*

MDPI

Review

Inhibitory Plasticity: From Molecules to Computation and Beyond

Daniela Gandolfi [1,2], Albertino Bigiani [1], Carlo Adolfo Porro [1] and Jonathan Mapelli [1,*]

1 Department of Biomedical, Metabolic and Neural Sciences and Center for Neuroscience and Neurotechnology, University of Modena and Reggio Emilia, Via Campi 287, 41125 Modena, Italy; daniela.gandolfi82@gmail.com (D.G.); albertino.bigiani@unimore.it (A.B.); carlo.porro@unimore.it (C.A.P)

2 Department of Brain and behavioral sciences, University of Pavia, 27100 Pavia, Italy

* Correspondence: jonathan.mapelli@unimore.it; Tel.: +39-059-205- 5459

Received: 31 January 2020; Accepted: 3 March 2020; Published: 6 March 2020

Abstract: Synaptic plasticity is the cellular and molecular counterpart of learning and memory and, since its first discovery, the analysis of the mechanisms underlying long-term changes of synaptic strength has been almost exclusively focused on excitatory connections. Conversely, inhibition was considered as a fixed controller of circuit excitability. Only recently, inhibitory networks were shown to be finely regulated by a wide number of mechanisms residing in their synaptic connections. Here, we review recent findings on the forms of inhibitory plasticity (IP) that have been discovered and characterized in different brain areas. In particular, we focus our attention on the molecular pathways involved in the induction and expression mechanisms leading to changes in synaptic efficacy, and we discuss, from the computational perspective, how IP can contribute to the emergence of functional properties of brain circuits.

Keywords: synaptic plasticity; inhibition; computational neuroscience; GABA; LTP; LTD

---

1. Introduction

The emergence of brain functions, from motion control to cognition and abstract thinking, is tightly bound to the ability of brain circuits to adjust synaptic connections [1]. For a long time, this capability was hypothesized to rely exclusively on the adaptability of excitatory synapses, assuming the substantial invariance of inhibitory connections. Only recently, a wide number of molecular and cellular mechanisms residing at inhibitory synapses and responsible for the emergence of complex brain states are beginning to be unraveled [2]. It is, in fact, evident that inhibitory synapses throughout the brain exhibit activity-dependent changes of their connectivity weights both in the form of long-term potentiation (LTP) and long-term depression (LTD). Nevertheless, the investigation of activity-dependent changes of inhibitory synapses has been traditionally prevented, or strongly limited by the wide number of GABAergic cell types and the consequent difficulty of isolating specific neuronal pathways [3]. The extraordinary variety of inhibitory interneurons and related connections expands the number of plasticity subtypes, which can be expressed. Importantly, the impact of IP on neuronal circuits, by acting on the overall neuronal excitability and on the possibility to further induce persistent forms of plasticity, can have important consequences on the brain functionality [4]. There is indeed increasing attention on the various forms of IP since circuit refinement induced by changes of the Excitatory/Inhibitory (E/I) balance can strongly influence learning and memory. In particular, the advent of sophisticated techniques either in the form of molecular, genetic, or electrophysiological and imaging approaches has started to allow precise dissection of microcircuits and of single synaptic types [5]. Moreover, long-term changes in inhibitory activity can induce pathological alterations of brain functions, while several neuropsychiatric disorders have been shown to be related to permanent GABAergic dysregulations [6,7].

In this article, we discuss how the various forms of IP shape brain circuits architectures and functions with particular attention on neuronal computation. We begin with a panoramic view on the variety of GABAergic cell types and circuits to continue with an analysis of the induction and expression mechanisms of inhibitory LTP and LTD. We then proceed with discussing the functional consequences of the aforementioned mechanisms, including computational models proposed to predict the effects of IP. Finally, we analyze the potential involvement of inhibitory plasticity in the emergence of pathological disorders to end with a few suggestions regarding the implementation of synaptic learning rules into artificial circuits.

2. Variety of Inhibitory Circuits in the Central Nervous System

Although it is widely accepted that inhibitory neurons are actively engaged in network computation by providing global stability to network dynamics, by controlling the degree of circuit synchronization and by controlling the timing of neuronal firing [8], it is still under debate the way fine-tuning of inhibitory connections participate into regulatory mechanisms of circuit dynamics. Certainly, information processing is strictly dependent on how excitation and inhibition are in balance with each other, engaging directional and recurrent wired networks, implementing computational functions like the expansion of the dynamic range of neuronal responses [8], the input separation through winner-take-all schemes [9], or spatial pattern separation through combinatorial operators [10]. Nevertheless, even the lack of a precise definition of the concept of balance strongly limits the possibility to fully understand the interplay between excitatory and inhibitory signaling. For instance, the temporal and spatial scales over which neuronal activity is tuned by the interplay between glutamatergic and GABAergic synapses affect the timing of spike generation [11–13] rather than the average firing rate [14] or the synchronization of local [15] and global networks [16].

One of the major obstacles in identifying the way inhibitory plasticity tunes the activity of brain circuits comes from the diversity of inhibitory interneurons in the CNS, which still leaves these cells the capacity to provide inhibition to a large variety of excitatory input classes [3]. Among neuronal categories, almost 20% is GABAergic [17] and show a wide variety of functional and molecular subtypes with variable locations and, interestingly, capacities for plasticity [2,18–20]. In some cases, even the classification of interneurons as exclusively inhibitory is technically difficult because, in early development, GABA can act as depolarizer [21], and into adulthood, some axo-axonic contacts may continue with this behavior [22].

Several attempts have been indeed made to classify cortical interneurons, for instance, the so-called "Petilla terminology" [23], by collecting features describing interneurons, classifies GABAergic cells following (i) morphological and (ii) molecular properties. A major effort has been made indeed to generate clusters of inhibitory neurons based on their gene expression. Although an attempt in classifying interneurons has been made by considering single properties independently, the amount of subtypes emerging from considering all the possible combinations increases dramatically. More in detail, recent works have identified at least 10 distinct classes of inhibitory neurons in the hippocampal circuits [24] with more than 30 subclasses. Although most of them show overlapping functional and computational properties, the analysis of the contribution of inhibitory neurons and plasticity in circuit computation has to deal with this heterogeneity. When focusing the attention only on cortical networks, the variety of GABAergic neurons has been investigated under the molecular, morphological, and functional points of view, as well as on the ability to undergo synaptic plasticity [3]. Furthermore, a critical issue regards the computational capacity of interneurons deriving from the organization of synaptic connectivity. For instance, chandelier cells, interneuron with the anatomical property of embracing the hillock of target neurons, can implement a simple modulation of the action potential generation in principal neurons (PNs) by exploiting synaptic contacts on the initial tract of axonal segments [25]. Conversely, basket cells provide inhibition to cell bodies and proximal dendrites of PNs.

The strategic location of efferents and the confined segregation of synapses allow controlling spike timing, oscillations, and integrative functions such as orientation selectivity or refinement of sensory maps [26] by exploiting peri-somatic innervation. Back to the molecular distinction, the neuronal subtypes identified in GABAergic interneuron classes can be, however, resumed into four main groups resulting from the non-overlapping neuronal clustering correlated to the expression of different markers: a) ParValbumin (PV), b) cholecystokinin (CCK) c) the co-transmitter SOMatostatin (SOM) and d) serotonin receptor 5-HT3A (5-HT). The main electrophysiological feature of PV cells is a fast-spiking firing enabling a powerful control of timing and rate of spike output from postsynaptic neurons [27]. Furthermore, PVs interneurons are the main actors in driving oscillations in cortical circuits [28,29]. The CCK interneurons, particular inhibitory subtypes with characteristics similar to PVs cells in terms of the anatomical organization of efferents, show regular firing providing fine control of the postsynaptic activity of PNs of both neocortical and hippocampal regions [30]. Even though CCK and PV neurons show differences, they share similar peri-somatic inhibitory properties, hence allowing the clustering in the same functional group. This aspect is particularly important in terms of the computational effects of IP because the morphological differences between neurons belonging to the same molecular class are responsible for the heterogeneity in the firing patterns. The peculiar electrophysiological characteristics such as membrane time constant, membrane capacity, and resistance, as well as leakage or conductance are critically involved in synaptic integration. Differently from PVs and CCKs, SOM neurons preferentially contact dendrites either on spines or shafts [31]. The main characteristic of SOM inhibitory interneurons are their action against the spatio-temporal diffusion of signals in dendrites [32]. By exerting their activity at dendritic level, SOM neurons regulate i) dendritic calcium fluxes, which are in turn involved into the induction mechanisms of LTP and LTD [33], ii) the insurgence of dendritic spikes driving neurons to somatic bursting activity and iii) other forms of stereotyped network activity [34]. The last class of inhibitory neurons (5-HT3A) can be subdivided into neuroglia form, inhibiting dendrites of excitatory neurons and thus massively suppressing circuit activity [35] and cells expressing vasoactive intestinal peptide (VIP) which exert their actions mainly versus other inhibitory circuits with paradoxical excitatory consequences on the overall network activity [36]. It appears evident that, despite the wide spectrum of molecular and morphological subtypes, inhibition is mainly exploited by the anatomical organization of afferent contacts, which can be alternatively perisomatic or dendritic. This distinction, in turn, leads to the determination of functional properties that can result in the emerge of the mechanisms leading to long-term plasticity induction and expression.

When considering the functional differences between GABAergic classes, it is evident that the shown diversity can be one of the main causes of the inhibitory capability to stabilize and finely regulate circuits activity [37]. Furthermore, since inhibitory neurons create a wide network of electrically and synaptically coupled cells with an exceptional variety of physiological and anatomical properties, it is widely accepted that inhibition cannot be merely considered as a regulator of circuit excitability [38,39]. Additionally, the reciprocal and recurrent broad connections between inhibitory and excitatory neurons are ideal to condition large circuits areas [40]. Several GABAergic neurons, in fact, widely connect excitatory cells while locally contact inhibitory neurons exploiting an interplay between excitation and inhibition in broad neural circuits [41], which is essential to maintain the circuit balance that favors neural computation [42].

3. Induction and Expression Mechanisms

Given the diversity of inhibitory classes, it can be envisaged that inhibitory plasticity presents heterogeneous molecular and functional characteristics accordingly. Interestingly, as observed for excitatory synapses, IP was shown to occur as changes in the presynaptic release, in postsynaptic $GABA_A$ receptors ($GABA_A$Rs) activity or in mixed forms [43]. Modifies at the presynaptic side require retrograde signaling that can persistently modulate GABA release [44], whereas purely postsynaptic mechanisms involve alterations of GABA receptors machinery [45]. The modulation of neurotransmitter vesicles release from presynaptic boutons is mainly triggered by heterosynaptic mechanisms, thus requiring non-GABAergic stimuli [44,46] from nearby synapses, and actually, the activation of inhibitory fibers is indeed not required. In order for this to happen, signals must be communicated to the presynaptic terminals following the postsynaptic induction, which can involve a wide series of mechanisms. The easiest and the most used strategy is via diffusible molecules acting as retrograde messengers [47]. Alternatively, the glutamate released during the induction process can be directly spread toward GABAergic terminals to induce changes in vesicles release through the activation of presynaptic receptors [48]. In the majority of the reported cases, heterosynaptic inhibitory plasticity was induced by high-frequency or theta-burst stimulation of excitatory axonal terminals [49]. The first forms of potentiation presynaptically expressed through the glutamatergic action of excitatory fibers were reported in the primary visual cortex and in the cerebellum. The high-frequency stimulation of layer 4 induced LTP of GABAergic inhibitory postsynaptic potentials (IPSPs) in layer 5 pyramidal neurons [50], through a not well-identified mechanism involving N-methyl-D-aspartate (NMDA) receptors in both the pre- and postsynaptic components. Interestingly, this form of IP remains one of the most investigated forms of plasticity given its importance in the determination of the E/I balance in the developing visual cortex [51]. Similarly, the stimulation of cerebellar glutamatergic climbing fibers, bringing the teaching error for motor learning to occur, induces a calcium-dependent long-lasting potentiation of inhibitory postsynaptic currents (IPSCs) in Purkinje cells mediated by molecular layer interneuron [52]. In the subsequent years, heterosynaptic LTP at the inhibitory connections has been discovered in several other brain areas, including the Ventral Tegmental Area [44], lamina I of the spinal cord [53], neonatal hippocampus [54] and basolateral amygdala [55]. Interestingly, heterosynaptic inhibitory potentiation shows strong similarities with the classical form of LTP discovered in the hippocampus in the early 70s: synapse specificity, associativity, calcium signaling, and dependence on NMDA receptors activation [56]. As in the case of presynaptic excitatory LTP [57], the potentiation of inhibitory synapses reported so far requires signals to be conveyed to presynaptic terminals following postsynaptic induction (Figure 1). The retrograde messengers, a class of molecules produced in the postsynaptic cell in an activity-dependent manner and traveling backward, are essential to modulate neurotransmitter release, therefore, allowing the expression of LTP. Various pathways involving retrograde messengers and participating in the triggering of inhibitory plasticity have been identified; however, two main molecules modulating heterosynaptic LTP are more recurrently found in different brain regions: i) the diffusible nitric oxide (NO) [44,46] and the brain-derived neurotrophic factor (BDNF) [58,59]. The NO originates in the postsynaptic compartment following calcium entry (Figure 1B). The activation of the Nitric Oxide Synthase (NOS) following postsynaptic calcium rise catalyzes NO, which can freely diffuse in the extracellular matrix by virtue of its gaseous nature. The NO then triggers cGMP, and eventually, other molecular targets in the presynaptic boutons [60]. The BDNF is involved in the regulation of neurogenesis, activity-dependent synaptic plasticity, and in other non-neuronal mechanisms [61]. It binds to its high affinity tyrosine kinase B (TrkB) receptor to activate transduction cascades crucial for early gene expression [62]. The BDNF signal cascade can be triggered by several mechanisms (Figure 1C), including calcium influx through voltage-dependent channels [63] and $GABA_B$ receptors activation [64]. In an alternative way to the retrograde diffusion of postsynaptically synthesized molecules, IP could be triggered, as in cerebellar stellate cells, by the direct activation of NMDA receptors (Figure 1D) on presynaptic GABAergic terminals in response to the glutamate released from parallel fibers [65]. Similarly, in the frontal cortex of developing rats, the calcium influx through NMDA-Rs opening caused by glutamate diffusion from nearby synapses is sufficient to trigger the increase of GABA release [48]. It should also be noted that a particular mechanism observed for inhibitory LTP has been characterized in the dorsomedial hypothalamus. The activation of CKK receptors by the exposure to neuromodulators combined with the concomitant activation of metabotropic glutamate receptors (mGluRs) induces the release of Adenosine triphosphate (ATP) by surrounding astrocytes acting on presynaptic receptors and in turn triggering a prolonged increase in GABA release [66].

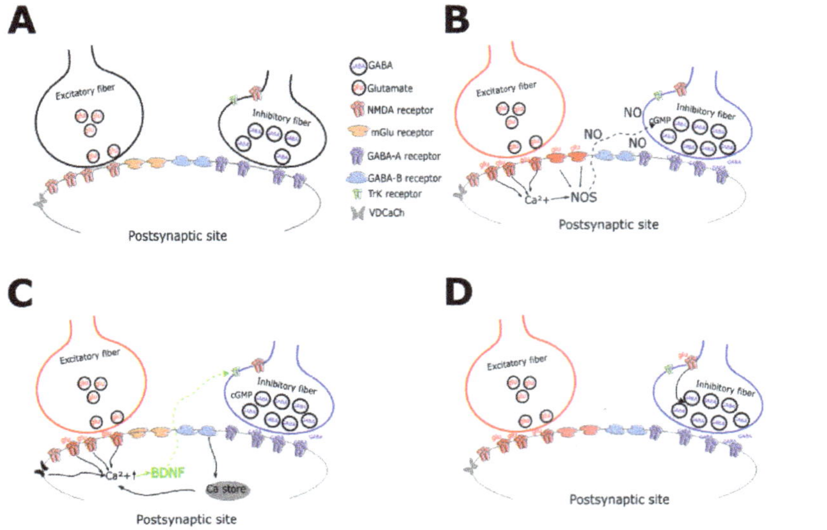

Figure 1. Schematic diagram collecting mechanisms underlying presynaptic LTP. A. Excitatory and inhibitory fibers contact, by releasing glutamate (glu) and GABA (GABA), a postsynaptic neuron expressing both ionotropic (mglu) and metabotropic receptors (GABA-B). B The repetitive release of glutamate triggers calcium entry through NMDA receptors in the postsynaptic terminals. The activation of Nitric Oxide (NO) synthase (NOS) induces the retrograde diffusion of NO which, in turn, activates cyclic-GMP potentiating vesicles release. C. Similarly to B, the repetitive glutamate release causes postsynaptic intracellular calcium rise in response to i) Voltage-dependent Calcium Channels (VDCaCh) opening, ii) NMDA receptors opening, or iii) mGlu receptors activation causing release from intracellular stores. Calcium increase triggers the retrograde diffusion of the Brain-Derived Neurotrophic Factor (BDNF) potentiating GABA release via Tyrosine Kinase-1 (TRK) receptors activation. D. The diffusion of glutamate in the extrasynaptic space can directly activate presynaptic NMDA-Rs favoring the potentiation of GABA release.

As in the case of i-LTP, the identified forms of i-LTD presynaptically expressed depend either on the direct NMDA-Rs activation in the GABAergic terminals [67] (Figure. 2) or through the retrograde diffusion of endocannabinoid (eCB) [49] (Figure 2D). These molecules, in response to afferent fibers stimulation, move from the post- to the presynaptic terminal triggering LTD induction. The eCB-dependent i-LTD, which is widely expressed throughout the brain [68–71], often requires the spread of glutamate from nearby excitatory synapses to activate metabotropic receptors (mGluR) as a trigger (Figure 2D). Interestingly, this form of plasticity does not necessarily involve calcium influx in the postsynaptic terminals, as demonstrated for the mGluR-dependent inhibitory LTD (i-LTD) in the amygdala [72]. Conversely, hippocampal i-LTD mediated by eCB is triggered by interneuron activity likely bringing calcium increase through voltage-dependent calcium channels, which in turn induces the simultaneous enhancement of calcineurin activity and the consequent reduction of the adenylyl cyclase-protein kinase A (PKA) transduction cascade leading to a long-term decrease of GABA release [73]. Additionally, other factors may actually contribute to modulate the retrograde diffusion of eCB. For instance, the dopamine receptors type 2 (D2R) were shown to suppress GABA release in the prefrontal cortex [74] and in the Ventral Tegmental Area [75] through a coactivation of the D2R and Cannabinoid Receptors by an increase of endogenous dopamine levels. As in the case of inhibitory presynaptic LTP, also the depression of GABA release can be induced by the direct activation of presynaptic NMDA-Rs by glutamate released from excitatory synapses in the next proximity [46,76]. Interestingly, our recent results show that the cerebellar inhibitory synapse between Golgi and Granule cells simultaneously exploits these mechanisms. The theta-burst protocol delivered to the excitatory mossy fibers can bidirectionally modulate GABA release through glutamate diffusion inducing LTP through the retrograde diffusion of nitric oxide toward GABAergic synapses or, alternative, LTD can be triggered by presynaptic activation of NMDA receptors [46].

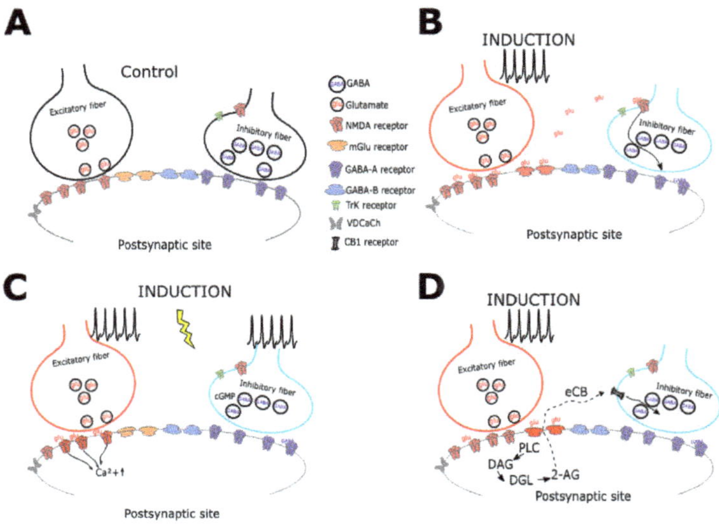

Figure 2. Schematic diagram collecting mechanisms underlying presynaptic LTD. A. Excitatory and inhibitory fibers contact, by releasing glutamate (glu) and GABA (GABA), a postsynaptic neuron expressing both ionotropic (mglu) and metabotropic receptors (GABA-B). B The diffusion of glutamate in the extrasynaptic space can directly activate presynaptic NMDA-Rs inducing the depression of GABA release. C. The coactivation of glutamatergic and GABAergic ionotropic receptors by simultaneous stimulation of excitatory and inhibitory fibers can lead to the depression of GABA release via a not well-identified mechanism. D. The activation of metabotropic glutamate receptors following repetitive excitatory stimulation triggers intracellular signal cascade, typically involving Phospholipase-C (PLC), diacylglycerol (DAG), Diacylglycerol lipase (DGL) and the 2-Arachidonoylglycerol (2-AG) endocannabinoid (eCB). This class of molecules can freely diffuse in the extracellular space acting as a retrograde messenger to activate specific cannabinoid receptors (CB) onto the GABAergic terminal that trigger the depression of vesicles release via different pathways.

Inhibitory plasticity can also be induced through mechanisms requiring the direct activity of GABAergic afferents. One of these homosynaptic mechanisms has been described in the primary visual cortex, where the firing of a presynaptic neuron paired with the activation of a postsynaptic pyramidal neuron can induce inhibitory plasticity [77,78]. The mechanisms subtending homosynaptic plasticity can depend on calcium changes, as in the case of star GABAergic connections between fast-spiking and star pyramidal neurons in the visual cortex during visual deprivation [77]. Nevertheless, other areas such as neocortex show inhibitory plasticity strongly correlated to calcium influx elicited by paired action potential during induction protocols [78] (Figure 2C). Additionally, it has been shown that a shift in the chloride transporter altering the driving force for GABAergic currents can occur in hippocampal neurons in response to coincidence activation of pre and postsynaptic activity [79].

Changes in inhibitory strength can also be associated with purely postsynaptic expression through a large variety of mechanisms (Figure. 3) [80], as it happens for excitatory synapses [81]. GABAergic weights can be adjusted postsynaptically by bidirectional changes in channels functionality. The ionotropic $GABA_A$ receptors, in response to specific patterns of induction requiring calcium influx following postsynaptic firing activity, can be phosphorylated by different kinases (e.g., PKC, CaMKII,

Src, and PKA) [82] (Figure 3B). The high-frequency firing in neocortical pyramidal neurons drives LTP of perisomatic inhibition via calcium entry through R-type voltage-gated calcium channels [83], which can be reverted to depression requiring calcium via L-type channels during hyperpolarization [84]. A similar form of postsynaptic plasticity was reported to occur in cerebellar Purkinje cells, where the potentiation of GABA release is triggered by repetitive postsynaptic discharge [52]. Interestingly, the increase of perisomatic inhibition by cerebellar basket cells through an increase of receptors trafficking is reported to be also triggered by the sole excitatory activation of climbing fibers [52]. Another form of postsynaptic plasticity was shown to occur in the hippocampal circuits and is expressed as an increase in the expression level of the scaffold protein for $GABA_A$ receptors gephyrin [85]. The availability of this molecule at GABAergic synapses is regulated by its state of phosphorylation [86] and is responsible for the induction of postsynaptic LTP that alters $GABA_A$-Rs dynamics. Moreover, the continuous cycling of $GABA_A$-Rs insertion and removal, together with movements of lateral diffusion at synaptic surface regulates synaptic functionality [87]. In addition, the number of $GABA_A$-Rs may change in response to receptor trafficking regulation. Inhibitory responses can be, therefore, bidirectionally modulated by alternatively acting on the exocytosis and endocytosis cycling [88]. Finally, also in the case of postsynaptic mechanisms, postsynaptic changes of intracellular concentrations of membrane-permeable ions can contribute to GABergic signaling [89].

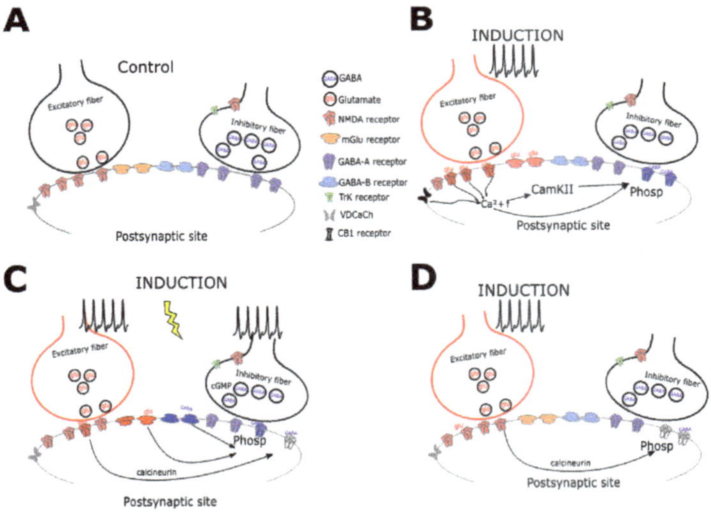

Figure 3. Schematic diagram collecting the mechanisms underlying postsynaptic plasticity. A. Excitatory and inhibitory fibers contact, by releasing glutamate (glu) and GABA (GABA), a postsynaptic neuron expressing both ionotropic (mglu) and metabotropic receptors (GABA-B). B. The activation of glutamatergic synapses can lead to an increase of postsynaptic intracellular calcium concentration either through NMDA-Rs or VDCaChs opening. Calcium increase can directly act on proteins phosphorylation (phosp) or can mediate CamKII activation leading to phosphorylation as well. Postsynaptic GABA-A receptors can thus increase their efficacy, can switch from silent to active state, or can move in the postsynaptic membrane. C. The simultaneous activation of metabotropic glutamatergic and metabotropic GABAergic receptors can lead to the phosphorylation required for the potentiation of ionotropic receptors activity. D. Conversely, the reduction of GABA-A-Rs receptors activity, either in the form of receptors silencing or in the sliding away from the postsynaptic density, can be induced by NMDA receptors opening following glutamate released. The protein phosphatase calcineurin or alternatively, the increase of intracellular calcium concentration in the postsynaptic neurons through NMDA or VDCaChs are the mediators of postsynaptic i-LTD.

The activity of extrasynaptic receptors mediates an alternative form of synaptic inhibition. The impact of this form of tonic inhibition is critically related to its impact on membrane conductance and membrane potential with time constants considerably lower than the one of receptors located in the terminals [90]. The tonic inhibition has been shown to undergo several forms of plasticity too; however, in most cases, glutamatergic signaling is essential to trigger persistent changes. In the hippocampus, kainate receptors activation triggers LTP of tonic inhibition [91], whereas persistent potentiation can be induced by block or genetic deletion of NMDA receptors [92] whilst depression is triggered by the activation of NMDA receptors [93]. Furthermore, tonic inhibition can be regulated by the direct activation of CB1 receptors [94] by retrograde diffusion of NO [95] or, alternatively by the direct activation of muscarinic acetylcholine receptors [96]. The fast synaptic inhibition related to the activity of $GABA_A$-Rs has been well characterized and described. Conversely, the functional alternative to the action of ionotropic $GABA_A$-Rs is the slow inhibition mediated by the metabotropic $GABA_B$ receptors. Although the biochemical signaling engaged by $GABA_B$-Rs is characterized in detail as well as their role in shaping neuronal activity, it is not much understood whether they can undergo plastic changes. Nevertheless, reports of persistent changes in $GABA_B$-Rs activity have been shown to occur in the hippocampus [97] and in lateral habenula [98], albeit the cellular and molecular mechanisms underlying this form of long-term plasticity need further investigation. It should also be noted that, as in the case of glutamatergic synapses, where the activity of G protein-activated inwardly rectifying K^+ (GIRK) channels has been shown to induce LTP in cultured hippocampal neurons [99], in a recent work, Sanchez-Rodriguez and colleagues showed that GIRK channels are implicated in the expression of inhibitory LTP in the hippocampal circuit in vivo and, importantly, these mechanisms are impaired by the presence of amyloid-β (Aβ), raising attention on the implication of inhibitory plasticity in neurodegenerative diseases [100].

4. Learning Rules and Computational Consequences of Inhibitory Plasticity

Unlike excitatory synapses, learning rules for inhibitory plasticity have not yet been extensively investigated [101]. However, among the variety of rules, the spike-timing-dependent plasticity (STDP) correlating the reciprocal timing of pre and postsynaptic firing [102] has been encoded for some GABAergic synapses [4,103]. The first reported evidence dates back to 2001 when Holmgren and Zilberter showed that in the neocortex, when action potentials in the presynaptic inhibiting interneuron are timely proximal to postsynaptic pyramidal cell firing, GABAergic synapses undergo LTD [78]. Conversely, if presynaptic spikes and postsynaptic firing are distant enough, synaptic weights are potentiated in a calcium-dependent way. After a few works demonstrating in hippocampal circuits the presence of a few variants of inhibitory STDP [104–106], in 2006, Haas and colleagues showed that in stellate cells of rat entorhinal cortex presynaptic incoming before postsynaptic spikes trigger strengthening of inhibitory synapses while the reserve leads to depression [107]. Surprisingly, the maximal efficacy changes did not perfectly match the pre-post coincidence, instantiating a time-shift with important consequences in the circuit computational rules. The authors, in fact, by using a mathematical model, showed that STDP exploits the clustering of neurons within the circuit providing flexible and dynamic organization of neuronal circuitry in a region where the uncontrolled spread of excitation often leads to epileptic foci. Similarly, in pyramidal neurons of the mouse auditory cortex, STDP is exploited by inhibitory conductance. The maximal effect was observed when pre and postsynaptic spikes were temporally proximal, with a relatively large time window ($\approx$ 10 ms), and independently from their reciprocal order [108]. Given the tendency to show potentiation with paired activity and the requirements of NMDA-Rs activation for the induction processes, this form of plasticity seems to be finalized to silencing network activity in response to diffuse circuit activation [108]. Furthermore, the auditory cortex also shows STDP in the PNs mediated by $GABA_B$-Rs with LTD induced by presynaptic before postsynaptic spikes. The fact that the sign of plasticity can be reversed during development makes this form of plasticity a suitable candidate for disinhibition during the auditory critical period [109]. Differently, electrophysiological recordings in the somatosensory cortex revealed that the coupling between prolonged postsynaptic bursts with single presynaptic spikes in temporal proximity leads to GABAergic depression, whereas potentiation was observed when presynaptic spikes were presented well beyond the end of the burst [73]. This mechanism has been suggested to participate in the sharpening of significant sensory patterns. In hippocampal neurons, the sole presynaptic spikes lead to depression of synaptic conductance while the coupling of pre and postsynaptic activity generated changes in local chloride reversal potential with the same sign [79]. At the circuit level, this effect appears to weaken the inhibitory strength leading the system toward a critical large excitatory/inhibitory balance with substantial reverberations on network activity [110]. Nevertheless, the firing activity of other neurons in the circuit [111], as well as membrane potential value during induction [48], could play a significant role in shaping synaptic changes by affecting the amplitude and direction of plasticity.

The precise identification of the functional role of inhibitory plasticity is still an open issue. Nonetheless, the recurrent leitmotif regarding GABAergic plasticity is the maintenance of a constant E/I balance in a circuit that can compensate for changes in the excitatory driving force triggered by plasticity at glutamatergic synapses. This homeostatic regulation can be obtained by reducing both feed-forward inhibition and excitation [112]. Alternatively, increasing the excitability of inhibitory interneurons following the potentiation of glutamatergic synapses can balance the circuit functioning, as it was shown in hippocampal circuits [113]. In the somatosensory cortex, the recruitment of inhibitory cells by the activity of pyramidal neurons can contribute to finely regulate cortical excitability through the sensitivity and the dynamic range of recurrent inhibition [114]. Furthermore, the tight correlation between fast inhibition and excitation allows the fine regulation and the balancing of neuronal circuits, either in the form of spontaneous or evoked firing activity [115]. Nevertheless, although the maintenance of the E/I balance seems essential for the correct circuit activity, changes in the excitatory to inhibitory balance could play a key role in receptive field organization [116] and sensory learning [117]. Importantly, since the exact value of E/I can be adjusted on different setpoints according to the brain region, the circuit activation by input stimuli can alternatively lead to the suppression or to the potentiation of excitatory output. It has recently been shown, in fact, that in neocortical circuits, the persistent increase of GABAergic synapses can impact output firing through a decreased spike probability and increased timing [118]. Furthermore, we have recently shown in cerebellar cortex that bidirectional plasticity of inhibitory circuits [46] contributes to control the spatial and temporal pattern activity of excitatory granule cells by sharpening center-surround structures [119] and by finely regulating the timing of first spike output through subthreshold integration processes [120].

From a purely computational perspective, the power of the brain is traditionally linked to the complex connectivity of neuronal networks, whereas single neurons are considered as linear integrators and thresholding devices. It is indeed clear that a wide series of non-linear mechanisms converting synaptic input into output firing is employed by single neurons to process information. These mechanisms include synaptic noise, inhibitory conductance, and notably synaptic plasticity [121]. As a general rule, the analysis of the effects of plasticity on neuronal computation has been mainly focused on excitatory circuits; nonetheless, recent discoveries on the involvement of IP require to deepen the investigation of its effects on network computation. Regardless of molecular subtypes, the prominent characteristic of inhibitory plasticity for its consequences on network computation is the architectural organization of inhibitory afferents. In particular, single neuron computation is strongly modulated by peri-somatic inhibition, which in turn exerts a critical additive or subtractive effect on the input-output ratio (I/O) when potentiated or depressed [121]. The specific peri-somatic targeting onto excitatory cortical neurons strongly influences the insurgence of network oscillations, which can be traced back to cognitive and sensory functions [122]. It has been suggested that the bidirectional modulation of peri-somatic inhibition by LTP and LTD could alternatively entrain single neurons in synchronous activity or decoupling oscillatory events [123], thus favoring the coordination of neurons sharing common functional properties. Conversely, despite dendritic inhibition is a crucial determinant for synaptic integration and underlies lateral inhibition during sensory tasks, as shown by evidence in cortical circuits [124], the impact on network computation of the plasticity of these synapses has been poorly investigated.

In order to analyze the impact of IP onto neuronal network computation, researchers tend to employ circuits models with a limited number of parameters to control. The simplest network is composed of basic computational units, for instance, integrate and fire neurons, endowed with excitatory and inhibitory synapses randomly connected with a sparse architecture [125]. In such a model, the balance between excitation and inhibition controls a wide range of circuit dynamics, including average firing rate, and the way neurons respond collectively to inputs [125]. By assuming these circuits constraints, Vogels and colleagues in 2011 demonstrated that the implementation of an STDP rule at inhibitory synapses with strong potentiation in case of coincident pre- and postsynaptic spikes, allows inhibition to approach output firing rates to a target value [126]. The final target value strongly depends on the ratio between LTP and LTD, while deviations from setpoint are suppressed by a contrary reaction. The resulting effect is, therefore, the stabilization of firing rates whenever the incoming of repeated and persistent excitatory inputs tend to disrupt the E/I balance [127]. If network connectivity is organized with clusters of excitatory units, the E/I ratio is extremely sensitive to the wiring architecture. This, in turn, brings the network to a winner-takes-all behavior. The implementation of inhibitory plasticity in such a context, by compensating changes in firing rates, prevents groups of neurons from dominating the network [128]. Moreover, it has also been demonstrated that, in a network where connections are implemented with realistic connectivity patterns, and single neuron firing rates are sparse throughout the network, synaptic weights can be dynamically adjusted by inhibitory plasticity to equalize E/I balance changes [129]. The increasing complexity encountered in more organized circuits models like multiple layered feedforward networks strongly limits the capacity of inhibitory plasticity to maintain the E/I ratio. Nevertheless, Haas and colleagues showed that in the entorhinal cortex the strengthening of inhibitory connections can block the propagation of excitatory waves, ensuring network stability [107]. Inhibitory plasticity has also been suggested to favor the selection of specific feedforward pathways either by altering or by maintaining the balance between excitatory inputs and inhibitory signals [130]. In 2019 Wilmes and Clopath published a work where, by using a spiking model of layer 2/3 primary visual cortex, they showed that IPs play a major role in adjusting stimulus representation by storing information about reward stimuli. This model allowed to demonstrate that IP is essential to increase stimulus representation by triggering excitatory plasticity [131]. Moreover, in a recent study Soloduchin and Shamir showed that in a network model composed of two neuronal populations reciprocally inhibited, the implementation of a simple STDP rule for inhibitory synapses can bring the network to rhythmic activity by itself [132], confirming the hypothesis that, spontaneous oscillation can be entrained by modulation of peri-somatic inhibition on principal excitatory neurons [122]. Finally, it has been proposed that the spatial tuning patterns showing invariance and selectivity observed, for instance, in hippocampal place cells could be the result of excitatory and inhibitory plasticity. The combination, in fact, of the two mechanisms leads to localized activity invariant to different spatial dimensions [133].

Synaptic plasticity is also thought to be the cellular and molecular counterpart of learning and memory. In particular, memories that can be recalled by contextual cues or commands must involve plasticity mechanisms to be exploited [134]. In Hopfield networks, a circuit model assembled with recurrent connectivity and particularly suitable to implement associative memory, groups of silent and active neurons recruited by recalling inputs are used to describe memories [135]. Given the mixture of excitatory and inhibitory connections in such a network, it can be envisaged that IP has a leading role in creating and exploiting memories recall. In a different perspective, Maas and colleagues proposed that memories are represented by networks through the pathway generated by the population activity of the whole circuit, instead of activating a bunch of neurons [136]. In networks with strong and random excitatory recurrence, inhibitory plasticity stabilizes circuit dynamics similar to what happens in the motor cortex during the execution of limb movements. These networks, in fact, amplify the activity states that can be used to execute movement patterns [137]. By using a supervised learning scheme for feedforward and recurrent connections, Gilra and Gestner showed that IPs could efficiently accomplish linear, non-linear, or chaotic dynamics, as well as motor coordination dynamics [138]. Similarly, the implementation of STDP rules at different sites in a cerebellar like structure allowed to implement an efficient adaptive scheme capable of motor learning performance [139]. Finally, the specificity of inhibitory feedback sustaining grid cell organization has been suggested to require IP for the generation of grid cell population (Table 1) [140].

Table 1. Mechanisms of inhibitory plasticity and functional consequences.

Sign of Plasticity	Molecular Mechanism	Brain Region/Neuron	Site of Expression	Computation/Functional Significance	Refs
LTP	$GABA_B$ receptor dependent, BDNF signaling	Visual cortex/Neonatal hippocampus	Presynaptic	Critical period plasticity/ E/I balancing	[54,104,105,141]
LTP	Postsynaptic NMDA, retrograde NO	VTA, Basolateral amygdala Cerebellum,	Presynaptic	Reward modulation/ spatio temporal pattern sharpening/ shaping conditioned fear response	[44,46,142]
LTP	Postsynaptic calcium, retrograde BDNF	Hippocampus	Presynaptic	Associative memory formation	[143]
LTP	Presynaptic NMDA	Cerebellum	Presynaptic	Motor learning regulation	[65,144]
LTP	Postsynaptic mGluR and retrograde NO	Lamina I spinal cord	Presynaptic	Signal to noise regulation	[53]
LTP	Postsynaptic calcium/NMDA	Deep cerebellar nuclei	Presynaptic	Regulation of spike firing for motor coordination	[145,146]
LTP	Postsynaptic NMDA and CamKII	Medial prefrontal cortex	Postsynaptic	Local regulation of E/I at cellular level	[147]
LTP	Postsynaptic Calcium/ CamKII	Cerebellum Purkinje cell	Postsynaptic	Regulation of output firing patterns	[52,141,148]
LTP	$GABA_B$/ mGluR	Hippocampal CA1	Postsynaptic	Reinforcement of rhythmic activity	[149]
LTP	Presynaptic firing paired with mild depolarization	Developing visual cortex	Postsynaptic	Regulating critical period for ocular dominance	[76]
LTP	Calcium influx receptor phosphorilation	Neocortex	postsynaptic	E/I balancing	[83–85]
LTP	Postsynaptic NMDA and calcium rise	Lateral amygdala	Postsynaptic	Processing stimuli during fear conditioning	[150]
LTP	Postsynaptic NMDA, L type calcium channels	Auditory cortex	Postsynaptic	Normalizing E/I and remodeling auditory map	[108]
LTD	mGlur, retrograde eCB	Hippocampus, amygdala, Visual cortex, prefrontal cortex	Presynaptic	Changes of E/I / extinction of aversive memories/ regulation of development in critical period	[49,69,72,75]
LTD	$GABA_A$ activation and postsynaptic NMDA	Neonatal hippocampus	Presynaptic	Regulation of synapse formation and maturation	[151]
LTD	Presynaptic NMDA	Cerebellum, visual cortex	Presynaptic	Spatio-temporal sharpening sensory information	[46,77]
LTD	Postsynaptic NMDA and mediated by calcineurin	hippocampus	postsynaptic	Disinhibit excitatory circuits	[152]
LTD	Postsynaptic calcium and protein phosphatase	Deep cerebellar nuclei	postsynaptic	Modulation of spontaneous cerebellar firing for motor coordination	[153]
LTD	Dopamine mediated eCBN signaling	Ventral tegmental area	postsynaptic	Regulation of addiction mechanisms	[75,154]

5. Perspectives and Concluding Remarks

The analysis of inhibitory plasticity presented so far is limited to the molecular characterization of the mechanisms underlying IPs together with the effects at the circuit level. However, a univocal determination of the roles for the IPs in circuit functions is still lacking, and evidences have been collected regarding the consequences of inhibitory plasticity at the integrative level. It is well known in fact that alterations of inhibitory circuits can contribute to the induction of neurological disorders. In particular, the regulation of the E/I balance, which is inherently bound to IP impacts the induction of excitatory LTP and LTD and can indeed shift the threshold required to switch between the two plastic conditions. It has recently been observed, in fact, that the inhibitory LTP, by modulating E/I balance, can effectively restore the hippocampal excitatory LTD preventing memory impairment related to Aβ protein accumulation [155]. Similarly, the disruption of the E/I balance in neocortical circuits is one of the most accredited explanation for the insurgence of the autism spectrum disorder (ASD), which can be therefore bound to inhibitory plasticity [156,157]. Again, schizophrenic patients often show disturbances in the GABAergic neurotransmission of the dorsolateral prefrontal cortex [158]. In particular, alterations in the peri-somatic regulation of pyramidal neurons lead to a reduced capacity of synchronizing gamma-band activity. Furthermore, one of the hypotheses on the etiology of epileptogenesis regards the hyperactivity of GABAergic neurons masking hyperexcitability activated by alterations of inhibitory strength triggered by insult of injury [159]. This cascade mechanism, responsible for the occurrence of the temporal lobe epilepsy, could be activated by overexpression or disruption of inhibitory plasticity. Finally, patients with Parkinson's disease were shown to display downregulation of GABAergic activity in the afferents to basal ganglia. The reduction of neurological symptoms observed in response to deep brain stimulation, one of the most efficient therapeutic treatments, can be attributed to the triggering of GABAergic LTP allowing the recovery from the aforementioned GABA downregulation [160].

Besides the obvious interest in the biomedical and clinical fields, the analysis of brain functions and particularly of synaptic connections also arouse great interest in computer science and, more generally, in the field of neuromorphic electronic and artificial intelligence. The goal in fact to build large artificial neural networks with vast amounts of computing elements has rendered the task of creating low consuming artificial synapses a high priority. The visionary idea of electrical elements behaving like synapses called memristors is now becoming true [161]. These physical devices can effectively behave like synaptic elements because of their capacity to reproduce synaptic features and plastic mechanisms such as STDP [162] or heterosynaptic plasticity [163]. Memristors can indeed be assembled to mimic neural functions and reproduce neuronal behavior [164] or perform autonomous complex learning tasks [165]. Nowadays, several physical elements have been proposed and adopted to reproduce synaptic learning; however, none of them has been used to mimic plastic behaviors of inhibitory synapses. There is therefore increasing attention on the role of the various forms of IP in circuit computation and on the further possibility to introduce such behaviors into electronic circuits performing complex tasks. It can be envisaged in fact that the implementation of unsupervised learning rules in electronic synapses could entrain artificial circuits to perform autonomous behavior [165]. The growing expansion of the neuromorphic field and of the brain-inspired computation requires to implement devices exploiting the properties of both excitatory and inhibitory synapses. The encoding then of learning rules in artificial synapses and in electronic circuits, paving the way to the next generation of neuromorphic devices, is opening promising perspectives for a series of applications with clinical relevance such as neuroprosthesis or with a high social impact for daily lives driving futuristic artificial intelligence machines.

Author Contributions: Writing—original draft preparation, D.G. and J.M.; writing—review and editing, D.G. A.B. and J.M.; Funding acquisition, C.A.P. and J.M. All authors have read and agreed to the published version of the manuscript.

Funding: This research was funded by: Italian Ministry of Education, University and Research (MIUR): Dipartimenti di Eccellenza Program (2018–2022)—Dept. of Biomedical, Metabolic and Neural Sciences, University of Modena and Reggio Emilia to C.A.P. and by the University of Modena and Reggio Emilia: FAR 2017 to J.M.

Conflicts of Interest: "The authors declare no conflict of interest." The funders had no role in the design of the design of the study; in the collection, analyses, or interpretation of data; in the writing of the manuscript, or in the

References

1. Siegelbaum, S.A.; Kandel, E.R. Learning-related synaptic plasticity: LTP and LTD. *Curr. Opin. Neurobiol.* **1991**, *1*, 113–120. [CrossRef]
2. Maffei, A. The many forms and functions of long term plasticity at GABAergic synapses. *Neural Plast.* **2011**, 254724. [CrossRef] [PubMed]
3. Allene, C.; Lourenço, J.; Bacci, A. The neuronal identity bias behind neocortical GABAergic plasticity. *Trends Neurosci.* **2015**, *38*, 524–534. [CrossRef] [PubMed]
4. Vogels, T.P.; Froemk e, R.C.; Doyon, N.; Gilson, M.; Haas, J.S.; Liu, R.; Maffei, A.; Miller, P.; Wierenga, C.J.; Woodin, M.A.; et al. Inhibitory synaptic plasticity: Spike timing-dependence and putative network function. *Front. Neural Circuits* **2013**, *18*, 7. [CrossRef] [PubMed]
5. Scanziani, M.; Häusser, M. Electrophysiology in the age of light. *Nature* **2009**, *461*, 930–939. [CrossRef]
6. Lewis, D.A.; Hashimoto, T. Deciphering the disease process of schizophrenia: The contribution of cortical GABA neurons. *Neuron* **2011**, *72*, 231–243.
7. Isaacson, J.S.; Scanziani, M. How inhibition shapes cortical activity. *Trends Neurosci.* **2015**, *38*, 524–534. [CrossRef]
8. Pouille, F.; Marin-Burgin, A.; Adesnik, H.; Atallah, B.V.; Scanziani, M. Input normalization by global feedforward inhibition expands cortical dynamic range. *Nat. Neurosci.* **2009**, *12*, 1577–1585. [CrossRef]
9. Tepper, J.M.; Koós, T.; Wilson, C.J. GABAergic microcircuits in the neostriatum. *Trends Neurosci.* **2004**, *27*, 662–669. [CrossRef]
10. Mapelli, J.; Gandolfi, D.; D'Angelo, E. Combinatorial responses controlled by synaptic inhibition in the cerebellum granular layer. *J. Neurophysiol.* **2010**, *103*, 250–261. [CrossRef]
11. Wehr, M.; Zador, A.M. Balanced inhibition underlies tuning and sharpens spike timing in auditory cortex. *Nature* **2003**, *426*, 442–446. [CrossRef] [PubMed]
12. Okun, M.; Lampl, I. Instantaneous correlation of excitation and inhibition during ongoing and sensory-evoked activities. *Nat. Neurosci.* **2008**, *11*, 535–537. [CrossRef]
13. Wilent, W.B.; Contreras, D. Dynamics of excitation and inhibition underlying stimulus selectivity in rat somatosensory cortex. *Nat. Neurosci.* **2005**, *8*, 1364–1370. [CrossRef] [PubMed]
14. Xue, M.; Atallah, B.V.; Scanziani, M. Equalizing excitation-inhibition ratios across visual cortical neurons. *Nature* **2014**, 596–600. [CrossRef] [PubMed]
15. Chiu, C.Q.; Lur, G.; Morse, T.M.; Carnevale, N.T.; Ellis-Davies, G.C.; Higley, M.J. Compartmentalization of GABAergic inhibition by dendritic spines. *Science* **2013**, *340*, 759–762. [CrossRef] [PubMed]
16. D'Angelo, E.; Solinas, S.; Mapelli, J.; Gandolfi, D.; Mapelli, L.; Prestori, F. The cerebellar Golgi cell and spatiotemporal organization of granular layer activity. *Front. Neural Circuits* **2013**, *7*, 93. [CrossRef]
17. Sahara, S.; Yanagawa, Y.; O'Leary, D.D.; Stevens, C.F. The fraction of cortical GABAergic neurons is constant from near the start of cortical neurogenesis to adulthood. *J. Neurosci.* **2012**, *32*, 4755–4761. [CrossRef]
18. Markram, H.; Toledo-Rodriguez, M.; Wang, Y.; Gupta, A.; Silberberg, G.; Wu, C. D Interneurons of the neocortical inhibitory system. *Nat. Rev. Neurosci.* **2004**, *5*, 793–807. [CrossRef]
19. Méndez, P.; Bacci, A. Assortment of GABAergic plasticity in the cortical interneuron melting pot. *Neural Plast.* **2011**, *2011*, 976856. [CrossRef]
20. Taniguchi, H. Genetic dissection of GABAergic neural circuits in mouse neocortex. *Front. Cell Neurosci.* **2014**, *8*, 8. [CrossRef]
21. Ben-Ari, Y.; Gaiarsa, J.L.; Tyzio, R.; Khazipov, R. GABA: A pioneer transmitter that excites immature neurons and generates primitive oscillations. *Physiol Rev.* **2007**, *87*, 1215–1284. [CrossRef] [PubMed]
22. Szabadics, J.; Varga, C.; Molnár, G.; Oláh, S.; Barzó, P.; Tamás, G. Excitatory effect of GABAergic axo-axonic cells in cortical microcircuits. *Science* **2006**, *311*, 233–235. [CrossRef] [PubMed]

23. DeFelipe, J.; López-Cruz, P.L.; Benavides-Piccione, R.; Bielza, C.; Larrañaga, P.; Anderson, S.; Burkhalter, A.; Cauli, B.; Fairén, A.; Feldmeyer, D.; et al. New insights into the classification and nomenclature of cortical GABAergic interneurons. *Nat. Rev. Neurosc.* **2013**, *14*, 202–216. [CrossRef] [PubMed]
24. Tasic, B.; Menon, V.; Nguyen, T.N.; Kim, T.K.; Jarsky, T.; Yao, Z.; Levi, B.; Gray, L.T.; Sorensen, S.A.; Dolbeare, T.; et al. Adult mouse cortical cell taxonomy revealed by single cell transcriptomics. *Nat. Neurosci.* **2016**, *19*, 335–346. [CrossRef]
25. Wefelmeyer, W.; Cattaert, D.; Burrone, J. Activity-dependent mismatch between axo-axonic synapses and the axon initial segment controls neuronal output. *Proc. Natl. Acad. Sci. USA* **2015**, *112*, 9757–9762. [CrossRef]
26. Somogyi, P.; Klausberger, T. Defined types of cortical interneurone structure space and spike timing in the hippocampus. *J. Physiol.* **2005**, *562*, 9–26. [CrossRef]
27. Pouille, F.; Scanziani, M. Enforcement of temporal fidelity in pyramidal cells by somatic feed-forward inhibition. *Science* **2001**, *293*, 1159–1163. [CrossRef]
28. Sohal, V.S.; Zhang, F.; Yizhar, O.; Deisseroth, K. Parvalbumin neurons and gamma rhythms enhance cortical circuit performance. *Nature* **2009**, *459*, 698–702. [CrossRef]
29. Cardin, J.A.; Carlén, M.; Meletis, K.; Knoblich, U.; Zhang, F.; Deisseroth, K.; Tsai, L.H.; Moore, C.I. Driving fast-spiking cells induces gamma rhythm and controls sensory responses. *Nature* **2009**, *459*, 663–667. [CrossRef]
30. Whissell, P.D.; Bang, J.Y.; Khan, I.; Xie, Y.F.; Parfitt, G.M.; Grenon, M.; Plummer, N.W.; Jensen, P.; Bonin, R.P.; Kim, J.C. Selective Activation of Cholecystokinin-Expressing GABA (CCK-GABA) Neurons Enhances Memory and Cognition. *eNeuro* **2019**, *6*, 1. [CrossRef]
31. Pelkey, K.A.; Chittajallu, R.; Craig, M.T.; Tricoire, L.; Wester, J.C.; McBain, C.J. Hippocampal GABAergic Inhibitory Interneurons. *Physiol Rev.* **2017**, *97*, 1619–1747. [CrossRef] [PubMed]
32. Mullner, F.E.; Wierenga, C.J.; Bonhoeffer, T. Precision of inhibition: Dendritic inhibition by individual GABAergic synapses on hippocampal pyramidal cells is confined in space and time. *Neuron* **2015**, *87*, 576–589. [CrossRef] [PubMed]
33. Hayama, T.; Noguchi, J.; Watanabe, S.; Takahashi, N.; Hayashi-Takagi, A.; Ellis-Davies, G.C.; Matsuzaki, M.; Kasai, H. GABA promotes the competitive selection of dendritic spines by controlling local Ca2+ signaling. *Nat. Neurosci.* **2013**, *16*, 1409–1416. [CrossRef] [PubMed]
34. Urban-Ciecko, J.; Barth, A.L. Somatostatin-expressing neurons in cortical networks. *Nat. Rev. Neurosci.* **2016**, *17*, 401–409. [CrossRef]
35. Overstreet-Wadiche, L.; McBain, C.J. Neurogliaform cells in cortical circuits. *Nat. Rev. Neurosci.* **2015**, *16*, 458–468. [CrossRef] [PubMed]
36. Pfeffer, C.K.; Xue, M.; He, M.; Huang, Z.J.; Scanziani, M. Inhibition of inhibition in visual cortex: The logic of connections between molecularly distinct interneurons. *Nat. Neurosci.* **2013**, *16*, 1068–1076. [CrossRef]
37. Miao, C.; Cao, Q.; Moser, M.B.; Moser, E.I. Parvalbumin and Somatostatin Interneurons Control Different Space-Coding Networks in the Medial Entorhinal Cortex. *Cell* **2017**, *171*, 507–521.e17. [CrossRef]
38. Kawaguchi, Y.; Kubota, Y. GABAergic cell subtypes and their synaptic connections in rat frontal cortex. *Cereb Cortex* **1997**, *7*, 476–486. [CrossRef]
39. Galarreta, M.; Hestrin, S. A network of fast-spiking cells in the neocortex connected by electrical synapses. *Nature* **1999**, *402*, 72–75. [CrossRef]
40. Song, H.F.; Yang, G.R.; Wang, X.J. Training Excitatory-Inhibitory Recurrent Neural Networks for Cognitive Tasks: A Simple and Flexible Framework. *PLoS Comput. Biol.* **2016**, *12*, e1004792. [CrossRef]
41. Shu, Y.; Hasenstaub, A.; McCormick, D.A. Turning on and off recurrent balanced cortical activity. *Nature.* **2003**, *423*, 288–293. [CrossRef]
42. Haider, B.; Krause, M.R.; Duque, A.; Yu, Y.; Touryan, J.; Mazer, J.A.; McCormick, D.A. Synaptic and network mechanisms of sparse and reliable visual cortical activity during nonclassical receptive field stimulation. *Neuron* **2010**, *65*, 107–121. [CrossRef]
43. Gaiarsa, J.; Caillard, O.; Ben-Ari, Y. Long-term plasticity at GABAergic and glycinergic synapses: Mechanisms and functional significance. *Trends Neurosci.* **2002**, *25*, 564–570. [CrossRef]
44. Nugent, F.S.; Penick, E.C.; Kauer, J.A. Opioids block long-term potentiation of inhibitory synapses. *J. Nature* **2007**, *446*, 1086–1090. [CrossRef]
45. Petrini, E.M.; Barberis, A. Diffusion dynamics of synaptic molecules during inhibitory postsynaptic plasticity. *Front. Cell Neurosci.* **2014**, *8*, 300.

46. Mapelli, J.; Gandolfi, D.; Vilella, A.; Zoli, M.; Bigiani, A. Heterosynaptic GABAergic plasticity bidirectionally driven by the activity of pre- and postsynaptic NMDA receptors. *Proc. Natl. Acad. Sci. USA* **2016**, *113*, 9898–9903. [CrossRef]
47. Regehr, W.G.; Carey, M.R.; Best, A.R. Activity-dependent regulation of synapses by retrograde messengers. *Neuron* **2009**, *63*, 154–170. [CrossRef]
48. Mathew, S.S.; Hablitz, J.J. Presynaptic NMDA receptors mediate IPSC potentiation at GABAergic synapses in developing rat neocortex. *PLoS ONE* **2011**, *6*, e17311. [CrossRef]
49. Chevaleyre, V.; Castillo, P.E. Heterosynaptic LTD of hippocampal GABAergic synapses: A novel role of endocannabinoids in regulating excitability. *Neuron* **2003**, *38*, 461–472. [CrossRef]
50. Komatsu, Y.; Iwakiri, M. Long-term modification of inhibitory synaptic transmission in developing visual cortex. *Neuroreport* **1993**, *4*, 907–910. [CrossRef]
51. Maffei, A.; Nelson, S.B.; Turrigiano, G.G. Selective reconfiguration of layer 4 visual cortical circuitry by visual deprivation. *Nat. Neurosci.* **2004**, *7*, 1353–1359. [CrossRef] [PubMed]
52. Kano, M.; Rexhausen, U.; Dreesen, J.; Konnerth, A. Synaptic excitation produces a long-lasting rebound potentiation of inhibitory synaptic signals in cerebellar Purkinje cells. *Nature* **1992**, *356*, 601–604. [CrossRef] [PubMed]
53. Fenselau, H.; Heinke, B.; Sandkühler, J. Heterosynaptic long-term potentiation at GABAergic synapses of spinal lamina I neurons. *J. Neurosci.* **2011**, *31*, 17383–17391. [CrossRef] [PubMed]
54. Caillard, O.; Ben-Ari, Y.; Gaiarsa, J.L. Long-term potentiation of GABAergic synaptic transmission in neonatal rat hippocampus. *J. Physiol.* **1999**, *518 (Pt 1)*, 109–119. [CrossRef]
55. Braga, M.F.; Aroniadou-Anderjaska, V.; Xie, J.; Li, H. Bidirectional modulation of GABA release by presynaptic glutamate receptor 5 kainate receptors in the basolateral amygdala. *J. Neurosci.* **2003**, *23*, 442–452. [CrossRef]
56. Bliss, T.V.; Lomo, T. Long-lasting potentiation of synaptic transmission in the dentate area of the anaesthetized rabbit following stimulation of the perforant path. *J. Physiol.* **1973**, *232*, 331–356. [CrossRef]
57. Nieus, T.; Sola, E.; Mapelli, J.; Saftenku, E.; Rossi, P.; D'Angelo, E. LTP regulates burst initiation and frequency at mossy fiber-granule cell synapses of rat cerebellum: Experimental observations and theoretical predictions. *J. Neurophysiol.* **2006**, *95*, 686–699. [CrossRef]
58. Baldelli, P.; Novara, M.; Carabelli, V.; Hernandez-Guijo, J.M.; Carbone, E. BDNF up-regulates evoked GABAergic transmission in developing hippocampus by potentiating presynaptic N- and P/Q-type Ca2+ channels signaling. *Eur. J. Neurosci.* **2002**, *16*, 2297–2310. [CrossRef]
59. Liu, Y.; Zhang, L.I.; Tao, H.W. Heterosynaptic scaling of developing GABAergic synapses: Dependence on glutamatergic input and developmental stage. *J. Neurosci.* **2007**, *27*, 5301–5312. [CrossRef]
60. Wood, J.; Garthwaite, J. Models of the diffusional spread of nitric oxide: Impli- cations for neural nitric oxide signalling and its pharmacological properties. *Neuropharmacology* **1994**, *33*, 1235–1244. [CrossRef]
61. Bathina, S.; Das, U.N. Brain-derived neurotrophic factor and its clinical implications. *Arch. Med. Sci.* **2015**, *11*, 1164–1178.
62. Gandolfi, D.; Cerri, S.; Mapelli, J.; Polimeni, M.; Tritto, S.; Fuzzati-Armentero, M.T.; Bigiani, A.; Blandini, F.; Mapelli, L.; D'Angelo, E. Activation of the CREB/c-Fos Pathway during Long-Term Synaptic Plasticity in the Cerebellum Granular Layer. *Front. Cell Neurosci.* **2017**, *11*, 184. [CrossRef] [PubMed]
63. Xu, H.; Kotak, V.C.; Sanes, P.H. Normal hearing is required for the emergence of long-lasting inhibitory potentiation in cortex. *J. Neurosci.* **2010**, *30*, 331–341. [CrossRef] [PubMed]
64. Komatsu, Y. GABAB receptors, monoamine receptors, and postsynaptic inositol trisphosphate-induced Ca2+ release are involved in the induction of long-term potentiation at visual cortical inhibitory synapses. *J. Neurosci.* **1996**, *16*, 6342–6352. [CrossRef]
65. Liu, S.J.; Lachamp, P. The activation of excitatory glutamate receptors evokes a long-lasting increase in the release of GABA from cerebellar stellate cells. *J. Neurosci.* **2006**, *26*, 9332–9339. [CrossRef]
66. Crosby, K.M.; Murphy-Royal, C.; Wilson, S.A.; Gordon, G.R.; Bains, J.S.; Pittman, Q.J. Cholecystokinin Switches the Plasticity of GABA Synapses in the Dorsomedial Hypothalamus via Astrocytic ATP Release. *J. Neurosci.* **2018**, *38*, 8515–8525. [CrossRef]
67. Castillo, P.E.; Chiu, C.Q.; Carroll, R.C. Long-term plasticity at inhibitory synapses. *Curr. Opin. Neurobiol.* **2011**, *21*, 328–338. [CrossRef]

68. Marsicano, G.; Wotjak, C.T.; Azad, S.C.; Bisogno, T.; Rammes, G.; Cascio, M.G.; Hermann, H.; Tang, J.; Hofmann, C.; Zieglgänsberger, W.; et al. The endogenous cannabinoid system controls extinction of aversive memories. *Nature* **2002**, *418*, 530–534. [CrossRef]
69. Adermark, L.; Lovinger, D.M. Frequency-dependent inversion of net striatal output by endocannabinoid-dependent plasticity at different synaptic inputs. *J. Neurosci.* **2009**, *29*, 1375–1380. [CrossRef]
70. Henneberger, C.; Redman, S.J.; Grantyn, R. Cortical efferent control of subcortical sensory neurons by synaptic disinhibition. *Cereb Cortex* **2007**, *17*, 2039–2049. [CrossRef]
71. Jiang, B.; Huang, S.; de Pasquale, R.; Millman, D.; Song, L.; Lee, H.K.; Tsumoto, T.; Kirkwood, A. The maturation of GABAergic transmission in visual cortex requires endocannabinoid- mediated LTD of inhibitory inputs during a critical period. *Neuron* **2010**, *66*, 248–259. [CrossRef] [PubMed]
72. Azad, S.C.; Monory, K.; Marsicano, G.; Cravatt, B.F.; Lutz, B.; Zieglgansberger, W.; Rammes, G. Circuitry for associative plasticity in the amygdala involves endocannabinoid signaling. *J. Neurosci.* **2004**, *24*, 9953–9961. [CrossRef] [PubMed]
73. Ahumada, J.; Fernández de Sevilla, D.; Couve, A.; Buño, W.; Fuenzalida, M. Long-term depression of inhibitory synaptic transmission induced by spike-timing dependent plasticity requires coactivation of endocannabinoid and muscarinic receptors. *Hippocampus* **2013**, *23*, 1439–1452. [CrossRef] [PubMed]
74. Chiu, C.Q.; Puente, N.; Grandes, P.; Castillo, P.E. Dopaminergic modulation of endocannabinoid-mediated plasticity at GABAergic synapses in the prefrontal cortex. *J. Neurosci.* **2010**, *30*, 7236–7248. [CrossRef]
75. Pan, B.; Hillard, C.J.; Liu, Q.S. D2 dopamine receptor activation facilitates endocannabinoid-mediated long-term synaptic depression of GABAergic synaptic transmission in midbrain dopamine neurons via cAMP-protein kinase A signaling. *J. Neurosci.* **2008**, *28*, 14018–14030. [CrossRef]
76. Lien, C.C.; Mu, Y.; Vargas-Caballero, M.; Poo, M.M. Visual stimuli-induced LTD of GABAergic synapses mediated by presynaptic NMDA receptors. *Nat. Neurosci.* **2006**, *9*, 372–380. [CrossRef]
77. Maffei, A.; Nataraj, K.; Nelson, S.B.; Turrigiano, G.G. Potentiation of cortical inhibition by visual deprivation. *Nature* **2006**, *443*, 81–84. [CrossRef]
78. Holmgren, C.D.; Zilberter, Y. Coincident spiking activity induces long-term changes in inhibition of neocortical pyramidal cells. *J. Neurosci.* **2001**, *21*, 8270–8277. [CrossRef]
79. Woodin, M.A.; Ganguly, K.; Poo, M.M. Coincident pre- and postsynaptic activity modifies GABAergic synapses by postsynaptic changes in Cl− transporter activity. *Neuron* **2003**, *39*, 807–820. [CrossRef]
80. Barberis, A. Postsynaptic plasticity of GABAergic synapses. *Neuropharmacology* **2019**, 107643. [CrossRef]
81. Malenka, R.C.; Nicoll, R.A. Learning and memory. Never fear, LTP is hear. *Nature* **1997**, *390*, 552–553. [PubMed]
82. Kittler, J.T.; Moss, S.J. Modulation of GABAA receptor activity by phosphorylation and receptor trafficking: Implications for the efficacy of synaptic inhibition. *Curr. Opin. Neurobiol.* **2003**, *13*, 341–347. [CrossRef]
83. Kurotani, T.; Yamada, K.; Yoshimura, Y.; Crair, M.C.; Komatsu, Y. State-dependent bidirectional modification of somatic inhibition in neocortical pyramidal cells. *Neuron* **2008**, *57*, 905–916. [CrossRef] [PubMed]
84. Chiu, C.Q.; Barberis, A.; Higley, M.J. Preserving the balance: Diverse forms of long-term GABAergic synaptic plasticity. *Nat. Rev. Neurosci.* **2019**, *20*, 272–281. [CrossRef]
85. Petrini, E.M.; Ravasenga, T.; Hausrat, T.J.; Iurilli, G.; Olcese, U.; Racine, V.; Sibarita, J.B.; Jacob, T.C.; Moss, S.J.; Benfenati, F.; et al. Synaptic recruitment of gephyrin regulates surface GABAA receptor dynamics for the expression of inhibitory LTP. *Nat. Commun.* **2014**, *5*, 3921.
86. Tyagarajan, S.K.; Ghosh, H.; Yévenes, G.E.; Nikonenko, I.; Ebeling, C.; Schwerdel, C.; Sidler, C.; Zeilhofer, H.U.; Gerrits, B.; Muller, D.; et al. Regulation of GABAergic synapse formation and plasticity by GSK3beta-dependent phosphorylation of gephyrin. *Proc. Natl. Acad. Sci. USA* **2011**, *108*, 379–384. [CrossRef]
87. Bannai, H.; Levi, S.; Schweizer, C.; Inoue, T.; Launey, T.; Racine, V.; Sibarit, J.B.; Mikoshiba, K.; Triller, A. Activity-dependent tuning of inhibitory neurotransmission based on GABAAR diffusion dynamics. *Neuron* **2009**, *62*, 670–682. [CrossRef]
88. Arancibia-Carcamo, I.L.; Kittler, J.T. Regulation of GABA(A) receptor membrane trafficking and synaptic localization. *Pharmacol. Ther.* **2009**, *123*, 17–31.
89. Fiumelli, H.; Woodin, M.A. Role of activity-dependent regulation of neuronal chloride homeostasis in development. *Curr. Opin. Neurobiol.* **2007**, *17*, 81–86. [CrossRef]

90. Farrant, M.; Nusser, Z. Variations on an inhibitory theme: Phasic and tonic activation of GABAA receptors. *Nat. Rev. Neurosci.* **2005**, *6*, 215–229. [CrossRef]
91. Jiang, L.; Kang, D.; Kang, J. Potentiation of tonic GABAergic inhibition by activation of postsynaptic kainate receptors. *Neuroscience* **2015**, *298*, 448–454. [CrossRef] [PubMed]
92. Gu, X.; Zhou, L.; Lu, W. An NMDA receptor-dependent mechanism underlies inhibitory synapse development. *Cell Rep.* **2016**, *14*, 471–478. [CrossRef] [PubMed]
93. Jaenisch, N.; Liebmann, L.; Guenther, M.; Hübner, C.A.; Frahm, C.; Witte, O.W. Reduced tonic inhibition after stroke promotes motor performance and epileptic seizures. *Sci. Rep.* **2016**, *6*, 26173.
94. Sigel, E.; Baur, R.; Rácz, I.; Marazzi, J.; Smart, T.G.; Zimmer, A.; Gertsch, J. The major central endocannabinoid directly acts at GABAA receptors. *Proc. Natl. Acad. Sci. USA* **2011**, *108*, 18150–18155. [CrossRef]
95. Gasulla, J.; Calvo, D.J. Enhancement of tonic and phasic GABAergic currents following nitric oxide synthase inhibition in hippocampal CA1 pyramidal neurons. *Neurosci. Lett.* **2015**, *590*, 29–34.
96. Dominguez, S.; Fernandez de Sevilla, D.; Buno, W. Muscarinic long-term enhancement of tonic and phasic GABAA inhibition in rat CA1 pyramidal neurons. *Front. Cell Neurosci.* **2016**, *10*, 244.
97. Chandler, K.E.; Princivalle, A.P.; Fabian-Fine, R.; Bowery, N.G.; Kullmann, D.M.; Walker, M.C. Plasticity of GABAB receptor-mediated heterosynaptic interactions at mossy fibers after status epilepticus. *J. Neurosci.* **2003**, *23*, 11382–11391. [CrossRef]
98. Lecca, S.; Trusel, M.; Mameli, M. Footshock-induced plasticity of GABAB signalling in the lateral habenula requires dopamine and glucocorticoid receptors. *Synapse* **2017**, *71*, e21948. [CrossRef]
99. Chung, H.J.; Ge, W.P.; Qian, X.; Wiser, O.; Jan, Y.N.; Jan, L.Y. G protein-activated inwardly rectifying potassium channels mediate depotentiation of long-term potentiation. *PNAS* **2009**, *106*, 635–640. [CrossRef]
100. Sánchez-Rodríguez, I.; Gruart, A.; Delgado-García, J.M.; Jiménez-Díaz, L.; Navarro-López, J.D. Role of GirK Channels in Long-Term Potentiation of Synaptic Inhibition in an In Vivo Mouse Model of Early Amyloid-β Pathology. *Int. J. Mol. Sci.* **2019**, *20*, 1168. [CrossRef]
101. Abbott, L.; Nelson, S. Synaptic plasticity: Taming the beast. *Nat. Neurosci.* **2000**, *3*, 1178–1183. [CrossRef] [PubMed]
102. Markram, H.; Gerstner, W.; Sjöström, P.J. A history of spike-timing-dependent plasticity. *Front. Synaptic Neurosci.* **2011**, *3*, 1–24.
103. Hennequin, G.; Agnes, E.J.; Vogels, T.P. Inhibitory Plasticity: Balance, Control, and Codependence. *Ann. Rev. Neurosci.* **2017**, *25*, 557–579. [CrossRef] [PubMed]
104. Fiumelli, H.; Cancedda, L.; Poo, M.M. Modulation of GABAergic transmission by activity via postsynaptic Ca2+-dependent regulation of KCC2 function. *Neuron* **2005**, *48*, 773–786. [CrossRef]
105. Gubellini, P.; Ben-Ari, Y.; Gaiarsa, J.L. Activity- and age-dependent GABAergic synaptic plasticity in the developing rat hippocampus. *Eur. J. Neurosci.* **2001**, *14*, 1937–1946. [CrossRef]
106. Huang, C.S.; Shi, S.H.; Ule, J.; Ruggiu, M.; Barker, L.A.; Darnell, R.B.; Jan, Y.N.; Jan, L.Y. Common molecular pathways mediate long-term potentiation of synaptic excitation and slow synaptic inhibition. *Cell* **2005**, *123*, 105–118. [CrossRef]
107. Haas, J.S.; Nowotny, T.; Abarbanel, H.D.I. Spike-timing-dependent plasticity of inhibitory synapses in the entorhinal cortex. *J. Neurophysiol.* **2006**, *96*, 3305–3313. [CrossRef]
108. D'amour, J.; Froemke, R. Inhibitory and excitatory spike-timing-dependent plasticity in the auditory cortex. *Neuron* **2015**, *86*, 514–528.
109. Vickers, E.D.; Clark, C.; Osypenko, D.; Fratzl, A.; Kochubey, O.; Bettler, B.; Schneggenburger, R. Parvalbumin-Interneuron Output Synapses Show Spike-Timing-Dependent Plasticity that Contributes to Auditory Map Remodeling. *Neuron* **2018**, *99*, 720–735.e6. [CrossRef]
110. Turrigiano, G. Homeostatic synaptic plasticity: Local and global mechanisms for stabilizing neuronal function. *Cold Spring Harb. Perspect. Biol.* **2012**, *4*, a005736.
111. House, D.R.; Elstrott, J.; Koh, E.; Chung, J.; Feldman, D.E. Parallel regulation of feedforward inhibition and excitation during whisker map plasticity. *Neuron* **2011**, *72*, 819–831. [CrossRef] [PubMed]
112. Wang, L.; Maffei, A. Inhibitory plasticity dictates the sign of plasticity at excitatory synapses. *J. Neurosci.* **2014**, *34*, 1083–1093. [CrossRef] [PubMed]
113. Campanac, E.; Gasselin, C.; Baude, A.; Rama, S.; Ankri, N.; Debanne, D. Enhanced intrinsic excitability in basket cells maintains excitatory-inhibitory balance in hippocampal circuits. *Neuron* **2013**, *77*, 712–722. [CrossRef] [PubMed]

114. Kapfer, C.; Glickfeld, L.L.; Atallah, B.V.; Scanziani, M. Supralinear increase of recurrent inhibition during sparse activity in the somatosensory cortex. *Nat. Neurosci.* **2007**, *10*, 743–753. [CrossRef] [PubMed]
115. Atallah, B.V.; Scanziani, M. Instantaneous modulation of gamma oscillation frequency by balancing excitation with inhibition. *Neuron* **2009**, *62*, 566–577. [CrossRef] [PubMed]
116. Froemke, R.C.; Merzenich, M.M.; Schreiner, C.E. A synaptic memory trace for cortical receptive field plasticity. *Nature* **2007**, *450*, 425–429. [CrossRef]
117. Gambino, F.; Holtmaat, A. Spike-timing-dependent potentiation of sensory surround in the somatosensory cortex is facilitated by deprivation-mediated disinhibition. *Neuron* **2012**, *75*, 490–502. [CrossRef]
118. Lourenço, J.; Pacioni, S.; Rebola, N.; van Woerden, G.M.; Marinelli, S.; DiGregorio, D.; Bacci, A. Non-associative potentiation of perisomatic Inhibition alters the temporal coding of neocortical layer 5 pyramidal neurons. *PLoS Biol.* **2014**, *12*, e1001903. [CrossRef]
119. Gandolfi, D.; Mapelli, J.; D'Angelo, E. Long-Term Spatiotemporal Reconfiguration of Neuronal Activity Revealed by Voltage-Sensitive Dye Imaging in the Cerebellar Granular Layer. *Neural Plas* **2015**, *2015*, 284986. [CrossRef]
120. Mapelli, J.; Gandolfi, D.; Giuliani, E.; Prencipe, F.P.; Pellati, F.; Barbieri, A.; D'Angelo, E.; Bigiani, A. The effect of desflurane on neuronal communication at a central synapse. *PLoS ONE* **2015**, *10*, e0123534. [CrossRef]
121. Silver, R.A. Neuronal Arithmetic. *Nat. Rev. Neurosci.* **2010**, *11*, 474–489. [CrossRef] [PubMed]
122. Freund, T.F. Interneuron diversity series: Rhythm and mood in perisomatic inhibition. *Trends Neurosci.* **2003**, *26*, 489–495. [CrossRef]
123. Buszaki, G. Neural syntax: Cell assemblies, synapsembles, and reader. *Neuron* **2010**, *68*, 362–385. [CrossRef] [PubMed]
124. Adesnik, H.; Bruns, W.; Taniguchi, H.; Huang, Z.J.; Scanziani, M. A neural circuit for spatial summation in visual cortex. *Nature* **2012**, *490*, 226–231. [CrossRef] [PubMed]
125. Tsodyks, M.V.; Sejnowski, T. Rapid state switching in balanced cortical network models. *Netw. Comput. Neural Syst.* **1995**, *6*, 111–124. [CrossRef]
126. Vogels, T.P.; Sprekeler, H.; Zenke, F.; Clopath, C.; Gerstner, W. Inhibitory plasticity balances excitation and inhibition in sensory pathways and memory networks. *Science* **2011**, *334*, 1569–1573. [CrossRef]
127. Luz, Y.; Shamir, M. Balancing feed-forward excitation and inhibition via Hebbian inhibitory synaptic plasticity. *PLOS Comput. Biol.* **2012**, *8*, e1002334.
128. Litwin-Kumar, A.; Doiron, B. Slow dynamics and high variability in balanced cortical networks with clustered connections. *Nat. Neurosci.* **2012**, *15*, 1498–1505.
129. Landau, I.; Egger, R.; Dercksen, V.; Oberlaender, M.; Sompolinsky, H. The impact of structural heterogeneity on excitation–inhibition balance in cortical networks. *Neuron* **2016**, *92*, 1106–1121. [CrossRef]
130. Vogels, T.P.; Abbott, L.F. Gating multiple signals through detailed balance of excitation and inhibition in spiking networks. *Nat. Neurosci.* **2009**, *12*, 483–491.
131. Wilmes, K.A.; Clopath, C. Inhibitory microcircuits for top-down plasticity of sensory representations. *Nat. Commun.* **2019**, *10*, 5055. [CrossRef] [PubMed]
132. Soloduchin, S.; Shamir, M. Rhythmogenesis evolves as a consequence of long-term plasticity of inhibitory synapses. *Sci. Rep.* **2018**, *8*, 13050.
133. Weber, S.N.; Sprekeler, H. Learning place cells, grid cells and invariances with excitatory and inhibitory plasticity. *Elife* **2018**, *7*, e34560. [CrossRef] [PubMed]
134. Stelly, C.E.; Pomrenze, M.B.; Cook, J.B.; Morikawa, H. Repeated social defeat stress enhances glutamatergic synaptic plasticity in the VTA and cocaine place conditioning. *Elife* **2016**, *4*, e15448. [CrossRef] [PubMed]
135. Hopfield, J.J. Neural networks and physical systems with emergent collective computational abilities. *PNAS* **1982**, *79*, 2554–2558. [CrossRef] [PubMed]
136. Maass, W.; Natschläger, T.; Markram, H. Real-time computing without stable states: A new framework for neural computation based on perturbations. *Neural Comput.* **2002**, *14*, 2531–2560. [CrossRef]
137. Hennequin, G.; Vogels, T.P.; Gerstner, W. Optimal control of transient dynamics in balanced networks supports generation of complex movements. *Neuron* **2014**, *82*, 1394–1406. [CrossRef]
138. Gilra, A.; Gerstner, W. Predicting non-linear dynamics by stable local learning in a recurrent spiking neural network. *Elife* **2017**, *6*, e28295. [CrossRef]
139. Luque, N.R.; Garrido, J.A.; Naveros, F.; Carrillo, R.R.; D'Angelo, E.; Ros, E. Distributed Cerebellar Motor Learning: A Spike-Timing-Dependent Plasticity Model. *Front. Comput Neurosci.* **2016**, *10*, 17. [CrossRef]

140. Widloski, J.; Fiete, I. A model of grid cell development through spatial exploration and spike time– dependent plasticity. *Neuron* **2014**, *83*, 481–495.
141. Inagaki, T.; Begum, T.; Reza, F.; Horibe, S.; Inaba, M.; Yoshimura, Y.; Komatsu, Y. Brain-derived neurotrophic factor-mediated retrograde signaling required for the induction of long-term potentiation at inhibitory synapses of visual cortical pyramidal neurons. *Neurosci. Res.* **2008**, *61*, 192–200. [CrossRef] [PubMed]
142. Lange, M.D.; Doengi, M.; Lesting, J.; Pape, H.C.; Jüngling, K. Heterosynaptic long-term potentiation at interneuron-principal neuron synapses in the amygdala requires nitric oxide signalling. *J. Physiol.* **2012**, *590*, 131–143. [CrossRef] [PubMed]
143. Sivakumaran, S.; Mohajerani, M.H.; Cherubini, E. At immature mossy-fiber-CA3 synapses, correlated presynaptic and postsynaptic activity persistently enhances GABA release and network excitability via BDNF and cAMP-dependent PKA. *J. Neurosci.* **2009**, *29*, 2637–2647. [CrossRef] [PubMed]
144. Duguid, I.C.; Smart, T.G. Retrograde activation of presynaptic NMDA receptors enhances GABA release at cerebellar interneuron–Purkinje cell synapses. *Nat. Neurosci.* **2004**, *7*, 525–533. [CrossRef]
145. He, Q.; Duguid, I.; Clark, B.; Panzanelli, P.; Patel, B.; Thomas, P.; Fritschy, J.M.; Smart, T.G. Interneuron- and GABA(A) receptor-specific inhibitory synaptic plasticity in cerebellar Purkinje cells. *Nat. Commun.* **2015**, *16*, 7364. [CrossRef]
146. Ouardouz, M.; Sastry, B.R. Mechanisms underlying LTP of inhibitory synaptic transmission in the deep cerebellar nuclei. *J. Neurophysiol.* **2000**, *84*, 1414–1421. [CrossRef]
147. Chiu, C.Q.; Martenson, J.S.; Yamazaki, M.; Natsume, R.; Sakimura, K.; Tomita, S.; Tavalin, S.J.; Higley, M.J. Input-Specific NMDAR-Dependent Potentiation of Dendritic GABAergic Inhibition. *Neuron* **2018**, *97*, 368–377.e3. [CrossRef]
148. Kawaguchi, S.Y.; Hirano, T. Sustained structural change of GABA(A) receptor-associated protein underlies long-term potentiation at inhibitory synapses on a cerebellar Purkinje neuron. *J. Neurosci.* **2007**, *27*, 6788–6799. [CrossRef]
149. Patenaude, C.; Chapman, C.A.; Bertrand, S.; Congar, P.; Lacaille, J.C. GABAB receptor- and metabotropic glutamate receptor-dependent cooperative long-term potentiation of rat hippocampal GABAA synaptic transmission. *J. Physiol.* **2003**, *553 (Pt 1)*, 155–167. [CrossRef]
150. Bauer, E.P.; LeDoux, J.E. Heterosynaptic long-term potentiation of inhibitory interneurons in the lateral amygdala. *J. Neurosci.* **2004**, *24*, 9507–9512. [CrossRef]
151. Caillard, O.; Ben-Ari, Y.; Gaïarsa, J.L. Mechanisms of induction and expression of long-term depression at GABAergic synapses in the neonatal rat hippocampus. *J. Neurosci.* **1999**, *19*, 7568–7777. [CrossRef] [PubMed]
152. Lu, Y.M.; Mansuy, I.M.; Kandel, E.R.; Roder, J. Calcineurin-mediated LTD of GABAergic inhibition underlies the increased excitability of CA1 neurons associated with LTP. *Neuron* **2000**, *26*, 197–205. [CrossRef]
153. Morishita, W.; Sastry, B.R. Postsynaptic mechanisms underlying long-term depression of GABAergic transmission in neurons of the deep cerebellar nuclei. *J. Neurophysiol.* **1996**, *76*, 59–68. [CrossRef]
154. Pan, B.; Hillard, C.J.; Liu, Q.S. Endocannabinoid signaling mediates cocaine-induced inhibitory synaptic plasticity in midbrain dopamine neurons. *J. Neurosci.* **2008**, *28*, 1385–1397. [CrossRef] [PubMed]
155. Sánchez-Rodríguez, I.; Djebari, S.; Temprano-Carazo, S.; Vega-Avelaira, D.; Jiménez-Herrera, R.; Iborra-Lázaro, G.; Yajeya, J.; Jiménez-Díaz, L.; Navarro-López, J.D. Hippocampal long-term synaptic depression and memory deficits induced in early amyloidopathy are prevented by enhancing G-protein-gated inwardly rectifying potassium channel activity. *J. Neurochem.* **2019**, e14946.
156. Gogolla, N.; Leblanc, J.J.; Quast, K.B.; Südhof, T.C.; Fagiolini, M.; Hensch, T.K. Common circuit defect of excitatory-inhibitory balance in mouse models of autism. *J. Neurodev. Disord.* **2009**, *1*, 172–181. [CrossRef]
157. Davenport, E.C.; Szulc, B.R.; Drew, J.; Taylor, J.; Morgan, T.; Higgs, N.F.; López-Doménech, G.; Kittler, J.T. Autism and Schizophrenia-Associated CYFIP1 Regulates the Balance of Synaptic Excitation and Inhibition. *Cell Rep.* **2019**, *26*, 2037–2051.e6. [CrossRef]
158. Lewis, D.A.; Hashimoto, T. Deciphering the disease process of schizophrenia: The contribution of cortical GABA neurons. *Int. Rev. Neurobiol.* **2007**, *78*, 109–131.
159. Scharfman, H.E.; Brooks-Kayal, A.R. Is plasticity of GABAergic mechanisms relevant to epileptogenesis? *Adv. Exp. Med. Biol.* **2014**, *813*, 133–150.
160. Milosevic, L.; Gramer, R.; Kim, T.H.; Algarni, M.; Fasano, A.; Kalia, S.K.; Hodaie, M.; Lozano, A.M.; Popovic, M.R.; Hutchison, W.D. Modulation of inhibitory plasticity in basal ganglia output nuclei of patients with Parkinson's disease. *Neurobiol. Dis.* **2019**, *124*, 46–56. [CrossRef]

161. Li, Q.; Serb, A.; Prodromakis, T.; Xu, H. A memristor SPICE model accounting for synaptic activity dependence. *PLoS ONE* **2015**, *10*, e0120506. [CrossRef] [PubMed]
162. Serrano-Gotarredona, T.; Masquelier, T.; Prodromakis, T.; Indiveri, G.; Linares-Barranco, B. STDP and STDP variations with memristors for spiking neuromorphic learning systems. *Front. Neurosci.* **2013**, *7*, 2. [CrossRef] [PubMed]
163. Yang, Y.; Chen, B.; Lu, W.D. Memristive physically evolving networks enabling the emulation of heterosynaptic plasticity. *Adv. Mater.* **2015**, *27*, 7720–7727. [CrossRef] [PubMed]
164. Abu-Hassan, K.; Taylor, J.D.; Morris, P.G.; Donati, E.; Bortolotto, Z.A.; Indiveri, G.; Paton, J.F.R.; Nogaret, A. Optimal solid state neuron. *Nat. Commun.* **2019**, *10*, 5309. [CrossRef] [PubMed]
165. Serb, A.; Bill, J.; Khiat, A.; Berdan, R.; Legenstein, R.; Prodromakis, T. Unsupervised learning in probabilistic neural networks with multi-state metal-oxide memristive synapses. *Nat. Commun.* **2016**, *7*, 12611. [CrossRef] [PubMed]

International Journal of
Molecular Sciences

MDPI

Article

Visualizing the Synaptic and Cellular Ultrastructure in Neurons Differentiated from Human Induced Neural Stem Cells—An Optimized Protocol

Philipp Capetian [1,*], Lorenz Müller [1], Jens Volkmann [1], Manfred Heckmann [2], Süleyman Ergün [3] and Nicole Wagner [3]

1 Department of Neurology, University Hospital Würzburg, 97080 Würzburg, Germany
2 Institute of Physiology, Department of Neurophysiology, University of Würzburg, 97070 Würzburg, Germany
3 Institute of Anatomy and Cell Biology, University of Würzburg, 97070 Würzburg, Germany
* Correspondence: capetian_p@ukw.de

Received: 31 December 2019; Accepted: 28 February 2020; Published: 2 March 2020

Abstract: The size of the synaptic subcomponents falls below the limits of visible light microscopy. Despite new developments in advanced microscopy techniques, the resolution of transmission electron microscopy (TEM) remains unsurpassed. The requirements of tissue preservation are very high, and human post mortem material often does not offer adequate quality. However, new reprogramming techniques that generate human neurons in vitro provide samples that can easily fulfill these requirements. The objective of this study was to identify the culture technique with the best ultrastructural preservation in combination with the best embedding and contrasting technique for visualizing neuronal elements. Two induced neural stem cell lines derived from healthy control subjects underwent differentiation either adherent on glass coverslips, embedded in a droplet of highly concentrated Matrigel, or as a compact neurosphere. Afterward, they were fixed using a combination of glutaraldehyde (GA) and paraformaldehyde (PFA) followed by three approaches (standard stain, Ruthenium red stain, high contrast en-bloc stain) using different combinations of membrane enhancing and contrasting steps before ultrathin sectioning and imaging by TEM. The compact free-floating neurospheres exhibited the best ultrastructural preservation. High-contrast en-bloc stain offered particularly sharp staining of membrane structures and the highest quality visualization of neuronal structures. In conclusion, compact neurospheres growing under free-floating conditions in combination with a high contrast en-bloc staining protocol, offer the optimal preservation and contrast with a particular focus on visualizing membrane structures as required for analyzing synaptic structures.

Keywords: transmission electron microscopy; human neurons; induced neural stem cells; synapse; synaptic vesicles; high contrast

1. Introduction

Neuronal synapses relay and transfer signals between cells and are key components of neural processing. The size of their sub-components (transmitter-filled vesicles, active zone, synaptic cleft, pre- and postsynaptic membranes) fall below the diffraction limits of visible light and thus conventional microscopy. Despite the advent of "superresolution" light microscopic techniques (e.g., direct stochastic optical reconstruction microscopy (dSTORM)), the possibility to visualize the biological membranes and the preservation of the cellular ultrastructure by transmission electron microscopy (TEM) remains unsurpassed [1]. However, for an optimal imaging quality allowing the resolution of structures in the nanometer range, high requirements for tissue preservation have to be met. Aside from rare suitable brain biopsies or surgical specimens, histological studies of human central nervous system (CNS) diseases almost entirely rely on post mortem samples. Due to varying degrees of tissue degradation up until fixation, pathological changes can easily be obscured and achieving sufficient sample quality remains a challenging task [2]. On the other hand, animal models allow perfect control over the modalities of sacrifice, perfusion, and tissue sampling to obtain the best ultrastructural quality. However, genetic animal models can only be obtained for diseases with known mutations that exclude disorders with complex modes of inheritance (e.g., essential tremor [3]). Reprogramming of easily obtainable human cell types, such as fibroblasts, to either pluripotent stem cells [4], followed by differentiation into neurons [5] or to neural stem cells/neurons directly [6,7] allows the derivation of neurons from patients with a wide spectrum of diseases in vitro. Neurons in cell culture can readily be fixed and processed in ways suitable for TEM [8]. We previously published a rather uncomplicated, straight forward protocol based on plasmid transfection that provides directly reprogrammed human induced neural stem cells (iNSC) [9]. These cells can be cultured for many passages and differentiated into neurons or into astrocytes within one month. We, therefore, consider this protocol as a quite accessible way of obtaining neural cell types in vitro from healthy donors or patients. In this scientific study, we explored the ultrastructural preservation of neurons differentiated from iNSC under different culture conditions followed by fixation and TEM imaging of relevant neural structures. Furthermore, we compared different contrasting protocols in their ability to provide optimal visualization of the synaptic apparatus and other neuronal cell components.

2. Results

2.1. *Differentiation of iNSC under Three Distinct Culture Conditions*

Differentiation of neural stem cells into adult neurons can be achieved under various culture conditions. We tested three of them to find out which one offered the best ultrastructural preservation for TEM.

Adherent differentiation on coated coverslips started from 2D-cultured iNSC (Figure 1A). After reaching confluency, differentiation was initiated, and during the following four weeks, cells with a higher cytoplasma/nucleus ratio formed a basal layer with a dense network of neurite-sprouting cells on top (Figure 1B). In a previous study, we identified the first type as astrocytes and the second as neurons [9].

Aggregated iNSC in Matrigel remained at the embedding site and formed a dense fiber network spanning the entire droplet, which was several millimeters in size (Figure 1C).

In u-shaped wells, iNSC aggregated within one to two weeks to a single compact sphere (Figure 1D). After the addition of a differentiation medium, the size of the spheres remained constant. In contrast to the considerable size of the Matrigel droplets, individual neurospheres remained below 1 mm in diameter. No discernible changes in the spheres were visible upon differentiation.

After re-plating neurospheres on coated glass coverslips, attachment of the spheres and outgrowth of cells could be observed (Figure 1E). Immunofluorescence could identify both microtubule-associated protein 2 (MAP2)-positive neurons as well as glial fibrillary acidic protein (GFAP)-positive astrocytes with a proportion of 5:1 (neurons:astrocytes).

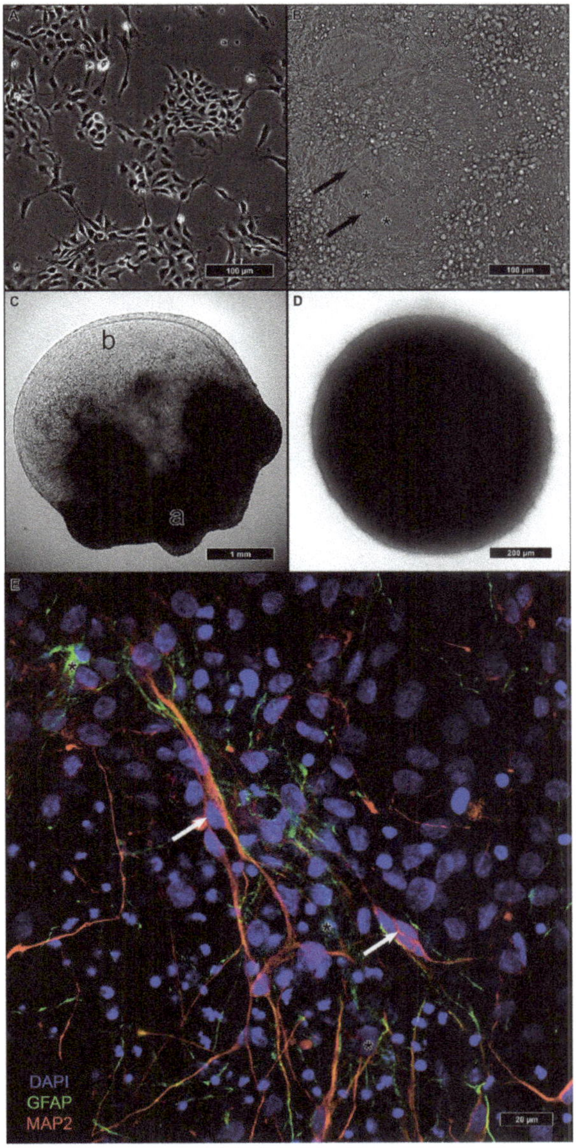

Figure 1. The three differentiation modalities in culture: induced neural stem cells (iNSC) growing adherently as a homogenous monolayer (**A**). After one month of adherent differentiation, neurite spreading cells (arrows) are placed on scattered layers of flat, presumably glial cells (asterisks) (**B**). Neural aggregates embedded in Matrigel, reaching several millimeters in size, kept their original polarity: spheres did not migrate away from their original site and form an apical part (a), while outgrowing neurites formed a dense network in the basal (b) parts of the aggregates (**C**). Neurospheres full of tightly packed cells appeared homogenously with a smooth surface, their size not exceeding 1 mm (**D**). Neurospheres replated on coverslips for outgrowth and incubated for immunofluorescence exhibited microtubule-associated protein 2 (MAP2)-positive neuronal (red, arrows) and glial fibrillary acidic protein (GFAP)-positive astrocytic cells (green, asterisks) in a proportion of roughly 5:1 (neurons: astrocytes) (**E**).

2.2. Ultrastructural Preservation under Three Different Culture Conditions

INSC, differentiated under the described culture conditions, underwent fixation, contrastation, and embedding following standard protocols.

Cells on coverslips, despite remaining adherent during differentiation, tended to lift off during the preparation for TEM. When specimens were imaged by TEM, only processes containing intermediate filaments, and thus, most likely belonging to the astrocytic basal layer, remained (Figure 2A, Supplementary Figure S1). The ultrastructure was decently preserved, but no traces of neuronal cells (such as neurofilaments or synapses) could be found.

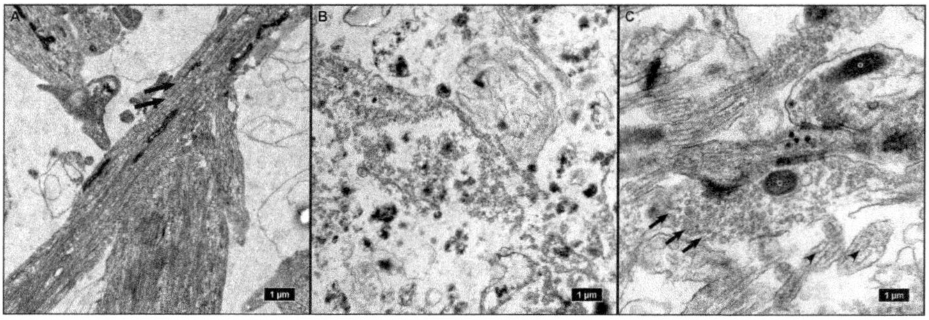

Figure 2. Ultrastructure of the three differentiation modalities: Adherent differentiation on coverslips resulted in the loss of all neuronal cells after sample preparation for transmission electron microscopy (TEM). Only astrocytic processes with intermediate filaments (arrows) were preserved (**A**). Spheres embedded in Matrigel exhibited a fragmented ultrastructure with low preservation of integrity (**B**). Only neurospheres provided preservation of neuronal elements, such as synapses, with vesicles (arrows), and synaptic mitochondria (asterisks), and neurites with neurofilaments (arrowheads) (**C**).

iNSC embedded and differentiated in Matrigel droplets exhibited inferior preservation of subcellular structures. Membranes and intracellular elements obviously had lost their integrity during sample processing for TEM and appeared fragmented (Figure 2B).

The only culture condition that provided sufficient ultrastructural preservation for TEM analysis was the neurosphere culture. Neurospheres remained tightly packed during fixation and embedding. The high cell density inside the spheres allowed screening and visualization of a high number of features in a small area (Figure 2C).

Since only neurospheres provided a sufficient ultrustructural quality after fixation and embedding, we settled on this differentiation method for further analyses.

2.3. Comparison of Three EM Preparation Protocols

Neuronal tissue staining for analysis of synaptic connections by electron microscopy requires optimal ultrastructural preservation in combination with strong deposition of heavy metal compounds into the biological membranes that outline neuronal processes, including axons and dendrites as well as synaptic vesicles. Besides combined primary fixation using a combination of glutaraldehyde and formaldehyde, several other parameters, including pH, osmolarity, and temperature of the washing buffer and primary fixative, are important for the success of ultrastructural preservation of neuronal tissues. In all three protocols, we used cacodylate buffer or phosphate buffer which have both been shown to be highly suitable for the preservation of neuronal tissue.

To enhance membrane contrast, standard staining protocols, including our standard-stain protocol, mostly used a combination of Osmium (Os) tetroxide (OsO_4), and uranyl acetate (UA).

In the second protocol, we added the inorganic dye Ruthenium red (ammoniated ruthenium oxy-chloride) to OsO_4 to enhance the staining, as it has been shown that when used in combination the two compounds react to form ruthenium tetroxide, which reacts with several cellular components resulting in enhanced contrast of diverse tissues [10,11].

Our third protocol was based on a study published by Deerinck and colleagues (Deerinck et al., 2010), which was designed primarily to emphasize the contrast of cellular membranes for serial block-face electron microscopy. Our high contrast en-bloc staining protocol combined subsequent steps after primary aldehyde fixation, including ferrocyanide-reduced osmium tetroxide postfixation, thiocarbohydrazide-osmium liganding (OTO), and subsequent uranyl acetate and en bloc lead aspartate staining. As Ca^{2+} ions are known to enhance membrane preservation and staining, $CaCl_2$ was included in a number of steps.

These three EM preparation protocols (standard stain, Ruthenium red stain, and high contrast en-bloc stain) were compared with respect to ultrastructural preservation and optimal visualization of biological membranes of the synaptic apparatus and other neuronal cell components.

All relevant subcellular structures were clearly discernible by all three protocols. However, unlike the neurospheres processed by the high contrast en-bloc stain, biological membranes were often not clearly visible in the neurospheres processed by the other two protocols (Figure 3–panel 1). We noticed that the chromatin was weakly stained in the high contrast en-bloc stain-processed specimens, likely due to the presence of membrane enhancing reagents during sample preparation (Figure 3–panel 1).

Detailed analysis of subcellular components in the differentially processed samples revealed that the continuity of nuclear envelope, and nuclear pores were preserved to a lesser extent in standard- or Ruthenium red stain-treated samples, whereas optimal preservation was achieved in the high contrast en-bloc stain-treated samples (Figure 3–panel 2). Membranes were essentially parallel to each other and showed no breaks and nuclear pores were clearly visible. No artificial dilation of the intermembraneous space of the nuclear envelope (Figure 3–panel 2), the rough endoplasmic reticulum (Figure 3–panel 3), or the golgi apparatus (Figure 3–panel 4) was seen in the high contrast en-bloc stain-treated samples compared to standard stain-treated samples. Dilation was seen to a lesser extent in Ruthenium red stain-treated samples (Figure 3–panel 2, panel 3, panel 4). However, ribosomes of the outer nuclear membrane, rough endoplasmic reticulum as well as free ribosomes could only be seen clearly in standard- or Ruthenium red stain-treated samples (Figure 3–panel 2, panel 3). Visualization of free ribosomes could be improved using post-staining of TEM sections with prolonged incubation with UAR (uranyl acetate replacement stain, Supplementary Figure S2). Analysis of all samples processed according to the three different protocols revealed sufficient preservation of mitochondria with only mild shrinkage or swelling observed in the neurospheres (Figure 3–panel 5). However, the double membranes and the cristae of mitochondria were only visible as continuous and undilated structures in the high contrast en-bloc stain-treated samples (Figure 3–panel 5).

The particularly enhanced membrane contrast in high contrast en-bloc stain-treated neurospheres led to optimal preservation and discernability of axonal and dendritic processes in the neuropil (Figure 4–panel 1, panel 2). In all three protocols, neurotubules were preserved, but neuronal membrane visualization was highly improved by high contrast en-bloc staining (Figure 4–panel 2). In addition, this method offers high-quality ultrastructural preservation and excellent membrane staining of synaptic connections, although the postsynaptic density was stained less intensely (Figure 5E′) compared to those processed with standard- or Ruthenium red stain (Figure 4–panel 3). Since UA has been implied in labeling proteins of the postsynaptic density similar to heterochromatin, we speculate that the weaker staining of these structures in our high contrast en-bloc stain-processed neurospheres is a result of interference with other membrane enhancing reagents present during sample preparation.

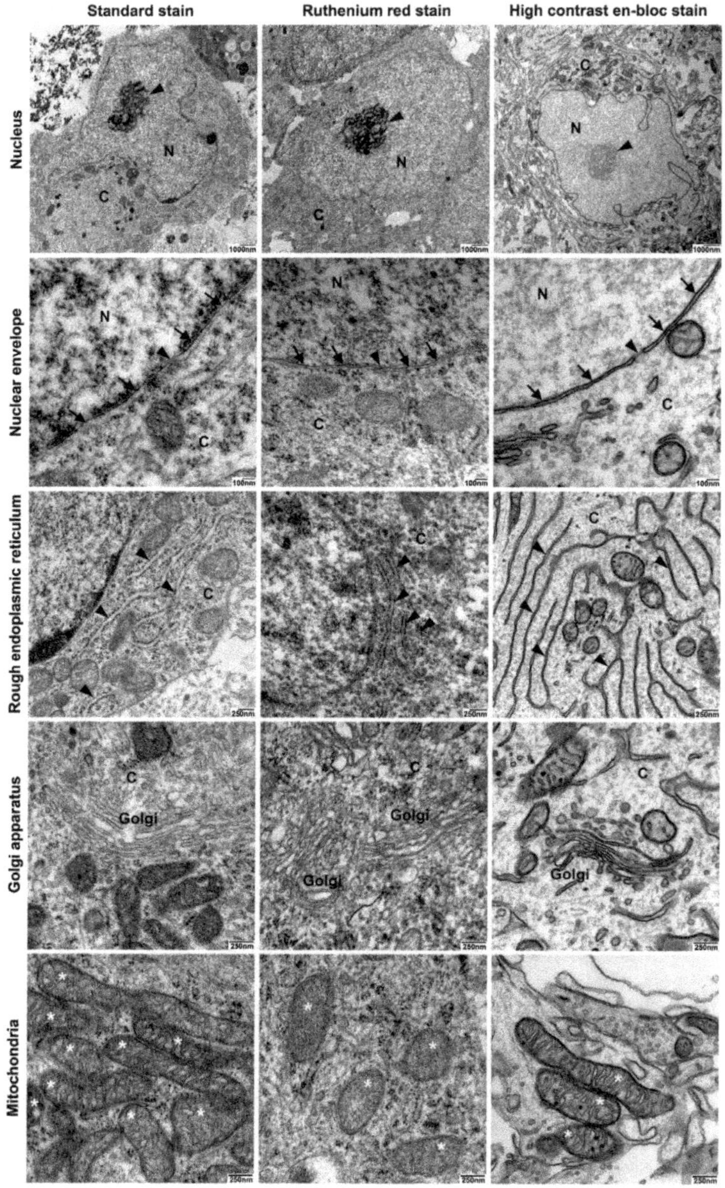

Figure 3. Comparison of the three fixation and staining protocols: Neuronal soma. Three TEM preparation protocols were compared with respect to ultrastructural preservation and optimal visualization of biological membranes of neuronal somata and cellular organelles: Transmission electron micrographs from cultured iNSC-derived neurospheres prepared using standard TEM staining protocol (standard stain, left column), Ruthenium red staining protocol (Ruthenium red stain, middle column), or high contrast electron microscopic staining protocol (high contrast en-bloc stain, right column). Panel 1: Neuronal soma. Soma showing cell nuclei (N) and nucleoli (arrowhead). Note the strong contrast of cellular membranes after sample processing with high contrast en-bloc staining protocol. C: Cytoplasm. Scale bars: 1000 nm. Panel 2: Nuclear envelope and nuclear pores. Preservation of the nuclear envelope (arrows) was improved by sample processing with high contrast en-bloc stain,

which resulted in intact and essentially parallel double membranes with clearly visible nuclear pores (arrowhead). N: nucleus, C: Cytoplasm. Scale bars: 100 nm. Panel 3: Rough endoplasmic reticulum. Optimal preservation and visualization of membranes of rough endoplasmic reticulum with flattened cisternae uniformly arranged in long profiles (arrowheads) after sample processing with high contrast en-bloc stain. Note that ribosomes are less visible after sample processing with high contrast en-bloc stain when compared to samples processed with standard stain or Ruthenium red stain. C: Cytoplasm. Scale bars: 250 nm. Panel 4: Golgi apparatus. Optimal preservation and visualization of golgi membranes with flattened cisternae uniformly arranged in long profiles and golgi vesicles after sample processing with high contrast en-bloc stain. C: Cytoplasm. Scale bars: 250 nm. Panel 5: Mitochondria (marked by asterisks). Optimal preservation and visualization of mitochondria with clearly visible outer double membrane, intact cristae, dense matrix, and mitochondrial granules after sample processing with high contrast en-bloc stain. Scale bars: 250 nm.

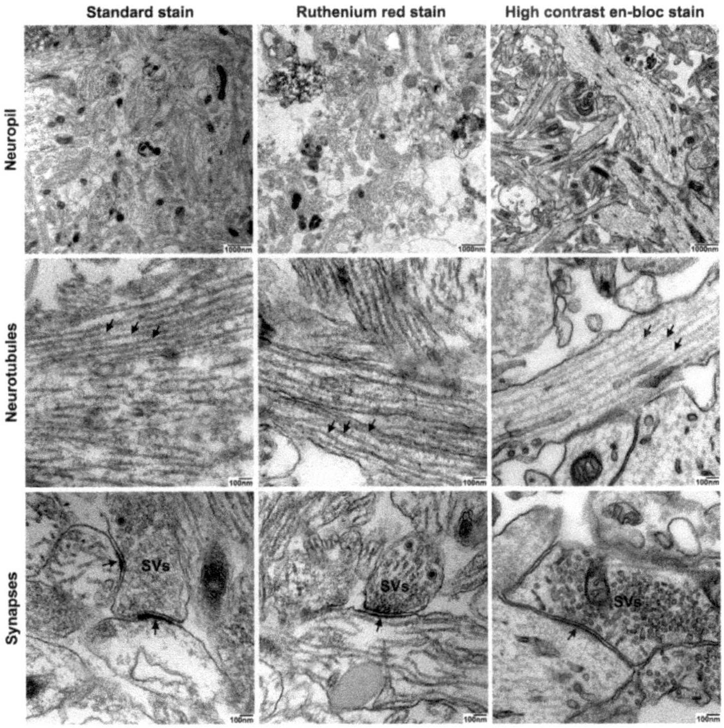

Figure 4. Comparison of the three fixation and staining protocols: Neuropil. Three TEM preparation protocols were compared with respect to ultrastructural preservation and optimal visualization of neuropil: Transmission electron micrographs from cultured iNSC-derived neurospheres prepared using standard TEM staining protocol (standard stain, left column), Ruthenium red staining protocol (Ruthenium red stain, middle column), or high contrast electron microscopic staining protocol (high contrast en-bloc stain, right column). Panel 1: Neuropil. Overview of neuronal processes. Note the strong contrast of cellular membranes after sample processing with high contrast en-bloc stain. Scale bars: 1000 nm. Panel 2: Neurotubules. Higher magnification of neuropil showing preservation of neurotubles (arrowheads) in neuronal processes after sample preparation using the different protocols. Scale bars: 100 nm. Panel 3: Synapses. Optimal preservation and visualization of membranes of the pre- and postsynapse as well as synaptic vesicles (SVs) after sample processing with high contrast en-bloc stain. Note the optimal preservation of the postsynaptic density (arrow) after sample processing with high contrast en-bloc stain. Scale bars: 100 nm.

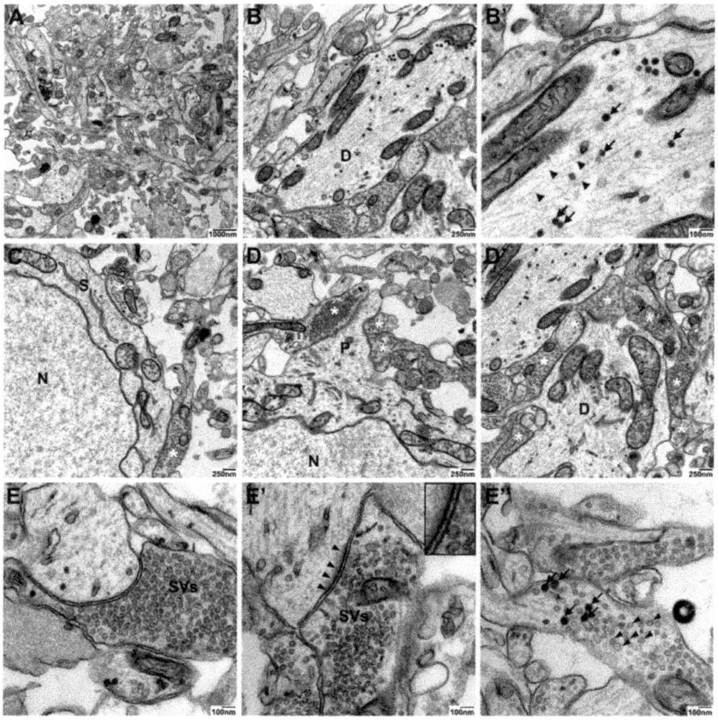

Figure 5. High contrast en-bloc staining protocol offers high-quality ultrastructural preservation and optimal membrane visualization in cultures neurospheres: Synapses. Transmission electron micrographs from cultured iNSC-derived neurospheres prepared using high contrast electron microscopic staining (high contrast en-bloc stain). Overview of neuronal processes (**A**). Single neuronal process (dendrite, **D**) at lower (**B**) and higher magnification (**B′**) showing neurofilaments (arrowheads) and dense core vesicles (arrow) with optimal preservation of cellular membranes. Axosomatic synapses (**C**) and axodendritic synapses (**D**, **D′**); presynaptic compartments. Examples of individual synapses (**E**–**E′**) showing optimal preservation of pre- and postsynaptic membranes, small synaptic cleft, postsynaptic density (**E′**, arrowheads), and synaptic vesicles (SVs). Enlarged view of the boxed region in E′ showing optimal preservation and staining of the lipid bilayers of pre- and postsynaptic membranes. Different types of synaptic vesicles could clearly be distinguished in neuronal processes showing clear (arrowheads) and dense core (arrows) vesicles (**E″**). N: nucleus, S: soma, P: process. Scale bars as indicated.

Detailed analyses of neuronal and synaptic structures in high contrast en-bloc stain-processed neurospheres revealed high-quality ultrastructural preservation and excellent membrane staining of dendritic processes with neurofilaments, dense core vesicles as well as axosomatic and –dendritic synapses (Figure 5A–D). A particular strength of the high contrast en-bloc stain was observed for visualizing subcomponents of the synaptic apparatus: The very clear membrane contrast made the pre- and postsynaptic membrane with the interjacent synaptic cleft easily discernible (Figure 5E and E′). Different types of synaptic vesicles (clear vs. dense core), docked vesicles at the presynaptic membrance, or free vesicles of the resting pool (Figure 5E–E″) could be clearly visualized.

Features of advanced synaptic maturity could be observed in the analyzed samples: All synapses appeared as entirely or nearly filled with synaptic vesicles. We could never observe synapses with single or no vesicles inside (Figure 6A). Of all synaptic contacts, 75% appeared as asymmetric synapses (postsynaptic membrane appearing thicker and more contrasted than the presynatic) (Figure 6B). As already described, most synapses were either axo-dendritic or axo-somatic, sporadically synaptic contacts could also be found on protrusions from the main dendrite, presumably representing spines, but without smooth endoplasmatic reticulum present inside the protrusions (Figure 6C). Occasionally, cells were fixed in the very moment of fusion between a synaptic vesicle and presynaptic membrane as a morphological correlate of synaptic transmission (Figure 6D).

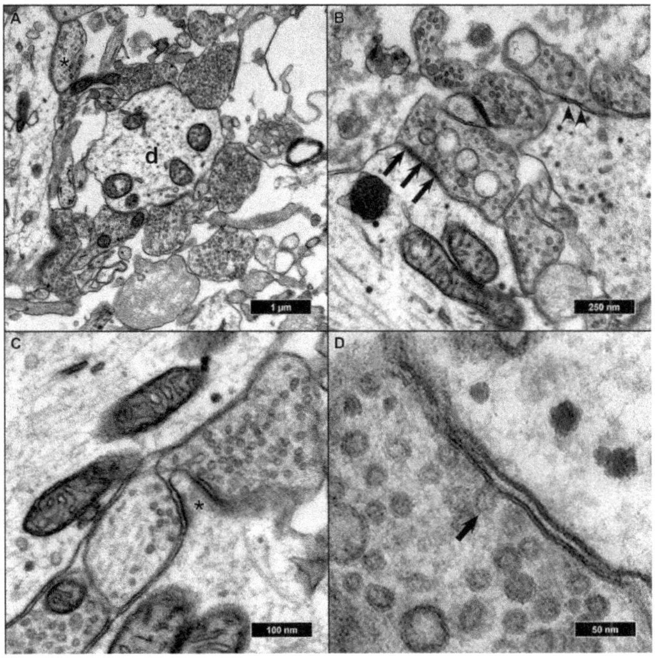

Figure 6. EM micrographs of features associated with synaptic maturity: (**A**) Synaptic boutons terminating at a dendrite (d). All entirely filled with synaptic vesicles but one (asterisk) which showed a sparser filling and thus exhibited a lesser state of maturity. (**B**) The majority of synapses were asymmetric (arrows), and some of them symmetric (arrowheads). (**C**) Protrusions resembling dendritic spines (asterisk) with surrounding synaptic boutons, but absent spine apparatus. (**D**) Synaptic bouton with a vesicle in the moment of fusion with the presynaptic membrane (arrow) as a sign of synaptic functionality.

3. Discussion

For decades visualizing synaptic structures and studying their morphology has been a particular strength of the TEM. In the past, the general notion has been that synaptic transmission is highly conserved [12] and thus was studied in a wider range of model organisms both invertebrate and vertebrate. However, the presence of distinctive features in the human neuromuscular synapse [13] and striking differences in the postsynaptic human proteome in comparison to mouse [14] challenge this notion. Therefore, there are obvious reasons to study synaptic neurotransmission in human neurons, but the aforementioned limitations of sufficiently preserved human CNS tissue challenge these endeavors. Reprogramming techniques can provide human neurons of healthy controls and patients with different kinds of diseases in vitro [15]. After a sufficient time span of differentiation, a functional synaptic network is established [9]. There have been studies in the past, which demonstrated the presence of synaptic connections between human neurons derived from reprogrammed stem cells by electron microscopy on a proof-of-principle basis [8]. However, a systematic comparison of differentiation and TEM preparation protocols concerning optimal visualization of neuronal elements of these cells has not been performed.

The first (and until today, most employed) reprogramming paradigm is the reprogramming towards pluripotent stem cells [4]. Most studies that had a more detailed look into the synapse and neuronal network formation of human neurons in vitro used this cell type [16]. These cells and their pluripotency are maintained by complex and work-intensive protocols requiring almost daily media change and manual removal of spontaneously differentiated cells. For deriving mature neurons, multi-step protocols, combining neural induction, regional patterning, and terminal differentiation, have to be followed that can easily take several months [15]. While this allows a certain enrichment of desired cell types, a 100% pure cell type is never achieved. Adherent differentiation on coated coverslips has been the standard approach leading to immature synapse formation after less than one week and spontaneous synaptic activity after roughly one month [17]. However, a full maturity (e.g., formation of synapses on spines) is possibly not achieved. Neuralizing and differentiating pluripotent stem cells as 3D aggregates (organoids) results in a maturation over months and is more likely to result in a mature synaptic network [18]. In our personal opinion, a culture technique requiring many months until specimens can be studied poses severe challenges to planning and performing experiments (especially if replicates are required).

The alternative is the generation of induced neurons (iN) from somatic cells: Overexpressing transcription factors associated with neuronal identity can reprogram non-neural cells to neurons [19]. The time needed for the reprogramming process to be completed is of similar length as the time needed for neural differentiation from induced pluripotent stem cells (iPSC) [20]. However, since iN are post-mitotic, the number derived is rather small and cannot be increased by proliferation. Thus, the reprogramming process has to be repeated for every set of new experiments.

A good compromise, in our opinion, is induced neural stem cells (iNSC) employed in this study: Derived from somatic cells by plasmid-based transfection, they are proliferative for at least 25 passages but cultivating them requires much less intervention and they are not prone to spontaneous differentiation [9,21]. Differentiation is simply initiated by a change in cell culture media and the addition of three recombinant growth factors and one small molecule. The timing until acquiring a certain level of maturity is not different from the other two methods described. The biggest disadvantage is that these cells are not responsive to patterning cues, meaning their regional identity cannot be altered. They exhibit a quite stable mix of neuronal subtypes (60% upper layer cortical layer, 20% GABAergic, 20% dopaminergic). Therefore, they might not be the first choice when it comes to obtaining a specific cell type, but their strength lies more in providing a good mixture of different neuronal subtypes in one dish.

Another important question in this context is the maturity and functionality of the synaptic network derived from reprogrammed stem cells. Neurons derived from iPSC might appear mature by morphology or specific protein expression after a couple of weeks in culture, yet the formation of functional networks usually takes more than a month, and still not all electrophysiological features associated with them might be present [16]. The same seems to apply to synaptic contacts. Morphologically, synapses begin their existence as mere contacts between two neuronal membranes. Later synaptic vesicles fill the presynaptic bouton, the pre- and postsynaptic membrane becomes more and more defined (resulting in an increasing number of asymmetric synapses with a thicker postsynaptic membrane) and an active synapse capable of neurotransmitter release by fusion of the synaptic vesicles with the presynaptic membrane has come into existence [22,23]. Certain neurons form dendritic spines, highly dynamic structures for multiple synaptic contacts [24]. The timing and sequence of synaptogenesis have already been studied in the human fetus decades ago, but only recently it has been acknowledged that neurons derived from human stem cells in vitro exhibit a different timing in synaptogenesis and certain features associated with synaptic maturity (e.g., spine formation) might be absent altogether [17]. As usual, when dealing with stem cell-derived neurons, much is dependent on the individual protocol employed. The neurospheres from iNSC we employed in our study exhibited a particularly mature phenotype: The majority of synaptic boutons were densely filled with synaptic vesicles, and the majority exhibited an asymmetric morphology and due to the superior membrane contrast of the high contrast en-bloc stain, individual fusions between vesicles and the presynaptic membrane could be observed. We even observed membrane protrusions that could be dendritic spines, a feature that has only be observed in iPSC derived cerebral organoids after many months in culture [25]. However, concerning this finding, uncertainty remained. We did not perform serial block-face scanning electron microscopy for three-dimensional reconstruction of the protrusions, which is the method of choice for unequivocally visualizing dendritic spines. Furthermore, we did not find a spine apparatus (smooth endoplasmatic reticulum inside the spine) in any protrusion. Not all spines contain a synaptic apparatus, but the presence of it is generally considered a sign of maturity [26]. We would, therefore, consider the presence of possibly still immature spines with no signs of full maturity.

The iNSC employed in this study were cultured adherently on Matrigel. In a previous publication, we induced differentiation of iNSC into mature neurons on glass coverslips as it simplified staining and fluorescence microscopy, transfer to recording chambers, etc. [9]. As we observed in the past, iNSC generated using our protocol differentiate into both neurons and astrocytes with the neurons exhibiting the tendency to aggregate on top of the astrocytes (Figure 1B). Presumably, the numerous pipetting steps leading to repeated shear stress resulted in a detachment of the neuronal layer leaving only the astrocytic basal layer.

Aggregating iNSC into neurospheres and embedding them into Matrigel droplets followed a protocol similar to the formation of cerebral organoids [18]. However, instead of forming a compact neuroepithelial layer with ventricle-like cavities, differentiating iNSC remained stationary and extended neurites throughout the aggregates. In contrast to iPSC-derived embryoid bodies that served as seeds for the cerebral organoids, our iNSC were different from early neuroepithelial precursors. We can only speculate that the comparatively large size of the aggregates of a few millimeters in conjunction with being mainly composed of a neurite fiber network, led to insufficient preservation for TEM.

Only the densely packed and rather small neurospheres reliably provided a sufficient ultrastructure. Neurospheres were generated by culturing iNSC under growth conditions until single spheres formed [27,28]. The growth of the spheres ceased after the induction of differentiation. Thus, the size of the spheres was mainly determined by the cell number seeded per well. This provided a good compromise between a small size allowing sufficient penetration of the fixatives and solutions during the embedding process and safe handling under the naked eye. Furthermore, the small size, yet a high density of cell bodies and processes, simplified screening during TEM analysis for relevant structures.

Besides the ultrastructural preservation, reliably identifying structures of interest by TEM requires high contrast with a sharp delineation. Neuronal structures that usually receive the most attention in TEM studies are either membrane-rich as synapses and mitochondria or filamentous, such as neurofilaments or neurotubuli. The standard- and Ruthenium red-stainings appeared "grainy", and the contours of membranes were sometimes hard to discriminate from surrounding structures. A better contrast was achieved for structures rich in DNA or RNA (nucleus, ribosomes) or proteins (filaments and tubuli). The high contrast en-block stain was primarily designed for enhancing membrane contrast in mammalian tissue in serial block-face scanning electron microscopy (SBF-SEM) [29]. Although developed for a different EM method, preserving and contrasting membranes proved useful for TEM as well. En-bloc contrasting protocols were optimized for better penetration of larger tissue samples and superseded the contrasting of ultra-thin cut sections on grids which is prone to non-specific deposition of the contrasting heavy metals [30]. In contrast to the other two staining protocols tested here, $CaCl_2$ (which is known to improve the stability of lipid bilayers [31]) was added to the fixative as well as to some of the contrasting steps. The increased contrast of membranes with this protocol was the result of a combination of different staining principles established in numerous individual studies: Ferrocyanide reduced osmium post-fixation in combination with cacodylate buffer, partially extracted the cytoplasmatic ground substance and mitochondrial matrix while strongly binding to membranes [32]. Insufficient preservation of the lipid bilayer structure by the aforementioned method was compensated by a downstream osmium–thiocarbohydrazide–osmium (OTO) step [33]. Using the classical contrasting agent uranyl–acetate (highly toxic and nowadays replaced by samarium and gadolinium [34]) and lead aspartate in pre-embedding en-bloc staining of wet tissue overcame their tendency to form contaminating precipitates when applied post-embedding [35,36].

To our knowledge, this is the first systematic comparison of different protocols for the visualization of neurons derived from human reprogrammed cells by TEM and the first example of en-bloc staining techniques employed on three-dimensional cellular aggregates. Consequently, the paramount properties of en-bloc staining in the preservation and contrasting of neuronal membranes work for in vitro specimens as well.

4. Materials and Methods

4.1. Derivation and Proliferation of iNSC

Detailed protocols describing the reprogramming and culture of iNSC from human fibroblast cultures have been published before [9,21]. In short, fibroblast cultures from 2 healthy donors who gave informed consent following the requirements and positive votum (AZ12-219) of the ethics committee of the University of Lübeck, Germany, were transfected with three polycistronic plasmids overexpressing transcription factors associated with pluripotency (Oct3/4, Sox2, Klf4, L-myc, Lin28) as well as a small hairpin RNA directed against p53 [37]. The presence of Epstein–Barr virus (EBV)-derived origin of viral replication/Epstein–Barr virus nuclear antigen 1 (oriP/EBNA-1) ensured the DNA amplification during cell cycles and extended the presence of plasmids in the cells. After switching the standard fibroblast medium (DMEM, 10% fetal calf serum, 1% Glutamax, 1% Antibiotic/Antimycotic, all from Thermo Fisher Scientific, Waltham, USA) to a commercially available neural induction medium (STEMCELL Technologies, Vancouver, Canada). Neural colonies began to emerge after about two to three weeks, were mechanically picked, and replated on Matrigel (Thermo Fisher Scientific, Waltham, USA) coated culture dishes. The growth medium was changed to neural progenitor medium (STEMCELL Technologies, Vancouver, Canada), confluent cultures split by accutase (Thermo Fisher Scientific, Waltham, USA) digestion and replated in a ratio of 1:10 for continuous proliferation.

4.2. Differentiation of iNSC

For differentiation, iNSC were split and viable cells determined by staining with 1:10 Trypan Blue 0.4% (Thermo Fisher Scientific, Waltham, MA, USA) and counting with a Neubauer improved counting chamber (Glaswarenfabrik Karl Hecht GmbH & Co KG, Sondheim vor der Rhön, Germany). INSC were plated and differentiated following three different protocols. All experiments were performed with the two iNSC lines in duplicates.

4.2.1. Adherent Differentiation on Glass Coverslips

Twelve-millimeter glass coverslips (Glaswarenfabrik Karl Hecht GmbH & Co KG, Sondheim vor der Rhön, Germany) were surface treated by 65% sulphuric acid (Carl Roth, Karlsruhe, Germany) overnight and heated to 180° for 8 h. Once put inside 24-well culture plates (Sigma–Aldrich, St. Louis, MI, USA), we performed coating overnight with a poly-D-lysine (Sigma–Aldrich, St. Louis, USA) solution (0.075 mg/mL in 0.1 M borate buffer) at room temperature and overnight by a laminin (Sigma–Aldrich, St. Louis, USA) solution (5 µg/mL in PBS) at 37 °C. One hundred and fifty thousand cells per well were plated in a neural progenitor medium with the addition of 1 µM of the rho kinase inhibitor Y-27632 (STEMCELL Technologies, Vancouver, Canada). After iNSC reached confluence, differentiation started by switching to a neural differentiation medium, composed as follows: DMEM/F12:Neurobasal 1:1, 1% N2, 2% B27 (Gibco by Life Technologies, Vancouver, BC, Canada), 20 ng/mL brain-derived neurotrophic factor (BDNF), 10 ng/mL glial cell line-derived neurotrophic factor (GDNF), 10 ng/mL insulin-like growth factor 1 (IGF-1) (all from PeproTech, Rocky Hill, NJ, USA), 0,5mM dibutyril cyclic adenosine–monophosphate (dbcAMP, EnzoLife Sciences, Farmingdale, NY, USA) and 10 μM of the Notch-pathway inhibitor DAPT (Tocris, Ellisville, MO, USA). Differentiation continued for four weeks.

4.2.2. Free Floating Differentiation of Aggregates Embedded in Matrigel

The second differentiation protocol was adapted from a protocol of cerebral organoid formation [18]: Fifty thousand cells per well were plated inside a u-shaped ultra-low attachment 96-well-plated (Thermo Fisher Scientific, Waltham, MA, USA) in neural progenitor medium. Cells aggregated into spheres for 7 days, were removed from the wells and embedded in groups of 5 inside 50 μL droplets of ice-cold Matrigel (Thermo Fisher Scientific, Waltham, MA, USA) on a sheet of sterilized parafilm (Bemis, Neenah, WI, USA). After polymerization for 30′ at 37 °C, the droplets were removed from the film, transferred into ultra-low attachment 6-well plates (Thermo Fisher Scientific, Waltham, MA, USA) and differentiation induced by differentiation medium for four weeks.

4.2.3. Free Floating Differentiation of Neurospheres

This method corresponded to the method described under 4.2.1 except that cells were kept and differentiated inside the ultra-low attachment 96-well plates for four weeks.

4.2.4. Replating and Immunofluorescence of Neurospheres

After four weeks of differentiation, neurospheres were put onto coverslips coated as described under 4.2.1 in the same medium as described above. Neurospheres attached to the coated surface and cells started migrating out. After one week, spheres and cells were fixed with 4% Paraformaldehyde (Carl Roth, Karlsruhe, Germany) and washed with phosphate-buffered saline (Thermo Fisher Scientific, Waltham, MA, USA). After blocking of unspecific binding sites with 5% donkey serum in PBS with 0.1% Triton X-100 (Thermo Fisher Scientific, Waltham, MA, USA), primary antibodies (MAP2, Millipore, Cat. No. MAB378, 1:500 and GFAP Zytomed Systems Cat. No. RBK037 1:500) were incubated overnight at 4 °C in 1% donkey serum in PBS with 0.1% Triton X-100. The next day specimens were washed for 3 × 15 min in PBS + 0.1% Triton X-100, followed by an incubation of the secondary antibodies (donkey anti-mouse CF568 and donkey anti-rabbit CF488, both Sigma–Aldrich, St. Louis, USA) for 2 h at room temperature. After another washing step for 3 × 15 min, nuclear staining wash performed by adding 1 μg/mL DAPI (Thermo Fisher Scientific, Waltham, USA) in PBS to the cells for 10 min and washing for 3 × 5 min with PBS. Coverslips were embedded on microscope slides (Glaswarenfabrik Karl Hecht GmbH & Co KG, Sondheim vor der Rhön, Germany) with ProLong Glass (Thermo Fisher Scientific, Waltham, MA, USA), and stored at 4 °C in the dark. Imaging was performed with an Olympus FV1000 confocal laser scanning microscope (Olympus, Tokyo, Japan). Six high-power fields (60× magnification) were taken and analyzed.

4.3. Specimen Preparation for Transmission Electron Microscopy

4.3.1. Standard Electron Microscopic Preparation (Standard Stain)

Coverslips, aggregates in Matrigel, or neurospheres were washed to remove cell culture medium and fixed in 0.1 M cacodylate buffer pH 7.4 containing 2.5% glutaraldehyde, 4% formaldehyde (fresh from paraformaldehyde) for 3 h at room temperature (RT) and left overnight in a fixative at 4 °C. After washing for 3 × 10 min in cold PBS (phosphate-buffered saline, pH 7.2), samples were subsequently fixed for 1 hr with 1% osmium tetroxide (buffered with PBS, pH 7.2). They were washed 3 x 10 min in PBS and 1 x 10 min in ddH_2O at RT. Afterward, specimens were incubated in aqueous UAR-EMS (4%, Uranyl acetate replacement stain, Electron Microscopy Sciences, Hatfield, USA). Washed 3 × 10 min in ddH_2O at RT and dehydrated in an ascending ethanol series using solutions of 30%, 50%, 70%, 90%, 96%, 100%, 100% ethanol 10 min each. They were incubated two times in propylene oxide (PO) for 30 min each before incubation in a mixture of PO and Epon812 (1:1) overnight. The following day, the Epon-PO mixture was substituted with pure Epon812 and samples were incubated for 2 h in

Epon812. Specimens were embedded in Epon812 and kept at 60 °C for 48hrs. For ultrathin sections, 70 nm thick ultrathin sections were cut with an ultramicrotome (Ultracut E, Reichert Jung, Germany) and collected on copper or nickel grids. Sections were post-stained with aqueous UAR-EMS (2,5%, Uranyl acetate replacement stain, Electron Microscopy Sciences, Hatfield, USA) and 0.2% lead citrate and finally analyzed with an LEO AB 912 transmission electron microscope (Carl Zeiss Microscopy GmbH, Germany).

4.3.2. Electron Microscopic Preparation Using Ruthenium Red (Ruthenium Red Stain)

To provide enhanced contrast staining of neurospheres compared to conventional electron microscopic protocols, the polycationic stain Ruthenium red was used in combination with OsO4 staining as follows: Neurospheres were washed to remove cell culture medium and fixed in 0.1 M cacodylate buffer pH 7.4 containing 2.5% glutaraldehyde, 0.01% Ruthenium red for 3 h at RT and left overnight in fixative at 4 °C. After washing for 3 × 5 min in 0.1 M cacodylate buffer, neurospheres were subsequently fixed for 1 hr with 2% osmium tetroxide (buffered with 0.1 M cacodylate buffer, pH 7.2). Neurospheres were washed 3 × 5 min 0.1 M cacodylate buffer and 1 × 5 min in ddH_2O at RT. Neurospheres were incubated in aqueous UAR-EMS (4%, Uranyl acetate replacement stain, Electron Microscopy Sciences, Hatfield, PA, USA), washed 3 × 5 min in ddH_2O at RT and dehydrated in an ascending methanol series using solutions of 25%, 50%, 70%, 80%, 95%, 95%, 100%, 100% methanol 10 min each. Neurospheres were incubated two times in Ethoxy propanol for 10 min each and incubated for 2 h in Epon812. Specimens were embedded in Epon812 and kept at 60 °C for 48hrs. For ultrathin sections, 70 nm thick ultrathin sections were cut with an ultramicrotome (Ultracut E, Reichert Jung, Germany) and collected on copper or nickel grids. Sections were post-stained with aqueous UAR-EMS (2,5%, Uranyl acetate replacement stain, Electron Microscopy Sciences, Hatfield, USA) and 0.2% lead citrate and finally analyzed with an LEO AB 912 transmission electron microscope (Carl Zeiss Microscopy GmbH, Germany).

4.3.3. High Contrast Electron Microscopic Preparation (High Contrast En-Bloc Stain)

To specifically enhance the staining of neuronal membranes, neurospheres were prepared using a modification of the National Center for Microscopy and Imaging Research (NCMIR) protocol [29], which was initially developed for the preparation of biological specimens for serial block-face scanning electron microscopy. Briefly, we rinsed neurospheres to remove cell culture medium and fixed in 0.12 M phosphate buffer (PB buffer) pH 7.5 containing 1.0% glutaraldehyde, 1.0% formaldehyde (fresh from paraformaldehyde), 0.002% $CaCl_2$ and 2% sacharose for 3 h at RT and left overnight in a fixative at 4 °C. Subsequently, neurospheres were washed in 0.15 mM sodium cacodylate buffer (pH 7.4) containing 2 mM $CaCl_2$ and incubated for 60 min on ice in a reduced osmium solution containing 2% osmium tetroxide, 1.5% potassium ferrocyanide, 2 mM $CaCl_2$ in 0.15 mM sodium cacodylate buffer (pH 7.4). Neurospheres were washed in ddH_2O at room temperature (RT) 5 × 5 min, followed by incubation in 1% thiocarbohydrazide (TCH) solution for 25 min at RT. Neurospheres were washed in ddH2O at RT, 5 × 5 min each and incubated in 2% osmium tetroxide in ddH_20 for 30 min at RT. Subsequently, neurospheres were washed 5 × 5 min at RT in ddH_2O then incubated in aqueous 4% UAR-EMS. Neurospheres were washed 3 × 3 min in ddH_2O at RT. Prior incubation with lead aspartate solution, neurospheres were washed 2 × 3 min in ddH_2O at 60 °C, and then subjected to en bloc Walton's lead aspartate staining [36] and placed in a 60 °C oven for 30 min. Neurospheres were washed 5 × 5 min in ddH_2O at RT and dehydrated using ice-cold solutions of freshly prepared 30%, 50%, 70%, 90%, 100%, 100% ethanol (anhydrous), 100% acetone (anhydrous) for 10 min each, then placed in anhydrous ice-cold acetone and left at RT for 10 min. Neurospheres were placed in 100% acetone at RT for 10 min. During this time, Epon812 was prepared. The resin was mixed thoroughly and samples were placed into 25% Epon:75% acetone for 2 h, then into 50% Epon:50% acetone for 2 h and 75% Epon:25% acetone for 2 h. Neurospheres were placed in 100% Epon overnight. The next day, Epon was replaced with fresh Epon for 2 h and neurospheres were placed in beem capsules and incubated in a 60 °C oven for

48 h for resin polymerization. For ultrathin sections, 70 nm thick ultrathin sections were cut with an ultramicrotome (Ultracut E, Reichert Jung, Germany) and collected on copper or nickel gridsm, and finally analyzed with a LEO AB 912 transmission electron microscope (Carl Zeiss Microscopy GmbH, Germany).

Supplementary Materials: Supplementary Materials can be found at http://www.mdpi.com/1422-0067/21/5/1708/s1.

Author Contributions: Conceptualization: N.W., P.C. Methodology: L.M., N.W., P.C. Validation: N.W., P.C. Formal analysis: N.W., P.C. Investigation: L.M., N.W., P.C. Resources: J.V., M.H., S.E. Writing—original draft preparation: N.W., P.C. Writing—review and editing J.V., L.M., M.H., N.W., P.C., S.E. Project administration: L.M., N.W., P.C. Funding acquisition: J.V., L.M., M.H., S.E. All authors have read and agreed to the published version of the manuscript.

Funding: This research was funded by BMBF, grant number 01GM1514A, and the interdisciplinary center of clinical research (IZKF) of the University Hospital Würzburg, grant number Z3-61. This publication was supported by the Open Access Publication Fund of the University of Wuerzburg.

Acknowledgments: We like to thank Claudia Wirth and Johanna Mehling for their outstanding technical assistance that made this study possible. We thank Mario Vallon for careful reading of the manuscript.

Conflicts of Interest: The authors declare no conflict of interest.

Abbreviations

TEM	Transmission electron microscopy
GA	Glutaraldehyde
dSTORM	Direct stochastic optical reconstruction microscopy
EBNA-1	Epstein–Barr virus nuclear antigen 1
EBV	Epstein–Barr virus
GFAP	Glial fibrillary acidic protein
iNSC	Induced neural stem cells
MAP2	Microtubule associated protein 2
OriP	Origin of viral replication
PFA	Paraformaldehyde
PO	Propylene oxide
RT	Room temperature
UAR	Uranyl acetate replacement

References

1. Galbraith, C.G.; Galbraith, J.A. Super-resolution microscopy at a glance. *J. Cell Sci.* **2011**, *124*, 1607–1611. [CrossRef] [PubMed]
2. Lewis, A.J.; Genoud, C.; Pont, M.; van de Berg, W.D.; Frank, S.; Stahlberg, H.; Shahmoradian, S.H.; Al-Amoudi, A. Imaging of post-mortem human brain tissue using electron and X-ray microscopy. *Curr. Opin. Struct. Biol.* **2019**, *58*, 138–148. [CrossRef] [PubMed]
3. Jiménez-Jiménez, F.J.; Alonso-Navarro, H.; García-Martín, E.; Lorenzo-Betancor, O.; Pastor, P.; Agúndez, J.a.G. Update on genetics of essential tremor. *Acta Neurol. Scand.* **2013**, *128*, 359–371. [CrossRef] [PubMed]
4. Takahashi, K.; Tanabe, K.; Ohnuki, M.; Narita, M.; Ichisaka, T.; Tomoda, K.; Yamanaka, S. Induction of pluripotent stem cells from adult human fibroblasts by defined factors. *Cell* **2007**, *131*, 861–872. [CrossRef] [PubMed]
5. Chambers, S.M.; Fasano, C.A.; Papapetrou, E.P.; Tomishima, M.; Sadelain, M.; Studer, L. Highly efficient neural conversion of human ES and iPS cells by dual inhibition of SMAD signaling. *Nat. Biotechnol.* **2009**, *27*, 275–280. [CrossRef] [PubMed]
6. Kim, J.; Efe, J.A.; Zhu, S.; Talantova, M.; Yuan, X.; Wang, S.; Lipton, S.A.; Zhang, K.; Ding, S. Direct reprogramming of mouse fibroblasts to neural progenitors. *Proc. Natl. Acad. Sci.* **2011**, *108*, 7838–7843. [CrossRef]
7. Liu, X.; Li, F.; Stubblefield, E.A.; Blanchard, B.; Richards, T.L.; Larson, G.A.; He, Y.; Huang, Q.; Tan, A.-C.; Zhang, D.; et al. Direct reprogramming of human fibroblasts into dopaminergic neuron-like cells. *Cell Res.* **2011**. [CrossRef]

8. Kim, J.-E.; O'Sullivan, M.L.; Sanchez, C.A.; Hwang, M.; Israel, M.A.; Brennand, K.; Deerinck, T.J.; Goldstein, L.S.B.; Gage, F.H.; Ellisman, M.H.; et al. Investigating synapse formation and function using human pluripotent stem cell-derived neurons. *Proc. Natl. Acad. Sci. USA.* **2011**, *108*, 3005–3010. [CrossRef]
9. Capetian, P.; Azmitia, L.; Pauly, M.G.; Krajka, V.; Stengel, F.; Bernhardi, E.-M.; Klett, M.; Meier, B.; Seibler, P.; Stanslowsky, N.; et al. Plasmid-based generation of induced neural stem cells from adult human fibroblasts. *Front. Cell. Neurosci.* **2016**, *10*, 245. [CrossRef]
10. Luft, J.H. Ruthenium red and violet. I. Chemistry, purification, methods of use for electron microscopy and mechanism of action. *Anat. Rec.* **1971**, *171*, 347–368. [CrossRef]
11. Sobota, A.; Mrozińska, K.; Popov, V.I. Anionic domains on the cytoplasmic surface of the plasma membrane of Acanthamoeba castellanii and their relation to calcium-binding microregions. *Acta Protozoologica* **1997**, *36*, 187–196.
12. Hadley, D.; Murphy, T.; Valladares, O.; Hannenhalli, S.; Ungar, L.; Kim, J.; Bućan, M. Patterns of sequence conservation in presynaptic neural genes. *Genome Biol.* **2006**, *7*, R105. [CrossRef] [PubMed]
13. Jones, R.A.; Harrison, C.; Eaton, S.L.; Llavero Hurtado, M.; Graham, L.C.; Alkhammash, L.; Oladiran, O.A.; Gale, A.; Lamont, D.J.; Simpson, H.; et al. Cellular and Molecular Anatomy of the Human Neuromuscular Junction. *Cell Rep.* **2017**, *21*, 2348–2356. [CrossRef] [PubMed]
14. Bayés, À.; Collins, M.O.; Croning, M.D.R.; Lagemaat, L.N. van de; Choudhary, J.S.; Grant, S.G.N. Comparative Study of Human and Mouse Postsynaptic Proteomes Finds High Compositional Conservation and Abundance Differences for Key Synaptic Proteins. *PLoS ONE* **2012**, *7*, e46683. [CrossRef] [PubMed]
15. Capetian, P.; Stanslowsky, N.; Bernhardi, E.; Grütz, K.; Domingo, A.; Brüggemann, N.; Naujock, M.; Seibler, P.; Klein, C.; Wegner, F. Altered glutamate response and calcium dynamics in iPSC-derived striatal neurons from XDP patients. *Exp. Neurol.* **2018**, *308*, 47–58. [CrossRef] [PubMed]
16. Bradford, A.B.; McNutt, P.M. Importance of being Nernst: Synaptic activity and functional relevance in stem cell-derived neurons. *World J. Stem Cells* **2015**, *7*, 899–921. [CrossRef]
17. Wilson, E.S.; Newell-Litwa, K. Stem cell models of human synapse development and degeneration. *Mol. Biol. Cell* **2018**, *29*, 2913–2921. [CrossRef]
18. Lancaster, M.A.; Renner, M.; Martin, C.-A.; Wenzel, D.; Bicknell, L.S.; Hurles, M.E.; Homfray, T.; Penninger, J.M.; Jackson, A.P.; Knoblich, J.A. Cerebral organoids model human brain development and microcephaly. *Nature* **2013**, *501*, 373–379. [CrossRef]
19. Vierbuchen, T.; Ostermeier, A.; Pang, Z.P.; Kokubu, Y.; Südhof, T.C.; Wernig, M. Direct conversion of fibroblasts to functional neurons by defined factors. *Nature* **2010**, *463*, 1035–1041. [CrossRef]
20. Drouin-Ouellet, J.; Pircs, K.; Barker, R.A.; Jakobsson, J.; Parmar, M. Direct Neuronal Reprogramming for Disease Modeling Studies Using Patient-Derived Neurons: What Have We Learned? *Front. Neurosci.* **2017**, *11*, 530. [CrossRef]
21. Azmitia, L.; Capetian, P. Single-Step Plasmid Based Reprogramming of Human Dermal Fibroblasts to Induced Neural Stem Cells. *Methods Mol. Biol. Clifton NJ* **2018**, *1842*, 31–41.
22. Molliver, M.E.; Kostović, I.; van der Loos, H. The development of synapses in cerebral cortex of the human fetus. *Brain Res.* **1973**, *50*, 403–407. [CrossRef]
23. Petit, T.L.; LeBoutillier, J.C.; Alfano, D.P.; Becker, L.E. Synaptic development in the human fetus: A morphometric analysis of normal and Down's syndrome neocortex. *Exp. Neurol.* **1984**, *83*, 13–23. [CrossRef]
24. Yuste, R.; Bonhoeffer, T. Genesis of dendritic spines: Insights from ultrastructural and imaging studies. *Nat. Rev. Neurosci.* **2004**, *5*, 24–34. [CrossRef]
25. Quadrato, G.; Nguyen, T.; Macosko, E.Z.; Sherwood, J.L.; Yang, S.M.; Berger, D.; Maria, N.; Scholvin, J.; Goldman, M.; Kinney, J.; et al. Cell diversity and network dynamics in photosensitive human brain organoids. *Nature* **2017**, *545*, 48–53. [CrossRef]
26. Spacek, J.; Harris, K.M. Three-Dimensional Organization of Smooth Endoplasmic Reticulum in Hippocampal CA1 Dendrites and Dendritic Spines of the Immature and Mature Rat. *J. Neurosci.* **1997**, *17*, 190–203. [CrossRef]
27. Jensen, J.B.; Parmar, M. Strengths and limitations of the neurosphere culture system. *Mol. Neurobiol.* **2006**, *34*, 153–161. [CrossRef]
28. Pauly, M.G.; Krajka, V.; Stengel, F.; Seibler, P.; Klein, C.; Capetian, P. Adherent vs. Free-Floating Neural Induction by Dual SMAD Inhibition for Neurosphere Cultures Derived from Human Induced Pluripotent Stem Cells. *Front. Cell Dev. Biol.* **2018**, *6*, 3. [CrossRef]

29. Deerinck, T.J.; Bushong, E.A.; Thor, A.; Ellisman, M.H.; Deerinck, T.J.; Bushong, E.; Ellisman, M.; Deerinck, T.; Thor, A.; Thor, C.A. *NCMIR Methods for 3D EM: A New Protocol for Preparation of Biological Specimens for Serial Block Face Scanning Electron Microscopy*; Center for Research in Biological Systems and the National Center for Microscopy and Imaging Research, University of California: San Diego, CA, USA, 2010.
30. Hua, Y.; Laserstein, P.; Helmstaedter, M. Large-volume en-bloc staining for electron microscopy-based connectomics. *Nat. Commun.* **2015**, *6*, 1–7. [CrossRef]
31. Glauert, A.M.; Lucy, J.A. Electron microscopy of lipids: Effects of pH and fixatives on the appearance of a macromolecular assembly of lipid micelles in negatively stained preparations. *J. Microsc.* **1969**, *89*, 1–18. [CrossRef]
32. Neiss, W.F. Electron staining of the cell surface coat by osmium-low ferrocyanide. *Histochemistry* **1984**, *80*, 231–242. [CrossRef] [PubMed]
33. Willingham, M.C.; Rutherford, A.V. The use of osmium-thiocarbohydrazide-osmium (OTO) and ferrocyanide-reduced osmium methods to enhance membrane contrast and preservation in cultured cells. *J. Histochem. Cytochem. Off. J. Histochem. Soc.* **1984**, *32*, 455–460. [CrossRef] [PubMed]
34. Nakakoshi, M.; Nishioka, H.; Katayama, E. New versatile staining reagents for biological transmission electron microscopy that substitute for uranyl acetate. *J. Electron Microsc. (Tokyo)* **2011**, *60*, 401–407. [CrossRef] [PubMed]
35. Kopriwa, B.M. Block-staining tissues with potassium ferrocyanide-reduced osmium tetroxide and lead aspartate for electron microscopic radioautography. *J. Histochem. Cytochem. Off. J. Histochem. Soc.* **1984**, *32*, 552–554. [CrossRef] [PubMed]
36. Walton, J. Lead aspartate, an en bloc contrast stain particularly useful for ultrastructural enzymology. *J. Histochem. Cytochem. Off. J. Histochem. Soc.* **1979**, *27*, 1337–1342. [CrossRef]
37. Okita, K.; Matsumura, Y.; Sato, Y.; Okada, A.; Morizane, A.; Okamoto, S.; Hong, H.; Nakagawa, M.; Tanabe, K.; Tezuka, K.-I.; et al. A more efficient method to generate integration-free human iPS cells. *Nat. Methods* **2011**, *8*, 409–412. [CrossRef]

International Journal of *Molecular Sciences*

MDPI

Review

Neurons, Glia, Extracellular Matrix and Neurovascular Unit: A Systems Biology Approach to the Complexity of Synaptic Plasticity in Health and Disease

Ciro De Luca [1], Anna Maria Colangelo [2,3,*], Assunta Virtuoso [1], Lilia Alberghina [3] and Michele Papa [1,3]

1 Laboratory of Morphology of Neuronal Network, Department of Public Medicine, University of Campania "Luigi Vanvitelli", 80138 Napoli, Italy; delucaciro88@gmail.com (C.D.L.); assunta-1989@hotmail.it (A.V.); michele.papa@unicampania.it (M.P.)

2 Laboratory of Neuroscience "R. Levi-Montalcini", Dept. of Biotechnology and Biosciences, University of Milano-Bicocca, 20126 Milano, Italy

3 SYSBIO Centre of Systems Biology ISBE.ITALY, University of Milano-Bicocca, 20126 Milano, Italy; lilia.alberghina@gmail.com

* Correspondence: annamaria.colangelo@unimib.it

Received: 31 December 2019; Accepted: 20 February 2020; Published: 24 February 2020

Abstract: The synaptic cleft has been vastly investigated in the last decades, leading to a novel and fascinating model of the functional and structural modifications linked to synaptic transmission and brain processing. The classic neurocentric model encompassing the neuronal pre- and post-synaptic terminals partly explains the fine-tuned plastic modifications under both pathological and physiological circumstances. Recent experimental evidence has incontrovertibly added oligodendrocytes, astrocytes, and microglia as pivotal elements for synapse formation and remodeling (tripartite synapse) in both the developing and adult brain. Moreover, synaptic plasticity and its pathological counterpart (maladaptive plasticity) have shown a deep connection with other molecular elements of the extracellular matrix (ECM), once considered as a mere extracellular structural scaffold altogether with the cellular glue (i.e., glia). The ECM adds another level of complexity to the modern model of the synapse, particularly, for the long-term plasticity and circuit maintenance. This model, called tetrapartite synapse, can be further implemented by including the neurovascular unit (NVU) and the immune system. Although they were considered so far as tightly separated from the central nervous system (CNS) plasticity, at least in physiological conditions, recent evidence endorsed these elements as structural and paramount actors in synaptic plasticity. This scenario is, as far as speculations and evidence have shown, a consistent model for both adaptive and maladaptive plasticity. However, a comprehensive understanding of brain processes and circuitry complexity is still lacking. Here we propose that a better interpretation of the CNS complexity can be granted by a systems biology approach through the construction of predictive molecular models that enable to enlighten the regulatory logic of the complex molecular networks underlying brain function in health and disease, thus opening the way to more effective treatments.

Keywords: glia; tripartite synapse; synaptic plasticity; neurovascular unit; systems biology

1. Introduction

Neuronal synapses are, at a biochemical level, stations of electrochemical signaling between the dynamic circuits underlying the complex and deeply interconnected processes of motor and learning functions. Intricate as brain processing may seem, the synapse and its plasticity represent the anatomic and functional unit that can explain it. Hundreds of proteins form the synaptic elements, and their correct expression, structural organization, turnover, and reshaping capability are pivotal for the proper function of the central nervous system (CNS) [1,2].

The pivotal characteristic of the brain is the continuous and strategic ability to modify itself in an experience-based fashion. The subsequent behavior could rely on the strength of circuit transmission and the reinforcing of active synapses or the pruning of new ones. Although development and adulthood show different patterns of synaptic plasticity, this function is fundamental for brain homeostasis [3,4]. Single-cell or matrix contribution could not be easily dissected; however, in the last decades, numerous studies have emerged to enroll glia in the first paradigm shift model, the tripartite synapse including the astrocytes [5]. Indeed, neuronal activity can be controlled by astrocytes in their different and specialized morphologies (i.e., protoplasmic, fibrous, perivascular, and Bergman glia) [6,7]. Higher functions and distinctive neurological competences of the human brain have also been associated with differences between humans and other mammals regarding glial cells and their pattern of gene expression, cellular morphology, and peculiar calcium dynamics [8]. Moreover, it is now clear that oligodendrocytes and microglia also contribute to synaptic plasticity. Oligodendrocytes have shown the potential role of signaling transducers and builder of the extracellular environment [9]. Microglia, instead, in addition to their role of specialized resident macrophage of the CNS, has shown to interact with neurons, to assist their formation in the neural niche and to guide circuit integration and tuning (axonal growth, dendritic sprouting, synapse remodeling) [10].

Finally, the extracellular matrix (ECM), acting as a functional scaffold, represents almost one-fifth of the brain volume. The complex network constituted by proteoglycans, glycoproteins, and glycosaminoglycan sustains neuronal function and provides, together with structural support, a reservoir of trophic factors, signaling molecules, biochemical pathways and long-distance gradient-like communication between cellular components of the CNS [11,12].

The fourth compartment of the synapse is indeed a non-cellular element [13]. The resilience of CNS and synaptic plasticity in the critical period of development and in the adult brain depends on specialized forms of ECM, such as the interstitial matrix, the perineural nets (PNNs) and the basement membrane. Particularly, this last structure is important for the integration of the neurovascular unit (NVU) to obtain an overall model that could be used as a start point for a systems biology-based approach [14]. The implementation of protein–protein interactions involving all the cellular and non-cellular elements of the system can help the building of hub-spoke network maps [14] to design further experiments with translational purposes. Indeed, the ECM is involved in the bidirectional exchange of nutrients and metabolic products between CNS and systemic circulation. The specialized blood–brain barrier (BBB) is one of the finest *exempla* of integration among cellular compartments (glia, pericytes, endothelium) and the ECM, that can selectively permit the transmembrane active transport, the diffusion of molecules through tight junctions, and the selective loosening and remodeling of the BBB [15]. The matrix metalloproteinases (MMPs), as well as other proteases and their relative matrix receptors and regulators, can actively participate in the modulation of CNS circuitry response to various stimuli. In addition, they can mediate the immune system activation and the reshaping of the NVU [16]. This complex and emergent system is furthermore pivotal in the so-called glymphatic regulation, a novel physiological model to clear out wastes of the cellular metabolism from the CNS parenchyma through the dynamic exchange between cerebrospinal fluid (CSF) and the ECM via the NVU [17,18].

In consideration of the great complexity of the synapse organization (defined as penta-partite if we take into account ECM and NVU), here we aim to construct a model of the synapse that can be used for a systems biology modeling. This approach can help to gain new insights into pathogenetic mechanisms underlying complex molecular processes, such as cancer and neurodegenerative disorders. For instance, this strategy is being used to integrate computational models and metabolic flux analysis in cancer cells and make prediction of metabolic reprogramming underlying cancer cell growth [19]. Computational studies of networks of genes and pathways in Alzheimer's and Parkinson's disease (PD) were also effective in identifying functional and topological similarities and differences between the two pathologies [20]. In addition, a modeling strategy has been used to construct a map of pathogenetic processes and pathways involved in PD [21]. Submodules of this map are currently used to unravel specific pathways and their interconnection with interacting processes. For instance, based on experimental evidence, we are currently implementing a mathematical model that exploits the ROS management system and its connection with the metabolism, as well as the relevance of ROS-mitochondria remodeling in neuronal differentiation and maintenance of the neuronal phenotype, neuroprotection, and antigliosis [22,23]. A novel computational model could be used to develop differential neuroprosthetic stimulation modulating pain processing [24]. Once validated, these mathematical models can be useful to predict the impact of any perturbation (genetic or environmental) on the complex biological process(es) under investigation. This could have many positive outcomes in terms of drug discovery and personalized medicine, as it can favor the identification of effective targets for functional recovery.

Impairment of the complex multicellular and multimolecular synaptic system induces acute or chronic CNS pathologies due to the dysfunction of any of these synaptic components with the consequent domino effect. To better understand how to favor the maintenance of adaptive plasticity, it would be useful to construct molecular models able to enlighten the regulating logic of the complex molecular network, which belongs to different cellular and subcellular domains. To this end, we will discuss in detail (*i*) the interactions between cellular elements in the synaptic cleft, (*ii*) how glial cells can modulate synaptic plasticity, and (*iii*) the role of interstitial ECM and the NVU in both physiological (adaptive) and pathological (maladaptive) circumstances (Figure 1). For each cellular and molecular component, we will consider some of the main molecular pathways that should be taken into account when considering the entire system as an interconnected unit.

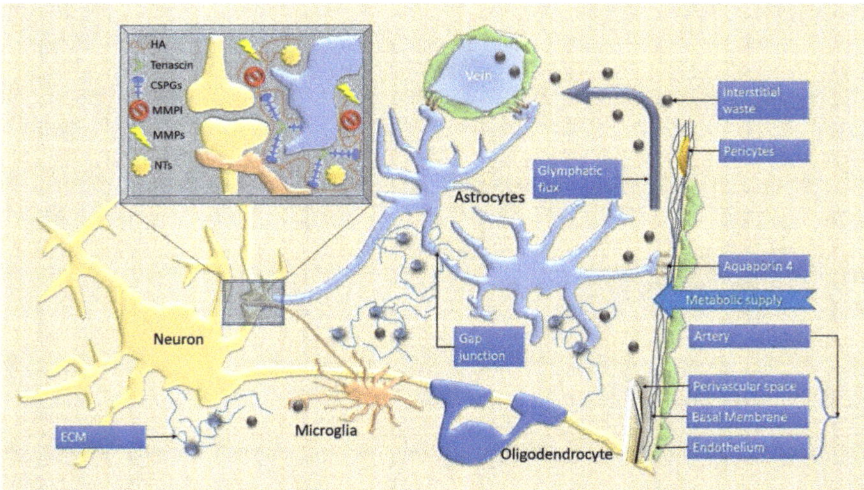

Figure 1. Schematic representation of the synaptic cleft. The main cellular and extracellular components implicated in both physiological and pathological synaptic changes are schematically represented. Oligodendrocytes ensure the correct myelination of the circuits. Microglia constantly scan the microenvironment and remove the debris. From the perivascular space of penetrating arteries, and the neurovascular unit (NVU) elements, fluid dynamics convey waste products toward perivenous spaces and control metabolic supply. The influx-efflux is regulated by Aquaporin-4 (AQP4) water channels densely expressed within astrocyte end-feet. The synaptic cleft magnified in the blue box in the high-left corner encompasses both glial and neuronal elements tightly connected through the extracellular matrix (ECM). The ECM functional scaffold composed of tenascin, hyaluronic acid (HA) and chondroitin sulfate proteoglycans (CSPGs) regulates the expression of neurotrophins (NTs), matrix metalloproteinases (MMPs), and their inhibitors (MMPI).

2. The Synaptic Cleft

Synaptic transmission is a highly specialized process. The punctual description of different types of neuronal cells with their various morpho-functional phenotypes goes beyond the aim of this review, thus only common features will be highlighted to describe the proposed system biology approach. Specific proteins organized in the synaptic cleft allow communication between neurons, rapidly and effectively through transmitter secretion [25]. The storage of transmitters inside the vesicles is a highly selective and energy-consuming task, with the employment of transporters, ion channels and the ATPase protonic pump that uses ATP to supply the proton gradient essential for vesicles loading with neurotransmitter [26]. The action potential is conducted through the axon voltage-gated channels leading to the increase of calcium concentration and the phosphorylation of synapsin that releases the tethering of vesicles to the cytoskeleton and permits the formation of the molecular machinery responsible for vesicle fusion with the cell membrane, the (soluble N-ethylmaleimide sensitive factor attachment protein -SNAP- receptor) SNARE complex [27].

The SNARE complex consists of (Vesicle-Associated Membrane Proteins) VAMPs, linked to the vesicular (v) membrane, the so-called v-SNAREs, and the cellular target (t)-SNAREs composed of the synaptosomal nerve-associated protein 25 (SNAP-25) and syntaxin-1, with a palmitoyl anchor to the pre-synaptic inside membrane [27]. VAMPs consist of seven recognized family members including VAMP1/2 (also known as synaptobrevin 1/2), VAMP4 and VAMP3/5/7/8 (also known as cellubrevin, myobrevin, Tetanus insensitive VAMP and endobrevin) [28]. To exert their role, these proteins are largely present on the cytoplasmic side of the vesicles and cellular membranes [29]. VAMP1/2 are the most abundant in the CNS, particularly expressed in neuronal vesicles, although also recognized on glandular secretory cells [27].

The fusion pore for vesicle secretion is the most controversial of the described mechanisms due to its heterogeneity and dynamicity. Indeed, various genes associated with the synapse have been implicated in neurological and psychiatric diseases, and their expression varies across brain areas, being modified by different cellular elements [30–32]. The difficulty of studying the complex protein–protein interactions is due to the inability of in vitro or in vivo conventional imaging to visualize multiple protein species in one intact sample with a high resolution of their sub-synaptic organization [33]. The probe-based imaging for sequential multiplexing (PRISM) methodology seems to be very versatile to obtain high resolution and dynamic visualization of multiple proteins interactions with reduced background fluorescence and the simultaneous immunostaining of an intact sample [34]. This technique could be useful to better investigate the synapse since it can screen protein interactions leading to maladaptive phenotypes and can count on multiple-level protein networks (e.g., 12 synaptic targets and 66 pair-wise synaptic co-localizations) in normal or perturbed cultures with high spatial or temporal resolution [34].

Synaptophysin (Syp) has been proposed as the initiator of the fusion pore, its role remaining however not utterly accepted [26,35]. The v-SNARE/t-SNARE interaction seems mandatory to generate the anchoring site for the pore formation [27]. These proteins are the target of the light chain of botulinum neurotoxins (BoNTs), the most toxic bacterial toxins produced by the anaerobic, spore-forming *Clostridium* (C.) species (i.e., *C. botulinum*, *C. butyricum*, and *C. baratii*) [36] The importance of SNARE proteins is practically the reason for the astonishing toxicity of BoNTs (median lethal dose (LD50): 1 ng/kg, intraperitoneally) [37]. The first considerations about the role of BoNTs in medicine were focused on the peripheral release of acetylcholine at the neuromuscular junction (NMJ) [38]. However, it was reported that subcutaneous administration of BoNTs could reduce synaptic transmission and vesicle release of neurotransmitters facilitating vasodilation and pain-sensitization (e.g., Calcitonin Gene-Related Peptide), thus interfering with not only the NMJ but also the trigeminovascular system [39] or with oxidative-stress production [40].

The release of vesicles is a fixed, all-or-none process. The increase of intracellular calcium concentration through voltage-gated ionotropic channels is essential to trigger quantal exocytosis [27]. The calcium-binding protein synaptotagmin is necessary to facilitate the binding of phospholipids on the cytoplasmic side of the membrane [41]. The described process allows the neurotransmitters to be released in the synaptic cleft; however, the complexity of CNS response could not rely on a simple all-or-none quantal exocytosis that is indeed finely-tuned by the glial cells (Figure 2) [41]. The most common synapse in the cerebral cortex (neocortex and hippocampal allocortex) is between the axon of a presynaptic cell and the dendrite of a postsynaptic cell (axodendritic synapses). The axodendritic synapses of some neurons are located on the spines, highly specialized structures, protruding from the dendritic trunk, localizing the specific connection and increasing the density of synaptic terminals [42]. Electron microscopy further recognized the presence in certain synapses (known as asymmetric or type I synapses) of electron-dense structures called postsynaptic densities (PSD). These PSD where recognized prevalently in excitatory synapses and encompass anchorage proteins to the cytoskeleton, postsynaptic receptors and associated signaling transducers (Figure 2) [43]. The hundreds of proteins inside the PSD are interconnected and function both as a scaffold and as transmitting complexes with the ability to interact with each other and form heteromeric structures.

The most represented protein of the PSD is known as synapse-associated protein 90 (SAP-90) or PSD-95 (based on its molecular weight). PSD-95 is part of the scaffold family proteins called membrane-associated guanylate guanyl kinases (MAGUK) and has been associated with the increase of dendritic spines, regulation of neurotransmitter receptors and synapse stabilization and plasticity [43]. Other scaffold proteins include Homer and SHANK (SH3 and multiple ankyrin-repeat domains) families [43].

On the other hand, symmetric or type II synapses have slight electron-dense postsynaptic structures and are mainly inhibitory [44]. This classical division is schematic and oversimplified, but useful to comprehend the functioning of the synaptic cleft in general. As stated before, a clear and utter understanding of protein–protein interaction at synaptic level needs to be further elucidated with novel methods that eventually will reveal a fine-tuned complexity, with multiple specialized forms of both immature or mature synapses depending on brain areas, functional state, involved neurotransmitters, and even pathological responses.

The presynaptic neurexins and their relative neuroligin ligands on the postsynaptic terminal are nonetheless important for the formation, maturation and stabilization of synapses and for the interaction with glial cells and the ECM (Figures 1 and 2). Neuroligins 1–4 seem to be unessential for synapse assembly in vivo, but they are pivotal for its maturation and proper functioning [41]. Their interactions with neurexins (α and β) affect both type I and type II synapses and the recruitment of other scaffolding proteins and receptors [45,46]. Mutations of these genes (NRXN1/2/3, NLGN1/3/4), together with SHANK family proteins, have been associated with autistic spectrum disorders (ASD) and schizophrenia [47].

Neurotransmitters, of course, are central in synaptic transmission, for stabilization of the forming synapse and to grant a bidirectional communication between neurons and other cell types (i.e., astrocytes, microglia, oligodendrocyte) through feedback systems in both adult and developing brains [48].

Finally, neurotrophins (NTs) are indeed the main regulator of synapse formation and function, as they strengthen robust and functional circuitries while preventing redundancies or pathological rewiring [49]. The NT family encompasses the nerve growth factor (NGF), the brain-derived neurotrophic factor (BDNF), NT-3, and NT-4/5 in mammalians (NT-6 and NT-7 in fishes) [50,51]. NT receptors are divided into two distinct classes. The p75 is the first receptor being identified. It shows a low affinity for all NT without a significant specificity and seems to be involved in the apoptotic pathway [52]. High-affinity receptors for NT are members of the tropomyosin receptor kinase (Trk) family. TrkA is the high-affinity receptor for NGF; NT-4/5 and BDNF are preferred ligands of TrkB, while NT3 binds to TrkC [51].

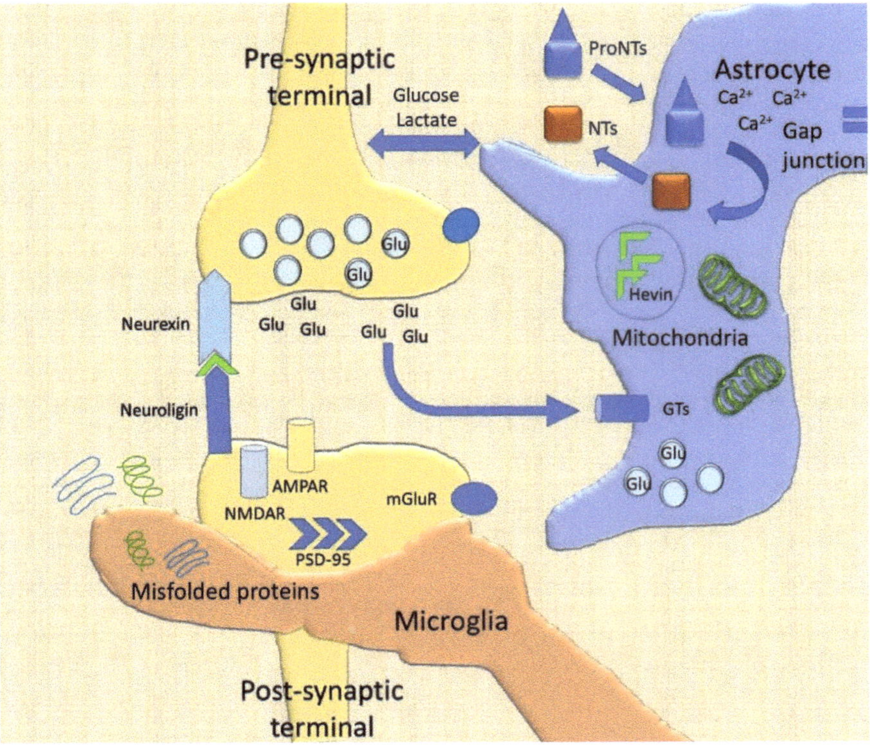

Figure 2. Cellular elements in the synaptic cleft. The schematic representation of a glutamatergic synapse highlights the role of molecular pre-synaptic (neurexins), post-synaptic (neuroligin) proteins and the astrocytic hevin in the stabilization of the cleft structure. The reuptake of neurotransmitters by glutamate transporters (GTs) is mainly provided by astrocytes. Moreover, astrocytes are responsible for the proneurotrophins (proNTs) alternative intracellular processing to active NTs and the metabolic coupling. The synaptic plasticity phenomena are widened by the astrocytic calcium waves, ensured through gap junction and by microglial trogocytosis (the partial engulfment of dendritic spines or axonal portions). The scavenger role of microglia is nonetheless necessary to avoid waste accumulation and synaptic failure.

All NTs share a very high amino acid homology (approximately 50%) and are initially synthetized as pre-pro-proteins in the endoplasmic reticulum. The amino terminal contains, in fact, the signal for intracellular transportation and is cleaved to obtain the proNT, which undergoes post-translational modification in the Golgi apparatus. ProNTs can be processed to NTs both intracellularly and extracellularly, with the consistent contribution of both astrocytes and ECM [53] (Figures 1 and 2). The intracellular pathway consists of proteolytic cleavage by the pro-protein convertase subtilisin kexin (PCSK), particularly the neuronal form PCSK1/2, which binds to specific recognition motifs of the pro-protein, leading to the mature NT that can be subsequently stored in vesicles and secreted [54,55]. Impaired extracellular proNTs processing and activation can compromise the stability of the synapse, as detailed below.

All NTs act as modulators of synaptic plasticity. TrkB activation has been shown to increase the density of presynaptic vesicles and the expression of vesicular VAMP1/2 and Syp, thus enhancing the exocytotic machinery. BDNF is practically ubiquitous in the CNS and is involved in neuronal survival, axon growth, cell morphology, induction of protein expression and adaptive plasticity, which is pivotal for both learning and memory formation [56]. Similarly, NGF has been shown to regulate mechanisms underlying development and energy homeostasis of NGF-dependent neurons [57], moreover to its well-established role in modulating synaptic plasticity of cholinergic neurons [58].

Levels of synaptic proteins have been investigated in models of neurological degenerative diseases by proteomic approaches to unravel their involvement in neurodegeneration. Dysfunctional levels of NTs, as well as SNAP-25, Syntaxin-1, Syp, PSD-95, MAGUK, and SHANK proteins, are involved in neurodegenerative diseases, such as Alzheimer's disease (AD), Parkinson's disease (PD), and dementia with Lewy bodies (DLB) [59]. Protein modification at the synaptic cleft are common pathways of neurodegeneration, preceding neuronal death and late-onset modifications in non-cell autonomous diseases. The complexity and similarity observed in the abovementioned neurodegenerative diseases corroborate the need for a systems biology approach [22,58]. The production of a holistic model that encompasses the complex, reductive, networks of proteins and metabolites belonging to distinct synaptic components (i.e., neurons, astrocytes, microglia, ECM and NVU), as well as to sub-cellular compartments (i.e., synaptic vesicles, mitochondria, etc.), can support a better understanding of cell–cell and cell–matrix-NVU interactions, as well as how these networks are affected under physiological or perturbed synaptic transmission conditions (adaptive and maladaptive plasticity) (Figure 1). For instance, such a model might allow predicting how the activation of a plasmatic protease, such as thrombin, could affect astrocytic and microglial activation, remodeling the ECM and ultimately influence the intracellular pathway of neuronal cells and proteins of the synaptic cleft (Figure 2).

3. Glial Cells

Synaptic plasticity is influenced by the three main non-neuronal resident cells: Astrocytes, microglia, and oligodendrocytes acting as a functional unit. The contribution of each cell-type will be described enlightening the peculiarities, possible networks within these elements and how they interact with the ECM and the NVU.

3.1. *Astrocytes*

Since the proposal of the tripartite synaptic model in 1999, astrocytes and their morphological variants and subtypes have been the most vastly investigated glial cells [5]. Indeed, astrocytes have a prominent role in synaptic plasticity: On one side, they, tightly enwrap neuronal cells and synapses (Figure 2) [60], and participate to the production and maintenance of the ECM; on the other hand, they are associated through their end-feet with the endothelium and pericytes in the NVU (Figure 1) [15]. In the penta-partite synaptic model, this cell-type could be considered as the cornerstone; the study of the related network with the other components could open a new field of experimental speculations.

The astrocytic influence on the synaptic cleft is required for CNS homeostasis [61]. For instance, the astrocytic Hevin interacts with the neurexin/neuroligin system to ensure synaptic clef stability (Figure 2) [62]. Astrocytes are hyperpolarized cells in the resting-state through the specific expression of inward rectifying potassium channels (Kir), particularly Kir 4.1 [63]. These channels finely expressed on the astrocytic membrane facing the synaptic cleft can reduce the potassium conductance during neuronal activity. The dysfunction of this mechanism, leading to improper neuronal membrane depolarization, has been found in various neurological diseases, such as epilepsy, multiple sclerosis (MS), amyotrophic lateral sclerosis (ALS), and Huntington's disease (HD) [64,65]. Another channel expressed on the astrocytic membrane is the aquaporin-4 (AQ-4) that is involved in neuromyelitis optic spectrum disorders (NMOSD) [64]. AQ-4 water channel has been described on the astrocytic end-feet, and associated with Kir channels to control the osmotic regulatory role of these cellular domains; in fact, potassium uptake during neuronal activity generates an osmotic imbalance [66]. Furthermore, AQ-4 is the pivotal player in the proposed glymphatic mechanism of ECM debris removal through NVU and CSF active directional filtration (Figure 1) [17,18,67]. Considering these proteins not only as peculiar in astrocytic regulation but as paramount in the entire plasticity process, could account for novel strategies of targeted therapies to improve, for instance, the clearance of misfolded proteins (e.g., AD, PD). On the other hand, the AQ4-NVU interface could explain the immune-mediated phenomenon of NMOSD and implement novel strategies for adaptive synaptic regulation following the lost immune privilege.

Astrocytes are also important in neurotransmitters reuptake (GABA and glutamate) through specific transporters, thus modulating the concentration of these mediators and confining them to the synaptic cleft (Figure 2). The main glutamate transporters for astrocytes are glutamate transporter 1 (GLT1) and glutamate/aspartate transporter (GLAST) that have been shown to prevent glutamate over-excitation observed in several pathological conditions, such as trauma or epilepsy [68,69]. The confinement of glutamate in the synaptic cleft avoids glutamate spillover, and its reuptake is adjuvated by neuronal transporters, the excitatory amino acid carrier 1 (EAAC1) [1,70]. The activation of extra-synaptic metabotropic glutamate receptors (Figure 2) is responsible for the modulation of excitatory postsynaptic currents (EPSCs) [71] that can modify temporal and local integration of synaptic currents, thus affecting synaptic plasticity. Inhibitory modulation, on the other hand, is due to the GABA transporters (GAT), particularly GAT1 and GAT3, the first one shared with neurons, the latter being able to regulate the astrocytic intracellular concentration of calcium [72]. Moreover, astrocytic GAT3 can enhance the release of purinergic mediators (ATP/adenosine) in the hippocampus [72]. Modification of GAT and astrocytic calcium dynamics have been reported in a rodent model of behavioral disorders [73].

Another important feature of astrocytes at the synapse is the calcium-mediated release of molecules, once thought to be exclusively neuronal, called gliotransmission (Figure 2) [74]. Although there are controversial data regarding the astrocytic expression of proteins necessary for vesicular release, the dynamic regulation of synaptic transmission through astrocytic activity has been extensively proved [75,76]. Among the molecules acting as gliotransmitters, glutamate, GABA, D-serine, ATP and adenosine have been shown to control the basal tone and threshold of synaptic activity, surpassing the all-or-none law of the action potential-mediated neurotransmitters release [77,78]. Finally, astrocytes participate in the secretion of proNTs and their processing in the ECM [14]. Moreover, NTs synthesized in neurons and secreted as proNTs into the ECM can be rapidly internalized into perineuronal astrocytes via p75 mediated endocytosis. After internalization, they can undergo a recycling or activation process [1] (Figure 2).

One single astrocyte can enwrap about 120,000 synapses in rodents. A human astrocyte might unsheathe from 270,000 to 2 million synapses [79,80] that can be both excitatory or inhibitory, thus encompassing different neuronal circuits and eventually integrating them [80], enhancing both short-term (STP) and long-term (LTP) potentiation, or decreasing the glutamatergic tone with GABA or the purinergic release [81,82]. The cannabinoid (CB) system, furthermore, seems to be a signaling pathway by which activated astrocytes release glutamate and enhance the synaptic strength with both short-term [83] and long-term plasticity [84]. D-serine is also released by CB activation and stimulates the N-methyl-D-aspartate glutamate receptor (NMDAR) contributing to hippocampal LTP, the proposed main mechanism for memory formation and maintenance [85]. Besides, the calcium-binding protein S100b produced by astrocytes can induce neuronal firing in both trigeminal sensory nucleus and prefrontal cortex and might regulate cognitive flexibility and neuronal oscillations [86]. Intriguing is even the role of astrocytes in the regulation of sleep: Adenosine accumulation is associated with sleep homeostasis [87] and the astrocytic release of ATP/adenosine and glutamate can induce a transition between wakefulness and sleep [88,89]. A fascinating pathway that connects astrocytes to the vascular system through the ECM, particularly during sleep, is the abovementioned glymphatic system [67]. This hypothesis considers experimental data conveying a possible CNS lymphatic-like system mediated by astrocytes that is responsible for ECM clearance of accumulated molecules and control of metabolic supplies [90]. A failure of the glymphatic system has been related to toxin accumulation, such as amyloid-β (Aβ) [91] and misfolded proteins in the perivascular space of murine models of type-2 diabetes and AD, although a direct correlation between these findings and the cognitive deficits associated with human dementia are still debated [92].

Astrocytes are interconnected with gap junctions formed by connexins (Cx), Cx43 and Cx30 being the most expressed [93,94]. The intercellular diffusion of ions, transmitters, and small molecules connects networks formed by hundreds of astrocytes, not with a mere neighboring principle but with functional purposes [95]. The cortical spreading depression (CSD), one of the main explanatory mechanisms of typical migraine aura, could depend on Cx-based astrocytic syncytia [96]. CSD, in fact, could involve waves of synchronous astrocytic activation (through their gap junctions), with a slow-propagating neuronal firing followed by reactive hyperpolarization in limited cortical area, thereby causing the transient neurological positive and negative focal symptoms characterizing the migraine aura [96]. As a clinical proof of concept, a gap junction inhibitor, tonabersat, was effective in both CSD inhibition and migraine aura frequency reduction [97]. Altered expression of Cx at the gap junction can also influence glucose and lactate supply and synaptic transmission [88,98]. Simple as it may sounds, there is still debate on the glucose/lactate coupling between astrocytes and neurons [99,100] (Figure 2).

Astrocytes are highly sensitive to plasticity processes and may act with structural changes involving their cytoskeleton and different expression of the glial fibrillary acidic protein (GFAP). These cells have a high motility rate of morpho-functional re-shaping in a timescale of minutes [101]. Sustained NMDAR activation can increase glutamate release from astrocytes even 1 h after neuronal stimuli, widening the time window for synaptic plasticity [102]. Astrocytes are practically able to store synaptic information over time and can, in turn, modulate late-onset synaptic plasticity of the same synaptic pathway or related neuronal circuits, reinforcing the interdependency of neuron-astrocyte processes. The features of astrocytes vary across brain areas and are dynamically connected to neuronal specialization, with their diversity being even more complex in different species [8,79].

All the discussed regulatory functions of astrocytes in synaptic plasticity make puzzling and intriguing the fact that these cells do not cover all synapses [60,103]. The extent of synapse sheathing varies from 15% for mossy fibers of granule cells to 85% for climbing fibers in the cerebellum and 90% for excitatory synapses in the somatosensory cortex of adult mice [104,105]. Differences between these synapses (as well as the relative involvement of other glial cells, nonresident-cells and matrix components in the maintenance of plasticity) corroborate the need for a multi-cellular interconnected model that could allow the understanding of a disease by analyzing the system as a functional unit. Moreover, it has been shown that synapse enwrapping can be dynamically induced by neuronal activity and by physiological conditions like nutritional state (satiety or starvation) [103,106,107].

3.2. Microglia

Microglia can rapidly sense differences and homeostatic perturbations by scanning other cells (astrocytes and neurons) and the brain ECM (Figure 1) [108], and surveying the environment for pathogens and autologous debris (phagocytic function) (Figure 2) as the first cellular line in the innate immune response [14,109–111]. In the last scenario, microglia can mediate the loosening of BBB and the secondary immune reaction [112]. All these functions are possible through a highly variable profile of gene expression and cell morphology [111]. In particular, microglial cells show the capability to actively assist not just the elimination (as thought considering their scavenger role), but practically the formation and/or relocation and reinforcement of synapses (e.g., maturation of excitatory synapses), by sensing the environmental pabulum for signaling molecules and re-wiring the circuitry following these biochemical instructions. Their role is ubiquitous in the CNS [10].

Their scavenger role has practically led to the strong parallelism between microglia and the macrophage system in other tissues. However, microglia is an ontogenically distinct population of the phagocyte system, with a different embryonic origin, compared to resident macrophages of other tissues [113]. For instance, there is not a continuous supply of microglia precursors from the general circulation (monocytes) and these cells renew themselves, slowly, in the mammalian brain maintaining a certain epigenetic memory of the environmental modifications [114,115]. Moreover, microglia bodies are relatively immobile, stretching highly dynamic elongations to scan the entire CNS in the timescale of hours [116]. This evidence and the aforementioned connection between microglia and the system gives the opportunity to enlighten the peculiar mechanisms that could be further studied to understand the main modifiers of synaptic plasticity. The classical macrophage-related categorization of M1 (pro-inflammatory phenotype) and M2 (resting or anti-inflammatory) phenotypes, is in contrast with transcriptomics studies showing multiple microglia responses with multifaceted polarization states [117,118]. The main activity of these cells does not seem to be the scavenging, as initially supposed. Their relevance to synaptic plasticity is linked to their capacity to sense the functional state of synapses, ECM and vascular compartment, and communicate with other resident cells [119]. This specialized function is based on high-density surface receptors (called sensome) to detect both physiological (cognitive stimulation, diet, physical exercise) or pathological stressors [108,120]. The main microglia receptors allow the communication with neurons, through the purinergic receptors family (e.g., P2XRs, P2YRs) [82] or the widely studied neuronal chemokine (fractalkine) receptors (CX3CR1) [121]. They also mediate the immune system and the NVU through the complement receptor CR3 [14], and activate the phagocytosis response by triggering the receptor expressed on myeloid cells 2 (TREM2) and DNAX-activation protein of 12 KDa (DAP12) [122,123].

These sensors have been shown to contribute to synaptic plasticity, neurogenesis, myelination, and blood vessel formation [124]. Concerning the synaptic plasticity, microglial processes establish contacts with neurites and synapses, particularly through the neuronal ligand fractalkine or the complement system C1q and C3, even if the exact mechanisms have not been completely elucidated [119, 125]. It has been suggested that microglia could enwrap small portions of axons (a mechanism called trogocytosis) to limit and guide their growth and to eliminate presynaptic structures [126] (Figure 2). It has also been debated whether microglia can phagocytose dendritic spines or can partially surround them [126,127]. The interposition of microglial cells in the synaptic cleft [128], and the ability of their filipodia to sustain dendritic spine formation and relocation on the dendritic shaft, are fascinating features of microglia-guided synaptic plasticity (Figure 1) [126]. Furthermore, it has been shown a close contact between the axon proximal segment and microglia, proposing an unknown mechanism for the formation, interruption, or elimination of synaptic connections [129].

Another paramount role during CNS development is the microglia-dependent synaptic elimination, essential for the correct wiring of the system, particularly studied in the visual cortex [119]. The interaction with the synaptic cleft has been supported by the evidence of PSD-95 in microglial specialized processes both in vitro and in the mouse cerebral and hippocampal cortex [122]. While the observation of trogocytosis has been demonstrated for the axonal portions, the engulfment of dendritic spines has been always shown as partial, by the endurance of a connection with the dendritic shaft through the neck of the spine [126]. Importantly, CXCR1 and CR3, pivotal in a previously proposed neuro-immune network [14] in the presynaptic region seem to be responsible, respectively, for synaptic and axonal enwrapping with an activity-dependent elimination in certain brain areas [130,131]. The postsynaptic modifications of dendritic spines could be indirect, with the enwrapping or interposition of microglial processes and functional deprivation, as speculated in a mechanism of interfering plasticity shown during microglia activation by lipopolysaccharide (LPS) [128]. Microglia can modulate synaptic plasticity by production and secretion of molecules, such as ECM components, NTs (i.e., BDNF), endo(e)CB, cytokines (e.g., tumor necrosis factor α -TNFα-), micro-ribonucleic acids (microRNAs) by surface-anchored release or the formation of extracellular vesicles (EV) [125]. Microglial BDNF acting through TrkB neuronal receptor seems to be involved in the switch of GABA from an excitatory to an inhibitory molecule in adult neurons [132]. The release of microglial BDNF is induced by P2 $\times$ 4 activation in a model of neuropathic pain; while it was shown that EVs containing eCB are released by rat microglia exposed to ATP [132,133]. Nonetheless, the secreted BDNF seems to be important for the presynaptic vesicular glutamate transporter 1 (vGLUT1) expression [134]. Furthermore, platelets are also able to release BDNF during pathological processes (such as stroke) that allows their activation in the NVU as well as glial activation [112]. These data associate the microglia with the previously described NTs survival network and the limitation of cell-based approaches considering neurons and astrocytes as the main regulators of NTs in CNS pathophysiology. Expression of NMDAR and AMPAR and their ratio could be modulated by the astrocyte/microglia crosstalk and by microglial TNFα release in the maladaptive plasticity [135]. Eventually, the activation of stress responses mediated by microglial CR3 seems to induce LTD [136].

Overall, these data show an interaction between the microglial system and multiple levels of synaptic plasticity. The mentioned pathways are strongly associated with the penta-partite model and pivotal in the development of systems biology approach to the adaptive/maladaptive tuning. The activation of both astrocytes and microglia in maladaptive plasticity and various neurological diseases, known as reactive gliosis [1,53,137], will not be discussed in this review.

3.3. Oligodendrocytes

Oligodendrocytes are the myelin-forming elements of the CNS. They develop from the oligodendrocytes precursors cells (OPC). OPCs are important in CNS development but, unlike other progenitors, remain abundant in the adulthood and maintain the ability to modify the state of the white matter both in physiological (learning, normal aging, experience-based system rewiring) and pathological conditions (e.g., MS and NMOSD) [138]. OPCs express the proteoglycan marker neural/glial antigen 2 (NG2) cells. The proliferation and differentiation of these cells are mediated by various growth factors and hormones. Platelet-derived growth factor (PDGF) seems to be the most powerful inducer of OPC proliferation mediated by neighboring cells (neurons and astrocytes). PDGF, however, acts in association with ECM molecules and their cellular integrins (the phosphorylated form of $\alpha_V\beta_3$ associated with PDGF receptor) to exert its mitogenic activity, with the relevant contribution of Tenascin-C and NG2 (Figure 1) [139]. These relations make OPCs relevant in the plasticity of the CNS with the ECM homeostasis.

The oligodendrocyte lineage proliferation, differentiation and survival, however, can be influenced also by neuregulins, NT-3 and NGF, further validating their role in synaptic plasticity [140,141]. Apart from their role in myelin and ECM homeostasis, novel interest is shown in the electrical signaling between oligodendrocytes and neurons. OPCs express a variety of neurotransmitter receptors, such as AMPAR, NMDAR, GABA, and acetylcholine (Ach) receptors [142] and voltage-gated channels (i.e., sodium, potassium, and calcium channels) that could in principle modulate the surrounding neuronal activity [138]. Moreover, OPCs are the only known glial cells to form synapses directly with glutamatergic neurons in both gray and white matter regions and with GABAergic neurons in the gray matter [143–145]. Unlike neurons, however, OPCs seem to participate only as postsynaptic terminals and their glutamatergic connections develop along with normal synapses in the surrounding neurons, while GABAergic signaling seems to gradually switch from synaptic to extra-synaptic [144]. The synaptic connection between OPCs and neurons appears to be counterintuitive, given the high mobility of OPC elements that continually reshape (forming and dissolving) synapses in their migratory pathway. The formation of these energetically expensive transient synapses allows the OPCs to monitor axonal activity in the neighboring neurons, guiding oligodendrocyte maturation and myelin synthesis. In fact, tetrodotoxin (TTX), a sodium channel blocker, can inhibit myelination [146], while stimulation of neural circuits even in adulthood can stimulate OPC proliferation and differentiation along the active pathways [147,148]. These synapses, however, can provide neurotransmitters, neurotrophic factors and integrins-mediated intercellular signaling, thus contributing to modulation of axonal outgrowth and/or neuronal excitability.

Oligodendrocytes, although maintaining the ability to regenerate, are long-lived cells, their lifespan apparently is independent from the brain area or the grade of axon myelinization [149] (Figure 1). New oligodendrocytes could replace myelin loss for physiological turnover and tend to accumulate with the time [150]. Myelination occurring during adulthood, however, seems to have different morphology, with a high density of shorter internodes [149]. The consequence is that the average internode length diminishes and the number of nodes per volume increases with aging, but the functional consequences of this remodeling are unknown and difficult to predict [151]. Behavioral phenomena should be expected in networks with rapid and synchronous information processing, such as auditory and visual systems [152]. In the auditory system, evidence suggests that synchronization on microseconds timescale is achieved through dynamic regulation of internode length and thickness [153]. Myelin synthesis as a proper synaptic plasticity mechanism can be triggered by a specific learning task, as demonstrated by myelin formation during motor-skill learning experiments [154] and the opposite was shown by the inhibition of a new motor-task acquisition by blocking new myelinization [147]. When myelination was blocked by conditional deletion of myelin regulatory factor, blocking the OPC differentiation into mature oligodendrocytes, animals were unable to learn new motor skills, but they were able to recall previous ones [147]. The role played by oligodendrocytes in synaptic plasticity has been neglected for decades, and substantially limited to demyelination processes [155]. A systems biology model of adaptive/maladaptive plasticity including this cellular component should help to obtain a more comprehensive framework of their involvement in acute (e.g., stroke) or chronic (e.g., corticobasal degeneration) CNS diseases.

4. ECM and NVU

Synaptic plasticity processes require structural and functional modifications, including shape, density, formation, and elimination of synapses [2]. The functional modifications have been mostly related to the neuron-glia interactions, like neurotransmitters secretion, receptors expression, activity-dependent synaptic plasticity (e.g., LTP, LTD, myelination). This physiological plasticity allows the circuitry to retain and reinforce the connections and store information [156,157]. The ECM and NVU, on the other hand, participate actively in the structural changes and mediate the bioavailability of nutrients, cytokines, molecular mediators (e.g., NT, transmitters, integrins ligands) that are fundamental for the correct sequence of functional modifications [14,112,157]. The major structural builder of these functional scaffolds could not be identified, as all CNS resident cells are involved in producing or re-shaping them both in health and diseases (Figure 1).

4.1. ECM

The molecular composition of ECM in the CNS has been deeply investigated to elucidate the intricate structure of this functional scaffold [11,12]. The major components of neural ECM are proteoglycans (e.g., brevican, neurocan), interacting with collagen, glycoproteins (e.g., tenascins) and hyaluronic acid (HA) synthesized in a different ratio by both glia and neurons (Figure 1). Unbiased mechanisms activated progressively during development and in physiological conditions in the adult brain, reshape the synaptic cleft through these ECM molecules.

Morpho-structural modifications of the ECM during CNS injury, aging and reactive astrocytosis have been studied for their implication in neuroinflammation, neurodegeneration and the maladaptive synaptic plasticity.

Elegant experiments were made to discriminate the astrocytic and neuronal ECM contribution using co-cultures of neurons and astrocytes obtained from quadruple knockout mice for Tenascin-C/R, brevican and neurocan. These co-cultures showed reduced production and organization of the ECM, with lowered neuronal activity. The integration of wild-type astrocytes did not rescue the neuronal phenotype [158] and co-cultures containing either quadruple knockout astrocytes or neurons showed a reduced number of synapses (after an initial transient increase of synaptic formation in co-cultures containing mutant astrocytes) [158]. These in-vitro data were partially complying with the in-vivo studies conducted with the same animal models, showing mild defects of ECM deposition and the replacement of the lost components with fibulin 1/2 during development [159]. Understanding the relevance of cell-type-specific ECM components could be fascinating, like the astrocytic hevin (or its antagonist SPARC) and thrombospondins, although all cellular elements seem to act all together as a functional unit in synaptic plasticity [3,160].

Furthermore, MMPs and ADAMTS (A Disintegrin And Metalloproteinase with thrombospondin motifs) proteases are secreted by both glia and neurons and can reshape the ECM in response to external or internal perturbations of the homeostasis [12]. These speculations have made the study of the ECM promising for novel therapeutic approaches, although still practically controversial, possibly for the delicate balance of this functional scaffold [161,162].

The PNNs, a specialized form of ECM, seem to be key elements in synaptic stability, creating a functionally permeable barrier that allows or limits the formation of new connections between neurons [156]. Furthermore, PNNs retain synaptic signaling molecules (e.g., semaphoring-plexin system) that in the adult brain prevent the formation of unfit neuronal circuitries [163]. ECM has a different composition in the developing brain with a high level of expression of matrix components reported before the postnatal synaptogenesis peak, suggesting the key role played by the interstitial matrix for the formation of immature synapses [164]. Aggrecan, also known as chondroitin sulfate proteoglycan (CSPG)-1, seems to be particularly expressed on the surface of neurons in correspondence to the loci of forming synapses [165].

Considering their role in synaptic stability, PNNs were shown as fundamental elements for memory maintenance [166]. CSPGs, major components of the PNNs, were increased in the amygdala of fear-conditioned animals and were associated with the protection of long-term memory traces, resistant to extinction. The formation of these memories seems to coincide with the organization of CSPGs into PNNs, related to the closure of a fragile period of the synaptic developmental plasticity, defined "critical period" [166].

The increased expression of CSPGs, Tenascin-C, Tenascin-R, and HA can regulate axon elongation mediated by reactive astrocytes [167,168]. CSPGs are selectively overexpressed in maladaptive plasticity. Astrocytes produce neurocan and phosphacan, following cerebral cortex injury, while brevican or versican expression is not increased [169]. Moreover, the induced modification of ECM through heparinase or chondroitinase ABC (ChABC) overcomes the limitation of neurite outgrowth following glial scar formation [169,170]. The specific expression of CPSGs seems to depend on the specific neurological disease. An overexpression of brevican in the frontal cortex was reported in AD [171]. Oligomers or fibrillary Aβ peptides (not the monomers) were able to bind brevican core protein [171]. Animal models of AD, when treated with ChABC, showed a certain degree of LTP function in the hippocampus and short-term memory formation with a higher synaptic density. The accumulation of brevican seems to be due to astrocytic deposition and leads to synaptic maladaptive plasticity [171].

The maturation of NTs (i.e., NGF and BDNF) can occur as abovementioned in the ECM. In this case, proNTs are processed by serine proteases like plasmin and MMPs, particularly MMP-7 and MMP-9 [55,172]. Plasmin, normally produced in its inactive form, the zymogen plasminogen, can be activated by the protease tissue plasminogen activator (tPA). Plasminogen seems to be exclusively expressed by neurons [173].

Cell adhesion molecules (CAM) are pivotal for the interaction between cellular elements and ECM. In particular synaptic CAM (SynCAM), neuroligins and hevin (involving both neurons and astrocytes) have demonstrated a role in the developmental synaptic plasticity, and more recently, also a member of the leucine-rich repeat transmembrane (LRRTM) proteins [174,175]. LRRTM members are transmembrane proteins, interacting with the ECM. They bind neurexins and induce presynaptic differentiation playing a role in the regulation of receptor composition [175,176]. The deficiency of LRRTM leads to the loss of different types of synapses with parallel impairment of pre- and post-synaptic components of the cleft [177,178]. LRRTM is able to bind neuroligin, but its role could influence different types of synaptic sprouting, through the different components of the surrounding ECM. However, the role of this protein family in ECM-mediated synaptic plasticity needs further investigation. Furthermore, the neural cell CAM (NCAM) is an important player in visual cortex development [179]. In particular, the visual stimuli could induce a polysialylation of NCAM (which account for 95% of CNS protein polysialylation) and enhance the homophilic interactions across the synapse [179]. Similar modifications are found in SynCAM, suggesting another possible mechanism to organize specialized synaptic composition [180]. The physiological function of polysialylation is to enhance hydration and volume size of the molecule, thereby increasing the distance between the cell membranes of polysialylated NCAM-carriers and following regulation of cell–cell interactions [181]. Specific patterns of polysialylation are characteristic of developmental and adult brains, with almost overlapping NCAM expression and a small fraction of SynCAM polysialylated after birth [181]. Moreover, polysialylation of SynCAM seems to be confined to few brain areas and could be found also on OPCs [181]. These data are in accordance with the role of OPCs in synapse maturation and ECM homeostasis.

4.2. NVU

Finally, we need to consider the complexity of the NVU. This component allows the CNS homeostasis, metabolic supply, and immunological privilege, being regulated by a fine tuning mode between the BBB elements (endothelial cells, astrocytic end-feet, pericytes, and the basal membrane), neurons, glia and interstitial ECM [14,182].

The immune privilege of the CNS applies to both brain and spinal cord (through the BBB homologous, blood-spinal cord barrier) [183] and can be maintained via the innate immune properties of resident microglia, constantly scanning the environment to detect endogenous perturbations or external pathogens and quickly restoring local homeostasis, thereby avoiding or confining the damage [184]. Neuroimmune regulators (NIRegs) are a group of signaling proteins that are expressed on both glia and neurons, and act to limit the immune activation [185]. Microglia remain in a resting state by interacting with NIRegs (e.g., CXCL1, CD200, and CD47) on other cells. NIRegs nonetheless inhibit complement activation through CD59, CD46 and factor H (FH) [185]. Furthermore, cytokine signaling can be physiologically modulated, by the constitutive levels of the suppressor of cytokines signaling (SOCS) that inhibits the Janus kinase (JAK)/signal transducer and activator of transcription (STAT) intracellular pathway [185].

Thrombin is one of the main activators of neuro-immune responses and is strictly linked to the NVU. This protease can exert both adaptive and maladaptive plasticity modifications on neurons and glia [186,187] depending on its concentration, by interacting with the proteinase-activated receptors (PARs). PARs are G-coupled receptors, with four recognized members [9,188] demonstrated on neurons, glia, endothelial, and immune cells. PAR-1 can be canonically activated by thrombin, activated coagulation factor X (FXa), MMP-1 and plasmin (also important as ECM reshaping proteins). The activation is mediated by proteolytic cleavage and the tethered ligand exposure [189].

PAR-1 canonical activation seems to be neurotoxic and is achieved with pathologically high concentrations of proteases [190] activating a guanosine triphosphatase (GTPase), the rat sarcoma protein (Ras), the related protein A (RhoA). PAR-1 biased agonism, instead, can be achieved with a controlled thrombin response, complexed with the activated protein C (aPC), which binds the endothelial PC receptor (EPCR). This induces a different proteolytic activation on PAR-1 stimulating another GTPase, the Ras-related C3 botulinum toxin substrate 1 (Rac-1), which interacts with βarrestin-2 and disheveled-2 [191]. Other PARs seem to show secondary functions, particularly PAR-3 which lacks intracellular domains [192], thus being unable to activate directly a G protein. Indeed, it could act as a cofactor and form an heteromeric complex with other members of the family to modulate intracellular transduction [193]. The complement factor C4a seems to act as an untethered ligand of both PAR-1 and PAR-4, leading to intracellular activation of phospholipase C (PLC) and intracellular calcium release [194], supporting the relevance of neuro-immune modulation of the NVU.

Serpins and thrombomodulin are also considered NIRegs and could reduce the toxicity of thrombin on the CNS by preventing the canonical activation of PAR-1 [195]. LPS administration can stimulate production in the hippocampus of microglial inflammatory factors and the expression of coagulation factors, probably depending on thrombin signaling activation [196]. The relevance of these factors has been proved in neurological and psychiatric diseases [14].

The role of pericytes, on the other hand, was largely ignored, although it was reported that a mutant mouse model with pericyte deficiency showed an significant increase of BBB permeability to both water and solutes of low and high molecular mass [197]. This increased permeability was due to endothelial trans-cytosis. Moreover, it has been shown that pericytes could alter gene expression patterns of the endothelium and induce the polarization of astrocytic processes into the end-feet surrounding the vessel [197]. Vascular changes involving pericytes and preceding the striatal and cortical changes were described also in a mouse model of HD, and changes of these cells were found in post-mortem HD human brains [198].

The CSF dynamic exchange of solutes with a directional flow from the arterial perivascular space into the brain parenchyma and the consequent venous drainage was recently added to the regulating functions of the NVU. In vivo studies demonstrated an active flow from the cisterna magna to the subendothelial space, regulated by astrocytic end-feet of the so-called glial-limiting membrane [17,92]. This process implicated in the CNS waste clearing (Figure 1), seems to be particularly active during sleep, with a cyclic increase of the interstitial space and lower noradrenergic tone mediated by the *locus coeruleus*. The increased ECM space alters the synaptic transmission and contributes with the arterial pump to the CSF influx and interstitial solute exchange during wakefulness-sleep rhythms [91] and possibly during sleep phases transitions. Although there are increasing data concerning the physiology of the glymphatic system and the involvement of the NVU in supporting its role in consciousness and CNS pathology (traumatic, vascular, autoimmune, or degenerative), a translational approach is still lacking [18,67,90,92,199,200].

5. Conclusions and Perspectives

Recent data suggest that a paradigm-shift in CNS studies is mandatory. To understand CNS complexity, novel experiments should focus on the functional cooperation between several cellular, sub-cellular and molecular components in synaptic plasticity. This can be achieved by a systems biology approach. Neurons, although paramount for synapse functioning, are not able to develop, reshape and reinforce the circuitry of the brain on their own.

The support of the glia is essential for trophic factors and neurotransmitters modulation, axon myelinization, and synapse re-localization and elimination. On the other hand, the structural scaffold of the ECM consistently regulated by all resident cells acting as a functional unit can be pivotal in both developmental and several physiological changes. Moreover, the ECM is modified by pathophysiological processes of CNS diseases due to its competence in NTs storage, axon guidance, circuitry protection and intercellular communication. This synaptic model also accounts for the role exerted by the NVU in the metabolic supply, BBB maintenance, glymphatic system, coagulation and immune-system intervention in both adaptive and maladaptive plasticity. NVU and its peculiar structure support selective and controlled exchange between CNS and the blood flow, with precise rules and the possibility to modulate synaptic transmission.

Here we have enlightened the state-of-the-art evidence of the main pathways that should be considered if we want to develop a comprehensive view of synaptic function under physiological and pathological modifications. The increasing amount of omics data (i.e., genomic, epigenomic, transcriptomic, proteomic, metabolomic, connectomic) gained us with an extremely complex picture of the molecular events underlying perturbed conditions linked to neurological and neurodegenerative disorders. On the other hand, we need to organize these big-data into a dynamic, integrative model that takes into account not only the molecular networks but also their relative distribution between cellular and sub-cellular elements. Indeed, modular systems biology is a strategy that allows structuring the information about complex biological processes to obtain modular and mathematical/computational models that may favor the identification of the key steps of the process, as well as the prediction of how the molecular events of the network will respond to specific perturbations of the system. The perspective is to be able to comprehend the regulatory logic of the complex molecular network, which belongs to different cellular and non-cellular domains (neurons, astrocytes, ECM, and NVU).

A clear understanding of these mechanisms, through an iterative process of computational and experimental validation, could lead to the design of new drugs and innovative effective treatments for neurological diseases.

Author Contributions: Conceptualization, C.D.L., A.M.C. and M.P.; writing—original draft preparation, C.D.L., A.V.; writing—review and editing, A.M.C., L.A. and M.P.; visualization, C.D.L.; supervision, M.P. All authors have read and agreed to the published version of the manuscript

Funding: This work was supported by grants from the Italian Minister of University and Research (MIUR) (PRIN2007 to AMC and MP; SYSBIO-Italian ROADMAP ESFRI Infrastructures to LA, MP and AMC; PRIN 2015-2015TM24JS_009 to MP; PRIN2017-2017XJ38A4_003-to MP; ALISEI-IVASCOMAR Italian National Cluster to AMC; and grant Dipartimenti di Eccellenza-2017 to the University of Milano-Bicocca Department of Biotechnology and Biosciences); Corbel: EU-H2020 [PID 2354] to AMC.

Conflicts of Interest: The authors declare no conflict of interests.

References

1. Papa, M.; De Luca, C.; Petta, F.; Alberghina, L.; Cirillo, G. Astrocyte-neuron interplay in maladaptive plasticity. *Neurosci. Biobehav. Rev.* **2014**, *42*, 35–54. [CrossRef]
2. Yamada, S.; Nelson, W.J. Synapses: Sites of cell recognition, adhesion, and functional specification. *Annu. Rev. Biochem.* **2007**, *76*, 267–294. [CrossRef] [PubMed]
3. Stogsdill, J.A.; Eroglu, C. The interplay between neurons and glia in synapse development and plasticity. *Curr. Opin. Neurobiol.* **2017**, *42*, 1–8. [CrossRef] [PubMed]
4. Papa, M.; Sellitti, S.; Sadile, A.G. Remodeling of neural networks in the anterior forebrain of an animal model of hyperactivity and attention deficits as monitored by molecular imaging probes. *Neurosci. Biobehav. Rev.* **2000**, *24*, 149–156. [CrossRef]
5. Araque, A.; Parpura, V.; Sanzgiri, R.P.; Haydon, P.G. Tripartite synapses: Glia, the unacknowledged partner. *Trends Neurosci.* **1999**, *22*, 208–215. [CrossRef]
6. Matyash, V.; Kettenmann, H. Heterogeneity in astrocyte morphology and physiology. *Brain Res. Rev.* **2010**, *63*, 2–10. [CrossRef]
7. Farmer, W.T.; Murai, K. Resolving Astrocyte Heterogeneity in the CNS. *Front. Cell. Neurosci.* **2017**, *11*, 300. [CrossRef]
8. Vasile, F.; Dossi, E.; Rouach, N. Human astrocytes: Structure and functions in the healthy brain. *Brain Struct. Funct.* **2017**, *222*, 2017–2029. [CrossRef]
9. Wang, Y.; Richter-Landsberg, C.; Reiser, G. Expression of protease-activated receptors (PARs) in OLN-93 oligodendroglial cells and mechanism of PAR-1-induced calcium signaling. *Neuroscience* **2004**, *126*, 69–82. [CrossRef]
10. Rodríguez-Iglesias, N.; Sierra, A.; Valero, J. Rewiring of Memory Circuits: Connecting Adult Newborn Neurons With the Help of Microglia. *Front. Cell Dev. Biol.* **2019**, *7*, 24. [CrossRef]
11. De Luca, C.; Papa, M. Looking Inside the Matrix: Perineuronal Nets in Plasticity, Maladaptive Plasticity and Neurological Disorders. *Neurochem. Res.* **2016**, *41*, 1507–1515. [CrossRef] [PubMed]
12. De Luca, C.; Papa, M. Chapter Five-Matrix Metalloproteinases, Neural Extracellular Matrix, and Central Nervous System Pathology. In *Progress in Molecular Biology and Translational Science*; Khalil, R.A., Ed.; Academic Press: Cambridge, MA, USA, 2017; Volume 148, pp. 167–202.
13. Dityatev, A.; Rusakov, D.A. Molecular signals of plasticity at the tetrapartite synapse. *Curr. Opin. Neurobiol.* **2011**, *21*, 353–359. [CrossRef] [PubMed]
14. De Luca, C.; Colangelo, A.M.; Alberghina, L.; Papa, M. Neuro-Immune Hemostasis: Homeostasis and Diseases in the Central Nervous System. *Front. Cell. Neurosci.* **2018**, *12*, 459. [CrossRef] [PubMed]
15. Liebner, S.; Dijkhuizen, R.M.; Reiss, Y.; Plate, K.H.; Agalliu, D.; Constantin, G. Functional morphology of the blood–brain barrier in health and disease. *Acta Neuropathol.* **2018**, *135*, 311–336. [CrossRef]
16. Daneman, R.; Prat, A. The blood–brain barrier. *Cold Spring Harb. Perspect. Biol.* **2015**, *7*, a020412. [CrossRef]
17. Iliff, J.J.; Lee, H.; Yu, M.; Feng, T.; Logan, J.; Nedergaard, M.; Benveniste, H. Brain-wide pathway for waste clearance captured by contrast-enhanced MRI. *J. Clin. Investig.* **2013**, *123*, 1299–1309. [CrossRef]
18. Davoodi-Bojd, E.; Ding, G.; Zhang, L.; Li, Q.; Li, L.; Chopp, M.; Zhang, Z.; Jiang, Q. Modeling glymphatic system of the brain using MRI. *NeuroImage* **2019**, *188*, 616–627. [CrossRef]

19. Gaglio, D.; Valtorta, S.; Ripamonti, M.; Bonanomi, M.; Damiani, C.; Todde, S.; Negri, A.S.; Sanvito, F.; Mastroianni, F.; Di Campli, A.; et al. Divergent in vitro/in vivo responses to drug treatments of highly aggressive NIH-Ras cancer cells: A PET imaging and metabolomics-mass-spectrometry study. *Oncotarget* **2016**, *7*, 52017–52031. [CrossRef]
20. Calderone, A.; Formenti, M.; Aprea, F.; Papa, M.; Alberghina, L.; Colangelo, A.M.; Bertolazzi, P. Comparing Alzheimer's and Parkinson's diseases networks using graph communities structure. *BMC Syst. Biol.* **2016**, *10*, 25. [CrossRef]
21. Fujita, K.A.; Ostaszewski, M.; Matsuoka, Y.; Ghosh, S.; Glaab, E.; Trefois, C.; Crespo, I.; Perumal, T.M.; Jurkowski, W.; Antony, P.M.A.; et al. Integrating pathways of Parkinson's disease in a molecular interaction map. *Mol. Neurobiol.* **2014**, *49*, 88–102. [CrossRef]
22. Colangelo, A.M.; Cirillo, G.; Alberghina, L.; Papa, M.; Westerhoff, H.V. Neural plasticity and adult neurogenesis: The deep biology perspective. *Neurotox. Res.* **2019**, *14*, 201–205. [CrossRef] [PubMed]
23. Martorana, F.; Foti, M.; Virtuoso, A.; Gaglio, D.; Aprea, F.; Latronico, T.; Rossano, R.; Riccio, P.; Papa, M.; Alberghina, L.; et al. Differential Modulation of NF-κB in Neurons and Astrocytes Underlies Neuroprotection and Antigliosis Activity of Natural Antioxidant Molecules. *Oxidative Med. Cell. Longev.* **2019**, *2019*, 8056904. [CrossRef] [PubMed]
24. Virtuoso, A.; Herrera-Rincon, C.; Papa, M.; Panetsos, F. Dependence of Neuroprosthetic Stimulation on the Sensory Modality of the Trigeminal Neurons Following Nerve Injury. Implications in the Design of Future Sensory Neuroprostheses for Correct Perception and Modulation of Neuropathic Pain. *Front. Neurosci.* **2019**, *13*, 389. [CrossRef]
25. Südhof, T.C. The synaptic vesicle cycle: A cascade of protein–protein interactions. *Nature* **1995**, *375*, 645–653. [CrossRef] [PubMed]
26. Valtorta, F.; Pennuto, M.; Bonanomi, D.; Benfenati, F. Synaptophysin: Leading actor or walk-on role in synaptic vesicle exocytosis? *BioEssays* **2004**, *26*, 445–453. [CrossRef]
27. Südhof, T.C. A molecular machine for neurotransmitter release: Synaptotagmin and beyond. *Nat. Med.* **2013**, *19*, 1227. [CrossRef] [PubMed]
28. Cupertino, R.B.; Kappel, D.B.; Bandeira, C.E.; Schuch, J.B.; da Silva, B.S.; Müller, D.; Bau, C.H.D.; Mota, N.R. SNARE complex in developmental psychiatry: Neurotransmitter exocytosis and beyond. *J. Neural Transm.* **2016**, *123*, 867–883. [CrossRef] [PubMed]
29. Woodman, P.G. The roles of NSF, SNAPs and SNAREs during membrane fusion. *Biochim. Biophys. Acta Mol. Cell Res.* **1997**, *1357*, 155–172. [CrossRef]
30. Gulsuner, S.; Walsh, T.; Watts, A.C.; Lee, M.K.; Thornton, A.M.; Casadei, S.; Rippey, C.; Shahin, H.; Consortium on the Genetics of Schizophrenia; PAARTNERS Study Group; et al. Spatial and temporal mapping of de novo mutations in schizophrenia to a fetal prefrontal cortical network. *Cell* **2013**, *154*, 518–529. [CrossRef]
31. Genovese, G.; Fromer, M.; Stahl, E.A.; Ruderfer, D.M.; Chambert, K.; Landén, M.; Moran, J.L.; Purcell, S.M.; Sklar, P.; Sullivan, P.F.; et al. Increased burden of ultra-rare protein-altering variants among 4,877 individuals with schizophrenia. *Nat. Neurosci.* **2016**, *19*, 1433–1441. [CrossRef]
32. Torres, V.I.; Vallejo, D.; Inestrosa, N.C. Emerging Synaptic Molecules as Candidates in the Etiology of Neurological Disorders. *Neural Plast.* **2017**, *2017*, 25. [CrossRef]
33. Wahlby, C.; Erlandsson, F.; Bengtsson, E.; Zetterberg, A. Sequential immunofluorescence staining and image analysis for detection of large numbers of antigens in individual cell nuclei. *Cytometry* **2002**, *47*, 32–41. [CrossRef] [PubMed]
34. Guo, S.-M.; Veneziano, R.; Gordonov, S.; Li, L.; Danielson, E.; Perez de Arce, K.; Park, D.; Kulesa, A.B.; Wamhoff, E.-C.; Blainey, P.C.; et al. Multiplexed and high-throughput neuronal fluorescence imaging with diffusible probes. *Nat. Commun.* **2019**, *10*, 4377. [CrossRef] [PubMed]
35. Gordon, S.L.; Cousin, M.A. The Sybtraps: Control of Synaptobrevin Traffic by Synaptophysin, α-Synuclein and AP-180. *Traffic* **2014**, *15*, 245–254. [CrossRef] [PubMed]
36. Stern, D.; Weisemann, J.; Le Blanc, A.; von Berg, L.; Mahrhold, S.; Piesker, J.; Laue, M.; Luppa, P.B.; Dorner, M.B.; Dorner, B.G.; et al. A lipid-binding loop of botulinum neurotoxin serotypes B, DC and G is an essential feature to confer their exquisite potency. *PLOS Pathog.* **2018**, *14*, e1007048. [CrossRef]
37. Gill, D.M. Bacterial toxins: A table of lethal amounts. *Microbiol. Rev.* **1982**, *46*, 86–94. [CrossRef]
38. Chen, S. Clinical uses of botulinum neurotoxins: Current indications, limitations and future developments. *Toxins* **2012**, *4*, 913–939. [CrossRef]

39. Kim, D.-W.; Lee, S.-K.; Ahnn, J. Botulinum Toxin as a Pain Killer: Players and Actions in Antinociception. *Toxins* **2015**, *7*, 2435–2453. [CrossRef]
40. Dini, E.; Mazzucchi, S.; De Luca, C.; Cafalli, M.; Chico, L.; Lo Gerfo, A.; Siciliano, G.; Bonuccelli, U.; Baldacci, F.; Gori, S. Plasma Levels of Oxidative Stress Markers, before and after BoNT/A Treatment, in Chronic Migraine. *Toxins* **2019**, *11*, 608. [CrossRef]
41. Südhof, T.C. Towards an Understanding of Synapse Formation. *Neuron* **2018**, *100*, 276–293. [CrossRef]
42. Nimchinsky, E.A.; Bernardo, L.S.; Svoboda, K. Structure and Function of Dendritic Spines. *Annu. Rev. Physiol.* **2002**, *64*, 313–353. [CrossRef] [PubMed]
43. Boeckers, T.M. The postsynaptic density. *Cell Tissue Res.* **2006**, *326*, 409–422. [CrossRef] [PubMed]
44. Sheng, M.; Kim, E. The Postsynaptic Organization of Synapses. *Cold Spring Harb. Perspect. Biol.* **2011**, *3*, a005678. [CrossRef] [PubMed]
45. Varoqueaux, F.; Aramuni, G.; Rawson, R.L.; Mohrmann, R.; Missler, M.; Gottmann, K.; Zhang, W.; Südhof, T.C.; Brose, N. Neuroligins Determine Synapse Maturation and Function. *Neuron* **2006**, *51*, 741–754. [CrossRef]
46. Heine, M.; Thoumine, O.; Mondin, M.; Tessier, B.; Giannone, G.; Choquet, D. Activity-independent and subunit-specific recruitment of functional AMPA receptors at neurexin/neuroligin contacts. *Proc. Natl. Acad. Sci. USA* **2008**, *105*, 20947–20952. [CrossRef]
47. Wang, J.; Gong, J.; Li, L.; Chen, Y.; Liu, L.; Gu, H.; Luo, X.; Hou, F.; Zhang, J.; Song, R. Neurexin gene family variants as risk factors for autism spectrum disorder. *Autism Res.* **2018**, *11*, 37–43. [CrossRef]
48. Sando, R.; Jiang, X.; Südhof, T.C. Latrophilin GPCRs direct synapse specificity by coincident binding of FLRTs and teneurins. *Science* **2019**, *363*, eaav7969. [CrossRef]
49. Lu, B.; Nagappan, G.; Lu, Y. BDNF and Synaptic Plasticity, Cognitive Function, and Dysfunction. In *Neurotrophic Factors*; Lewin, G.R., Carter, B.D., Eds.; Springer: Berlin/Heidelberg, Germany, 2014; pp. 223–250.
50. Chao, M.V.; Bothwell, M.A.; Ross, A.H.; Koprowski, H.; Lanahan, A.A.; Buck, C.R.; Sehgal, A. Gene transfer and molecular cloning of the human NGF receptor. *Science* **1986**, *232*, 518–521. [CrossRef]
51. Huang, E.J.; Reichardt, L.F. Neurotrophins: Roles in neuronal development and function. *Annu. Rev. Neurosci.* **2001**, *24*, 677–736. [CrossRef]
52. McMahon, S.; Murinson, B. CHAPTER 25—Therapeutic Potential of Neurotrophic Factors. In *From Neuroscience To Neurology*; Waxman, S., Ed.; Academic Press: Burlington, NJ, USA, 2005; pp. 419–431. [CrossRef]
53. Colangelo, A.M.; Cirillo, G.; Lavitrano, M.L.; Alberghina, L.; Papa, M. Targeting reactive astrogliosis by novel biotechnological strategies. *Biotechnol. Adv.* **2012**, *30*, 261–271. [CrossRef]
54. Longo, F.M.; Massa, S.M. Small-molecule modulation of neurotrophin receptors: A strategy for the treatment of neurological disease. *Nat. Rev. Drug Discov.* **2013**, *12*, 507–525. [CrossRef] [PubMed]
55. Bruno, M.A.; Cuello, A.C. Activity-dependent release of precursor nerve growth factor, conversion to mature nerve growth factor, and its degradation by a protease cascade. *Proc. Natl. Acad. Sci. USA* **2006**, *103*, 6735–6740. [CrossRef] [PubMed]
56. Park, H.; Poo, M.-M. Neurotrophin regulation of neural circuit development and function. *Nat. Rev. Neurosci.* **2013**, *14*, 7–23. [CrossRef] [PubMed]
57. Martorana, F.; Gaglio, D.; Bianco, M.R.; Aprea, F.; Virtuoso, A.; Bonanomi, M.; Alberghina, L.; Papa, M.; Colangelo, A.M. Differentiation by nerve growth factor (NGF) involves mechanisms of crosstalk between energy homeostasis and mitochondrial remodeling. *Cell Death Dis.* **2018**, *9*, 391. [CrossRef]
58. Alberghina, L.; Colangelo, A.M. The modular systems biology approach to investigate the control of apoptosis in Alzheimer's disease neurodegeneration. *BMC Neurosci.* **2006**, *7*, S2. [CrossRef]
59. Bereczki, E.; Branca, R.M.; Francis, P.T.; Pereira, J.B.; Baek, J.-H.; Hortobágyi, T.; Winblad, B.; Ballard, C.; Lehtiö, J.; Aarsland, D. Synaptic markers of cognitive decline in neurodegenerative diseases: A proteomic approach. *Brain* **2018**, *141*, 582–595. [CrossRef]
60. Witcher, M.R.; Park, Y.D.; Lee, M.R.; Sharma, S.; Harris, K.M.; Kirov, S.A. Three-dimensional relationships between perisynaptic astroglia and human hippocampal synapses. *Glia* **2010**, *58*, 572–587. [CrossRef]
61. Perea, G.; Navarrete, M.; Araque, A. Tripartite synapses: Astrocytes process and control synaptic information. *Trends Neurosci.* **2009**, *32*, 421–431. [CrossRef]

62. Singh, S.K.; Stogsdill, J.A.; Pulimood, N.S.; Dingsdale, H.; Kim, Y.H.; Pilaz, L.-J.; Kim, I.H.; Manhaes, A.C.; Rodrigues, W.S., Jr.; Pamukcu, A.; et al. Astrocytes Assemble Thalamocortical Synapses by Bridging NRX1α and NL1 via Hevin. *Cell* **2016**, *164*, 183–196. [CrossRef]
63. Seifert, G.; Hüttmann, K.; Binder, D.K.; Hartmann, C.; Wyczynski, A.; Neusch, C.; Steinhäuser, C. Analysis of Astroglial K+ Channel Expression in the Developing Hippocampus Reveals a Predominant Role of the Kir4.1 Subunit. *J. Neurosci.* **2009**, *29*, 7474. [CrossRef]
64. Dossi, E.; Vasile, F.; Rouach, N. Human astrocytes in the diseased brain. *Brain Res. Bull.* **2018**, *136*, 139–156. [CrossRef] [PubMed]
65. Jiang, R.; Diaz-Castro, B.; Looger, L.L.; Khakh, B.S. Dysfunctional Calcium and Glutamate Signaling in Striatal Astrocytes from Huntington's Disease Model Mice. *J. Neurosci.* **2016**, *36*, 3453. [CrossRef] [PubMed]
66. Larsen, B.R.; Assentoft, M.; Cotrina, M.L.; Hua, S.Z.; Nedergaard, M.; Kaila, K.; Voipio, J.; MacAulay, N. Contributions of the Na+/K+-ATPase, NKCC1, and Kir4.1 to hippocampal K+ clearance and volume responses. *Glia* **2014**, *62*, 608–622. [CrossRef] [PubMed]
67. Plog, B.A.; Nedergaard, M. The Glymphatic System in Central Nervous System Health and Disease: Past, Present, and Future. *Annu. Rev. Pathol. Mech. Dis.* **2018**, *13*, 379–394. [CrossRef] [PubMed]
68. Goodrich, G.S.; Kabakov, A.Y.; Hameed, M.Q.; Dhamne, S.C.; Rosenberg, P.A.; Rotenberg, A. Ceftriaxone Treatment after Traumatic Brain Injury Restores Expression of the Glutamate Transporter, GLT-1, Reduces Regional Gliosis, and Reduces Post-Traumatic Seizures in the Rat. *J. Neurotrauma* **2013**, *30*, 1434–1441. [CrossRef] [PubMed]
69. Lehre, K.P.; Danbolt, N.C. The Number of Glutamate Transporter Subtype Molecules at Glutamatergic Synapses: Chemical and Stereological Quantification in Young Adult Rat Brain. *J. Neurosci.* **1998**, *18*, 8751. [CrossRef]
70. Bianchi, M.G.; Bardelli, D.; Chiu, M.; Bussolati, O. Changes in the expression of the glutamate transporter EAAT3/EAAC1 in health and disease. *Cell. Mol. Life Sci. CMLS* **2014**, *71*, 2001–2015. [CrossRef]
71. Murphy-Royal, C.; Dupuis, J.P.; Varela, J.A.; Panatier, A.; Pinson, B.; Baufreton, J.; Groc, L.; Oliet, S.H.R. Surface diffusion of astrocytic glutamate transporters shapes synaptic transmission. *Nat. Neurosci.* **2015**, *18*, 219. [CrossRef]
72. Boddum, K.; Jensen, T.P.; Magloire, V.; Kristiansen, U.; Rusakov, D.A.; Pavlov, I.; Walker, M.C. Astrocytic GABA transporter activity modulates excitatory neurotransmission. *Nat. Commun.* **2016**, *7*, 13572. [CrossRef]
73. Yu, X.; Taylor, A.M.W.; Nagai, J.; Golshani, P.; Evans, C.J.; Coppola, G.; Khakh, B.S. Reducing Astrocyte Calcium Signaling In Vivo Alters Striatal Microcircuits and Causes Repetitive Behavior. *Neuron* **2018**, *99*, 1170–1187. [CrossRef]
74. Savtchouk, I.; Volterra, A. Gliotransmission: Beyond Black-and-White. *J. Neurosci.* **2018**, *38*, 14–25. [CrossRef] [PubMed]
75. Bohmbach, K.; Schwarz, M.K.; Schoch, S.; Henneberger, C. The structural and functional evidence for vesicular release from astrocytes in situ. *Brain Res. Bull.* **2018**, *136*, 65–75. [CrossRef] [PubMed]
76. Allen, N.J.; Eroglu, C. Cell Biology of Astrocyte-Synapse Interactions. *Neuron* **2017**, *96*, 697–708. [CrossRef] [PubMed]
77. De Pittà, M.; Brunel, N. Modulation of Synaptic Plasticity by Glutamatergic Gliotransmission: A Modeling Study. *Neural Plast.* **2016**, *2016*, 30. [CrossRef] [PubMed]
78. Panatier, A.; Vallée, J.; Haber, M.; Murai, K.K.; Lacaille, J.-C.; Robitaille, R. Astrocytes Are Endogenous Regulators of Basal Transmission at Central Synapses. *Cell* **2011**, *146*, 785–798. [CrossRef] [PubMed]
79. Oberheim, N.A.; Takano, T.; Han, X.; He, W.; Lin, J.H.C.; Wang, F.; Xu, Q.; Wyatt, J.D.; Pilcher, W.; Ojemann, J.G.; et al. Uniquely Hominid Features of Adult Human Astrocytes. *J. Neurosci.* **2009**, *29*, 3276. [CrossRef]
80. Bushong, E.A.; Martone, M.E.; Jones, Y.Z.; Ellisman, M.H. Protoplasmic Astrocytes in CA1 Stratum Radiatum Occupy Separate Anatomical Domains. *J. Neurosci.* **2002**, *22*, 183. [CrossRef]
81. Covelo, A.; Araque, A. Neuronal activity determines distinct gliotransmitter release from a single astrocyte. *eLife* **2018**, *7*, e32237. [CrossRef]
82. Cirillo, G.; Colangelo, A.M.; Berbenni, M.; Ippolito, V.M.; De Luca, C.; Verdesca, F.; Savarese, L.; Alberghina, L.; Maggio, N.; Papa, M. Purinergic Modulation of Spinal Neuroglial Maladaptive Plasticity Following Peripheral Nerve Injury. *Mol. Neurobiol.* **2015**, *52*, 1440–1457. [CrossRef]

83. Martin-Fernandez, M.; Jamison, S.; Robin, L.M.; Zhao, Z.; Martin, E.D.; Aguilar, J.; Benneyworth, M.A.; Marsicano, G.; Araque, A. Synapse-specific astrocyte gating of amygdala-related behavior. *Nat. Neurosci.* **2017**, *20*, 1540. [CrossRef]
84. Gómez-Gonzalo, M.; Navarrete, M.; Perea, G.; Covelo, A.; Martín-Fernández, M.; Shigemoto, R.; Luján, R.; Araque, A. Endocannabinoids Induce Lateral Long-Term Potentiation of Transmitter Release by Stimulation of Gliotransmission. *Cereb. Cortex* **2014**, *25*, 3699–3712. [CrossRef] [PubMed]
85. Robin, L.M.; Oliveira da Cruz, J.F.; Langlais, V.C.; Martin-Fernandez, M.; Metna-Laurent, M.; Busquets-Garcia, A.; Bellocchio, L.; Soria-Gomez, E.; Papouin, T.; Varilh, M.; et al. Astroglial CB1 Receptors Determine Synaptic D-Serine Availability to Enable Recognition Memory. *Neuron* **2018**, *98*, 935–944. [CrossRef] [PubMed]
86. Brockett, A.T.; Kane, G.A.; Monari, P.K.; Briones, B.A.; Vigneron, P.-A.; Barber, G.A.; Bermudez, A.; Dieffenbach, U.; Kloth, A.D.; Buschman, T.J.; et al. Evidence supporting a role for astrocytes in the regulation of cognitive flexibility and neuronal oscillations through the Ca2+ binding protein S100β. *PLoS ONE* **2018**, *13*, e0195726. [CrossRef]
87. Brown, R.E.; Basheer, R.; McKenna, J.T.; Strecker, R.E.; McCarley, R.W. Control of Sleep and Wakefulness. *Physiol. Rev.* **2012**, *92*, 1087–1187. [CrossRef] [PubMed]
88. Clasadonte, J.; Scemes, E.; Wang, Z.; Boison, D.; Haydon, P.G. Connexin 43-Mediated Astroglial Metabolic Networks Contribute to the Regulation of the Sleep-Wake Cycle. *Neuron* **2017**, *95*, 1365–1380. [CrossRef]
89. Poskanzer, K.E.; Yuste, R. Astrocytes regulate cortical state switching in vivo. *Proc. Natl. Acad. Sci. USA* **2016**, *113*, E2675. [CrossRef]
90. Lundgaard, I.; Lu, M.L.; Yang, E.; Peng, W.; Mestre, H.; Hitomi, E.; Deane, R.; Nedergaard, M. Glymphatic clearance controls state-dependent changes in brain lactate concentration. *J. Cereb. Blood Flow Metab.* **2016**, *37*, 2112–2124. [CrossRef]
91. Xie, L.; Kang, H.; Xu, Q.; Chen, M.J.; Liao, Y.; Thiyagarajan, M.; O'Donnell, J.; Christensen, D.J.; Nicholson, C.; Iliff, J.J.; et al. Sleep Drives Metabolite Clearance from the Adult Brain. *Science* **2013**, *342*, 373. [CrossRef]
92. Bacyinski, A.; Xu, M.; Wang, W.; Hu, J. The Paravascular Pathway for Brain Waste Clearance: Current Understanding, Significance and Controversy. *Front. Neuroanat.* **2017**, *11*. [CrossRef]
93. De Bock, M.; Leybaert, L.; Giaume, C. Connexin Channels at the Glio-Vascular Interface: Gatekeepers of the Brain. *Neurochem. Res.* **2017**, *42*, 2519–2536. [CrossRef]
94. Pannasch, U.; Freche, D.; Dallerac, G.; Ghezali, G.; Escartin, C.; Ezan, P.; Cohen-Salmon, M.; Benchenane, K.; Abudara, V.; Dufour, A.; et al. Connexin 30 sets synaptic strength by controlling astroglial synapse invasion. *Nat. Neurosci.* **2014**, *17*, 549–558. [CrossRef] [PubMed]
95. Pannasch, U.; Vargová, L.; Reingruber, J.; Ezan, P.; Holcman, D.; Giaume, C.; Syková, E.; Rouach, N. Astroglial networks scale synaptic activity and plasticity. *Proc. Natl. Acad. Sci. USA* **2011**, *108*, 8467. [CrossRef] [PubMed]
96. Charles, A.C.; Baca, S.M. Cortical spreading depression and migraine. *Nat. Rev. Neurol.* **2013**, *9*, 637–644. [CrossRef] [PubMed]
97. Hauge, A.W.; Asghar, M.S.; Schytz, H.W.; Christensen, K.; Olesen, J. Effects of tonabersat on migraine with aura: A randomised, double-blind, placebo-controlled crossover study. *Lancet Neurol.* **2009**, *8*, 718–723. [CrossRef]
98. Tarczyluk, M.A.; Nagel, D.A.; O'Neil, J.D.; Parri, H.R.; Tse, E.H.; Coleman, M.D.; Hill, E.J. Functional astrocyte-neuron lactate shuttle in a human stem cell-derived neuronal network. *J. Cereb. Blood Flow Metab.* **2013**, *33*, 1386–1393. [CrossRef]
99. Falkowska, A.; Gutowska, I.; Goschorska, M.; Nowacki, P.; Chlubek, D.; Baranowska-Bosiacka, I. Energy Metabolism of the Brain, Including the Cooperation between Astrocytes and Neurons, Especially in the Context of Glycogen Metabolism. *Int. J. Mol. Sci.* **2015**, *16*, 25959–25981. [CrossRef]
100. Alberini, C.M.; Cruz, E.; Descalzi, G.; Bessieres, B.; Gao, V. Astrocyte glycogen and lactate: New insights into learning and memory mechanisms. *Glia* **2018**, *66*, 1244–1262. [CrossRef]
101. Haber, M.; Zhou, L.; Murai, K.K. Cooperative Astrocyte and Dendritic Spine Dynamics at Hippocampal Excitatory Synapses. *J. Neurosci.* **2006**, *26*, 8881. [CrossRef]
102. Pirttimaki, T.M.; Hall, S.D.; Parri, H.R. Sustained Neuronal Activity Generated by Glial Plasticity. *J. Neurosci.* **2011**, *31*, 7637. [CrossRef]

103. Chung, W.-S.; Allen, N.J.; Eroglu, C. Astrocytes Control Synapse Formation, Function, and Elimination. *Cold Spring Harb. Perspect. Biol.* **2015**, *7*, a020370. [CrossRef]
104. Xu-Friedman, M.A.; Harris, K.M.; Regehr, W.G. Three-Dimensional Comparison of Ultrastructural Characteristics at Depressing and Facilitating Synapses onto Cerebellar Purkinje Cells. *J. Neurosci.* **2001**, *21*, 6666. [CrossRef] [PubMed]
105. Ventura, R.; Harris, K.M. Three-Dimensional Relationships between Hippocampal Synapses and Astrocytes. *J. Neurosci.* **1999**, *19*, 6897. [CrossRef] [PubMed]
106. Genoud, C.; Quairiaux, C.; Steiner, P.; Hirling, H.; Welker, E.; Knott, G.W. Plasticity of Astrocytic Coverage and Glutamate Transporter Expression in Adult Mouse Cortex. *PLOS Biol.* **2006**, *4*, e343. [CrossRef] [PubMed]
107. Theodosis, D.T.; Poulain, D.A.; Oliet, S.H.R. Activity-Dependent Structural and Functional Plasticity of Astrocyte-Neuron Interactions. *Physiol. Rev.* **2008**, *88*, 983–1008. [CrossRef] [PubMed]
108. Kettenmann, H.; Hanisch, U.-K.; Noda, M.; Verkhratsky, A. Physiology of Microglia. *Physiol. Rev.* **2011**, *91*, 461–553. [CrossRef]
109. Cirillo, G.; Colangelo, A.M.; De Luca, C.; Savarese, L.; Barillari, M.R.; Alberghina, L.; Papa, M. Modulation of Matrix Metalloproteinases Activity in the Ventral Horn of the Spinal Cord Re-stores Neuroglial Synaptic Homeostasis and Neurotrophic Support following Peripheral Nerve Injury. *PLoS ONE* **2016**, *11*, e0152750. [CrossRef]
110. De Luca, C.; Savarese, L.; Colangelo, A.M.; Bianco, M.R.; Cirillo, G.; Alberghina, L.; Papa, M. Astrocytes and Microglia-Mediated Immune Response in Maladaptive Plasticity is Differently Modulated by NGF in the Ventral Horn of the Spinal Cord Following Peripheral Nerve Injury. *Cell. Mol. Neurobiol.* **2016**, *36*, 37–46. [CrossRef]
111. Sierra, A.; Beccari, S.; Diaz-Aparicio, I.; Encinas, J.M.; Comeau, S.; Tremblay, M. Surveillance, Phagocytosis, and Inflammation: How Never-Resting Microglia Influence Adult Hippocampal Neurogenesis. *Neural Plast.* **2014**, *2014*, 15. [CrossRef]
112. De Luca, C.; Virtuoso, A.; Maggio, N.; Papa, M. Neuro-Coagulopathy: Blood Coagulation Factors in Central Nervous System Diseases. *Int. J. Mol. Sci.* **2017**, *18*, 2128. [CrossRef]
113. Ginhoux, F.; Greter, M.; Leboeuf, M.; Nandi, S.; See, P.; Gokhan, S.; Mehler, M.F.; Conway, S.J.; Ng, L.G.; Stanley, E.R.; et al. Fate Mapping Analysis Reveals That Adult Microglia Derive from Primitive Macrophages. *Science* **2010**, *330*, 841. [CrossRef]
114. Cheray, M.; Joseph, B. Epigenetics Control Microglia Plasticity. *Front. Cell. Neurosci.* **2018**, *12*, 243. [CrossRef] [PubMed]
115. Röszer, T. Understanding the Biology of Self-Renewing Macrophages. *Cells* **2018**, *7*, 103. [CrossRef] [PubMed]
116. Paris, I.; Savage, J.C.; Escobar, L.; Abiega, O.; Gagnon, S.; Hui, C.-W.; Tremblay, M.-È.; Sierra, A.; Valero, J. ProMoIJ: A new tool for automatic three-dimensional analysis of microglial process motility. *Glia* **2018**, *66*, 828–845. [CrossRef] [PubMed]
117. Gosselin, D.; Skola, D.; Coufal, N.G.; Holtman, I.R.; Schlachetzki, J.C.M.; Sajti, E.; Jaeger, B.N.; O'Connor, C.; Fitzpatrick, C.; Pasillas, M.P.; et al. An environment-dependent transcriptional network specifies human microglia identity. *Science* **2017**, *356*, eaal3222. [CrossRef]
118. Xue, J.; Schmidt, S.V.; Sander, J.; Draffehn, A.; Krebs, W.; Quester, I.; De Nardo, D.; Gohel, T.D.; Emde, M.; Schmidleithner, L.; et al. Transcriptome-Based Network Analysis Reveals a Spectrum Model of Human Macrophage Activation. *Immunity* **2014**, *40*, 274–288. [CrossRef]
119. Kettenmann, H.; Kirchhoff, F.; Verkhratsky, A. Microglia: New Roles for the Synaptic Stripper. *Neuron* **2013**, *77*, 10–18. [CrossRef]
120. Hanamsagar, R.; Bilbo, S.D. Environment matters: Microglia function and dysfunction in a changing world. *Curr. Opin. Neurobiol.* **2017**, *47*, 146–155. [CrossRef]
121. Luo, P.; Chu, S.F.; Zhang, Z.; Xia, C.Y.; Chen, N.H. Fractalkine/CX3CR1 is involved in the cross-talk between neuron and glia in neurological diseases. *Brain Res. Bull.* **2019**, *146*, 12–21. [CrossRef]
122. Filipello, F.; Morini, R.; Corradini, I.; Zerbi, V.; Canzi, A.; Michalski, B.; Erreni, M.; Markicevic, M.; Starvaggi-Cucuzza, C.; Otero, K.; et al. The Microglial Innate Immune Receptor TREM2 Is Required for Synapse Elimination and Normal Brain Connectivity. *Immunity* **2018**, *48*, 979–991. [CrossRef]

123. Roumier, A.; Béchade, C.; Poncer, J.-C.; Smalla, K.-H.; Tomasello, E.; Vivier, E.; Gundelfinger, E.D.; Triller, A.; Bessis, A. Impaired Synaptic Function in the Microglial KARAP/DAP12-Deficient Mouse. *J. Neurosci.* **2004**, *24*, 11421. [CrossRef]
124. Frost, J.L.; Schafer, D.P. Microglia: Architects of the Developing Nervous System. *Trends Cell Biol.* **2016**, *26*, 587–597. [CrossRef] [PubMed]
125. Szepesi, Z.; Manouchehrian, O.; Bachiller, S.; Deierborg, T. Bidirectional Microglia–Neuron Communication in Health and Disease. *Front. Cell. Neurosci.* **2018**, *12*, 323. [CrossRef] [PubMed]
126. Weinhard, L.; Di Bartolomei, G.; Bolasco, G.; Machado, P.; Schieber, N.L.; Neniskyte, U.; Exiga, M.; Vadisiute, A.; Raggioli, A.; Schertel, A.; et al. Microglia remodel synapses by presynaptic trogocytosis and spine head filopodia induction. *Nat. Commun.* **2018**, *9*, 1228. [CrossRef] [PubMed]
127. Kim, H.J.; Cho, M.H.; Shim, W.H.; Kim, J.K.; Jeon, E.Y.; Kim, D.H.; Yoon, S.Y. Deficient autophagy in microglia impairs synaptic pruning and causes social behavioral defects. *Mol. Psychiatry* **2017**, *22*, 1576–1584. [CrossRef]
128. Chen, Z.; Jalabi, W.; Hu, W.; Park, H.-J.; Gale, J.T.; Kidd, G.J.; Bernatowicz, R.; Gossman, Z.C.; Chen, J.T.; Dutta, R.; et al. Microglial displacement of inhibitory synapses provides neuroprotection in the adult brain. *Nat. Commun.* **2014**, *5*, 4486. [CrossRef] [PubMed]
129. Clark, K.C.; Josephson, A.; Benusa, S.D.; Hartley, R.K.; Baer, M.; Thummala, S.; Joslyn, M.; Sword, B.A.; Elford, H.; Oh, U.; et al. Compromised axon initial segment integrity in EAE is preceded by microglial reactivity and contact. *Glia* **2016**, *64*, 1190–1209. [CrossRef] [PubMed]
130. Lowery, R.L.; Tremblay, M.-E.; Hopkins, B.E.; Majewska, A.K. The microglial fractalkine receptor is not required for activity-dependent plasticity in the mouse visual system. *Glia* **2017**, *65*, 1744–1761. [CrossRef]
131. Schecter, R.W.; Maher, E.E.; Welsh, C.A.; Stevens, B.; Erisir, A.; Bear, M.F. Experience-Dependent Synaptic Plasticity in V1 Occurs without Microglial CX3CR1. *J. Neurosci.* **2017**, *37*, 10541. [CrossRef]
132. Ferrini, F.; De Koninck, Y. Microglia Control Neuronal Network Excitability via BDNF Signalling. *Neural Plast.* **2013**, *2013*, 11. [CrossRef]
133. Gabrielli, M.; Battista, N.; Riganti, L.; Prada, I.; Antonucci, F.; Cantone, L.; Matteoli, M.; Maccarrone, M.; Verderio, C. Active endocannabinoids are secreted on extracellular membrane vesicles. *EMBO Rep.* **2015**, *16*, 213–220. [CrossRef]
134. Parkhurst, C.N.; Yang, G.; Ninan, I.; Savas, J.N.; Yates, J.R., III; Lafaille, J.J.; Hempstead, B.L.; Littman, D.R.; Gan, W.-B. Microglia Promote Learning-Dependent Synapse Formation through Brain-Derived Neurotrophic Factor. *Cell* **2013**, *155*, 1596–1609. [CrossRef] [PubMed]
135. Pickering, M.; Cumiskey, D.; O'Connor, J.J. Actions of TNF-α on glutamatergic synaptic transmission in the central nervous system. *Exp. Physiol.* **2005**, *90*, 663–670. [CrossRef] [PubMed]
136. Zhang, J.; Malik, A.; Choi, H.B.; Ko, R.W.Y.; Dissing-Olesen, L.; MacVicar, B.A. Microglial CR3 Activation Triggers Long-Term Synaptic Depression in the Hippocampus via NADPH Oxidase. *Neuron* **2014**, *82*, 195–207. [CrossRef] [PubMed]
137. Burda, J.E.; Sofroniew, M.V. Reactive gliosis and the multicellular response to CNS damage and disease. *Neuron* **2014**, *81*, 229–248. [CrossRef]
138. Bergles, D.E.; Richardson, W.D. Oligodendrocyte Development and Plasticity. *Cold Spring Harb. Perspect. Biol.* **2015**, *8*, a020453. [CrossRef]
139. Baron, W.; Colognato, H.; Ffrench-Constant, C. Integrin-growth factor interactions as regulators of oligodendroglial development and function. *Glia* **2005**, *49*, 467–479. [CrossRef] [PubMed]
140. Ortega, M.C.; Bribian, A.; Peregrin, S.; Gil, M.T.; Marin, O.; de Castro, F. Neuregulin-1/ErbB4 signaling controls the migration of oligodendrocyte precursor cells during development. *Exp. Neurol.* **2012**, *235*, 610–620. [CrossRef]
141. Cohen, R.I.; Marmur, R.; Norton, W.T.; Mehler, M.F.; Kessler, J.A. Nerve growth factor and neurotrophin-3 differentially regulate the proliferation and survival of developing rat brain oligodendrocytes. *J. Neurosci.* **1996**, *16*, 6433–6442. [CrossRef]
142. Cahoy, J.D.; Emery, B.; Kaushal, A.; Foo, L.C.; Zamanian, J.L.; Christopherson, K.S.; Xing, Y.; Lubischer, J.L.; Krieg, P.A.; Krupenko, S.A.; et al. A transcriptome database for astrocytes, neurons, and oligodendrocytes: A new resource for understanding brain development and function. *J. Neurosci.* **2008**, *28*, 264–278. [CrossRef]
143. Ziskin, J.L.; Nishiyama, A.; Rubio, M.; Fukaya, M.; Bergles, D.E. Vesicular release of glutamate from unmyelinated axons in white matter. *Nat. Neurosci.* **2007**, *10*, 321–330. [CrossRef]

144. Maldonado, P.P.; Velez-Fort, M.; Angulo, M.C. Is neuronal communication with NG2 cells synaptic or extrasynaptic? *J. Anat.* **2011**, *219*, 8–17. [CrossRef] [PubMed]
145. Balia, M.; Velez-Fort, M.; Passlick, S.; Schafer, C.; Audinat, E.; Steinhauser, C.; Seifert, G.; Angulo, M.C. Postnatal down-regulation of the GABAA receptor gamma2 subunit in neocortical NG2 cells accompanies synaptic-to-extrasynaptic switch in the GABAergic transmission mode. *Cereb. Cortex* **2015**, *25*, 1114–1123. [CrossRef] [PubMed]
146. Demerens, C.; Stankoff, B.; Logak, M.; Anglade, P.; Allinquant, B.; Couraud, F.; Zalc, B.; Lubetzki, C. Induction of myelination in the central nervous system by electrical activity. *Proc. Natl. Acad. Sci. USA* **1996**, *93*, 9887–9892. [CrossRef] [PubMed]
147. McKenzie, I.A.; Ohayon, D.; Li, H.; de Faria, J.P.; Emery, B.; Tohyama, K.; Richardson, W.D. Motor skill learning requires active central myelination. *Science* **2014**, *346*, 318–322. [CrossRef]
148. Gibson, E.M.; Purger, D.; Mount, C.W.; Goldstein, A.K.; Lin, G.L.; Wood, L.S.; Inema, I.; Miller, S.E.; Bieri, G.; Zuchero, J.B.; et al. Neuronal activity promotes oligodendrogenesis and adaptive myelination in the mammalian brain. *Science* **2014**, *344*, 1252304. [CrossRef]
149. Young, K.M.; Psachoulia, K.; Tripathi, R.B.; Dunn, S.J.; Cossell, L.; Attwell, D.; Tohyama, K.; Richardson, W.D. Oligodendrocyte dynamics in the healthy adult CNS: Evidence for myelin remodeling. *Neuron* **2013**, *77*, 873–885. [CrossRef]
150. Sandell, J.H.; Peters, A. Effects of age on the glial cells in the rhesus monkey optic nerve. *J. Comp. Neurol.* **2002**, *445*, 13–28. [CrossRef]
151. Peters, A.; Kemper, T. A review of the structural alterations in the cerebral hemispheres of the aging rhesus monkey. *Neurobiol. Aging* **2012**, *33*, 2357–2372. [CrossRef]
152. Kimura, F.; Itami, C. Myelination and isochronicity in neural networks. *Front. Neuroanat.* **2009**, *3*, 12. [CrossRef]
153. Seidl, A.H.; Rubel, E.W.; Barria, A. Differential conduction velocity regulation in ipsilateral and contralateral collaterals innervating brainstem coincidence detector neurons. *J. Neurosci.* **2014**, *34*, 4914–4919. [CrossRef]
154. Sampaio-Baptista, C.; Khrapitchev, A.A.; Foxley, S.; Schlagheck, T.; Scholz, J.; Jbabdi, S.; DeLuca, G.C.; Miller, K.L.; Taylor, A.; Thomas, N.; et al. Motor skill learning induces changes in white matter microstructure and myelination. *J. Neurosci.* **2013**, *33*, 19499–19503. [CrossRef] [PubMed]
155. Lombardi, M.; Parolisi, R.; Scaroni, F.; Bonfanti, E.; Gualerzi, A.; Gabrielli, M.; Kerlero de Rosbo, N.; Uccelli, A.; Giussani, P.; Viani, P.; et al. Detrimental and protective action of microglial extracellular vesicles on myelin lesions: Astrocyte involvement in remyelination failure. *Acta Neuropathol.* **2019**, *138*, 987–1012. [CrossRef] [PubMed]
156. Corvetti, L.; Rossi, F. Degradation of Chondroitin Sulfate Proteoglycans Induces Sprouting of Intact Purkinje Axons in the Cerebellum of the Adult Rat. *J. Neurosci.* **2005**, *25*, 7150. [CrossRef] [PubMed]
157. Dzwonek, J.; Rylski, M.; Kaczmarek, L. Matrix metalloproteinases and their endogenous inhibitors in neuronal physiology of the adult brain. *FEBS Lett.* **2004**, *567*, 129–135. [CrossRef] [PubMed]
158. Geissler, M.; Gottschling, C.; Aguado, A.; Rauch, U.; Wetzel, C.H.; Hatt, H.; Faissner, A. Primary hippocampal neurons, which lack four crucial extracellular matrix molecules, display abnormalities of synaptic structure and function and severe deficits in perineuronal net formation. *J. Neurosci.* **2013**, *33*, 7742–7755. [CrossRef] [PubMed]
159. Rauch, U.; Zhou, X.H.; Roos, G. Extracellular matrix alterations in brains lacking four of its components. *Biochem. Biophys. Res. Commun.* **2005**, *328*, 608–617. [CrossRef]
160. Allen, N.J.; Bennett, M.L.; Foo, L.C.; Wang, G.X.; Chakraborty, C.; Smith, S.J.; Barres, B.A. Astrocyte glypicans 4 and 6 promote formation of excitatory synapses via GluA1 AMPA receptors. *Nature* **2012**, *486*, 410–414. [CrossRef]
161. Vandenbroucke, R.E.; Libert, C. Is there new hope for therapeutic matrix metalloproteinase inhibition? *Nat. Rev. Drug Discov.* **2014**, *13*, 904–927. [CrossRef]
162. Jablonska-Trypuc, A.; Matejczyk, M.; Rosochacki, S. Matrix metalloproteinases (MMPs), the main extracellular matrix (ECM) enzymes in collagen degradation, as a target for anticancer drugs. *J. Enzym. Inhib. Med. Chem.* **2016**, *31*, 177–183. [CrossRef]
163. Saroja, S.R.; Sase, A.; Kircher, S.G.; Wan, J.; Berger, J.; Höger, H.; Pollak, A.; Lubec, G. Hippocampal proteoglycans brevican and versican are linked to spatial memory of Sprague–Dawley rats in the morris water maze. *J. Neurochem.* **2014**, *130*, 797–804. [CrossRef]

164. Milev, P.; Maurel, P.; Chiba, A.; Mevissen, M.; Popp, S.; Yamaguchi, Y.; Margolis, R.K.; Margolis, R.U. Differential Regulation of Expression of Hyaluronan-Binding Proteoglycans in Developing Brain: Aggrecan, Versican, Neurocan, and Brevican. *Biochem. Biophys. Res. Commun.* **1998**, *247*, 207–212. [CrossRef] [PubMed]
165. Dino, M.R.; Harroch, S.; Hockfield, S.; Matthews, R.T. Monoclonal antibody Cat-315 detects a glycoform of receptor protein tyrosine phosphatase beta/phosphacan early in CNS development that localizes to extrasynaptic sites prior to synapse formation. *Neuroscience* **2006**, *142*, 1055–1069. [CrossRef] [PubMed]
166. Gogolla, N.; Caroni, P.; Lüthi, A.; Herry, C. Perineuronal Nets Protect Fear Memories from Erasure. *Science* **2009**, *325*, 1258. [CrossRef] [PubMed]
167. Apostolova, I.; Irintchev, A.; Schachner, M. Tenascin-R restricts posttraumatic remodeling of motoneuron innervation and functional recovery after spinal cord injury in adult mice. *J. Neurosci.* **2006**, *26*, 7849–7859. [CrossRef] [PubMed]
168. Costa, C.; Martínez-Sáez, E.; Gutiérrez-Franco, A.; Eixarch, H.; Castro, Z.; Ortega-Aznar, A.; Ramón y Cajal, S.; Montalban, X.; Espejo, C. Expression of semaphorin 3A, semaphorin 7A and their receptors in multiple sclerosis lesions. *Mult. Scler. J.* **2015**, *21*, 1632–1643. [CrossRef] [PubMed]
169. McKeon, R.J.; Jurynec, M.J.; Buck, C.R. The Chondroitin Sulfate Proteoglycans Neurocan and Phosphacan Are Expressed by Reactive Astrocytes in the Chronic CNS Glial Scar. *J. Neurosci.* **1999**, *19*, 10778. [CrossRef]
170. Bovolental, P.; Fernaud-Espinosa, I.; Mendez-Otero, R.; Nieto-Sampedro, M. Neurite Outgrowth Inhibitor of Gliotic Brain Tissue. Mode of Action and Cellular Localization, Studied with Specific Monoclonal Antibodies. *Eur. J. Neurosci.* **1997**, *9*, 977–989. [CrossRef]
171. Howell, M.D.; Bailey, L.A.; Cozart, M.A.; Gannon, B.M.; Gottschall, P.E. Hippocampal administration of chondroitinase ABC increases plaque-adjacent synaptic marker and diminishes amyloid burden in aged APPswe/PS1dE9 mice. *Acta Neuropathol. Commun.* **2015**, *3*, 54. [CrossRef]
172. Lessmann, V.; Brigadski, T. Mechanisms, locations, and kinetics of synaptic BDNF secretion: An update. *Neurosci. Res.* **2009**, *65*, 11–22. [CrossRef]
173. Taniguchi, Y.; Inoue, N.; Morita, S.; Nikaido, Y.; Nakashima, T.; Nagai, N.; Okada, K.; Matsuo, O.; Miyata, S. Localization of plasminogen in mouse hippocampus, cerebral cortex, and hypothalamus. *Cell Tissue Res.* **2011**, *343*, 303–317. [CrossRef]
174. Frei, J.A.; Stoeckli, E.T. SynCAMs—From axon guidance to neurodevelopmental disorders. *Mol. Cell. Neurosci.* **2017**, *81*, 41–48. [CrossRef] [PubMed]
175. Siddiqui, T.J.; Tari, P.K.; Connor, S.A.; Zhang, P.; Dobie, F.A.; She, K.; Kawabe, H.; Wang, Y.T.; Brose, N.; Craig, A.M. An LRRTM4-HSPG complex mediates excitatory synapse development on dentate gyrus granule cells. *Neuron* **2013**, *79*, 680–695. [CrossRef] [PubMed]
176. Roppongi, R.T.; Karimi, B.; Siddiqui, T.J. Role of LRRTMs in synapse development and plasticity. *Neurosci. Res.* **2017**, *116*, 18–28. [CrossRef] [PubMed]
177. Bhouri, M.; Morishita, W.; Temkin, P.; Goswami, D.; Kawabe, H.; Brose, N.; Sudhof, T.C.; Craig, A.M.; Siddiqui, T.J.; Malenka, R. Deletion of LRRTM1 and LRRTM2 in adult mice impairs basal AMPA receptor transmission and LTP in hippocampal CA1 pyramidal neurons. *Proc. Natl. Acad. Sci. USA* **2018**, *115*, E5382–E5389. [CrossRef] [PubMed]
178. Ko, J.; Soler-Llavina, G.J.; Fuccillo, M.V.; Malenka, R.C.; Sudhof, T.C. Neuroligins/LRRTMs prevent activity- and Ca2+/calmodulin-dependent synapse elimination in cultured neurons. *J. Cell Biol.* **2011**, *194*, 323–334. [CrossRef]
179. Di Cristo, G.; Chattopadhyaya, B.; Kuhlman, S.J.; Fu, Y.; Belanger, M.C.; Wu, C.Z.; Rutishauser, U.; Maffei, L.; Huang, Z.J. Activity-dependent PSA expression regulates inhibitory maturation and onset of critical period plasticity. *Nat. Neurosci.* **2007**, *10*, 1569–1577. [CrossRef]
180. Muhlenhoff, M.; Rollenhagen, M.; Werneburg, S.; Gerardy-Schahn, R.; Hildebrandt, H. Polysialic acid: Versatile modification of NCAM, SynCAM 1 and neuropilin-2. *Neurochem. Res.* **2013**, *38*, 1134–1143. [CrossRef]
181. Schnaar, R.L.; Gerardy-Schahn, R.; Hildebrandt, H. Sialic acids in the brain: Gangliosides and polysialic acid in nervous system development, stability, disease, and regeneration. *Physiol. Rev.* **2014**, *94*, 461–518. [CrossRef]
182. Muoio, V.; Persson, P.B.; Sendeski, M.M. The neurovascular unit—Concept review. *Acta Physiol.* **2014**, *210*, 790–798. [CrossRef]

183. Bartanusz, V.; Jezova, D.; Alajajian, B.; Digicaylioglu, M. The blood-spinal cord barrier: Morphology and Clinical Implications. *Ann. Neurol.* **2011**, *70*, 194–206. [CrossRef]
184. Lampron, A.; ElAli, A.; Rivest, S. Innate Immunity in the CNS: Redefining the Relationship between the CNS and Its Environment. *Neuron* **2013**, *78*, 214–232. [CrossRef] [PubMed]
185. Bedoui, Y.; Neal, J.W.; Gasque, P. The Neuro-Immune-Regulators (NIREGs) Promote Tissue Resilience; a Vital Component of the Host's Defense Strategy against Neuroinflammation. *J. Neuroimmune Pharmacol.* **2018**, *13*, 309–329. [CrossRef] [PubMed]
186. Lin, C.-C.; Lee, I.T.; Wu, W.-B.; Liu, C.-J.; Hsieh, H.-L.; Hsiao, L.-D.; Yang, C.-C.; Yang, C.-M. Thrombin Mediates Migration of Rat Brain Astrocytes via PLC, Ca2+, CaMKII, PKCα, and AP-1-Dependent Matrix Metalloproteinase-9 Expression. *Mol. Neurobiol.* **2013**, *48*, 616–630. [CrossRef] [PubMed]
187. Lin, H.; Trejo, J. Transactivation of the PAR1-PAR2 Heterodimer by Thrombin Elicits β-Arrestin-mediated Endosomal Signaling. *J. Biol. Chem.* **2013**, *288*, 11203–11215. [CrossRef] [PubMed]
188. Junge, C.E.; Lee, C.J.; Hubbard, K.B.; Zhang, Z.; Olson, J.J.; Hepler, J.R.; Brat, D.J.; Traynelis, S.F. Protease-activated receptor-1 in human brain: Localization and functional expression in astrocytes. *Exp. Neurol.* **2004**, *188*, 94–103. [CrossRef]
189. Flaumenhaft, R.; De Ceunynck, K. Targeting PAR1: Now What? *Trends Pharmacol. Sci.* **2017**, *38*, 701–716. [CrossRef]
190. Rajput, P.S.; Lyden, P.D.; Chen, B.; Lamb, J.A.; Pereira, B.; Lamb, A.; Zhao, L.; Lei, I.F.; Bai, J. Protease activated receptor-1 mediates cytotoxicity during ischemia using in vivo and in vitro models. *Neuroscience* **2014**, *281*, 229–240. [CrossRef]
191. Soh, U.J.K.; Trejo, J. Activated protein C promotes protease-activated receptor-1 cytoprotective signaling through β-arrestin and dishevelled-2 scaffolds. *Proc. Natl. Acad. Sci. USA* **2011**, *108*, E1372–E1380. [CrossRef]
192. Coughlin, S.R. Protease-activated receptors in hemostasis, thrombosis and vascular biology. *J. Thromb. Haemost. JTH* **2005**, *3*, 1800–1814. [CrossRef]
193. Ossovskaya, V.S.; Bunnett, N.W. Protease-activated receptors: Contribution to physiology and disease. *Physiol. Rev.* **2004**, *84*, 579–621. [CrossRef]
194. Wang, H.; Ricklin, D.; Lambris, J.D. Complement-activation fragment C4a mediates effector functions by binding as untethered agonist to protease-activated receptors 1 and 4. *Proc. Natl. Acad. Sci. USA* **2017**, *114*, 10948–10953. [CrossRef] [PubMed]
195. Niego, B.; Samson, A.L.; Petersen, K.U.; Medcalf, R.L. Thrombin-induced activation of astrocytes in mixed rat hippocampal cultures is inhibited by soluble thrombomodulin. *Brain Res.* **2011**, *1381*, 38–51. [CrossRef] [PubMed]
196. Shavit Stein, E.; Ben Shimon, M.; Artan Furman, A.; Golderman, V.; Chapman, J.; Maggio, N. Thrombin Inhibition Reduces the Expression of Brain Inflammation Markers upon Systemic LPS Treatment. *Neural Plast.* **2018**, *2018*, 8. [CrossRef]
197. Armulik, A.; Genove, G.; Mae, M.; Nisancioglu, M.H.; Wallgard, E.; Niaudet, C.; He, L.; Norlin, J.; Lindblom, P.; Strittmatter, K.; et al. Pericytes regulate the blood–brain barrier. *Nature* **2010**, *468*, 557–561. [CrossRef] [PubMed]
198. Padel, T.; Roth, M.; Gaceb, A.; Li, J.Y.; Bjorkqvist, M.; Paul, G. Brain pericyte activation occurs early in Huntington's disease. *Exp. Neurol.* **2018**, *305*, 139–150. [CrossRef] [PubMed]
199. Lee, H.; Xie, L.; Yu, M.; Kang, H.; Feng, T.; Deane, R.; Logan, J.; Nedergaard, M.; Benveniste, H. The Effect of Body Posture on Brain Glymphatic Transport. *J. Neurosci.* **2015**, *35*, 11034–11044. [CrossRef] [PubMed]
200. Petta, F.; De Luca, C.; Triggiani, M.; Casolaro, V. Fragments of truth: T-cell targets of polyclonal immunoglobulins in autoimmune diseases. *Curr. Opin. Pharmacol.* **2014**, *17*, 1–11. [CrossRef]

International Journal of
Molecular Sciences

MDPI

Review

Synaptic GluN2A-Containing NMDA Receptors: From Physiology to Pathological Synaptic Plasticity

Luca Franchini †, Nicolò Carrano †, Monica Di Luca and Fabrizio Gardoni *

Department of Pharmacological and Biomolecular Sciences, University of Milan, Via Balzaretti 9, 20133 Milan, Italy; luca.franchini@unimi.it (L.F.); nicolo.carrano@unimi.it (N.C.); monica.diluca@unimi.it (M.D.L.)

* Correspondence: fabrizio.gardoni@unimi.it; Tel.: +39-0250318374/314

† These authors contributed equally to this work.

Received: 29 January 2020; Accepted: 21 February 2020; Published: 24 February 2020

Abstract: N-Methyl-D-Aspartate Receptors (NMDARs) are ionotropic glutamate-gated receptors. NMDARs are tetramers composed by several homologous subunits of GluN1-, GluN2-, or GluN3-type, leading to the existence in the central nervous system of a high variety of receptor subtypes with different pharmacological and signaling properties. NMDAR subunit composition is strictly regulated during development and by activity-dependent synaptic plasticity. Given the differences between GluN2 regulatory subunits of NMDAR in several functions, here we will focus on the synaptic pool of NMDARs containing the GluN2A subunit, addressing its role in both physiology and pathological synaptic plasticity as well as the contribution in these events of different types of GluN2A-interacting proteins.

Keywords: Glutamate; NMDA receptors; dendritic spines; synaptic plasticity; brain disorders

1. Introduction

N-Methyl-D-Aspartate Receptors (NMDARs) are ligand- and voltage-gated ionotropic glutamate receptors (iGluRs) promoting Ca^{2+} and Na^{+} influx. In particular, the NMDAR pore channel is blocked by Mg^{2+} at resting potential and this event is removed by an adequate depolarization. Therefore, in the canonical view, activation of NMDARs represents a coincidence detector of presynaptic glutamate release and postsynaptic depolarization [1,2]. This view requires NMDARs be located postsynaptically [3]. Nonetheless, the presence of NMDARs in the presynapse has long been described in several brain areas, like cortex [4,5], hippocampus [6,7], and cerebellum [8,9], and also in the spinal cord [10]. Moreover, in parallel with the classic view of ion influx as the only NMDAR signaling, several studies pointed out a NMDAR metabotropic cascade independently of ion-flow. In particular, the NMDAR metabotropic activity results from a complex intracellular "signalosome", which is mostly involved in synaptic receptor endocytosis and trafficking [11–14], and depression of neurotransmission such as NMDAR-dependent Long-Term Depression (LTD) [15–19].

NMDARs are tetramers composed by two obligatory GluN1 subunits, associated with two regulatory subunits of GluN2-type and GluN3-type, expressed in several different isoforms (GluN2A-D and GluN3A-B) at different brain regions and at different developmental periods. According to the neurodevelopmental switch theory, GluN2B/GluN2D and GluN3A/GluN3B are more abundant in early developmental stages, whereas GluN2A/GluN2C are more expressed in mature ones [20,21].

GluN2A and GluN2B are by far the most abundant NMDAR regulatory subunits expressed in the mammalian brain [20–22]. GluN2A-containing NMDARs are highly expressed in the adult hippocampus and neocortex while in other brain areas such as the striatum the GluN2B-containing NMDARs are predominant. Notably, GluN2A-containing NMDARs are more localized at synaptic sites and enriched in the postsynaptic density (PSD) compared to extrasynaptic sites [22–24] and display a slower mobility compared to other GluN2-containing ones [25].

GluN2A confers unique channel properties to NMDARs containing this subunit, starting from high sensitivity to Mg^{2+} [26]. Chen et al. reported that GluN2A-containing NMDAR mediate large currents, as suggested by the peak current densities [27]. Moreover, this subunit grants the NMDAR a high channel open probability [27]. In postnatal cortical neurons, regardless their age, the presence of GluN2A mRNA was associated to faster NMDAR EPSC compared to negative cells [28]. GluN2 subunits are also the main regulators of the open/close state of the NMDAR. In this context, GluN2A-containing receptors have a reversible calcium-dependent inactivation, whereas other kinds of NMDARs, like the GluN2B-containing ones, do not significantly suffer for calcium-dependent inactivation [29]. Another important feature owned by GluN2A-containing NMDARs is the fast recovery from desensitization [30]. To summarize, GluN2A confers to NMDAR specific electrophysiological properties, like high channel opening probability and conductance, more predisposed to desensitize but with a faster recovery, making these receptors versatile modulators of synaptic activity.

Synaptic NMDARs mainly mediate pro-survival and synaptic plasticity pathways, whereas extrasynaptic NMDARs are mostly responsible for glutamate excitotoxicity and are detrimental for neuronal functions [23,31]. Furthermore, the balance in synaptic GluN2-type subunits is responsible for adequate glutamatergic neurotransmission, which is altered in several neurological disorders and which is linked to the pathophysiology of brain diseases [22,32,33]. Therefore, understanding the molecular mechanisms regulating NMDAR subunit composition at synapses, such as membrane insertion, and assembly and removal of specific regulatory subunits, represents a pharmacological challenge for the setting up of new therapeutic strategies for neurological disorders where NMDARs play a pivotal role [34,35]. In this review, we will focus on synaptic GluN2A-containing NMDARs, their role in synaptic plasticity and their contribution to pathological plasticity as observed in several brain disorders.

2. NMDAR Structure: Focus on the GluN2A Subunit

All NMDAR subunits contain an extracellular N-terminal domain (NTD), a transmembrane domain (TD), and an intracellular C-terminal domain (CTD). Within NTDs are located the agonist binding site (ABD) for glutamate (on GluN2 subunits) and glycine (on GluN1 and GluN3 subunits) [36]. Zn^{2+} ions are able to inhibit NMDAR currents probably by interacting with pore channel amino acids [37,38]; however, Zn^{2+} can also inhibit specifically GluN2A-containing NMDARs due to an interaction site located on GluN2A NTD. In particular, Zn^{2+} can relate allosterically with four residues (H42, H128, K233, and E266) [39] belonging to GluN2A NTD, therefore providing inhibition and reducing NMDAR currents [40–43]. GluN2A NTD is also responsible for receptor structural changes in a closed conformation due to extracellular protonation (i.e., pH variations), which were shown to be independent of plasma membrane potential [44–46] and agonist binding [47]. This event was shown to be synergistically promoted by other NMDAR inhibitors, such as Zn^{2+} [48]. Recently, R370W and P79R mutations of the *GRIN2A* (Glutamate Ionotropic Receptor NMDA Type Subunit 2A) gene, encoding for the GluN2A subunit, have been associated with augmented and reduced sensitivity to Zn^{2+}, respectively [49]. Finally, the NTD of the GluN2A subunit shows an endoplasmic reticulum (ER) retention signal, which is absent on GluN2B homolog [50] probably accounting for different delivery mechanisms to the synapse.

The TD displays an M2 loop involved in pore channel formation and responsible for Mg^{2+} interaction. In particular, two asparagine residues from M2 loops belonging to the GluN2 and the GluN1 subunits are responsible of the Mg^{2+} blockade [51]. The *C1845A* mutation on *GRIN2A* gene encodes for GluN2AN651K variant at the M2 loop leading to low Mg^{2+} blockade [51], decrease in Ca^{2+} permeability both for di-heteromeric and tri-heteromeric NMDARs, and to epileptic encephalopathy with cognitive impairment [52,53].

The long CTD of GluN2 subunits displays the lowest homology among GluN2 isoforms, as well as various length and binding partners [25,32,54]. GluN2 CTDs are necessary for receptor surface dynamics and activation of specific metabotropic signaling [32,55]. Protein–protein interactions at the GluN2A CTD are involved in the lower mobility of GluN2A-containing NMDARs at synapses compared to GluN2B-containing ones [25] (see below) [55]. GluN2 CTD as well as NTD were reported to affect receptor trafficking from the ER, indicating a complex modulation of the receptor delivery to target regions [56].

NMDAR can be modulated through subunit phosphorylation of different residues at the CTD, which represent also a well-demonstrated fast mechanism for activation or maintenance of synaptic plasticity events [57]. In particular, CTD phosphorylation is important for the endocytosis/removal of NMDARs, which occurs during synaptic plasticity events [58] and agonist binding [11–13,59]. Vissel and collaborators (2001) reported a use-dependent dephosphorylation of NMDAR associated with low amplitude recordings given by augmented endocytosis of the receptors mediated by AP2 and dynamin [11]. This event was dependent on agonist concentrations and time interval of treatments. Furthermore, they identified the GluN2A CTD portions responsible for the endocytosis, such as Tyr842 which is well known to interact with a consensus motif (μ2) on AP2, and additional residues between the 874-1464 fragment [11].

Selective GluN2 antagonists represent pharmacological tools to investigate GluN2 subunits role both in physiological and pathological conditions. As for the GluN2A subunit, the mostly used selective antagonist is NVP-AAM077, which interacts with ABD of GluN2A and Glu781 on GluN1 subunit [60]. The selectivity of NVP-AAM077 for GluN2A over GluN2B is, however, only approximately fivefold, suggesting a tight range of concentrations to be used [61]. This probably explains why the variability of the results obtained with NVP-AAM077 in different experimental systems, thus suggesting that this compound should be used cautiously to investigate GluN2A contribution to neuronal functions.

Researchers recently developed new promising compounds with different pharmacological profile, such as GluN2A Negative Allosteric Modulators (NAMs) and Positive Allosteric Modulators (PAMs). As for GluN2A-NAMs, the most studied is TCN-201, which interacts with the ABD lower portion called S2 of GluN2A and GluN1; in particular, Leu780 and Val783 on GluN2A, whereas Arg755 on GluN1 [62–64]. As for GluN2A-PAMs, scientists are making efforts in obtaining clean molecular profiles due to a PAM binding site shared between NMDARs and AMPA receptors (AMPARs) [54]. Among PAMs, GNE-0723 was the first to be identified with a high potency and brain penetrance but poor pharmacokinetic profile [65,66]. After modification of GNE-0723 structure into GNE-5729, almost fivefold increased selectivity against AMPAR was achieved, maintaining good GluN2A potency and specificity [66]. Overall, even if several advances have been performed in the last decade trying to identify novel compounds that are able to modulate selectively the GluN2A-containing NMDARs, additional work is needed to reach this goal.

3. GluN2A Subunit: Binding Partners

The GluN2A CTD binds a variety of synaptic proteins with different cellular functions. These protein–protein interactions play a pivotal role in the modulation of numerous properties of GluN2A-containing NMDARs, ranging from the regulation of downstream intracellular signaling to mechanisms involved in the synaptic retention of the receptor.

The GluN2A CTD domains responsible for the interaction with these partners sometimes overlap thus leading to the competition of different types of proteins for the interaction with the GluN2A subunit (see below). Importantly, it is known that some of these interactions depend on the activation state of the synapse. In particular, interactions at the GluN2A CTD are very often not static but dynamically regulated by synaptic activity and plasticity (Figure 1). Accordingly, different pools of GluN2A-containing NMDAR can be associated to different signalosome. Moreover, several protein–protein interactions described below have been shown to play a key role for Long-Term Potentiation (LTP) induction. Therefore, the knowledge of the interactors at the GluN2A-CTD and their binding mechanisms represent a helpful tool in understanding how GluN2A-containing NMDAR function is regulated.

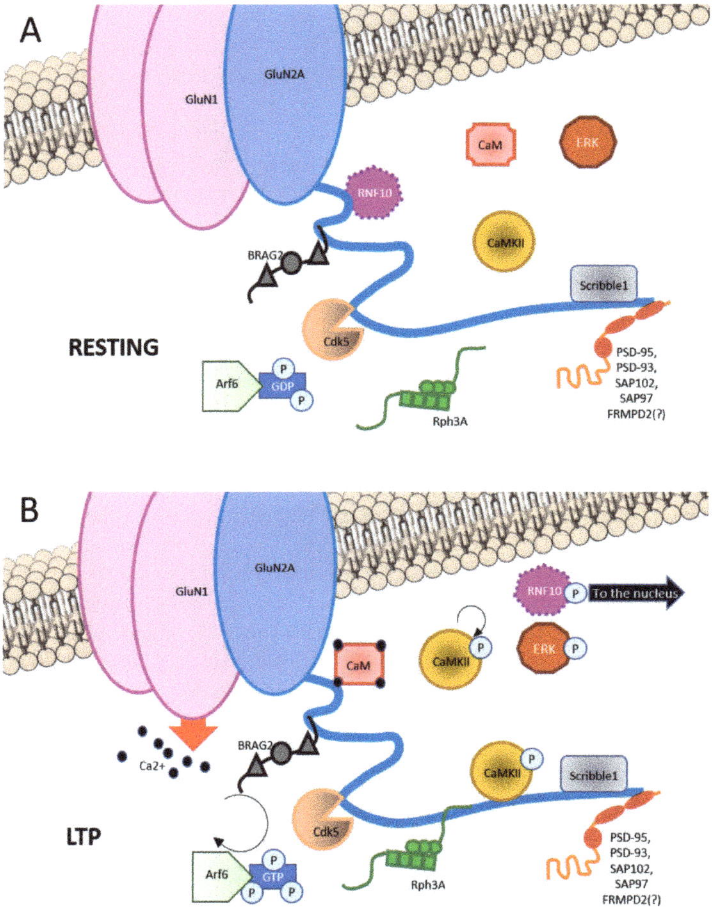

Figure 1. Graphic representation of GluN2A interactors at C-terminal domain (CTD). (**A**) In basal condition, GluN2A CTD binds RNF10 at aa991-1049; BRAG2 at aa1078-1117; Cdk5 at aa1218-1246; CaMKII at aa1389-1464; Scribble1 at 1458-1464; and PSD scaffolding proteins such as PSD-93, PSD-95, SAP102, and SAP97 at aa1461-1464. (**B**) Upon LTP induction, phosphorylation of RNF10 triggers RNF10 detachment and its nuclear translocation. Simultaneously, CaM binds to aa875-1029 of GluN2A. BRAG2 exchanges GDP with GTP to Arf6-GTPase, which is involved is vesicles recycle at the postsynapse, also with Scribble1. Rph3A interaction with GluN2A at aa1349-1389 increases. CaMKII is disinhibited and is recruited at the postsynapse. Ras-ERK downstream cascade is initiated.

Among GluN2A CTD interacting partners there are (1) scaffolding proteins, (2) synapse-to-nucleus messengers, (3) protein kinases, and (4) other proteins (see Figure 1).

3.1. Scaffolding Proteins

Several types of scaffolding proteins bind the distal domain of GluN2A CTD and these interactions mainly regulate GluN2A stability at synapses. Members of membrane-associated guanylate kinase (MAGUK) from the Discs Large subfamily (Dlg) such as PSD-95, PSD-93, SAP102, and SAP97 bind to the last three aa (1461–1464) of the GluN2A subunit and are responsible for GluN2A-containing NMDARs retention in PSD [67–70]. Similarly, the rab-effector protein Rabphilin3A (Rph3A) binds to GluN2A1349-1389 domain at synapses and promotes stabilization of GluN2A-containing NMDARs at the PSD, also thanks to the formation of a triple complex with PSD-95. Interestingly, the formation of this protein complex is triggered by LTP and disruption of Rph3A/GluN2A interaction through cell permeable peptides or Rph3A shRNA rapidly reduces synaptic levels of GluN2A-containing NMDARs and prevents LTP induction [71]. FRMPD2, a scaffolding protein typically present in polarized cells and neurons, was found to interact selectively with GluN2A subunit of NMDARs through its PDZ2. Again, this association is necessary for receptor stabilization in PSD and its disruption leads to reduced GluN2A synaptic levels in CA1 region of hippocampus [72]. Scribble1, a scaffolding protein involved in neural tube closure and presynaptic architecture was also described as a GluN2A binding partner at 1458–1464 domain promoting insertion at the PSD of GluN2A-containing NMDARs via interplay with AP2 [73].

3.2. *Synapse-to-Nucleus Messengers*

Long-distance NMDAR signaling to the nucleus contributes to activity-dependent regulation of gene expression that feeds back to synaptic function in health and disease. In addition to calcium signals, which represent a major route for communication of NMDAR activity to the nucleus, synaptonuclear protein messengers connect synapses and nucleus enabling bidirectional transfer of information. In particular, several synaptonuclear messengers are associated to NMDAR complex at the glutamatergic postsynaptic compartment and play a key role in the modulation of the LTP [74,75]. Ring Finger protein 10 (RNF10) binds the 991–1049 fragment of GluN2A subunit and migrates to the nucleus through the transporter importin-α after induction of LTP or synaptic NMDAR activation [76].

AIDA-1d was shown to be present in the PSD during resting conditions, and to migrate to the nucleus upon NMDAR activation where it regulates protein synthesis and nucleoli number [77]. An interaction site on GluN2 CTD has not been reported yet for AIDA-1d and it shows to preferentially colocalize and coimmunoprecipitate with GluN2B subunits over GluN2A, suggesting a promiscuous synapse-to-nucleus messenger. However, we cannot exclude AIDA-1d as potentially part of GluN2A-containing NMDAR signaling components. Nevertheless, it seems to mainly affect GluN2B-containing NMDARs assembly and trafficking to the synapse from the endoplasmic reticulum [77].

Extracellular regulated kinase (ERK) is involved in Mitogen-Activated Protein Kinase (MAPK) downstream signaling and a synapse-to-nucleus communication, which still needs to be clarified. Both GluN2A- and GluN2B-containing NMDAR can be found upstream of ERK activation, suggesting ERK as a promiscuous signaling mediator [55]. ERK was shown to be activated by only a subset of dendritic spines modulating CREB and Elk-1 activity [78]. However, there is no direct demonstration of ERK translocation to the nucleus as well as no direct interaction reported yet on GluN2 CTDs.

3.3. *Protein Kinases*

Several protein kinases play a relevant role at GluN2A-containing NMDAR complexes not only phosphorylating the GluN2A subunit, but also forming protein–protein interactions. Ca^{2+}/calmodulin-dependent protein Kinase II (CaMKII) interacts with 1389–1464 domain of GluN2A [79,80] and with different CTD portions of GluN2B subunit (839–1120/1120–1489) depending on the kinase phosphorylation state [81,82]. CaMKII is the most abundant protein at the glutamatergic PSD and is activated by Ca^{2+}-Calmodulin complexes driven by NMDAR opening. When this occurs, CaMKII autophosphorylates its autoinhibitory domain in T286, becoming active. The active conformation of CaMKII remains, even when the Ca^{2+} stimulus has subsided, therefore it is also named as "molecular memory" [83–85]. The interaction between GluN2A and this kinase was shown to be in an unphosphorylated state, but it is promoted by autophosphorylation of CaMKII [86]. Increased formation of GluN2A/CaMKII complex induces also a disruption of GluN2A/PSD-95 interaction thus leading to a dynamic modification of GluN2A-interacting proteins following NMDAR activation [79].

Cyclin-Dependent Kinase 5 (CdK5) binds GluN2A at fragment 1218–1246, regulating receptor recycle and degradation via coordination with the protein AP2 [87,88]. Of relevance, Cdk5-dependent phosphorylation at Ser1232 seems to be important for LTP maintenance in CA1 region of hippocampus [88].

Phospatidyl-inositol-3 kinase (PI3K) was reported to bind GluN2A- or GluN2B-containing NMDAR by interacting with GluN1 CTD. Activation of GluN2A containing NMDAR was shown to trigger PI3K signaling [89,90].

3.4. Other Proteins

Guanine nucleotide exchange factors (GEFs): Brefeldin A-resistant Arf guanine nucleotide exchange factor 2 (BRAG2) binds 1078–1117 fragment of GluN2A CTD and mediates exchange of GDP to GTP at ADP-rybosilation factors (Arf) [91]. However, also BRAG1 is able to interact with synaptic GluN2A-containing NMDAR complex, as reported previously by Sakagami [92].

Ras-guanine nucleotide-releasing factor 2 (Ras-GRF2) is a Ca^{2+}/Calmodulin (CaM) sensor mediating GluN2A-containing NMDAR signaling in mature neurons [93], in fact its expression is developmentally regulated [94]. In particular, Ras-GRF2 displays a GEF domain, a C-terminal cell division cycle 25 (CDC25) domain for Ras/ERK activation [95,96], and an N-terminal dbl homology (DH) domain activating Rac/p38 pathway [97,98]. Ras-GEF are important mediators of NMDARs late phase signaling, since they involve MAP kinases, ERK1/2 and also CREB phosphorylation which is important for gene expression [94]. However, the domain of interaction between Ras-GRF2 and GluN2A has not been investigated.

IQGAP1 serves as a scaffold protein for different signaling pathways involving B-Raf, Rac1, Lis1, Cdc42, ERK, and MEK. IQGAP1 was found to preferentially interact with GluN2A-PSD-95 complexes compared to GluN2B-containing ones, thus indicating a possible signaling interplay through IQGAP1 between different pools of NMDAR [99–102].

Calmodulin (CaM) interacts with CTD of GluN2A subunit in a calcium-dependent manner at the 875-1029 domain, through interaction with a tryptophan residue (1014), which is critical for protein–protein interaction [103]. Ca^{2+}/Calmodulin complexes can be also considered as a synapse-to-nucleus messengers of the γCaMKII and γCaMKI signaling, regulating gene expression and cognitive functions [104–106].

Bip, a recently discovered ER chaperone protein, is involved in the selective delivery of GluN2A-containing NMDAR at the synapse from the dendritic ER [107].

4. GluN2A Subunit and Synaptic Plasticity

The influence of the NMDAR subunit composition on synaptic plasticity has been subject of a vast number of studies, trying to explain the unsolved question on how different patterns of neuronal stimulation lead to opposite modulation of synaptic strength acting on the same type of receptor. It is commonly accepted that the different levels of calcium influx associated with different degrees of NMDAR activation polarize the direction of plasticity towards LTP or LTD, as they couple to different intracellular signaling pathways. However, it was proposed that distinct subpopulations of NMDARs could exert a finer level of regulation on synaptic potentiation or depression triggered by different activity patterns. In this regard, GluN2A subunit has been intensively studied, as GluN2A-containing NMDAR has particular channel properties that generate distinct calcium dynamics in the postsynapse. Moreover, as described above, GluN2A CTD allows unique intracellular molecular associations with proteins like kinases and phosphatases and synaptonuclear messengers that intrinsically direct the plasticity signaling [33,75,108]. Recently it was also reported a specific metabotropic function of GluN2A-containing NMDAR in mediating glycine-induced potentiation of AMPAR currents through the activation ERK1/2 signaling [109]. The synaptic abundance and motility of this subunit affects the basal state of the spine and provides another level of control on the possibility to induce synaptic strengthening or weakening. In particular, the CA3-CA1 synapse of the hippocampal neurocircuitry represents the most commonly used model for the study of the putative role of different GluN2 subunits in synaptic plasticity. Pharmacological and genetic manipulations, or the combination of the two approaches, have been used in combination with several plasticity protocols.

4.1. Role of GluN2A in LTP

In 2004, a couple of seminal ex vivo studies showed that the selective pharmacological inhibition of GluN2A by NVP-AAM077 blocks tetanus- and pairing-induced LTP in three- to four-week rat hippocampal neurons [110] and HFS-induced LTP in layer II and III cortical neurons [111], but not LTD. A few years later the findings were confirmed on hippocampal neurons using the same antagonist and age of rats [112] and also in two-week-old mice [113]. In vivo pharmacological studies also showed that both intrahippocampal infusion and intraperitoneal injection of NVP-AAM077 was able to abolish LTP induction [114]. The data in these studies seem to point out that GluN2A is a necessary element for LTP. However, the NVP-AAM077 concentration used in the aforementioned studies is relatively high and, as NVP-AAM077 affinity is just 5 times higher to GluN2A than GluN2B, it cannot be excluded that the block of LTP seen is uniquely GluN2A dependent. Moreover, this elevated concentration of NVP-AAM077 was able to induce a drastic reduction of the total NMDAR-mediated currents [110,112] that might have prevented the calcium influx necessary for LTP induction. Given that, the block of LTP induction seemed more likely to have arisen from a threshold effect on the total NMDAR-mediated currents than on the selective inhibition of GluN2A. In support of this theory, the group of Köhr found that in mice treated with a concentration of the antagonist that was titrated not to cause an excessive NMDAR-current reduction, pairing-induced LTP was not impaired [115]. In another study, the same low NVP-AAM077 concentration also did not abolish LTP induction by a theta burst protocol [116], but partially impaired tetanus-induced LTP [117]. Overall, pharmacological studies did not clarify the role of the GluN2A subunits in LTP, even if the data of these studies suggest that GluN2A might have a more relevant function in tetanus-induced LTP, since it resulted the most sensitive to NVP-AAM077 treatment.

Several lines of evidence from a genetic manipulation approach further supported the idea of an involvement of GluN2A in LTP, as in both mice with reduced synaptic GluN2A [118] or no GluN2A [119], LTP is altered, but can be restored though a multiple tetanic induction protocol [120]. A possible explanation for this functional recovery is that a stronger neuronal stimulation can restore the calcium influx that is usually supplied by GluN2A-containing NMDAR in high-frequency stimulation conditions. However, other studies put forward the idea that also GluN2A-interaction with molecular partners is important for synaptic strengthening. It has been shown that form of LTP dependent on the Ras-GRF2/Erk Map Kinase pathway can be sustained by GluN2A subunit-containing NMDARs alone [93], suggesting that unique interactions may occur. This idea seems to be confirmed by another study, in which a GluN2A C-terminal truncation mouse model showed impaired tetanus-induced LTP [121]. However, these mice have a reduction of the total levels of GluN2A, so it is possible that the LTP impairment is caused by the reduction in calcium influx (threshold effect) rather than the loss of the signaling downstream GluN2A. Also, in this case [121], and in another study [122], using more intense stimulation protocols overcomes the LTP induction deficit, probably thanks to the expected increase of calcium influx in the spine. To conclude, these experimental data indicate an involvement of GluN2A in LTP but there is no clear indication for a privileged or necessary role of this subunit in the process induction.

GluN2A has a bidirectional involvement in synaptic plasticity: it regulates and is regulated by it. In particular, LTP is a potent trigger of GluN2A upregulation, especially at synaptic levels. Several evidences seem to indicate that GluN2A rise in the postsynapse is associated to an early mobilization of NMDARs localized in non-synaptic pools and a later increase of total level of the protein. In adult rats HFS-induced LTP is coupled to waves of GluN2A increase in DG total homogenates: a rapid surge 20 min after LTP induction, then the subunit levels go back to the base levels 1 h and 4 h after LTP to go up again 24 h after induction [43]. LTP induced by HFS is also able to promote a GluN2B- to GluN2A-mediated current switch in newborn mice hippocampal slices, in a very short time window [123]. Accordingly, LTP induction increased GluN2A synaptic levels in both hippocampal organotypic cultures from neonatal rats [12] and in synaptosomal fractions of hippocampal slices from 6- to 8-week-old rats [124]. Similarly, Franchini et al. found also in primary hippocampal cultures a synaptic increase of GluN2A driven by chemically induced LTP protocol, with a concurrent increase of its CTD-binding partner Rph3A (see Figure 1). Interestingly, interfering with GluN2A/Rph3A interaction prevents the synaptic increase of GluN2A levels [125].

Baez and coworkers found that only after the establishment of an actual long-term potentiation (after 70 min), but not after 30 min, GluN2A levels were increased [126]. Even if these data seem to contrast with the previously described ones, the different results obtained may arise from the different model used (young adult rats) and the different cellular fraction analyzed (hippocampal homogenates and not synaptic fractions). In accordance, GluN2A mRNA dendritic levels and translation increased 30 min after NMDA stimulation [127], whereas LTP induction immediately increases local translation and membrane insertion of GluN2A and for the 30 min after the induction [128].

In conclusion, GluN2A seems to respond to the appropriate stimulus with a fast lateral mobilization from extra-synaptic to synaptic sites and, to sustain synaptic potentiation, with an increase of the local translation and surface expression at longer time points.

4.2. Role of GluN2A in LTD

As described above, the exact nature of the role of GluN2A in the LTP process is still not completely clear. Similarly, it is still not possible to make an educated guess on the exact role of GluN2A in LTD, as there are contradictory results on the matter. Some studies using the pharmacological approach revealed that a concentration of NVP-AAM077 that is able to impair LTP does not affect LTD in culture [129], in acute slices [110], or in vivo [130], suggesting that GluN2A is not necessary for LTD. However, there are other studies in opposition, in which the antagonist blocked LTD at the same concentration in which blocked LTP [112,113]. In another in vivo study, mice injected intraperitoneally with two different concentrations of NVP-AAM077 that were able to block LTP showed LTD impairments just at the higher concentration [114], suggesting that just a substantial reduction in NMDAR-mediated current and subsequent calcium entry is able to perturb LTD induction, whereas at low concentration of NVP-AAM077, the residual calcium influx is not able to trigger LTP but is still able to mediate LTD. Also, genetic approaches, providing mostly negative results, did not provide a convincing answer to the question. No LTD deficit was found in GluN2A knockout mice [131] and neither in case of direct [132] nor indirect GluN2A overexpression [133] at 1 Hz stimulation protocol, although at 3–5 Hz, GluN2A overexpression reduced LTD without affecting LTP [132]. It was demonstrated that NMDAR-mediated ion flux block could not stop LTD induction, suggesting the emerging idea that the basal state of the spine, rather than the amount of the calcium influx through NMDARs, is essential for LTD. However, to what extent GluN2A is involved in LTD, if through the intracellular association or the channel properties conferred to the NMDAR, it is still not determined, even if the data provided do not evidence a necessary role of the subunit in the process.

4.3. Role of GluN2A Interactors in Synaptic Plasticity

Despite GluN2A important role mediated by the particular channel properties it confers to the NMDAR, data seem to point out that many of the GluN2A specific effects, both in and out of the context of synaptic plasticity, may be due to the unique intracellular interactors that mediate GluN2A specific signaling pathways. First, as already described, some GluN2A CTD associated proteins are scaffolding proteins and mediate GluN2A membrane association and synaptic localization (See Figure 1). It was reported that Rph3A promotes stabilization of GluN2A-containing NMDARs in the postsynapse [71]. Franchini et al. recently published that Rph3A is targeted to the synapse after chemically induced LTP, which, in turn, drives the upregulation of GluN2A-containing NMDAR induced by LTP through the formation of the Rph3A-GluN2A-PSD-95 complex. Remarkably Rhp3A silencing in vitro blocks LTP induction. Interfering with Rph3A/GluN2A interaction produces the same effect and in vivo induces a strong memory deficit [125]. Other important interactors of GluN2A CTD that modulate its function in synaptic plasticity are protein kinases. Cdk5 can upregulate GluN2A function via phosphorylating GluN2A at Ser1232 in vitro and in vivo [87]. CaMKIIα competes with PSD-95 for binding to GluN2A [79]. This interaction occurs with non-phosphorylated CaMKII, but is positively regulated by kinase autophosphorylation [86]. CaMKII activity has been demonstrated to be necessary and sufficient for LTP induction (reviewed in [134]). Finally, GluN2A has the unique association with the synaptonuclear messenger RNF10 (see Figure 1) that has been demonstrated to mediate GluN2A-mediated LTP maintenance as well as LTP-dependent structural modifications of dendritic spines [76]. More recently Carrano et al. demonstrated a wider role of the GluN2A-associated synaptonuclear communication mediated by RNF10 in the regulation of dendritic arborization [135].

5. Role of GluN2A in Learning and Memory

Synaptic plasticity is considered the cellular and molecular basis for learning and memory formation [136,137] and several studies have demonstrated an important role for NMDARs in specific forms of learning and memory [138,139]. The specific involvement of GluN2A in learning and memory has been investigated through both pharmacological and genetical approaches. Genetic ablation of GluN2A and expression of a GluN2A lacking the CTD did not impair spatial tasks acquired over a number of days (the Morris water maze (MWM) and radial arm maze) [140]. Neto1 knockout mice, characterized by reduced level of GluN2A and LTP deficits, perform normally at MWM [118]. Pharmacological manipulations gave similar results: intraperitoneal injections of the GluN2A antagonist NVP-AAM077, able to block LTP, in vivo did not alter MWM acquisition or consolidation in rats [130]. However, interfering with GluN2A has been associated with short-term memory deficits. In fact, GluN2A-knockout or GluN2A C-terminus-deficient mice exhibit impaired spatial working memory [140]. Moreover, CA1 infusion of NVP-AAM077 impaired the performance on a delayed alternation T-maze task [141]. Other works suggested that GluN2A might participate in the rapid acquisition of context or object representations. Indeed, GluN2A knockout mice showed an impaired performance in a variant of a hippocampus-dependent contextual fear-conditioning with reduced context exposure before shock delivery [120]. Moreover, Neto1 knockout mice were impaired in a displaced object recognition task, but not in a novel object recognition task, suggesting a specific deficit in a hippocampal dependent test [118]. Similarly, Franchini et al. found that mice treated with a cell-permeable peptide that interferes with Rph3A-mediated GluN2A synaptic delivery also display memory defects in the object displacement task [125].

Thus, GluN2A does not seem necessary for incrementally acquired long-term memory tasks. It seems that GluN2A partially controls learning and memory: it is not necessary for long-term memory tasks but seems to affect short-term memory and the rapid acquisition of spatial information.

However, it is interesting to note that pharmacological and genetic manipulations of the GluN2B subunit produce similar cognitive alterations of the ones described by manipulating GluN2A, suggesting that a general reduction in NMDAR-mediated current might induce short-term memory deficits, rather than specifically interfering with GluN2A-containing NMDAR.

NMDAR has been subject of study also in the context of fear memory. Even if most of the studies support the idea that GluN2B is the subunit that is primarily involved in this kind of memory process [142–146], recently some studies have also pointed out an involvement of GluN2A-containing NMDARs. The group of Luo demonstrated that interfering with the activity-dependent insertion of GluN2A mediated by the ER chaperone Bip in the dorsal hippocampus was sufficient to impair the acquisition of fear conditioning test, suggesting that the interaction between the two proteins is crucial for fear memory formation [107]. Moreover, an increase of GluN2A/GluN2B ratio in basolateral amygdala after fear memory consolidation inhibits retrieval-dependent memory destabilization and modification of the fear memory trace, suggesting a neurobiological molecular explanation of why some memory events are resistant to variation and extinction [147]. Similarly to synaptic plasticity, the process of learning and memory also influences GluN2A. A hippocampal-dependent spatial memory task, MWM, was shown to upregulate GluN2A mRNA levels after the training session. Baez et al. showed an increase in GluN2A levels in the hippocampi in habituation of a new environment starting from 30 min after the test [126]. Interestingly, a similar upregulation of GluN2A was also found after each phase of object presentation of the novel object recognition task [24]. This evidence supports the idea that specific types of memory are supported by a contribution of GluN2A-containing NMDAR mediated neurotransmission and so the subunit is upregulated by the process itself.

6. Role of GluN2A in Pathological Plasticity

Modifications of the GluN2A-containing NMDARs have been implicated in several pathological conditions including among others cerebral ischemia [148], depression [149–156], anxiety [149,157], schizophrenia [158–161], and Huntington's disease [162–164]. In this review we decided to focus our attention on selected brain disorders in which GluN2A disfunctions is strictly correlated to altered synaptic plasticity, such as epilepsy, Alzheimer's disease, Parkinson's disease, Fragile-X Syndrome, and autism.

6.1. GluN2A in Epilepsy

Epilepsy represents a common disease in the population with complex pathophysiology leading to aberrant firing in brain circuits [165]. Several mutations of *GRIN2A* (encoding the GluN2A subunit) leading to NMDAR gain- or loss-of-function have been associated with different epilepsy-aphasia spectrum (EAS) such as benign epilepsy with centrotemporal spikes (BECTS), the Landau–Kleffner syndrome (LKS) and epileptic encephalopathy with continuous-spike-and-waves-during-slow-wave-sleep (CSWSS) [166–168]. A patient with the GluN2A P552R substitution presented seizures and intellectual disability [169]. This mutation affects the pre-M1 region of GluN2A leading to higher glutamate and glycine potency of the NMDAR, without impairments in receptor surface trafficking or expression. Hippocampal primary cultures transfected with this receptor variant showed increased neurotoxicity under synaptic stimulation conditions, which was rescued by memantine, underlining the contribution of this mutation to the clinical phenotype observed [170]. Other mutations associated with epilepsy, the GluN2A M817V as well as the L812M substitution which both reside in the pre-M4 region [171,172], increase agonist potency, channel mean open time and open probability while reducing endogenous negative modulator effects on receptor activity [173]. A GluN2A mutation at Tyr1387 was reported in a family displaying CSWSS and an autistic phenotype [167]. Interestingly, Tyr1387 is a well-known GluN2A phosphorylation site by Src Kinase, important for the modulation of the receptor activity [174].

Other *GRIN2A* mutations such as the GluN2A-I148S and R512H were associated with reduced surface NMDAR expression, activation time and current amplitude, when expressed in heterozygous or homozygous condition in HEK cells [175]. Overall, GluN2A-I148S and R512H can be considered dominant negative variants of GluN2A, as well as the I184T, C436R, R518H, T531M, V685G, and D731N [175–178]. Moreover, also the GluN2A-N651K variant previously described (see NMDAR structure paragraph) can be considered as a dominant negative form of the GluN2A subunit due to low Ca^{2+} influx in spite of low Mg^{2+} sensitivity [53]. Finally, there are some GluN2A mutations that are also present in GluN2B (equivalent sites) but give rise to different phenotypes. In particular, the GluN2A variants C436R, V685G, and D731N were shown to induce intractable epilepsy with developmental delay and general tonic-clonic seizures [167,176,177], whereas GluN2B ones (C436R, Q413G, and C461F) with intellectual disability and absence seizures [178–180].

However, the involvement of GluN2A subunit in epilepsy has been correlated not only to genetical variants, but also to altered trafficking, expression and balance with other NMDAR subunits for an adequate receptor functioning.

Injection of convulsive agents, such as 3-mercaptopropionic acid or 4-aminopyridine for several days, is able to increase hippocampal GluN2A levels [181,182]. Chen and colleagues demonstrated how selective antagonism of the GluN2 subunits is neuroprotective in the hippocampus of pilocarpine and kindling animal models of epilepsy. However, only GluN2A antagonism reduced epileptogenesis, seizure-induced mossy fiber sprouting, and the activity-dependent BDNF expression indicating a crucial role of this subunit in epilepsy [183]. Recently, activation of GPR40 (a Gq coupled GPCR) in the kainic-acid (KA) induced epilepsy model and PTZ-kindling model reduced susceptibility to SE, most probably reducing NMDAR surface localization by regulating the endocytosis of the receptor [184].

A characteristic of some epilepsy models is that a single application of convulsive agent (electrical stimulation, GABA-A receptor antagonist, and Kainic Acid) has a long-term impact on electrical activity of the model itself, even after a washout period. Abegg and colleagues revealed how the Bicuculline (GABA-A receptor antagonist) epilepsy model induces abnormal LTP-like phenotype by exaggerated potentiation of excitatory synapses [185]. In particular, they observed how NMDAR activity was responsible for synapse potentiation by increasing synaptic AMPAR insertion, then occluding LTP induction but enhancing LTD amplitude in CA1. In this study, they provided a probable explanation for impaired learning abilities after epileptic exposure [186–188]. However, more recent literature identified both increased [189–191] and decreased LTP responses, probably depending on experimental conditions and how efficiently the seizure-induced LTP-like condition was realized [192–194]. Finally, in a recent work Hanson et al. found that the seizure phenotype was ameliorated by the use of a Positive Allosteric Modulator of GluN2A in a mouse model of Dravet syndrome [195]. In this case, the activation of GluN2A-containing NMDAR was able to improve the brain oscillations and neuronal hypersynchrony, improving the epileptic discharges.

6.2. GluN2A in Autism Spectrum Disorder (ASD) and in Fragile X Syndrome (FXS)

Neurodevelopmental disorders characterized by reduced learning and social abilities are also called Intellectual Disabilities (IDs). Among IDs, ASD shows high variability in the severity of social-deficit interaction, communication, and stereotyped behaviors [196–198]. Genetic background as well as environmental exposure to different stressors play an impact on the disease emergence [199,200]. FXS is an X-linked disease characterized by IDs and often accompanied by autistic features. In particular, FXS is due to lack or altered expression in *FMR1* gene, encoding Fragile Mental Retardation Protein (FMRP), which binds mRNA [201] and regulates its transport and translation [202,203].

Among the several neuronal alterations detected in ASD and FXS, a convergent point is dysregulation in synaptic protein expression and activity, leading to altered intracellular signaling during neurotransmission. Interestingly, FMRP target mRNAs include both pre and postsynaptic proteins such as NMDARs, PSD-95, Shank1-3, Calcium channels, SNAP-25, and many others [204].

In particular, NMDAR and mGluR signaling modulating ERK/MAPK kinase activity have been associated with the pathophysiology of FXS [205]. Indeed, FXS-patient brain and Fmr1-KO mouse model show hyperactive ERK/MAPK pathway [206,207]. Furthermore, elevated activation of ERK was detected in Fmr1-KO mice and associated to overproduction of APPα derived from APP cleavage [208].

Synaptic plasticity events in different regions of Fmr1-KO animals are altered compared to WT. In particular, the LTP in CA1 region of hippocampus is impaired in animal models of FXS [209,210] as well as in other brain regions [211–215]. Importantly, Lundbye and colleagues recently reported the NMDAR-dependent LTP in CA1 of Fmr1-KO animals was restored to normal levels with selective inhibition of GluN2A subunit through different antagonists and negative allosteric modulators (NAM) or via GRIN2A deletion [216].

mGluR1/5-dependent LTD in CA1 was reported to be enhanced in Fmr1-KO animals compared to WT [216–219]. Interestingly, mGluR1/5 LTD was also recovered to basal levels via GluN2A inhibition [216].

A reduction of both LTP [217] and NMDAR-mediated LTD [220] were also recorded in perforant path-granule cells of Fmr1-KO animals. In agreement with the DG involvement in pattern separation aspects of learning [221], Fmr1-KO mice display no impairments in acquisition phase of Morris Water Maze [222,223], but have reduced context discrimination and altered memory extinction [222,224], which is thought to be dependent on NMDAR-LTD [225].

Recently, integrated transcriptome analysis revealed, among other genes, alterations of *GRIN2A* in the adult hippocampus of ASD mouse models [226], strengthening its role in the pathophysiology of neurodevelopmental disorders.

6.3. Pathological Synaptic Plasticity in Parkinson's Disease and Dystonia

A balanced cross-talk between dopaminergic and glutamatergic transmission in the striatum is essential for corticostriatal synaptic plasticity and motor activity [227]. In Parkinson's disease, the degeneration of nigrostriatal dopaminergic neurons leads to complex alterations of the basal ganglia pathways involving also impaired synaptic plasticity of striatal spiny projecting neurons (SPNs) [228]. SPNs express GluN2A- and GluN2B-containing NMDARs [229,230] and an alteration of GluN2A/GluN2B subunit ratio of the striatal synaptic NMDAR is a key element in the regulation of motor behavior and synaptic plasticity in PD [231,232]. Interestingly, distinct levels of dopamine denervation differentially alter striatal synaptic plasticity and NMDA receptor subunit composition [231]. The complete depletion of striatal DA induces in the PD rat model by 6OHDA, mimicking advanced stages of the disease, results in the loss of both LTP and LTD and an increase of GluN2A/GluN2B ratio at synapses [230,231]. However, an early PD model characterized by only a partial DA depletion and mild motor symptoms shows a normal LTD and an altered LTP associated with a selective increased expression of synaptic GluN2A NMDARs. The use of cell permeable peptides able to modulate GluN2A interaction with scaffolding elements and, consequently, a physiological NMDAR subunit composition led to a full rescue of synaptic plasticity and to an amelioration of the motor behavior [231]. Overall, these studies indicate that an uncorrected composition of NMDARs is a key element in both motor behavior and synaptic plasticity in experimental PD.

Notably, more recent studies suggest that also α-synuclein (α-syn) produces deficits in visuospatial learning and impairs LTP in striatal SPNs belonging to the direct and indirect basal ganglia pathways by altering the activity and synaptic localization of GluN2A-containing NMDARs [233]. In addition, α-syn impairs NMDAR-dependent LTP in striatal SPNs without inducing any effect on LTD [233], in agreement with previous reports showing that this latter form of synaptic plasticity in the striatum does not require NMDAR activation [228]. Similarly, also in hippocampal neurons α-syn oligomers impair LTP and modify basal synaptic transmission through an NMDAR-mediated mechanism [234]. Even if levodopa remains the major pharmacological approach for Parkinson's disease, patients treated for several years develop disabling motor complications known as levodopa-induced dyskinesia [235]. These events are strictly correlated to adaptive changes of the glutamatergic corticostriatal signaling and involve also an aberrant functioning of NMDARs. Accordingly, the use of the low-affinity noncompetitive NMDAR antagonist amantadine allows at least a short-term benefit in the treatment of dyskinesia [235,236]. Of relevance, chronic levodopa treatment restores bidirectional synaptic plasticity (LTD, LTP, and depotentiation) only in non-dyskinetic conditions, whereas both in animal models and in patients, dyskinesias are associated with the loss of depotentiation of previously induced LTP [210,237,238].

Several reports demonstrated an increased synaptic GluN2A/GluN2B ratio of NMDARs in dendritic spines of SPNs, both in animal models and in patients [230,232,239], paralleled by modifications in the association of GluN2-type subunits with scaffolding proteins. Interestingly, treatment with peptides disrupting GluN2A interaction with MAGUK proteins or with Rph3A demonstrated that a decrease in synaptic GluN2A-containing NMDAR is sufficient to induce a significant reduction in the severity of Levodopa-induced dyskinesia [240,241].

Alterations of synaptic plasticity have been reported as a key determinant in the pathophysiology of early-onset generalized torsion dystonia (DYT1), an autosomal dominant movement disorder [242]. Human studies identified modifications of neuronal processing and plasticity in dystonic patients [243] and an impairment of striatal plasticity has been shown in different DYT1 animal models. In particular, a recent study demonstrated in a DYT1 dystonia mouse model that LTP appeared prematurely in a critical developmental window in striatal SPNs, whereas LTD was never recorded [244]. Notably these functional alterations where closely related to an increase of postsynaptic GluN2A suggesting a "premature" GluN2A/GluN2B switch [244].

6.4. Dysfunction of Glutamatergic Synaptic Plasticity in Alzheimer's Disease

Dysfunction of the excitatory synapse and spine loss represent early functional and morphological events of Alzheimer's disease (AD). Several postsynaptic proteins of the glutamatergic synapses have been demonstrated to be a key target of the amyloid-β_{1-42} (Aβ_{1-42}) peptide, whose accumulation and deposition are the main hallmarks of AD. Notably, several studies indicate that Aβ_{1-42} is an important modulator of neuronal activity affecting in turn synaptic transmission and plasticity [245]. In particular, oligomeric Aβ_{1-42} can alter the activation of NMDARs and downstream signaling cascades promoting the loss of LTP, the induction of LTD, and synaptic loss [246–248].

The different contribution of GluN2A- and GluN2B-containing NMDARs to Aβ_{1-42}-induced synaptic toxicity has been widely evaluated. Several studies reported the role of the extra-synaptic GluN2B-containing NMDARs in these events [249–251]. Aβ_{1-42} oligomers induce intracellular Ca^{2+} overload particularly mediated by the GluN2A subunit and neuronal death that can be prevented by NMDAR antagonists [252]. In addition, another study showed that presenilin knockout mice models are characterized by an early increase in GluN2A-containing NMDARs at the PSD with a concomitant reduction at non-synaptic sites before synaptic loss [253]. Moreover, oligomeric Aβ_{1-42} caused selective loss of synaptic GluN2B responses, promoting a switch in subunit composition from GluN2B to GluN2A, a process normally observed during development [254]. As GluN2A subunits have been implicated in protective pathways, whereas GluN2B subunits appear to increase neuronal vulnerability [255], the early increase in GluN2A and decrease in GluN2B subunit-composed NMDARs activity may be an attempt to reduce Aβ_{1-42}-induced neuronal dysfunction. Finally, a very recent study provided a novel mechanistic explanation for Aβ_{1-42}-dependent aberrant plasticity induced oligomeric forms [256]. In particular, Marcello and co-workers demonstrated that oligomeric Aβ_{1-42} activates a pathway that requires neuronal activity and the involvement of the GluN2A-containing NMDARs. Moreover, Aβ_{1-42}-induced neuronal degeneration involves activation of extrasynaptic GluN2B-containing NMDARs followed by increased Tau phosphorylation, whereas spine loss can be mainly mediated by GluN2A-containing NMDARs signaling and active caspase-3 [257]. Noteworthy, among Aβ toxicity mechanisms, also NMDAR metabotropic signaling was shown to impact on synaptic functions. However, this process seems to mainly involve GluN2B-containing NMDARs as reported by Malinow's laboratory in 2013 [254]. However, Aβ treatment was also found to act indirectly on GluN2A subunit, increasing the synaptic GluN2A/GluN2B ratio [254] and so inducing a switch in NMDAR subunits similar to the neurodevelopmental one.

The impact of GluN2A subunit in AD has been recently taken into account also as a potential pharmacological target. In particular, GNE-0723, a novel positive allosteric modulator of GluN2A-containing NMDAR, was shown to ameliorate the cognitive deficits in a mouse model of AD (J20) [195], supporting the importance of the involvement of synaptic NMDAR neurotransmission in AD pathogenesis.

7. Conclusions

NMDARs have been intensively studied in the last decades because of their involvement in many aspects of neuronal transmission as well as learning and memory. It has been clearly demonstrated that NMDAR properties depend on its subunit composition and this review focused on the distinctive functions of synaptic GluN2A-containing NMDARs. Even if these NMDAR subtypes play a key role in both physiology and pathological synaptic plasticity, future studies aiming to understand their contribution to the pathophysiology of neurological disorders will require novel and specific approaches to overcome the limitations due to the high similarity among GluN2-type regulatory subunits. At present, genetic manipulation of GluN2A did not fully clarify its pathophysiological role and the available pharmacological tools are not specific enough to give a better answer. Understanding the role and the modulation of GluN2A protein–protein interactions at the CTD may provide new possible targets for pharmacological tools and be a future direction in which to investigate the unique role of this NMDAR subunit in the context of glutamatergic neurotransmission and synaptic plasticity.

Conflicts of Interest: The authors declare no conflicts of interest.

References

1. Seeburg, P.H.; Burnashev, N.; Köhr, G.; Kuner, T.; Sprengel, R.; Monyer, H. The NMDA Receptor Channel: Molecular Design of a Coincidence Detector. In *Proceedings of the 1993 Laurentian Hormone Conference*; Recent progress in hormone research; Academic Press: London, UK, 1995; pp. 19–34.
2. Mayer, M.L.; Westbrook, G.L.; Guthrie, P.B. Voltage-dependent block by Mg 2+ of NMDA responses in spinal cord neurones. *Nature* **1984**, *309*, 261–263. [CrossRef] [PubMed]
3. Duguid, I.; Sjöström, P.J. Novel presynaptic mechanisms for coincidence detection in synaptic plasticity. *Curr. Opin. Neurobiol.* **2006**, *16*, 312–322. [CrossRef] [PubMed]
4. Berretta, N.; Jones, R.S.G. Tonic facilitation of glutamate release by presynaptic N-methyl-D-aspartate autoreceptors in the entorhinal cortex. *Neuroscience* **1996**, *75*, 339–344. [CrossRef]
5. Sjöström, P.J.; Turrigiano, G.G.; Nelson, S.B. Neocortical LTD via coincident activation of presynaptic NMDA and cannabinoid receptors. *Neuron* **2003**, *39*, 641–654. [CrossRef]
6. Siegel, S.J.; Brose, N.; Janssen, W.G.; Gasic, G.P.; Jahn, R.; Heinemann, S.F.; Morrison, J.H. Regional, cellular, and ultrastructural distribution of N-methyl-D- aspartate receptor subunit 1 in monkey hippocampus. *Proc. Natl. Acad. Sci. USA* **1994**, *91*, 564–568. [CrossRef]
7. McGuinness, L.; Taylor, C.; Taylor, R.D.T.; Yau, C.; Langenhan, T.; Hart, M.L.; Christian, H.; Tynan, P.W.; Donnelly, P.; Emptage, N.J. Presynaptic NMDARs in the Hippocampus Facilitate Transmitter Release at Theta Frequency. *Neuron* **2010**, *68*, 1109–1127. [CrossRef]
8. Casado, M.; Dieudonné, S.; Ascher, P. Presynaptic N-methyl-D-aspartate receptors at the parallel fiber-Purkinje cell synapse. *Proc. Natl. Acad. Sci. USA* **2000**, *97*, 11593–11597. [CrossRef]
9. Bidoret, C.; Ayon, A.; Barbour, B.; Casado, M. Presynaptic NR2A-containing NMDA receptors implement a high-pass filter synaptic plasticity rule. *Proc. Natl. Acad. Sci. USA* **2009**, *106*, 14126–14131. [CrossRef]
10. Liu, H.; Wang, H.; Sheng, M.; Jan, L.Y.; Jan, Y.N.; Basbaum, A.I. Evidence for presynaptic N-methyl-D-aspartate autoreceptors in the spinal cord dorsal horn. *Proc. Natl. Acad. Sci. USA* **1994**, *91*, 8383–8387. [CrossRef]
11. Vissel, B.; Krupp, J.J.; Heinemann, S.F.; Westbrook, G.L. A use-dependent tyrosine dephosphorylation of NMDA receptors is independent of ion flux. *Nat. Neurosci.* **2001**, *4*, 587–596. [CrossRef]
12. Barria, A.; Malinow, R. Subunit-Specific NMDA Receptor Trafficking to Synapses. *Neuron* **2002**, *35*, 345–353. [CrossRef]
13. Nong, Y.; Huang, Y.Q.; Ju, W.; Kalia, L.V.; Ahmadian, G.; Wang, Y.T.; Salter, M.W. Glycine binding primes NMDA receptor internalization. *Nature* **2003**, *422*, 302–307. [CrossRef] [PubMed]
14. Ferreira, J.S.; Papouin, T.; Ladépêche, L.; Yao, A.; Langlais, V.C.; Bouchet, D.; Dulong, J.; Mothet, J.P.; Sacchi, S.; Pollegioni, L.; et al. Co-agonists differentially tune GluN2B-NMDA receptor trafficking at hippocampal synapses. *Elife* **2017**, *6*, e25492. [CrossRef] [PubMed]
15. Mayford, M.; Wang, J.; Kandel, E.R.; O'Dell, T.J. CaMKII regulates the frequency-response function of hippocampal synapses for the production of both LTD and LTP. *Cell* **1995**, *81*, 891–904. [CrossRef]
16. Scanziani, M.; Malenka, R.C.; Nicoll, R.A. Role of intercellular interactions in heterosynaptic long-term depression. *Nature* **1996**, *380*, 446–450. [CrossRef]
17. Carter, B.C.; Jahr, C.E. Postsynaptic, not presynaptic NMDA receptors are required for spike-timing-dependent LTD induction. *Nat. Neurosci.* **2016**, *19*, 1218–1224. [CrossRef]
18. Nabavi, S.; Kessels, H.W.; Alfonso, S.; Aow, J.; Fox, R.; Malinow, R. Metabotropic NMDA receptor function is required for NMDA receptor-dependent long-term depression. *Proc. Natl. Acad. Sci. USA* **2013**, *110*, 4027–4032. [CrossRef]

19. Dore, K.; Aow, J.; Malinow, R. Agonist binding to the NMDA receptor drives movement of its cytoplasmic domain without ion flow. *Proc. Natl. Acad. Sci. USA* **2015**, *112*, 14705–14710. [CrossRef]
20. Monyer, H.; Burnashev, N.; Laurie, D.J.; Sakmann, B.; Seeburg, P.H. Developmental and regional expression in the rat brain and functional properties of four NMDA receptors. *Neuron* **1994**, *12*, 529–540. [CrossRef]
21. Akazawa, C.; Shigemoto, R.; Bessho, Y.; Nakanishi, S.; Mizuno, N. Differential expression of five N-methyl-D-aspartate receptor subunit mRNAs in the cerebellum of developing and adult rats. *J. Comp. Neurol.* **1994**, *347*, 150–160. [CrossRef]
22. Paoletti, P.; Bellone, C.; Zhou, Q. NMDA receptor subunit diversity: Impact on receptor properties, synaptic plasticity and disease. *Nat. Rev. Neurosci.* **2013**, *14*, 383–400. [CrossRef] [PubMed]
23. Luo, T.; Wu, W.H.; Chen, B.S. NMDA receptor signaling: Death or survival? *Front. Biol. (Beijing)* **2011**, *6*, 468–476. [CrossRef] [PubMed]
24. Cercato, M.; Vàzquez, C.; Kornisiuk, E.; Aquirre, A.; Colettis, N.; Snitcofsky, M.; Jerusalinski, D.; Baez, M. GluN1 and GluN2A NMDA Receptor Subunits Increase in the Hippocampus during Memory Consolidation in the Rat. *Front. Behav. Neurosci.* **2017**, *13*, 242. [CrossRef] [PubMed]
25. Groc, L.; Heine, M.; Cousins, S.L.; Stephenson, F.A.; Lounis, B.; Cognet, L.; Choquet, D. NMDA receptor surface mobility depends on NR2A-2B subunits. *Proc. Natl. Acad. Sci. USA* **2006**, *103*, 18769–18774. [CrossRef]
26. Retchless, B.S.; Gao, W.; Johnson, J.W. A single GluN2 subunit residue controls NMDA receptor channel properties via intersubunit interaction. *Nat. Neurosci.* **2012**, *15*, 406-S2. [CrossRef]
27. Chen, N.; Luo, T.; Raymond, L.A. Subtype-dependence of NMDA receptor channel open probability. *J. Neurosci.* **1999**. [CrossRef]
28. Flint, A.C.; Maisch, U.S.; Weishaupt, J.H.; Kriegstein, A.R.; Monyer, H. NR2A subunit expression shortens NMDA receptor synaptic currents in developing neocortex. *J. Neurosci.* **1997**. [CrossRef]
29. Krupp, J.J.; Vissel, B.; Heinemann, S.F.; Westbrook, G.L. Calcium-dependent inactivation of recombinant N-methyl-D-aspartate receptors is NR2 subunit specific. *Mol. Pharmacol.* **1996**, *50*, 1680–1688.
30. Vicini, S.; Wang, J.F.; Li, J.H.; Zhu, W.J.; Wang, Y.H.; Luo, J.H.; Wolfe, B.B.; Grayson, D.R. Functional and pharmacological differences between recombinant N- methyl-D-aspartate receptors. *J. Neurophysiol.* **1998**, *79*, 555–566. [CrossRef]
31. Bell, K.F.S.; Hardingham, G.E. The influence of synaptic activity on neuronal health. *Curr. Opin. Neurobiol.* **2011**, *21*, 299–305. [CrossRef]
32. Hardingham, G. NMDA receptor C-terminal signaling in development, plasticity, and disease. *F1000Research* **2019**, *8*, 1547. [CrossRef] [PubMed]
33. Sun, Y.; Cheng, X.; Zhang, L.; Hu, J.; Chen, Y.; Zhan, L.; Gao, Z. The Functional and Molecular Properties, Physiological Functions, and Pathophysiological Roles of GluN2A in the Central Nervous System. *Mol. Neurobiol.* **2017**, *54*, 1008–1021. [CrossRef] [PubMed]
34. Baez, M.V.; Cercato, M.C.; Jerusalinsky, D.A. NMDA receptor subunits change after synaptic plasticity induction and learning and memory acquisition. *Neural Plast.* **2018**, *2018*. [CrossRef] [PubMed]
35. Gardoni, F.; Di Luca, M. Targeting glutamatergic synapses in Parkinson's disease. *Curr. Opin. Pharmacol.* **2015**, *20*, 24–28. [CrossRef]
36. Johnson, J.W.; Ascher, P. Glycine potentiates the NMDA response in cultured mouse brain neurons. *Nature* **1987**, *325*, 529–531. [CrossRef]
37. Christine, C.W.; Choi, D.W. Effect of zinc on NMDA receptor-mediated channel currents in cortical neurons. *J. Neurosci.* **1990**, *10*, 108–116. [CrossRef]
38. Legendre, P.; Westbrook, G.L. The inhibition of single N-methyl-D-aspartate-activated channels by zinc ions on cultured rat neurones. *J. Physiol.* **1990**, *429*, 429–449. [CrossRef]
39. Fayyazuddin, A.; Villarroel, A.; Le Goff, A.; Lerma, J.; Neyton, J. Four residues of the extracellular N-terminal domain of the NR2A subunit control high-affinity Zn2+ binding to NMDA receptors. *Neuron* **2000**, *25*, 683–694. [CrossRef]
40. Chen, N.; Moshaver, A.; Raymond, L.A. Differential sensitivity of recombinant N-methyl-D-aspartate receptor subtypes to zinc inhibition. *Mol. Pharmacol.* **1997**, *51*, 1015–1023. [CrossRef]
41. Mayer, M.L.; Vyklicky, L. The action of zinc on synaptic transmission and neuronal excitability in cultures of mouse hippocampus. *J. Physiol.* **1989**, *415*, 315–365. [CrossRef]

42. Paoletti, P.; Ascher, P.; Neyton, J. High-affinity zinc inhibition of NMDA NR1-NR2A receptors. *J. Neurosci.* **1997**, *17*, 5711–5725. [CrossRef] [PubMed]
43. Williams, J.M.; Mason-Parker, S.E.; Abraham, W.C.; Tate, W.P. Biphasic changes in the levels of N-methyl-D-aspartate receptor-2 subunits correlate with the induction and persistence of long-term potentiation. *Mol. Brain Res.* **1998**, *60*, 21–27. [CrossRef]
44. Traynelis, S.F.; Cull-Candy, S.G. Pharmacological properties and H+ sensitivity of excitatory amino acid receptor channels in rat cerebellar granule neurones. *J. Physiol.* **1991**, *443*, 727–763. [CrossRef] [PubMed]
45. Traynelis, S.F.; Cull-Candy, S.G. Proton inhibition of N-methyl-D-aspartate receptors in cerebellar neurons. *Nature* **1990**, *345*, 347–350. [CrossRef]
46. Velíšek, L.; Dreier, J.P.; Stanton, P.K.; Heinemann, U.; Moshé, S.L. Lowering of extracellular pH suppresses low-Mg2+-induces seizures in combined entorhinal cortex-hippocampal slices. *Exp. Brain Res.* **1994**, *101*, 44–52. [CrossRef]
47. Banke, T.G.; Dravid, S.M.; Traynelis, S.F. Protons trap NR1/NR2B NMDA receptors in a nonconducting state. *J. Neurosci.* **2005**, *25*, 42–51. [CrossRef]
48. Zhang, J.B.; Chang, S.; Xu, P.; Miao, M.; Wu, H.; Zhang, Y.; Zhang, T.; Wang, H.; Zhang, J.; Xie, C.; et al. Structural Basis of the Proton Sensitivity of Human GluN1-GluN2A NMDA Receptors. *Cell Rep.* **2018**, *25*, 3582–3590. [CrossRef]
49. Serraz, B.; Grand, T.; Paoletti, P. Altered zinc sensitivity of NMDA receptors harboring clinically-relevant mutations. *Neuropharmacology* **2016**, *109*, 196–204. [CrossRef]
50. Qiu, S.; Zhang, X.M.; Cao, J.Y.; Yang, W.; Yan, Y.G.; Shan, L.; Zheng, J.; Luo, J.H. An endoplasmic reticulum retention signal located in the extracellular amino-terminal domain of the NR2A subunit of N-methyl-D-aspartate receptors. *J. Biol. Chem.* **2009**, *284*, 20285–20298. [CrossRef]
51. Wollmuth, L.P.; Kuner, T.; Sakmann, B. Adjacent asparagines in the NR2-subunit of the NMDA receptor channel control the voltage-dependent block by extracellular Mg 2+. *J. Physiol.* **1998**, *506*, 13–32. [CrossRef]
52. Marwick, K.F.M.; Skehel, P.A.; Hardingham, G.E.; Wyllie, D.J.A. The human NMDA receptor GluN2AN615K variant influences channel blocker potency. *Pharmacol. Res. Perspect.* **2019**, *7*, e00495. [CrossRef] [PubMed]
53. Endele, S.; Rosenberger, G.; Geider, K.; Popp, B.; Tamer, C.; Stefanova, I.; Milh, M.; Kortüm, F.; Fritsch, A.; Pientka, F.K.; et al. Mutations in GRIN2A and GRIN2B encoding regulatory subunits of NMDA receptors cause variable neurodevelopmental phenotypes. *Nat. Genet.* **2010**, *42*, 1021–1026. [CrossRef] [PubMed]
54. Traynelis, S.F.; Wollmuth, L.P.; McBain, C.J.; Menniti, F.S.; Vance, K.M.; Ogden, K.K.; Hansen, K.B.; Yuan, H.; Myers, S.J.; Dingledine, R. Glutamate receptor ion channels: Structure, regulation, and function. *Pharmacol. Rev.* **2010**, *62*, 405–496. [CrossRef] [PubMed]
55. Sun, Y.; Xu, Y.; Cheng, X.; Chen, X.; Xie, Y.; Zhang, L.; Wang, L.; Hu, J.; Gao, Z. The differences between GluN2A and GluN2B signaling in the brain. *J. Neurosci. Res.* **2018**, *96*, 1430–1433. [CrossRef] [PubMed]
56. Hawkins, L.M.; Prybylowski, K.; Chang, K.; Moussan, C.; Stephenson, F.A.; Wenthold, R.J. Export from the endoplasmic reticulum of assembled N-methyl-D-aspartic acid receptors is controlled by a motif in the C terminus of the NR2 subunit. *J. Biol. Chem.* **2004**, *279*, 28903–28910. [CrossRef] [PubMed]
57. Chen, B.S.; Roche, K.W. Regulation of NMDA receptors by phosphorylation. *Neuropharmacology* **2007**, *53*, 362–368. [CrossRef] [PubMed]
58. Snyder, E.M.; Philpot, B.D.; Huber, K.M.; Dong, X.; Fallon, J.R.; Bear, M.F. Internalization of ionotropic glutamate receptors in response to mGluR activation. *Nat. Neurosci.* **2001**, *4*, 1079–1085. [CrossRef]
59. Li, B.; Chen, N.; Luo, T.; Otsu, Y.; Murphy, T.H.; Raymond, L.A. Differential regulation of synaptic and extrasynaptic NMDA receptors. *Nat. Neurosci.* **2002**, *5*, 833–834. [CrossRef]
60. Romero-Hernandez, A.; Furukawa, H. Novel mode of antagonist binding in NMDA receptors revealed by the crystal structure of the GluN1-GluN2A ligand-binding domain complexed to NVP-AAM077. *Mol. Pharmacol.* **2017**, *92*, 22–29. [CrossRef]
61. Frizelle, P.; Chen, P.; Wyllie, D. Equilibrium constants for (R)-[(S)-1-(4-bromo-phenyl)-ethylamino]-(2,3-dioxo-1,2,3,4-tetrahydroquinoxalin-5-yl)-methyl]-phosphonic acid (NVP-AAM077) acting at recombinant NR1NR2A and NR1NR2B N-methyl-D-aspartate receptors Im. *Mol. Pharmacol.* **2006**, *70*, 1022–1032. [CrossRef]
62. Schreiber, J.A.; Müller, S.L.; Westphälinger, S.E.; Schepmann, D.; Strutz-Seebohm, N.; Seebohm, G.; Wünsch, B. Systematic variation of the benzoylhydrazine moiety of the GluN2A selective NMDA receptor antagonist TCN-201. *Eur. J. Med. Chem.* **2018**, *158*, 259–269. [CrossRef] [PubMed]

63. Yi, F.; Mou, T.C.; Dorsett, K.N.; Volkmann, R.A.; Menniti, F.S.; Sprang, S.R.; Hansen, K.B. Structural Basis for Negative Allosteric Modulation of GluN2A-Containing NMDA Receptors. *Neuron* **2016**, *91*, 1316–1329. [CrossRef] [PubMed]
64. Hackos, D.H.; Lupardus, P.J.; Grand, T.; Chen, Y.; Wang, T.M.; Reynen, P.; Gustafson, A.; Wallweber, H.J.A.; Volgraf, M.; Sellers, B.D.; et al. Positive Allosteric Modulators of GluN2A-Containing NMDARs with Distinct Modes of Action and Impacts on Circuit Function. *Neuron* **2016**, *89*, 983–999. [CrossRef] [PubMed]
65. Volgraf, M.; Sellers, B.; Jiang, Y.; Wu, G.; Ly, C.; Villemure, E.; Pastor, R.; Yuen, P.; Lu, A.; Luo, X.; et al. Discovery of GluN2A-Selective NMDA Receptor Positive Allosteric Modulators (PAMs) Tuning Deactivation Kinetics via Structure-Based Design. *J. Med. Chem.* **2016**, *59*, 2760–2779. [CrossRef] [PubMed]
66. Villemure, E.; Volgraf, M.; Jiang, Y.; Wu, G.; Ly, C.Q.; Yuen, P.W.; Lu, A.; Luo, X.; Liu, M.; Zhang, S.; et al. GluN2A-selective pyridopyrimidinone series of nmdar positive allosteric modulators with an improved in vivo profile. *ACS Med. Chem. Lett.* **2016**, *8*, 84–89. [CrossRef] [PubMed]
67. Lim, I.A.; Hall, D.D.; Hell, J.W. Selectivity and promiscuity of the first and second PDZ domains of PSD-95 and synapse-associated protein 102. *J. Biol. Chem.* **2002**, *277*, 21697–21711. [CrossRef] [PubMed]
68. Kornau, H.C.; Schenker, L.T.; Kennedy, M.B.; Seeburg, P.H. Domain interaction between NMDA receptor subunits and the postsynaptic density protein PSD-95. *Science (80-.)* **1995**. [CrossRef]
69. Howard, M.A.; Elias, G.M.; Elias, L.A.B.; Swat, W.; Nicoll, R.A. The role of SAP97 in synaptic glutamate receptor dynamics. *Proc. Natl. Acad. Sci. USA* **2010**, *107*, 3805–3810. [CrossRef]
70. Sheng, M.; Sala, C. PDZ Domains and the Organization of Supramolecular Complexes. *Annu. Rev. Neurosci.* **2001**, *24*, 1–29. [CrossRef]
71. Stanic, J.; Carta, M.; Eberini, I.; Pelucchi, S.; Marcello, E.; Genazzani, A.A.; Racca, C.; Mulle, C.; Di Luca, M.; Gardoni, F. Rabphilin 3A retains NMDA receptors at synaptic sites through interaction with GluN2A/PSD-95 complex. *Nat. Commun.* **2015**, *6*, 10181. [CrossRef]
72. Lu, X.; Zhang, H.; Tian, X.; Wang, X. FRMPD2: A novel GluN2A-interacting scaffold protein in synaptic excitatory transmission. *Biochem. Biophys. Res. Commun.* **2018**.
73. Piguel, N.H.; Fievre, S.; Blanc, J.M.; Carta, M.; Moreau, M.M.; Moutin, E.; Pinheiro, V.L.; Medina, C.; Ezan, J.; Lasvaux, L.; et al. Scribble1/AP2 complex coordinates NMDA receptor endocytic recycling. *Cell Rep.* **2014**, *9*, 712–727. [CrossRef] [PubMed]
74. Panayotis, N.; Karpova, A.; Kreutz, M.R.; Fainzilber, M. Macromolecular transport in synapse to nucleus communication. *Trends Neurosci.* **2015**, *38*, 108–116. [CrossRef] [PubMed]
75. Marcello, E.; Di Luca, M.; Gardoni, F. Synapse-to-nucleus communication: From developmental disorders to Alzheimer's disease. *Curr. Opin. Neurobiol.* **2018**, *48*, 160–166. [CrossRef] [PubMed]
76. Dinamarca, M.C.; Guzzetti, F.; Karpova, A.; Lim, D.; Mitro, N.; Musardo, S.; Mellone, M.; Marcello, E.; Stanic, J.; Samaddar, T.; et al. Ring finger protein 10 is a novel synaptonuclear messenger encoding activation of NMDA receptors in hippocampus. *Elife* **2016**, *5*, e12430. [CrossRef] [PubMed]
77. Jordan, B.A.; Fernholz, B.D.; Khatri, L.; Ziff, E.B. Activity-dependent AIDA-1 nuclear signaling regulates nucleolar numbers and protein synthesis in neurons. *Nat. Neurosci.* **2007**. [CrossRef]
78. Zhai, S.; Ark, E.D.; Parra-Bueno, P.; Yasuda, R. Long-distance integration of nuclear ERK signaling triggered by activation of a few dendritic spines. *Science (80-.)* **2013**. [CrossRef]
79. Gardoni, F.; Schrama, L.; Kamal, A.; Gispen, W.; Cattabeni, F.; Di Luca, M. Hippocampal synaptic plasticity involves competition between Ca2+calmodulin-dependent protein kinase II and postsynaptic density 95 for binding to the NR2A subunit of the NMDA receptor. *J. Neurosci.* **2001**, *21*, 1501–1509. [CrossRef]
80. Gardoni, F.; Bellone, C.; Cattabeni, F.; Di Luca, M. Protein Kinase C Activation Modulates α-Calmodulin Kinase II Binding to NR2A Subunit of N-Methyl-D-Aspartate Receptor Complex. *J. Biol. Chem.* **2001**, *273*, 20689–20692. [CrossRef]
81. Strack, S.; Colbran, R.J. Autophosphorylation-dependent targeting of calcium calmodulin-dependent protein kinase II by the NR2B subunit of the N-methyl-D-aspartate receptor. *J. Biol. Chem.* **1998**, *273*, 20689–20692. [CrossRef]
82. Bayer, K.U.; Schulman, H. Regulation of signal transduction by protein targeting: The case for CaMKII. *Biochem. Biophys. Res. Commun.* **2001**, *289*, 917–923. [CrossRef] [PubMed]
83. Braun, A.P.; Schulman, H. The Multifunctional Calcium/Calmodulin-Dependent Protein Kinase: From Form to Function. *Annu. Rev. Physiol.* **1995**, *57*, 417–445. [CrossRef] [PubMed]

84. Miller, S.G.; Kennedy, M.B. Regulation of brain Type II Ca2+ calmodulin-dependent protein kinase by autophosphorylation: A Ca2+-triggered molecular switch. *Cell* **1986**, *44*, 861–870. [CrossRef]
85. Colbran, R.J.; Brown, A.M. Calcium/calmodulin-dependent protein kinase II and synaptic plasticity. *Curr. Opin. Neurobiol.* **2004**, *14*, 318–327. [CrossRef] [PubMed]
86. Gardoni, F.; Schrama, L.H.; Van Dalen, J.J.W.; Gispen, W.H.; Cattabeni, F.; Di Luca, M. αCaMKII binding to the C-terminal tail of NMDA receptor subunit NR2A and its modulation by autophosphorylation. *FEBS Lett.* **1999**, *456*, 394–398. [CrossRef]
87. Wang, J.; Liu, S.H.; Fu, Y.P.; Wang, J.H.; Lu, Y.M. Cdk5 activation induces hippocampal CA1 cell death by directly phosphorylating NMDA receptors. *Nat. Neurosci.* **2003**, *6*, 1039–1047. [CrossRef]
88. Li, B.S.; Sun, M.K.; Zhang, L.; Takahashi, S.; Ma, W.; Vinade, L.; Kulkarni, A.B.; Brady, R.O.; Pant, H.C. Regulation of NMDA receptors by cyclin-dependent kinase-5. *Proc. Natl. Acad. Sci. USA* **2001**, *98*, 12742–12747. [CrossRef]
89. Hisatsune, C.; Umemori, H.; Mishina, M.; Yamamoto, T. Phosphorylation-dependent interaction of the N-methyl-D-aspartate receptor ε2 subunit with phosphatidylinositol 3-kinase. *Genes to Cells* **1999**, *4*, 657–666. [CrossRef]
90. Lee, F.J.S.; Xue, S.; Pei, L.; Vukusic, B.; Chéry, N.; Wang, Y.; Wang, Y.T.; Niznik, H.B.; Yu, X.M.; Liu, F. Dual regulation of NMDA receptor functions by direct protein-protein interactions with the dopamine D1 receptor. *Cell* **2002**, *111*, 219–230. [CrossRef]
91. Elagabani, M.N.; Brisevac, D.; Kintscher, M.; Pohle, J.; Köhr, G.; Schmitz, D.; Kornau, H.C. Subunit-selective N-Methyl-D-aspartate (NMDA) receptor signaling through brefeldin a-resistant arf guanine nucleotide exchange factors BRAG1 and BRAG2 during synapse maturation. *J. Biol. Chem.* **2016**, *291*, 9105–9118. [CrossRef]
92. Sakagami, H.; Sanda, M.; Fukaya, M.; Miyazaki, T.; Sukegawa, J.; Yanagisawa, T.; Suzuki, T.; Fukunaga, K.; Watanabe, M.; Kondo, H. IQ-ArfGEF/BRAG1 is a guanine nucleotide exchange factor for Arf6 that interacts with PSD-95 at postsynaptic density of excitatory synapses. *Neurosci. Res.* **2008**, *60*, 199–212. [CrossRef]
93. Li, S.; Tian, X.; Hartley, D.M.; Feig, L.A. Distinct roles for Ras-guanine nucleotide-releasing factor 1 (Ras-GRF1) and Ras-GRF2 in the induction of long-term potentiation and long-term depression. *J. Neurosci.* **2006**, *26*, 1721–1729. [CrossRef] [PubMed]
94. Tian, X.; Gotoh, T.; Tsuji, K.; Lo, E.H.; Huang, S.; Feig, L.A. Developmentally regulated role for Ras-GRFs in coupling NMDA glutamate receptors to Ras, Erk and CREB. *EMBO J.* **2004**, *23*, 1567–1575. [CrossRef] [PubMed]
95. Fam, N.P.; Fan, W.T.; Wang, Z.; Zhang, L.J.; Chen, H.; Moran, M.F. Cloning and characterization of Ras-GRF2, a novel guanine nucleotide exchange factor for Ras. *Mol. Cell. Biol.* **1997**, *17*, 1396–1406. [CrossRef] [PubMed]
96. Shou, C.; Farnsworth, C.L.; Neel, B.G.; Feig, L.A. Molecular cloning of cDNAs encoding a guanine-nucleotide-releasing factor for Ras p21. *Nature* **1992**, *358*, 351–354. [CrossRef] [PubMed]
97. Fan, W.T.; Koch, C.A.; De Hoog, C.L.; Fam, N.P.; Moran, M.F. The exchange factor Ras-GRF2 activates Ras-dependent and Rac-dependent mitogen-activated protein kinase pathways. *Curr. Biol.* **1998**, *8*, 935–938. [CrossRef]
98. Innocenti, M.; Zippel, R.; Brambilla, R.; Sturani, E. CDC25(Mm)/Ras-GRF1 regulates both Ras and Rac signaling pathways. *FEBS Lett.* **1999**, *460*, 357–362. [CrossRef]
99. Ren, J.G.; Li, Z.; Sacks, D.B. IQGAP1 integrates Ca2+calmodulin and B-Raf signaling. *J. Biol. Chem.* **2008**, *283*, 22972–22982. [CrossRef]
100. Reiner, O.; Sapoznik, S.; Sapir, T. Lissencephaly 1 linking to multiple diseases: Mental retardation, neurodegeneration, schizophrenia, male sterility, and more. *NeuroMolecular Med.* **2006**, *8*, 547–565. [CrossRef]
101. Kholmanskikh, S.S.; Koeller, H.B.; Wynshaw-Boris, A.; Gomez, T.; Letourneau, P.C.; Ross, M.E. Calcium-dependent interaction of Lis1 with IQGAP1 and Cdc42 promotes neuronal motility. *Nat. Neurosci.* **2006**, *9*, 50–57. [CrossRef]
102. Ryu, J.; Futai, K.; Feliu, M.; Weinberg, R.; Sheng, M. Constitutively active Rap2 transgenic mice display fewer dendritic spines, reduced extracellular signal-regulated kinase signaling, enhanced long-term depression, and impaired spatial learning and fear extinction. *J. Neurosci.* **2008**, *28*, 8178–8188. [CrossRef] [PubMed]
103. Bajaj, G.; Hau, A.M.; Hsu, P.; Gafken, P.R.; Schimerlik, M.I.; Ishmael, J.E. Identification of an atypical calcium-dependent calmodulin binding site on the C-terminal domain of GluN2A. *Biochem. Biophys. Res. Commun.* **2014**, *44*, 588–594. [CrossRef] [PubMed]

104. Ma, H.; Groth, R.D.; Cohen, S.M.; Emery, J.F.; Li, B.; Hoedt, E.; Zhang, G.; Neubert, T.A.; Tsien, R.W. γcaMKII shuttles Ca2+/CaM to the nucleus to trigger CREB phosphorylation and gene expression. *Cell* **2014**, *159*, 281–294. [CrossRef] [PubMed]
105. Cohen, S.M.; Suutari, B.; He, X.; Wang, Y.; Sanchez, S.; Tirko, N.N.; Mandelberg, N.J.; Mullins, C.; Zhou, G.; Wang, S.; et al. Calmodulin shuttling mediates cytonuclear signaling to trigger experience-dependent transcription and memory. *Nat. Commun.* **2018**, *9*. [CrossRef] [PubMed]
106. Cohen, S.M.; Ma, H.; Kuchibhotla, K.V.; Watson, B.O.; Buzsáki, G.; Froemke, R.C.; Tsien, R.W. Excitation-Transcription Coupling in Parvalbumin-Positive Interneurons Employs a Novel CaM Kinase-Dependent Pathway Distinct from Excitatory Neurons. *Neuron* **2016**, *90*, 292–307. [CrossRef] [PubMed]
107. Zhang, X.M.; Yan, X.Y.; Zhang, B.; Yang, Q.; Ye, M.; Cao, W.; Qiang, W.B.; Zhu, L.J.; Du, Y.L.; Xu, X.X.; et al. Activity-induced synaptic delivery of the GluN2A-containing NMDA receptor is dependent on endoplasmic reticulum chaperone Bip and involved in fear memory. *Cell Res.* **2015**, *25*, 818–836. [CrossRef]
108. Malenka, R.C.; Bear, M.F. LTP and LTD: An embarrassment of riches. *Neuron* **2004**, *44*, 5–21. [CrossRef]
109. Li, L.J.; Hu, R.; Lujan, B.; Chen, J.; Zhang, J.J.; Nakano, Y.; Cui, T.Y.; Liao, M.X.; Chen, J.C.; Man, H.Y.; et al. Glycine potentiates AMPA receptor function through metabotropic activation of GluN2A-containing NMDA receptors. *Front. Mol. Neurosci.* **2016**, *9*. [CrossRef]
110. Liu, L.; Wong, T.P.; Pozza, M.F.; Lingenhoehl, K.; Wang, Y.; Sheng, M.; Auberson, Y.P.; Wang, Y.T. Role of NMDA Receptor Subtypes in Governing the Direction of Hippocampal Synaptic Plasticity. *Science (80-.)* **2004**, *304*, 1021–1024. [CrossRef]
111. Massey, P.V.; Johnson, B.E.; Moult, P.R.; Auberson, Y.P.; Brown, M.W.; Molnar, E.; Collingridge, G.L.; Bashir, Z.I. Differential roles of NR2A and NR2B-containing NMDA receptors in cortical long-term potentiation and long-term depression. *J. Neurosci.* **2004**, *24*, 7821–7828. [CrossRef]
112. Li, R.; Huang, F.S.; Abbas, A.K.; Wigström, H. Role of NMDA receptor subtypes in different forms of NMDA-dependent synaptic plasticity. *BMC Neurosci.* **2007**, *8*, 55. [CrossRef] [PubMed]
113. Bartlett, T.E.; Bannister, N.J.; Collett, V.J.; Dargan, S.L.; Massey, P.V.; Bortolotto, Z.A.; Fitzjohn, S.M.; Bashir, Z.I.; Collingridge, G.L.; Lodge, D. Differential roles of NR2A and NR2B-containing NMDA receptors in LTP and LTD in the CA1 region of two-week old rat hippocampus. *Neuropharmacology* **2007**, *52*, 60–70. [CrossRef] [PubMed]
114. Fox, C.J.; Russell, K.I.; Wang, Y.T.; Christie, B.R. Contribution of NR2A and NR2B NMDA subunits to bidirectional synaptic plasticity in the hippocampus in vivo. *Hippocampus* **2006**, *16*, 907–915. [CrossRef] [PubMed]
115. Berberich, S.; Punnakkal, P.; Jensen, V.; Pawlak, V.; Seeburg, P.H.; Hvalby, Ø.; Köhr, G. Lack of NMDA receptor subtype selectivity for hippocampal long-term potentiation. *J. Neurosci.* **2005**, *25*, 6907–6910. [CrossRef]
116. Romberg, C.; Raffel, J.; Martin, L.; Sprengel, R.; Seeburg, P.H.; Rawlins, J.N.P.; Bannerman, D.M.; Paulsen, O. Induction and expression of GluA1 (GluR-A)-independent LTP in the hippocampus. *Eur. J. Neurosci.* **2009**, *29*, 1141–1152. [CrossRef]
117. Izumi, Y.; Auberson, Y.P.; Zorumski, C.F. Zinc modulates bidirectional hippocampal plasticity by effects on NMDA receptors. *J. Neurosci.* **2006**, *26*, 7181–7188. [CrossRef]
118. Ng, D.; Pitcher, G.M.; Szilard, R.K.; Sertié, A.; Kanisek, M.; Clapcote, S.J.; Lipina, T.; Kalia, L.V.; Joo, D.; McKerlie, C.; et al. Neto1 is a novel CUB-domain NMDA receptor-interacting protein required for synaptic plasticity and learning. *PLoS Biol.* **2009**, *7*, e41. [CrossRef]
119. Sakimura, K.; Kutsuwada, T.; Ito, I.; Manabe, T.; Takayama, C.; Kushiya, E.; Yagi, T.; Shinichi, A.; Inoue, Y.; Sugiyama, H.; et al. Reduced hippocampal LTP and spatial learning in mice lacking NMDA receptor ɛ1 subunit. *Nature* **1995**, *373*, 151–155. [CrossRef]
120. Kiyama, Y.; Manabe, T.; Sakimura, K.; Kawakami, F.; Mori, H.; Mishina, M. Increased Thresholds for Long-Term Potentiation and Contextual Learning in Mice Lacking the NMDA-type Glutamate Receptor e1 Subunit. *J. Neurosci.* **1998**, *18*, 6704–6712. [CrossRef]
121. Köhr, G.; Jensen, V.; Koester, H.J.; Mihaljevic, A.L.A.; Utvik, J.K.; Kvello, A.; Ottersen, O.P.; Seeburg, P.H.; Sprengel, R.; Hvalby, Ø. Intracellular Domains of NMDA Receptor Subtypes Are Determinants for Long-Term Potentiation Induction. *J. Neurosci.* **2003**, *23*, 10791–10799. [CrossRef]

122. Moody, T.D.; Watabe, A.M.; Indersmitten, T.; Komiyama, N.H.; Grant, S.G.N.; O'Dell, T.J. β-adrenergic receptor activation rescues theta frequency stimulation-induced LTP deficits in mice expressing C-terminally truncated NMDA receptor GluN2A subunits. *Learn. Mem.* **2011**, *18*, 118–127. [CrossRef] [PubMed]
123. Bellone, C.; Nicoll, R.A. Rapid Bidirectional Switching of Synaptic NMDA Receptors. *Neuron* **2007**, *55*, 779–785. [CrossRef]
124. Grosshans, D.R.; Clayton, D.A.; Coultrap, S.J.; Browning, M.D. Analysis of glutamate receptor surface expression in acute hippocampal slices. *Sci. STKE* **2002**, *2002*, pl8. [CrossRef] [PubMed]
125. Franchini, L.; Stanic, J.; Ponzoni, L.; Mellone, M.; Carrano, N.; Musardo, S.; Zianni, E.; Olivero, G.; Marcello, E.; Pittaluga, A.; et al. Linking NMDA Receptor Synaptic Retention to Synaptic Plasticity and Cognition. *iScience* **2019**, *19*, 927–939. [CrossRef] [PubMed]
126. Baez, M.V.; Oberholzer, M.V.; Cercato, M.C.; Snitcofsky, M.; Aguirre, A.I.; Jerusalinsky, D.A. NMDA Receptor Subunits in the Adult Rat Hippocampus Undergo Similar Changes after 5 Minutes in an Open Field and after LTP Induction. *PLoS ONE* **2013**, *8*, e55244. [CrossRef]
127. Udagawa, T.; Swanger, S.A.; Takeuchi, K.; Kim, J.H.; Nalavadi, V.; Shin, J.; Lorenz, L.J.; Zukin, R.S.; Bassell, G.J.; Richter, J.D. Bidirectional Control of mRNA Translation and Synaptic Plasticity by the Cytoplasmic Polyadenylation Complex. *Mol. Cell* **2012**, *47*, 253–266. [CrossRef]
128. Swanger, S.A.; He, Y.A.; Richter, J.D.; Bassell, G.J. Dendritic GluN2A synthesis mediates activity-induced NMDA receptor insertion. *J. Neurosci.* **2013**, *33*, 8898–8908. [CrossRef]
129. Gerkin, R.C.; Lau, P.-M.; Nauen, D.W.; Wang, Y.T.; Bi, G.-Q. Modular Competition Driven by NMDA Receptor Subtypes in Spike-Timing-Dependent Plasticity. *J. Neurophysiol.* **2007**, *97*, 2851–2862. [CrossRef]
130. Ge, Y.; Dong, Z.; Bagot, R.C.; Howland, J.G.; Phillips, A.G.; Wong, T.P.; Wang, Y.T. Hippocampal long-term depression is required for the consolidation of spatial memory. *Proc. Natl. Acad. Sci. USA* **2010**, *107*, 16697–16702. [CrossRef]
131. Longordo, F.; Kopp, C.; Mishina, M.; Luján, R.; Lüthi, A. NR2A at CA1 synapses is obligatory for the susceptibility of hippocampal plasticity to sleep loss. *J. Neurosci.* **2009**, *29*, 9026–9041. [CrossRef]
132. Cui, Z.; Feng, R.; Jacobs, S.; Duan, Y.; Wang, H.; Cao, X.; Tsien, J.Z. Increased NR2A:NR2B ratio compresses long-term depression range and constrains long-term memory. *Sci. Rep.* **2013**, *3*, 1036. [CrossRef] [PubMed]
133. Tang, T.T.T.; Yang, F.; Chen, B.S.; Lu, Y.; Ji, Y.; Roche, K.W.; Lu, B. Dysbindin regulates hippocampal LTP by controlling NMDA receptor surface expression. *Proc. Natl. Acad. Sci. USA* **2009**, *106*, 21395–21400. [CrossRef] [PubMed]
134. Lisman, J.; Schulman, H.; Cline, H. The molecular basis of CaMKII function in synaptic and behavioural memory. *Nat. Rev. Neurosci.* **2002**, *3*, 175–190. [CrossRef] [PubMed]
135. Carrano, N.; Samaddar, T.; Brunialti, E.; Franchini, L.; Marcello, E.; Ciana, P.; Mauceri, D.; Di Luca, M.; Gardoni, F. The Synaptonuclear Messenger RNF10 Acts as an Architect of Neuronal Morphology. *Mol. Neurobiol.* **2019**, *56*, 7583–7593. [CrossRef]
136. Bliss, T.V.P.; Collingridge, G.L. A synaptic model of memory: Long-term potentiation in the hippocampus. *Nature* **1993**, *36*, 31–39. [CrossRef]
137. Martin, S.J.; Grimwood, P.D.; Morris, R.G.M. Synaptic Plasticity and Memory: An Evaluation of the Hypothesis. *Annu. Rev. Neurosci.* **2000**, *23*, 649–711. [CrossRef]
138. Morris, R.G.M.; Anderson, E.; Lynch, G.S.; Baudry, M. Selective impairment of learning and blockade of long-term potentiation by an N-methyl-D-aspartate receptor antagonist, AP5. *Nature* **1986**, *319*, 774–776. [CrossRef]
139. Tsien, J.Z.; Huerta, P.T.; Tonegawa, S. The essential role of hippocampal CA1 NMDA receptor-dependent synaptic plasticity in spatial memory. *Cell* **1996**, *87*, 1327–1338. [CrossRef]
140. Bannerman, D.M.; Niewoehner, B.; Lyon, L.; Romberg, C.; Schmitt, W.B.; Taylor, A.; Sanderson, D.J.; Cottam, J.; Sprengel, R.; Seeburg, P.H.; et al. NMDA receptor subunit NR2A is required for rapidly acquired spatial working memory but not incremental spatial reference memory. *J. Neurosci.* **2008**, *28*, 3623–3630. [CrossRef]
141. Zhang, X.H.; Liu, S.S.; Yi, F.; Zhuo, M.; Li, B.M. Delay-dependent impairment of spatial working memory with inhibition of NR2B-containing NMDA receptors in hippocampal CA1 region of rats. *Mol. Brain* **2013**. [CrossRef]
142. Goodfellow, M.J.; Abdulla, K.A.; Lindquist, D.H. Neonatal Ethanol Exposure Impairs Trace Fear Conditioning and Alters NMDA Receptor Subunit Expression in Adult Male and Female Rats. *Alcohol. Clin. Exp. Res.* **2016**, *40*, 309–318. [CrossRef] [PubMed]

143. Zhan, J.Q.; Zheng, L.L.; Chen, H.B.; Yu, B.; Wang, W.; Wang, T.; Ruan, B.; Pan, B.X.; Chen, J.R.; Li, X.F.; et al. Hydrogen sulfide reverses aging-associated amygdalar synaptic plasticity and fear memory deficits in rats. *Front. Neurosci.* **2018**, *12*. [CrossRef] [PubMed]
144. Chen, Y.F.; Chen, Z.X.; Wang, R.H.; Shi, Y.W.; Xue, L.; Wang, X.G.; Zhao, H. Knockdown of CLC-3 in the hippocampal CA1 impairs contextual fear memory. *Prog. Neuro-Psychopharmacology Biol. Psychiatry* **2019**, *89*, 132–145. [CrossRef] [PubMed]
145. Jacobs, S.; Cui, Z.; Feng, R.; Wang, H.; Wang, D.; Tsien, J.Z. Molecular and genetic determinants of the NMDA receptor for superior learning and memory functions. *PLoS ONE* **2014**, *9*. [CrossRef]
146. Gupta-Agarwal, S.; Jarome, T.J.; Fernandez, J.; Lubin, F.D. NMDA receptor- and ERK-dependent histone methylation changes in the lateral amygdala bidirectionally regulate fear memory formation. *Learn. Mem.* **2014**, *21*, 351–362. [CrossRef]
147. Holehonnur, R.; Phensy, A.J.; Kim, L.J.; Milivojevic, M.; Vuong, D.; Daison, D.K.; Alex, S.; Tiner, M.; Jones, L.E.; Kroener, S.; et al. Increasing the GluN2A/GluN2B ratio in neurons of the mouse basal and lateral amygdala inhibits the modification of an existing fear memory trace. *J. Neurosci.* **2016**, *36*, 9490–9504. [CrossRef]
148. Sun, Y.; Wang, L.; Gao, Z. Identifying the role of GluN2A in cerebral ischemia. *Front. Mol. Neurosci.* **2017**. [CrossRef]
149. Boyce-Rustay, J.M.; Holmes, A. Genetic inactivation of the NMDA receptor NR2A subunit has anxiolytic- and antidepressant-like effects in mice. *Neuropsychopharmacology* **2006**, *31*, 2405–2414. [CrossRef]
150. Taniguchi, S.; Nakazawa, T.; Tanimura, A.; Kiyama, Y.; Tezuka, T.; Watabe, A.M.; Katayama, N.; Yokoyama, K.; Inoue, T.; Izumi-Nakaseko, H.; et al. Involvement of NMDAR2A tyrosine phosphorylation in depression-related behaviour. *EMBO J.* **2009**, *28*, 3717–3729. [CrossRef]
151. Han, X.; Shao, W.; Liu, Z.; Fan, S.; Yu, J.; Chen, J.; Qiao, R.; Zhou, J.; Xie, P. ITRAQ-based quantitative analysis of hippocampal postsynaptic density-associated proteins in a rat chronic mild stress model of depression. *Neuroscience* **2015**, *298*, 220–292. [CrossRef]
152. Feyissa, A.M.; Chandran, A.; Stockmeier, C.A.; Karolewicz, B. Reduced levels of NR2A and NR2B subunits of NMDA receptor and PSD-95 in the prefrontal cortex in major depression. *Prog. Neuro-Psychopharmacology Biol. Psychiatry* **2009**, *33*, 70–75. [CrossRef] [PubMed]
153. Karolewicz, B.; Szebeni, K.; Gilmore, T.; Maclag, D.; Stockmeier, C.A.; Ordway, G.A. Elevated levels of NR2A and PSD-95 in the lateral amygdala in depression. *Int. J. Neuropsychopharmacol.* **2009**, *12*, 143–153. [CrossRef] [PubMed]
154. Kaut, O.; Schmitt, I.; Hofmann, A.; Hoffmann, P.; Schlaepfer, T.E.; Wüllner, U.; Hurlemann, R. Aberrant NMDA receptor DNA methylation detected by epigenome-wide analysis of hippocampus and prefrontal cortex in major depression. *Eur. Arch. Psychiatry Clin. Neurosci.* **2015**, *265*, 331–341. [CrossRef] [PubMed]
155. Francija, E.; Petrovic, Z.; Brkic, Z.; Mitic, M.; Radulovic, J.; Adzic, M. Disruption of the NMDA receptor GluN2A subunit abolishes inflammation-induced depression. *Behav. Brain Res.* **2019**, *359*, 550–559. [CrossRef] [PubMed]
156. Sun, H.; Guan, L.; Zhu, Z.; Li, H. Reduced levels of NR1 and NR2A with depression-like behavior in different brain regions in prenatally stressed juvenile offspring. *PLoS ONE* **2013**, *8*. [CrossRef] [PubMed]
157. Sun, H.; Jia, N.; Guan, L.; Su, Q.; Wang, D.; Li, H.; Zhu, Z. Involvement of NR1, NR2A different expression in brain regions in anxiety-like behavior of prenatally stressed offspring. *Behav. Brain Res.* **2013**, *257*, 1–7. [CrossRef]
158. Itokawa, M.; Yamada, K.; Yoshitsugu, K.; Toyota, T.; Suga, T.; Ohba, H.; Watanabe, A.; Hattori, E.; Shimizu, H.; Kumakura, T.; et al. A microsatellite repeat in the promoter of the N-methyl-D-aspartate receptor 2A subunit (GRIN2A) gene suppresses transcriptional activity and correlates with chronic outcome in schizophrenia. *Pharmacogenetics* **2003**, *13*, 271–278. [CrossRef]
159. Pinacho, R.; Villalmanzo, N.; Roca, M.; Iniesta, R.; Monje, A.; Haro, J.M.; Meana, J.J.; Ferrer, I.; Gill, G.; Ramos, B. Analysis of Sp transcription factors in the postmortem brain of chronic schizophrenia: A pilot study of relationship to negative symptoms. *J. Psychiatr. Res.* **2013**, *47*, 926–934. [CrossRef]
160. Beneyto, M.; Meador-Woodruff, J.H. Lamina-specific abnormalities of NMDA receptor-associated postsynaptic protein transcripts in the prefrontal cortex in schizophrenia and bipolar disorder. *Neuropsychopharmacology* **2008**, *33*, 2175–2186. [CrossRef]

161. Bitanihirwe, B.K.Y.; Lim, M.P.; Kelley, J.F.; Kaneko, T.; Woo, T.U.W. Glutamatergic deficits and parvalbumin-containing inhibitory neurons in the prefrontal cortex in schizophrenia. *BMC Psychiatry* **2009**, *9*. [CrossRef]
162. Arning, L.; Kraus, P.H.; Valentin, S.; Saft, C.; Andrich, J.; Epplen, J.T. NR2A and NR2B receptor gene variations modify age at onset in Huntington disease. *Neurogenetics* **2005**, *6*, 25–28. [CrossRef] [PubMed]
163. Luthi-Carter, R.; Apostol, B.L.; Dunah, A.W.; DeJohn, M.M.; Farrell, L.A.; Bates, G.P.; Young, A.B.; Standaert, D.G.; Thompson, L.M.; Cha, J.H.J. Complex alteration of NMDA receptors in transgenic Huntington's disease mouse brain: Analysis of mRNA and protein expression, plasma membrane association, interacting proteins, and phosphorylation. *Neurobiol. Dis.* **2003**, *14*, 624–636. [CrossRef] [PubMed]
164. Ali, N.J.; Levine, M.S. Changes in expression of N-methyl-D-aspartate receptor subunits occur early in the R6/2 mouse model of Huntington's disease. *Dev. Neurosci.* **2006**, *28*, 230–238. [CrossRef] [PubMed]
165. Devinsky, O.; Vezzani, A.; O'rien, T.J.; Jette, N.; Scheffer, I.E.; de Curtis, M.; Perucca, P. Epilepsy. *Nat. Rev. Dis. Prim.* **2018**, *3*, 18024.
166. Lemke, J.R.; Lal, D.; Reinthaler, E.M.; Steiner, I.; Nothnagel, M.; Alber, M.; Geider, K.; Laube, B.; Schwake, M.; Finsterwalder, K.; et al. Mutations in GRIN2A cause idiopathic focal epilepsy with rolandic spikes. *Nat. Genet.* **2013**, *45*, 1067–1072. [CrossRef]
167. Lesca, G.; Rudolf, G.; Bruneau, N.; Lozovaya, N.; Labalme, A.; Boutry-Kryza, N.; Salmi, M.; Tsintsadze, T.; Addis, L.; Motte, J.; et al. GRIN2A mutations in acquired epileptic aphasia and related childhood focal epilepsies and encephalopathies with speech and language dysfunction. *Nat. Genet.* **2013**, *45*, 1061–1066. [CrossRef]
168. Carvill, G.L.; Regan, B.M.; Yendle, S.C.; O'Roak, B.J.; Lozovaya, N.; Bruneau, N.; Burnashev, N.; Khan, A.; Cook, J.; Geraghty, E.; et al. GRIN2A mutations cause epilepsy-aphasia spectrum disorders. *Nat. Genet.* **2013**, *45*, 1073–1076. [CrossRef]
169. De Ligt, J.; Willemsen, M.H.; Van Bon, B.W.M.; Kleefstra, T.; Yntema, H.G.; Kroes, T.; Vulto-van Silfhout, A.T.; Koolen, D.A.; De Vries, P.; Gilissen, C.; et al. Diagnostic exome sequencing in persons with severe intellectual disability. *N. Engl. J. Med.* **2012**, *367*, 1921–1929. [CrossRef]
170. Ogden, K.K.; Chen, W.; Swanger, S.A.; McDaniel, M.J.; Fan, L.Z.; Hu, C.; Tankovic, A.; Kusumoto, H.; Kosobucki, G.J.; Schulien, A.J.; et al. Molecular Mechanism of Disease-Associated Mutations in the Pre-M1 Helix of NMDA Receptors and Potential Rescue Pharmacology. *PLoS Genet.* **2017**, *13*, e1006536. [CrossRef]
171. Pierson, T.M.; Yuan, H.; Marsh, E.D.; Fuentes-Fajardo, K.; Adams, D.R.; Markello, T.; Golas, G.; Simeonov, D.R.; Holloman, C.; Tankovic, A.; et al. GRIN2A mutation and early-onset epileptic encephalopathy: personalized therapy with memantine. *Ann. Clin. Transl. Neurol.* **2014**, *1*, 190–198. [CrossRef]
172. Yuan, H.; Hansen, K.B.; Zhang, J.; Mark Pierson, T.; Markello, T.C.; Fajardo, K.V.F.; Holloman, C.M.; Golas, G.; Adams, D.R.; Boerkoel, C.F.; et al. Functional analysis of a de novo GRIN2A missense mutation associated with early-onset epileptic encephalopathy. *Nat. Commun.* **2014**, *5*, 3251. [CrossRef] [PubMed]
173. Chen, W.; Tankovic, A.; Burger, P.B.; Kusumoto, H.; Traynelis, S.F.; Yuan, H. Functional evaluation of a de novo GRIN2A mutation identified in a patient with profound global developmental delay and refractory epilepsy. *Mol. Pharmacol.* **2017**, *91*, 317–330. [CrossRef] [PubMed]
174. Salter, M.W.; Kalia, L.V. SRC kinases: A hub for NMDA receptor regulation. *Nat. Rev. Neurosci.* **2004**, *5*, 317–328. [CrossRef] [PubMed]
175. Sibarov, D.A.; Bruneau, N.; Antonov, S.M.; Szepetowski, P.; Burnashev, N.; Giniatullin, R. Functional properties of human NMDA receptors associated with epilepsy-related mutations of GluN2A subunit. *Front. Cell. Neurosci.* **2017**, 155. [CrossRef]
176. Lemke, J.R.; Geider, K.; Helbig, K.L.; Heyne, H.O.; Schütz, H.; Hentschel, J.; Courage, C.; Depienne, C.; Nava, C.; Heron, D.; et al. Delineating the GRIN1 phenotypic spectrum: A distinct genetic NMDA receptor encephalopathy. *Neurology* **2016**, *86*, 2171–2178. [CrossRef]
177. Swanger, S.A.; Chen, W.; Wells, G.; Burger, P.B.; Tankovic, A.; Bhattacharya, S.; Strong, K.L.; Hu, C.; Kusumoto, H.; Zhang, J.; et al. Mechanistic Insight into NMDA Receptor Dysregulation by Rare Variants in the GluN2A and GluN2B Agonist Binding Domains. *Am. J. Hum. Genet.* **2016**, *99*, 1261–1280. [CrossRef]
178. Xu, X.X.; Luo, J.H. Mutations of N-Methyl-D-Aspartate Receptor Subunits in Epilepsy. *Neurosci. Bull.* **2018**, *34*, 549–565. [CrossRef]
179. Allen, A.S.; Berkovic, S.F.; Cossette, P.; Delanty, N.; Dlugos, D.; Eichler, E.E.; Epstein, M.P.; Glauser, T.; Goldstein, D.B.; Han, Y.; et al. De novo mutations in epileptic encephalopathies. *Nature* **2013**, *501*, 217–221.

180. Adams, D.R.; Yuan, H.; Holyoak, T.; Arajs, K.H.; Hakimi, P.; Markello, T.C.; Wolfe, L.A.; Vilboux, T.; Burton, B.K.; Fajardo, K.F.; et al. Three rare diseases in one Sib pair: RAI1, PCK1, GRIN2B mutations associated with Smith-Magenis Syndrome, cytosolic PEPCK deficiency and NMDA receptor glutamate insensitivity. *Mol. Genet. Metab.* **2014**, *113*, 161–170. [CrossRef]
181. Gori, M.B.; Girardi, E. 3-Mercaptopropionic acid-induced repetitive seizures increase glun2a expression in rat hippocampus: A potential neuroprotective role of cyclopentyladenosine. *Cell. Mol. Neurobiol.* **2013**, *33*, 803–813. [CrossRef]
182. Borbély, S.; Dobó, E.; Czégé, D.; Molnár, E.; Bakos, M.; Szucs, B.; Vincze, A.; Világi, I.; Mihály, A. Modification of ionotropic glutamate receptor-mediated processes in the rat hippocampus following repeated, brief seizures. *Neuroscience* **2009**, *159*, 358–368. [CrossRef] [PubMed]
183. Chen, Q.; He, S.; Hu, X.L.; Yu, J.; Zhou, Y.; Zheng, J.; Zhang, S.; Zhang, C.; Duan, W.H.; Xiong, Z.Q. Differential roles of NR2A- and NR2B-containing NMDA receptors in activity-dependent brain-derived neurotrophic factor gene regulation and limbic epileptogenesis. *J. Neurosci.* **2007**, *27*, 542–552. [CrossRef] [PubMed]
184. Yang, Y.; Tian, X.; Xu, D.; Zheng, F.; Lu, X.; Zhang, Y.; Ma, Y.; Li, Y.; Xu, X.; Zhu, B.; et al. GPR40 modulates epileptic seizure and NMDA receptor function. *Sci. Adv.* **2018**, *4*, eaau2357. [CrossRef] [PubMed]
185. Abegg, M.H.; Savic, N.; Ehrengruber, M.U.; McKinney, R.A.; Gähwiler, B.H. Epileptiform activity in rat hippocampus strengthens excitatory synapses. *J. Physiol.* **2004**, *554*, 439–448. [CrossRef]
186. Cain, D.P.; Hargreaves, E.L.; Boon, F.; Dennison, Z. An examination of the relations between hippocampal long-term potentiation, kindling, afterdischarge, and place learning in the water maze. *Hippocampus* **1993**, *3*, 153–163. [CrossRef]
187. McNamara, R.K.; Kirkby, R.D.; DePape, G.E.; Skelton, R.W.; Corcoran, M.E. Differential effects of kindling and kindled seizures on place learning in the morris water maze. *Hippocampus* **1993**, *3*, 149–152. [CrossRef]
188. Rutten, A.; Van Albada, M.; Silveira, D.C.; Cha, B.H.; Liu, X.; Hu, Y.N.; Cilio, M.R.; Holmes, G.L. Memory impairment following status epilepticus in immature rats: Time-course and environmental effects. *Eur. J. Neurosci.* **2002**, *16*, 501–513. [CrossRef]
189. Guli, X.; Tokay, T.; Kirschstein, T.; Köhling, R. Status Epilepticus Enhances Depotentiation after Fully Established LTP in an NMDAR-Dependent but GluN2B-Independent Manner. *Neural Plast.* **2016**, *2016*. [CrossRef]
190. O'Leary, H.; Bernard, P.B.; Castano, A.M.; Benke, T.A. Enhanced long term potentiation and decreased AMPA receptor desensitization in the acute period following a single kainate induced early life seizure. *Neurobiol. Dis.* **2016**, *87*, 134–144. [CrossRef]
191. Müller, L.; Tokay, T.; Porath, K.; Köhling, R.; Kirschstein, T. Enhanced NMDA receptor-dependent LTP in the epileptic CA1 area via upregulation of NR2B. *Neurobiol. Dis.* **2013**, *54*, 183–193. [CrossRef]
192. Kryukov, K.A.; Kim, K.K.; Magazanik, L.G.; Zaitsev, A.V. Status epilepticus alters hippocampal long-term synaptic potentiation in a rat lithium-pilocarpine model. *Neuroreport* **2016**, *27*, 1191–1195. [CrossRef] [PubMed]
193. Zhou, J.L.; Shatskikh, T.N.; Liu, X.; Holmes, G.L. Impaired single cell firing and long-term potentiation parallels memory impairment following recurrent seizures. *Eur. J. Neurosci.* **2007**, *25*, 3667–3677. [CrossRef] [PubMed]
194. Suárez, L.M.; Cid, E.; Gal, B.; Inostroza, M.; Brotons-Mas, J.R.; Gómez-Domínguez, D.; de la Prida, L.M.; Solís, J.M. Systemic Injection of Kainic Acid Differently Affects LTP Magnitude Depending on its Epileptogenic Efficiency. *PLoS ONE* **2012**, *7*, e48128. [CrossRef] [PubMed]
195. Hanson, J.E.; Ma, K.; Elstrott, J.; Weber, M.; Saillet, S.; Khan, A.S.; Simms, J.; Liu, B.; Kim, T.A.; Yu, G.Q.; et al. GluN2A NMDA Receptor Enhancement Improves Brain Oscillations, Synchrony, and Cognitive Functions in Dravet Syndrome and Alzheimer's Disease Models. *Cell Rep.* **2020**, *30*, 381–396.e4. [CrossRef]
196. Bagni, C.; Zukin, R.S. A Synaptic Perspective of Fragile X Syndrome and Autism Spectrum Disorders. *Neuron* **2019**, *101*, 1070–1088. [CrossRef]
197. Bhat, S.; Acharya, U.R.; Adeli, H.; Bairy, G.M.; Adeli, A. Autism: Cause factors, early diagnosis and therapies. *Rev. Neurosci.* **2014**, *25*, 841–850. [CrossRef]
198. Volkmar, F.R.; McPartland, J.C. From Kanner to DSM-5: Autism as an Evolving Diagnostic Concept. *Annu. Rev. Clin. Psychol.* **2014**, *10*, 193–212. [CrossRef]
199. Park, H.R.; Lee, J.M.; Moon, H.E.; Lee, D.S.; Kim, B.N.; Kim, J.; Kim, D.G.; Paek, S.H. A short review on the current understanding of autism spectrum disorders. *Exp. Neurobiol.* **2016**, *25*, 1–13. [CrossRef]

200. Waye, M.M.Y.; Cheng, H.Y. Genetics and epigenetics of autism: A Review. *Psychiatry Clin. Neurosci.* **2018**, *72*, 228–244. [CrossRef]
201. Bagni, C.; Tassone, F.; Neri, G.; Hagerman, R. Fragile X syndrome: Causes, diagnosis, mechanisms, and therapeutics. *J. Clin. Invest.* **2012**, *122*, 4314–4322. [CrossRef]
202. Bassell, G.J.; Warren, S.T. Fragile X Syndrome: Loss of Local mRNA Regulation Alters Synaptic Development and Function. *Neuron* **2008**, *60*, 201–214. [CrossRef] [PubMed]
203. De Rubeis, S.; Bagni, C. Fragile X mental retardation protein control of neuronal mRNA metabolism: Insights into mRNA stability. *Mol. Cell. Neurosci.* **2010**, *43*, 43–50. [CrossRef] [PubMed]
204. Pasciuto, E.; Bagni, C. SnapShot: FMRP mRNA targets and diseases. *Cell* **2014**, *158*, 1446–1446.e1. [CrossRef] [PubMed]
205. Zoghbi, H.Y.; Bear, M.F. Synaptic dysfunction in neurodevelopmental disorders associated with autism and intellectual disabilities. *Cold Spring Harb. Perspect. Biol.* **2012**, *4*, a009886. [CrossRef] [PubMed]
206. Sawicka, K.; Pyronneau, A.; Chao, M.; Bennett, M.V.L.; Zukin, R.S. Elevated erk/p90 ribosomal S6 kinase activity underlies audiogenic seizure susceptibility in Fragile X mice. *Proc. Natl. Acad. Sci. USA* **2016**, *113*, E6290–E6297. [CrossRef] [PubMed]
207. Wang, X.; Snape, M.; Klann, E.; Stone, J.G.; Singh, A.; Petersen, R.B.; Castellani, R.J.; Casadesus, G.; Smith, M.A.; Zhu, X. Activation of the extracellular signal-regulated kinase pathway contributes to the behavioral deficit of fragile x-syndrome. *J. Neurochem.* **2012**, *121*, 672–679. [CrossRef]
208. Pasciuto, E.; Ahmed, T.; Wahle, T.; Gardoni, F.; D'Andrea, L.; Pacini, L.; Jacquemont, S.; Tassone, F.; Balschun, D.; Dotti, C.G.; et al. Dysregulated ADAM10-Mediated Processing of APP during a Critical Time Window Leads to Synaptic Deficits in Fragile X Syndrome. *Neuron* **2015**, *87*, 382–398. [CrossRef]
209. Lauterborn, J.C.; Rex, C.S.; Kramár, E.; Chen, L.Y.; Pandyarajan, V.; Lynch, G.; Gall, C.M. Brain-derived neurotrophic factor rescues synaptic plasticity in a mouse model of fragile X syndrome. *J. Neurosci.* **2007**, *27*, 10685–10694. [CrossRef]
210. Lee, H.Y.; Ge, W.P.; Huang, W.; He, Y.; Wang, G.X.; Rowson-Baldwin, A.; Smith, S.J.; Jan, Y.N.; Jan, L.Y. Bidirectional regulation of dendritic voltage-gated potassium channels by the fragile X mental retardation protein. *Neuron* **2011**, *72*, 630–642. [CrossRef]
211. Paradee, W.; Melikian, H.E.; Rasmussen, D.L.; Kenneson, A.; Conn, P.J.; Warren, S.T. Fragile X mouse: Strain effects of knockout phenotype and evidence suggesting deficient amygdala function. *Neuroscience* **1999**, *94*, 185–192. [CrossRef]
212. Zhao, M.G.; Toyoda, H.; Ko, S.W.; Ding, H.K.; Wu, L.J.; Zhuo, M. Deficits in trace fear memory and long-term potentiation in a mouse model for fragile X syndrome. *J. Neurosci.* **2005**, *25*, 7385–7392. [CrossRef] [PubMed]
213. Desai, N.S.; Casimiro, T.M.; Gruber, S.M.; Vanderklish, P.W. Early postnatal plasticity in neocortex of Fmr1 knockout mice. *J. Neurophysiol.* **2006**, *96*, 1734–1745. [CrossRef] [PubMed]
214. Meredith, R.M.; Holmgren, C.D.; Weidum, M.; Burnashev, N.; Mansvelder, H.D. Increased Threshold for Spike-Timing-Dependent Plasticity Is Caused by Unreliable Calcium Signaling in Mice Lacking Fragile X Gene Fmr1. *Neuron* **2007**, *54*, 627–638. [CrossRef]
215. Shang, Y.; Wang, H.; Mercaldo, V.; Li, X.; Chen, T.; Zhuo, M. Fragile X mental retardation protein is required for chemically-induced long-term potentiation of the hippocampus in adult mice. *J. Neurochem.* **2009**, *111*, 635–646. [CrossRef] [PubMed]
216. Lundbye, C.J.; Toft, A.K.H.; Banke, T.G. Inhibition of GluN2A NMDA receptors ameliorates synaptic plasticity deficits in the Fmr1−/y mouse model. *J. Physiol.* **2018**, *596*, 5017–5031. [CrossRef] [PubMed]
217. Yun, S.H.; Trommer, B.L. Fragile X mice: Reduced long-term potentiation and N-Methyl-D-Aspartate receptor-mediated neurotransmission in dentate gyrus. *J. Neurosci. Res.* **2011**, *89*, 176–182. [CrossRef]
218. Huber, K.M.; Gallagher, S.M.; Warren, S.T.; Bear, M.F. Altered synaptic plasticity in a mouse model of fragile X mental retardation. *Proc. Natl. Acad. Sci. USA* **2002**, *99*, 7746–7750. [CrossRef]
219. Bear, M.F.; Huber, K.M.; Warren, S.T. The mGluR theory of fragile X mental retardation. *Trends Neurosci.* **2004**, *27*, 370–377. [CrossRef]
220. Eadie, B.D.; Cushman, J.; Kannangara, T.S.; Fanselow, M.S.; Christie, B.R. NMDA receptor hypofunction in the dentate gyrus and impaired context discrimination in adult Fmr1 knockout mice. *Hippocampus* **2012**, *22*, 241–254. [CrossRef]

221. McHugh, T.J.; Jones, M.W.; Quinn, J.J.; Balthasar, N.; Coppari, R.; Elmquist, J.K.; Lowell, B.B.; Fanselow, M.S.; Wilson, M.A.; Tonegawa, S. Dentate gyrus NMDA receptors mediate rapid pattern separation in the hippocampal network. *Science (80-.)* **2007**, *317*, 94–99. [CrossRef]
222. Eadie, B.D.; Zhang, W.N.; Boehme, F.; Gil-Mohapel, J.; Kainer, L.; Simpson, J.M.; Christie, B.R. Fmr1 knockout mice show reduced anxiety and alterations in neurogenesis that are specific to the ventral dentate gyrus. *Neurobiol. Dis.* **2009**, *36*, 361–373. [CrossRef] [PubMed]
223. D'Hooge, R.; Nagels, G.; Franck, F.; Bakker, C.E.; Reyniers, E.; Storm, K.; Kooy, R.F.; Oostra, B.A.; Willems, P.J.; De Deyn, P.P. Mildly impaired water maze performance in male Fmr1 knockout mice. *Neuroscience* **1997**, *76*, 367–376. [CrossRef]
224. Dölen, G.; Osterweil, E.; Rao, B.S.S.; Smith, G.B.; Auerbach, B.D.; Chattarji, S.; Bear, M.F. Correction of Fragile X Syndrome in Mice. *Neuron* **2007**, *56*, 955–962. [CrossRef] [PubMed]
225. Dalton, G.L.; Wang, Y.T.; Floresco, S.B.; Phillips, A.G. Disruption of AMPA receptor endocytosis impairs the extinction, but not acquisition of learned fear. *Neuropsychopharmacology* **2008**, *33*, 2416–2426. [CrossRef]
226. Duan, W.; Wang, K.; Duan, Y.; Chu, X.; Ma, R.; Hu, P.; Xiong, B. Integrated Transcriptome Analyses Revealed Key Target Genes in Mouse Models of Autism. *Autism Res.* **2019**. [CrossRef]
227. Gardoni, F.; Bellone, C. Modulation of the glutamatergic transmission by Dopamine: A focus on Parkinson, Huntington and Addiction diseases. *Front. Cell. Neurosci.* **2015**, *9*. [CrossRef]
228. Calabresi, P.; Picconi, B.; Tozzi, A.; Di Filippo, M. Dopamine-mediated regulation of corticostriatal synaptic plasticity. *Trends Neurosci.* **2007**, *30*, 211–219. [CrossRef]
229. Dunah, A.W.; Wang, Y.; Yasuda, R.P.; Kameyama, K.; Huganir, R.L.; Wolfe, B.B.; Standaert, D.G. Alterations in subunit expression, composition, and phosphorylation of striatal N-methyl-D-aspartate glutamate receptors in a rat 6-hydroxydopamine model of Parkinson's disease. *Mol. Pharmacol.* **2000**, *57*, 342–352.
230. Gardoni, F.; Picconi, B.; Ghiglieri, V.; Polli, F.; Bagetta, V.; Bernardi, G.; Cattabeni, F.; Di Luca, M.; Calabresi, P. A critical interaction between NR2B and MAGUK in L-DOPA induced dyskinesia. *J. Neurosci.* **2006**, *26*, 2914–2922. [CrossRef]
231. Paillé, V.; Picconi, B.; Bagetta, V.; Ghiglieri, V.; Sgobio, C.; Di Filippo, M.; Viscomi, M.T.; Giampà, C.; Fusco, F.R.; Gardoni, F.; et al. Distinct levels of dopamine denervation differentially alter striatal synaptic plasticity and NMDA receptor subunit composition. *J. Neurosci.* **2010**, *30*, 14182–14193. [CrossRef]
232. Mellone, M.; Stanic, J.; Hernandez, L.F.; Iglesias, E.; Zianni, E.; Longhi, A.; Prigent, A.; Picconi, B.; Calabresi, P.; Hirsch, E.C.; et al. NMDA receptor GluN2A/GluN2B subunit ratio as synaptic trait of levodopa-induced dyskinesias: From experimental models to patients. *Front. Cell. Neurosci.* **2015**, *9*, 245. [CrossRef] [PubMed]
233. Durante, V.; De Iure, A.; Loffredo, V.; Vaikath, N.; De Risi, M.; Paciotti, S.; Quiroga-Varela, A.; Chiasserini, D.; Mellone, M.; Mazzocchetti, P.; et al. Alpha-synuclein targets GluN2A NMDA receptor subunit causing striatal synaptic dysfunction and visuospatial memory alteration. *Brain* **2019**, *142*, 1365–1385. [CrossRef] [PubMed]
234. Diógenes, M.J.; Dias, R.B.; Rombo, D.M.; Vicente Miranda, H.; Maiolino, F.; Guerreiro, P.; Näsström, T.; Franquelim, H.G.; Oliveira, L.M.A.; Castanho, M.A.R.B.; et al. Extracellular alpha-synuclein oligomers modulate synaptic transmission and impair LTP via NMDA-receptor activation. *J. Neurosci.* **2012**, *32*, 11750–11762. [CrossRef] [PubMed]
235. Bastide, M.F.; Meissner, W.G.; Picconi, B.; Fasano, S.; Fernagut, P.O.; Feyder, M.; Francardo, V.; Alcacer, C.; Ding, Y.; Brambilla, R.; et al. Pathophysiology of L-dopa-induced motor and non-motor complications in Parkinson's disease. *Prog. Neurobiol.* **2015**, *132*, 96–168. [CrossRef]
236. Perez-Lloret, S.; Rascol, O. Efficacy and safety of amantadine for the treatment of l-DOPA-induced dyskinesia. *J. Neural Transm.* **2018**, *125*, 1237–1250. [CrossRef]
237. Picconi, B.; Centonze, D.; Håkansson, K.; Bernardi, G.; Greengard, P.; Fisone, G.; Cenci, M.A.; Calabresi, P. Loss of bidirectional striatal synaptic plasticity in L-DOPA-induced dyskinesia. *Nat. Neurosci.* **2003**, *6*, 501–506. [CrossRef]
238. Prescott, I.A.; Liu, L.D.; Dostrovsky, J.O.; Hodaie, M.; Lozano, A.M.; Hutchison, W.D. Lack of depotentiation at basal ganglia output neurons in PD patients with levodopa-induced dyskinesia. *Neurobiol. Dis.* **2014**, *71*, 24–33. [CrossRef]
239. Hallett, P.J.; Dunah, A.W.; Ravenscroft, P.; Zhou, S.; Bezard, E.; Crossman, A.R.; Brotchie, J.M.; Standaert, D.G. Alterations of striatal NMDA receptor subunits associated with the development of dyskinesia in the MPTP-lesioned primate model of Parkinson's disease. *Neuropharmacology* **2005**, *48*, 503–516. [CrossRef]

240. Gardoni, F.; Sgobio, C.; Pendolino, V.; Calabresi, P.; Di Luca, M.; Picconi, B. Targeting NR2A-containing NMDA receptors reduces L-DOPA-induced dyskinesias. *Neurobiol. Aging* **2012**, *33*, 2138–2144. [CrossRef]
241. Stanic, J.; Mellone, M.; Napolitano, F.; Racca, C.; Zianni, E.; Minocci, D.; Ghiglieri, V.; Thiolat, M.L.; Li, Q.; Longhi, A.; et al. Rabphilin 3A: A novel target for the treatment of levodopa-induced dyskinesias. *Neurobiol. Dis.* **2017**, *108*, 54–64. [CrossRef]
242. Quartarone, A.; Hallett, M. Emerging concepts in the physiological basis of dystonia. *Mov. Disord.* **2013**, *28*, 958–967. [CrossRef] [PubMed]
243. Quartarone, A.; Rizzo, V.; Terranova, C.; Morgante, F.; Schneider, S.; Ibrahim, N.; Girlanda, P.; Bhatia, K.P.; Rothwell, J.C. Abnormal sensorimotor plasticity in organic but not in psychogenic dystonia. *Brain* **2009**, *132*, 2871–2877. [CrossRef] [PubMed]
244. Maltese, M.; Stanic, J.; Tassone, A.; Sciamanna, G.; Ponterio, G.; Vanni, V.; Martella, G.; Imbriani, P.; Bonsi, P.; Mercuri, N.B.; et al. Early structural and functional plasticity alterations in a susceptibility period of DYT1 dystonia mouse striatum. *Elife* **2018**, *7*, e33331. [CrossRef] [PubMed]
245. Kamenetz, F.; Tomita, T.; Hsieh, H.; Seabrook, G.; Borchelt, D.; Iwatsubo, T.; Sisodia, S.; Malinow, R. APP Processing and Synaptic Function. *Neuron* **2003**, *37*, 925–937. [CrossRef]
246. Hsieh, H.; Boehm, J.; Sato, C.; Iwatsubo, T.; Tomita, T.; Sisodia, S.; Malinow, R. AMPAR Removal Underlies Aβ-Induced Synaptic Depression and Dendritic Spine Loss. *Neuron* **2006**, *52*, 831–843. [CrossRef]
247. Kim, J.H.; Anwyl, R.; Suh, Y.H.; Djamgoz, M.B.A.; Rowan, M.J. Use-dependent effects of amyloidogenic fragments of β-amyloid precursor protein on synaptic plasticity in rat hippocampus in vivo. *J. Neurosci.* **2001**, *21*, 1327–1333. [CrossRef]
248. Li, S.; Hong, S.; Shepardson, N.E.; Walsh, D.M.; Shankar, G.M.; Selkoe, D. Soluble Oligomers of Amyloid β Protein Facilitate Hippocampal Long-Term Depression by Disrupting Neuronal Glutamate Uptake. *Neuron* **2009**, *62*, 788–801. [CrossRef]
249. Hu, N.W.; Klyubin, I.; Anwy, R.; Rowan, M.J. GluN2B subunit-containing NMDA receptor antagonists prevent Aβ-mediated synaptic plasticity disruption in vivo. *Proc. Natl. Acad. Sci. USA* **2009**, *106*, 20504–20509. [CrossRef]
250. Li, S.; Jin, M.; Koeglsperger, T.; Shepardson, N.E.; Shankar, G.M.; Selkoe, D.J. Soluble a β oligomers inhibit long-term potentiation through a mechanism involving excessive activation of extrasynaptic NR2B-containing NMDA receptors. *J. Neurosci.* **2011**, *31*, 6627–6638. [CrossRef]
251. Rönicke, R.; Mikhaylova, M.; Rönicke, S.; Meinhardt, J.; Schröder, U.H.; Fändrich, M.; Reiser, G.; Kreutz, M.R.; Reymann, K.G. Early neuronal dysfunction by amyloid β oligomers depends on activation of NR2B-containing NMDA receptors. *Neurobiol. Aging* **2011**, *32*, 2219–2228. [CrossRef]
252. Texidó, L.; Martín-Satué, M.; Alberdi, E.; Solsona, C.; Matute, C. Amyloid β peptide oligomers directly activate NMDA receptors. *Cell Calcium* **2011**, *49*, 184–190. [CrossRef] [PubMed]
253. Aoki, C.; Lee, J.; Nedelescu, H.; Ahmed, T.; Ho, A.; Shen, J. Increased levels of NMDA receptor NR2A subunits at pre- and postsynaptic sites of the hippocampal CA1: An early response to conditional double knockout of presenilin 1 and 2. *J. Comp. Neurol.* **2009**, *517*, 512–523. [CrossRef] [PubMed]
254. Kessels, H.W.; Nabavi, S.; Malinow, R. Metabotropic NMDA receptor function is required for β-amyloid-induced synaptic depression. *Proc. Natl. Acad. Sci. USA* **2013**, *110*, 4033–4038. [CrossRef]
255. Liu, Y.; Tak, P.W.; Aarts, M.; Rooyakkers, A.; Liu, L.; Ted, W.L.; Dong, C.W.; Lu, J.; Tymianski, M.; Craig, A.M.; et al. NMDA receptor subunits have differential roles in mediating excitotoxic neuronal death both in vitro and in vivo. *J. Neurosci.* **2007**, *27*, 2846–2857. [CrossRef] [PubMed]
256. Marcello, E.; Musardo, S.; Vandermeulen, L.; Pelucchi, S.; Gardoni, F.; Santo, N.; Antonucci, F.; Di Luca, M. Amyloid-β Oligomers Regulate ADAM10 Synaptic Localization Through Aberrant Plasticity Phenomena. *Mol. Neurobiol.* **2019**, *56*, 7136–7143. [CrossRef]
257. Tackenberg, C.; Grinschgl, S.; Trutzel, A.; Santuccione, A.C.; Frey, M.C.; Konietzko, U.; Grimm, J.; Brandt, R.; Nitsch, R.M. NMDA receptor subunit composition determines beta-amyloid-induced neurodegeneration and synaptic loss. *Cell Death Dis.* **2013**, *4*, e608. [CrossRef]

International Journal of
Molecular Sciences

MDPI

Article

Developmental Time Course of SNAP-25 Isoforms Regulate Hippocampal Long-Term Synaptic Plasticity and Hippocampus-Dependent Learning

Katisha R. Gopaul [1,†], Muhammad Irfan [1,2,†], Omid Miry [1], Linnea R. Vose [1], Alexander Moghadam [1], Galadu Subah [1], Tomas Hökfelt [2], Christina Bark [2,*] and Patric K. Stanton [1,*]

1 Department of Cell Biology & Anatomy, New York Medical College, Valhalla, NY 10595, USA; krgopaul@gmail.com (K.R.G.); irfanjadoon@icloud.com (M.I.); omidm7@stanford.edu (O.M.); linnea316@gmail.com (L.R.V.); alexander_moghadam@nymc.edu (A.M.); galadu_subah@nymc.edu (G.S.)

2 Department of Neuroscience, Karolinska Institutet, 171 77 Stockholm, Sweden; Toaffmas.Hokfelt@ki.se

* Correspondence: christina.bark@ki.se (C.B.); patric_stanton@nymc.edu (P.K.S.); Tel. +46-085-248-6984 (C.B.); +1-914-594-4883 (P.K.S.)

† These authors contributed equally to this study.

Received: 31 December 2019; Accepted: 18 February 2020; Published: 20 February 2020

Abstract: SNAP-25 is essential to activity-dependent vesicle fusion and neurotransmitter release in the nervous system. During early development and adulthood, SNAP-25 appears to have differential influences on short- and long-term synaptic plasticity. The involvement of SNAP-25 in these processes may be different at hippocampal and neocortical synapses because of the presence of two different splice variants, which are developmentally regulated. We show here that the isoform SNAP-25a, which is expressed first developmentally in rodent brain, contributes to developmental regulation of the expression of both long-term depression (LTD) and long-term potentiation (LTP) at Schaffer collateral-CA1 synapses in the hippocampus. In one month old mice lacking the developmentally later expressed isoform SNAP-25b, Schaffer collateral-CA1 synapses showed faster release kinetics, decreased LTP and enhanced LTD. By four months of age, SNAP-25b-deficient mice appeared to have compensated for the lack of the adult SNAP-25b isoform, now exhibiting larger LTP and no differences in LTD compared to wild type mice. Interestingly, learning a hippocampus-dependent task reversed the reductions in LTP, but not LTD, seen at one month of age. In four month old adult mice, learning prevented the compensatory up-regulation of LTD that we observed prior to training. These findings support the hypothesis that SNAP-25b promotes stronger LTP and weakens LTD at Schaffer collateral-CA1 synapses in young mice, and suggest that compensatory mechanisms can reverse alterations in synaptic plasticity associated with a lack of SNAP-25b, once mice reach adulthood.

Keywords: SNARE proteins; long-term potentiation; long-term depression; learning and memory; cognition; Schaffer collateral-CA1 synapses

---

1. Introduction

Synaptosomal Associated Protein of 25kDa (SNAP-25) is one of three core SNARE (Soluble N-ethylmaleimide sensitive factor Attachment protein REceptor) proteins that participate in the fusion of the neurotransmitter-containing vesicles with the plasma membrane [1]. These three SNARE proteins interact to form a trimeric core-complex, and in the process overcome the energy barrier for plasma membrane fusion [2]. Protein–protein interactions between the SNARE fusion proteins, and with additional ancillary proteins, can modify the properties and alter vesicle fusion [3,4]. The other core SNARE proteins, synaptobrevin 2 or VAMP2 and syntaxin 1A, are anchored to the membrane via their C-terminal domains. In contrast, SNAP-25 tethers to the membrane via palmitoylation, which is a reversible and dynamic process [5–7], and participates in the formation of coiled–coil quaternary SNARE complexes, where alpha helices of all three proteins wrap around each other. In the nervous system, SNAP-25 plays a major role in neurotransmitter release [8], and during development, SNAP-25 has been suggested to play a role in promoting neurite outgrowth [9,10].

Alternative splicing is a mechanism that increases the diversity profile of protein coding genes. More than 95% of multiexon genes in higher eukaryotes go through alternative splicing [11–13]. SNAP-25 utilizes this process to produce two variants, SNAP-25a, which is the predominant isoform in neurons in mice during embryonic and early postnatal (PN) development, and SNAP-25b, which becomes the dominant isoform as central synapses mature into adulthood. During mRNA formation, the splicing machinery splices in either exon 5a or exon 5b, corresponding to SNAP-25a or SNAP-25b proteins, respectively. While the differences that affect SNAP-25a and SNAP-25b functionality are still being elucidated, there are clear phenotypic differences in mice expressing only SNAP-25a versus predominantly SNAP-25b [14,15]. The two splice variants differ by only nine out of 206 amino acids [16–19]. However, SNAP-25a is the major isoform peripherally in endocrine and neuroendocrine excitable cells throughout life and certain hypothalamic structures, but this still needs further thorough investigation [18,20,21]. Moreover, the differences between SNAP-25a and SNAP-25b span regions of the N-terminal amphipathic helix and the cysteine-rich linker region, spacing the two (N- and C-terminal) helices [19,22]. Likely, the different SNAP-25 isoforms change the tertiary structure of the core SNARE complex and thereby modify the affinities of SNARE-binding partner proteins. Additionally, SNAP-25 isoforms differentially interact with accessory proteins such as Munc18-1 and $G\beta\gamma$ subtypes [23].

Within the first week of life, in the mouse hippocampal formation, SNAP-25a is the dominant isoform that participates in vesicle fusion and development [18,24]. As mice mature, SNAP-25b and overall SNAP-25 levels increase dramatically in the entire brain. By three weeks of age, the level of SNAP-25b is more than 50% greater than SNAP-25a in the hippocampal formation. Additionally, between weeks 3 and 8, likely the time of synaptic maturation, there is a dramatic increase in the total amount of SNAP-25.

During the first two weeks of development, when SNAP-25a is the predominant isoform in brain, rodents exhibit predominantly, or entirely, metabotropic glutamate receptor (mGluR)-dependent forms of long-term depression (LTD) of synaptic transmission, while long-term potentiation (LTP) has yet to appear [25–28]. The timing of this change in relative levels of SNAP-25a to SNAP-25b mirrors the developmental upregulation of N-methyl-D-aspartate receptors (NMDARs), suggesting that SNAP-25a may play a greater role in the expression of mGluR-dependent LTD, while SNAP-25b may be more important for expression of NMDAR-dependent LTD and LTP. It is possible, but still undetermined, whether the switch in expression from SNAP-25a to SNAP-25b confers this ability to express LTD and LTP in older animals.

We have previously reported that, at four weeks of age, SNAP-25b-deficient mice demonstrate significantly reduced LTP and less ability to discriminate between intensities of presynaptic stimuli [29]. To evaluate the role of SNAP-25 isoforms on later developmental stages of synaptic transmission and plasticity, we studied this same strain of SNAP-25b-deficient mice, using one month and four month old animals, to compare and assess the early developmental and later adult impact of the absence of SNAP-25b. Normally, the switch in SNAP-25 mRNA isoforms happens during postnatal weeks 1–3 in the hippocampal formation, so in one month old wild type (WT) mice, a complete switch of the SNAP-25 proteins has occurred. We compared the magnitude of hippocampal LTP and LTD in in vitro hippocampal slices, and learning acquisition and memory retention in vivo, at these different ages, to assess the early developmental and long-term adult impact of the absence of SNAP-25b on synaptic plasticity and cognitive function.

2. Results

2.1. Vesicular Release Probability is Increased in One Month Old SNAP-25b-Deficient Mice

SNAP-25a and SNAP-25b are functionally different in their ability to facilitate exocytosis following association with the other core SNARE proteins in the SNARE complex [30]. Therefore, we analyzed whether SNAP-25a-only expressing mice exhibited functional differences in vesicular transmitter release properties by direct imaging of the release kinetics of the styryl dye FM1-43 using two-photon laser scanning microscopy. We compared SNAP-25b-deficient mice to wild type (WT) controls and assessed the effects of the lack of SNAP-25b on release kinetics at glutamatergic Schaffer collateral-CA1 terminals in hippocampal slices.

To directly assess neurotransmitter release probability, Schaffer collateral presynaptic terminal vesicles were loaded with FM1-43, and the time course of fluorescent destaining was monitored in response to stimulus-evoked release of the dye using two-photon laser scanning microscopy. Schaffer collateral axons were stimulated with 0.1 Hz frequency to evoke neurotransmitter release. Schaffer collateral terminals in field CA1 of hippocampal slices from SNAP-25b-deficient mice showed significantly faster neurotransmitter release kinetics compared to the WT controls (Figure 1A), as measured by a faster rate of fluorescence decay of FM1-43. Fitting of destaining curves (Figure 1B) with a single exponential decay function Y = (Y0-NS)*exp(-K*X)+NS (Y = total vesicle brightness, Y0 = brightness at time 0, X = time, NS = brightness after end of stimulation period, K = rate constant in inverse units of X, half-life = 0.69/K) revealed significantly smaller half-life (13.7) and smaller decay time constant tau (0.05) values for SNAP-25b-deficient mice compared to WT half-life (23.3) and tau (0.031). Comparison of the two fits with extra sum-of-squares *F* test for the dissociation time constants (K) showed significant difference between SNAP-25b-deficient and wild type slices ($p < 0.01$).

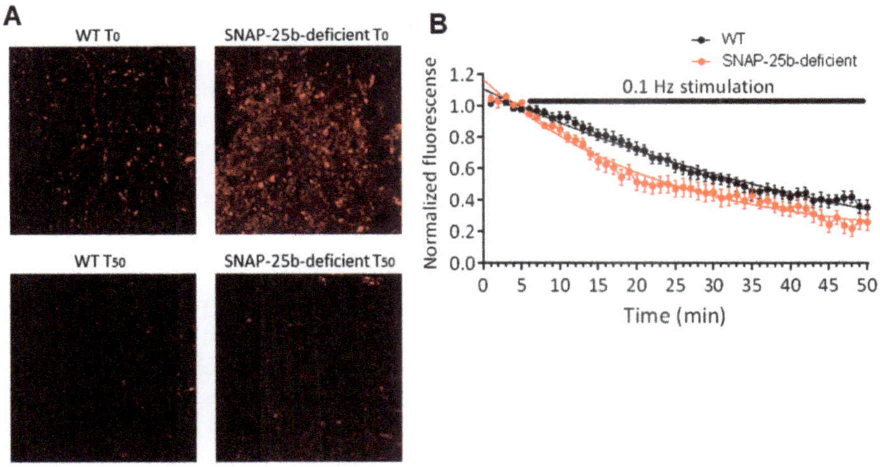

Figure 1. Presynaptic vesicular release from Schaffer collateral terminals in one month old SNAP-25b-deficient mice. (**A**) Representative fluorescent images of hippocampal field CA1 stratum radiatum in slices from WT and SNAP-25b-deficient mice prior to synaptic stimulation (T_0) and after 50 min of 2Hz stimulation (T_{50}). Bright puncta are presynaptic Schaffer collateral terminals. (**B**) Fluorescence decay of vesicular dye FM1-43 from Schaffer collateral-CA1 presynaptic boutons elicited by 0.1 Hz stimulation (black bar). SNAP-25b-deficient mice (filled circles, $n = 20$ puncta from 4 slices) showed a significantly faster rate of FM1-43 decay compared to wild type mice (open circles, $n = 18$ puncta from 6 slices). Each point is mean ± SEM of 'n' puncta.

2.2. *Long-Term Potentiation and Long-Term Depression of Synaptic Strength*

A high-frequency theta-burst stimulus (TBS) was used to induce stimulus-dependent LTP (sLTP) at Schaffer collateral-CA1 synapses in the hippocampal slices from one and four month old mice. At one month of age, sLTP was reduced in SNAP-25b-deficient mice compared to age-matched WT littermate controls (Figure 2A). The reduced expression of LTP suggests an important role for the missing SNAP-25b isoform in the expression of LTP. However, there clearly is LTP retained that does not require presence of SNAP-25b.

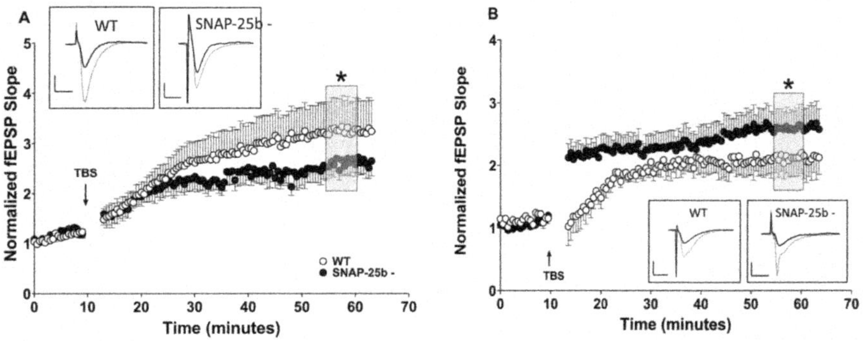

Figure 2. Alterations in stimulus-evoked LTP in SNAP-25b-deficient mice of both sexes at one and four months of age. (**A**) Theta burst stimulation (TBS)-induced LTP (burst data points not shown) at one month in hippocampal Schaffer collateral synapses. Arrows indicate application of TBS in each panel. Young SNAP-25b deficient mice expressed lower sLTP (filled circles, $n = 13$) compared to age-matched controls (open circles, $n = 12$). (* $p < 0.01$; Student's t-test). Each point is normalized to the averaged baseline and is mean ± SEM of n slices. (**B**) Time course of LTP induced by TBS in littermate controls (open circles, $n = 9$) and SNAP-25b-deficient mice (filled circles, $n = 9$) in slices from 4 month old mice. Older SNAP-25b-deficient mice show enhanced LTP with an altered PTP compared to littermate controls (* $p < 0.01$; Student's t-test). Each point is normalized to the averaged baseline and is mean ± SEM of n slices. LTP was calculated as mean of grey box, compared to pre-TBS normalized baseline. Insets: Representative Schaffer collateral-evoked fEPSPs in WT and SNAP-25b deficient slices immediately before (black traces) and 50 min after (grey traces TBS stimulation, with calibration bars in (**A**) 1mV/10msec, in (**B**) 2mV/10msec.

In contrast to slices from one month old mice, at four months of age SNAP-25b-deficient mice displayed LTP that was significantly larger than LTP in WT control mice (Figure 2B). In addition to expressing augmented LTP, four month old SNAP-25b deficient mice exhibited larger rapid post-tetanic potentiation of fEPSPs (PTP) shortly after induction of high-frequency stimulation (HFS), in contrast to the reduced PTP in one month old mice. While one month old mice lacking SNAP-25b displayed reduced LTP, the older animals, exhibited enhanced LTP compared to age-matched controls. This is clear evidence for presynaptic compensatory mechanisms that upregulate LTP in absence of the switch from SNAP-25a to SNAP-25b.

Due to the essential roles posited for both LTP and LTD in learning and memory, we also examined stimulus-evoked LTD of synaptic transmission (sLTD) in mice lacking SNAP-25b at one and four months of age. At one month, but not four months, of age, hippocampal slices from SNAP-25b-deficient mice showed enhanced sLTD induced by a low-frequency stimulation (LFS; 2Hz/10min), compared to littermate controls (Figure 3A). Taken together, our LTP and LTD data in one month old animals suggest that SNAP-25a favors the expression of sLTD over sLTP in the developing brain, consistent with a previous study [29] in which we showed that LTP at Schaffer collateral-CA1 synapses is reduced in both male and female SNAP-25b-deficient mice at one month of age. In contrast, four month old

SNAP-25b deficient mice displayed sLTD that did not differ from their littermate controls (Figure 3B). These results also suggest that compensatory effects can restore both LTP and LTD to control levels and balance in adult mice lacking SNAP-25b.

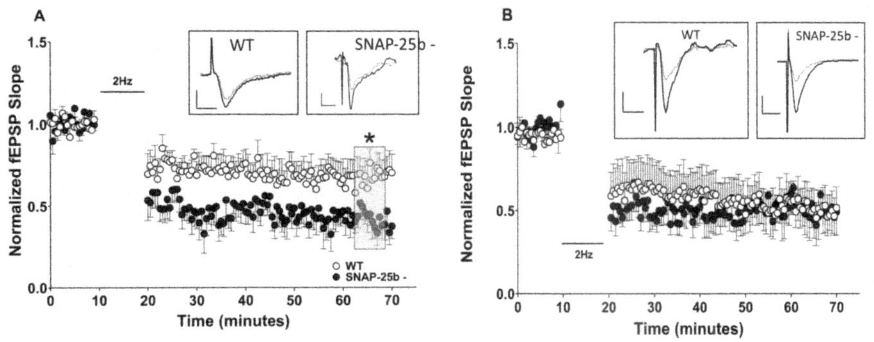

Figure 3. Changes in LTD at Schaffer collateral-CA1 synapses in hippocampal slices from one and four month old SNAP-25b-deficient mice of both sexes. (**A**) LTD at one month of age at Schaffer collateral-CA1 synapses induced by a 2Hz LFS. SNAP-25b-deficient mice (filled circles, $n = 8$) exhibited significantly greater LTD compared to controls (open circles, $n = 7$). Bar indicates application of LTD stimulus (2Hz/10min, data points not shown). (* $p < 0.01$; Student's t-test). Each point is normalized to the averaged baseline and is mean ± SEM of n slices. (**B**) After prolonged absence of SNAP-25b, four month old SNAP-25b-deficient mice (filled circles, $n = 8$) displayed LTD similar to WT littermates (open circles, $n = 8$). Each point is normalized to the averaged baseline and is mean ± SEM of n slices. LTD was calculated as mean of grey box, compared to pre-TBS normalized baseline. Insets: Representative Schaffer collateral-evoked fEPSPs in WT and SNAP-25b deficient slices immediately before (black traces) and 50 min after (grey traces) LBS stimulation, with all calibration bars 1mV/10msec.

2.3. Place Avoidance Spatial Learning

A recent study from our group found, in an active avoidance spatial learning test, that the same one month old SNAP-25b-deficient mice showed impaired learning acquisition rates, and slower extinction of learning once acquired, suggesting weaker initial learning and greater flexibility for relearning new contingencies [29]. Our current findings suggest that compensatory mechanisms have returned LTD and LTP to near control levels by four months of age in mice lacking the adult SNAP-25b isoform, leading us to test whether behavioral phenotypes found in younger mice lacking SNAP-25b are also compensated in adulthood.

To evaluate the behavioral phenotypes associated with a lack of SNAP-25b throughout developmental into adulthood, we used an active place avoidance assay developed by Fenton and colleagues [31]. In this task, rodents are placed on a turning metal grid platform and, using spatial cues, must learn to move to avoid a shock given when the animal enters one segment of the circular field (Figure 4A). Littermate controls and SNAP-25b-deficient mice were subject to multiday habituation, training trials, extinction and conflict discrimination as described [31]. In contrast to deficits in this learning, as previously observed in one month old mutant mice, adult mice lacking SNAP-25b showed no differences in the initial learning phase in days 1–3 (Figure 4B). However, after extinction (day 4), when the shock zone was shifted 180° from its initial position, SNAP-25b-deficient mice entered the new shock zone fewer times than controls, suggesting a more rapid relearning of the new shock zone location. This may reflect a behavioral learning flexibility. No difference in anxiety-like behavior was detected in four month old SNAP-25b deficient mice (Figure 4C), using an elevated plus maze. Increased anxiety observed at one month in our previous study was no longer detected in older SNAP-25b-deficient mice. Finally, motor function as assessed by total path length was not different in SNAP-25b-deficient mice and WT littermates (Figure 4D). These data indicate that adult

SNAP-25b-deficient mice have compensated for the impairments in synaptic plasticity and learning at one month of age, consistent with the shift from a reduction in LTP at one month of age, to larger LTP at four months of age, and the 'renormalization' of the magnitude of LTD reached by adulthood.

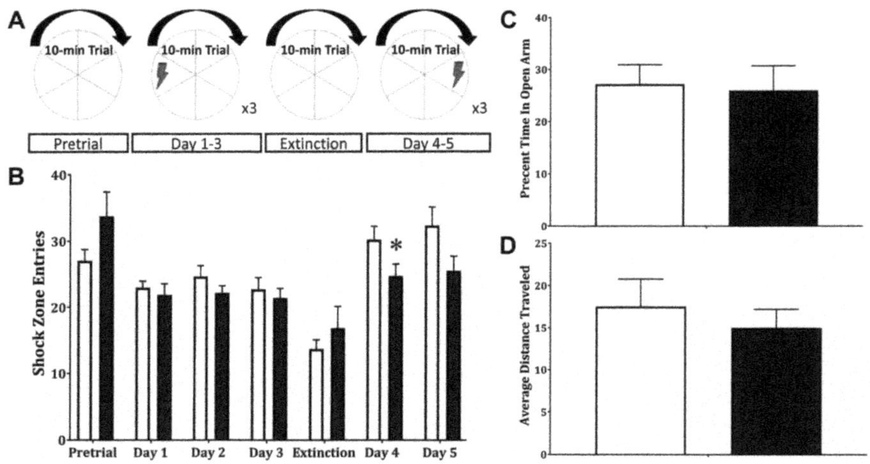

Figure 4. Active avoidance initially shows altered cognitive flexibility in adult SNAP-25b-deficient mice of both sexes with no difference in anxiety-like behaviors or locomotion. (**A**): Active avoidance training schedule. Pretrial and Extinction: 1–10 min trial without shock. Initial learning: 3 days of 3–10 min trails/day with a left shock zone. Conflict learning: 2 days of 3–10 min trials/day with a 180° shifted shock region. (**B**): Active avoidance assay average daily counts of shock zone entries. Four month old SNAP-25b-deficient mice (black bars, $n = 12$) learned to avoid the original shock zone as well as controls (white bars, $n = 17$). After extinction, four month old SNAP-25b-deficient mice initially learned faster and entered the new shock zone fewer times than WT littermates (day 4; $p < 0.05$; one-way ANOVA for repeated measures). (**C**): The same four month old mice deficient in SNAP-25b (black bar, $n = 12$) showed no difference in time spent in the open arm of the elevated plus maze (EPM) relative to controls (white bar, $n = 17$). Therefore, no difference in anxiety-like behavior was detected. (**D**): These older SNAP-25b-deficient mice (black bar, $n = 12$), compared to WT littermates (white bar, $n = 17$), also showed no difference in general locomotion, indicated by total distance traveled during EPM assessment. Each point is mean + SEM of n animals.

2.4. *Effects of Learning Acquisition on Long-Term Synaptic Plasticity*

In a cohort of mice that went through active place avoidance behavioral testing, sLTP and sLTD were assessed post-training, to determine the impact of hippocampal-dependent training of place avoidance on hippocampal synaptic transmission and long-term activity-dependent plasticity in SNAP-25b-deficient mice. Once mice completed all five days of the training paradigm, they were sacrificed and used for slice electrophysiology recordings. In contrast to untrained mice, one month old SNAP-25b-deficient mice exhibited enhanced sLTP compared to littermate controls (Figure 5A). Prior to training, the same SNAP-25b-deficient mouse line had displayed reduced LTP at Schaffer collateral-CA1 synapses. While the response to HFS post-training was larger in SNAP-25b-deficient mice at one month of age, the magnitude of sLTP was not altered by behavioral training in four month old SNAP-25b deficient mice (Figure 5B).

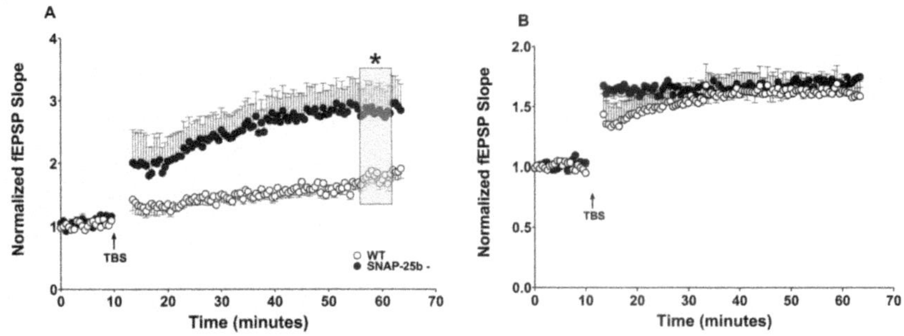

Figure 5. Hippocampal dependent training alters LTP in adolescent SNAP-25b-deficient mice of both sexes. (**A**): Time course of TBS (arrows) used to induce LTP in younger mice. After completing an active avoidance assay, adolescent mice lacking SNAP-25b showed enhanced LTP (filled circles, $n = 7$) relative to controls (open circles, $n = 11$). (* $p < 0.01$; Student's t-test). Each point is normalized to the averaged baseline and is mean ± SEM of n slices; (**B**): TBS LTP (arrows) recorded from area CA1 in four month old mice that had completed a multiday active avoidance training. SNAP-25b-deficient mice (filled circles, $n = 18$) expressed LTP at levels similar to controls (open circles, $n = 18$). However, SNAP-25b deficient mice did show significantly larger PTP. Each point is normalized to the averaged baseline and is mean ± SEM of n slices.

Unlike adolescent mice lacking SNAP-25b, sLTD was altered in four month old adult SNAP-25b-deficient mice compared to WT littermate controls only after the passive avoidance spatial learning task (Figure 6B). Prior to training, four month old SNAP-25b-deficient mice did not show altered sLTD, while younger mice showed the same enhancement in sLTD relative to controls both before and after passive avoidance training. These data show that, in one month old mice lacking SNAP-25b, reductions in LTP and increases in LTD are both present, while by four months of age, compensatory mechanisms have returned LTD levels to normal. Nevertheless, the effect of learning acquisition on adult mice expressing only the juvenile SNAP-25a isoform was to upregulate both LTP and LTD at Schaffer collateral-CA1 synapses.

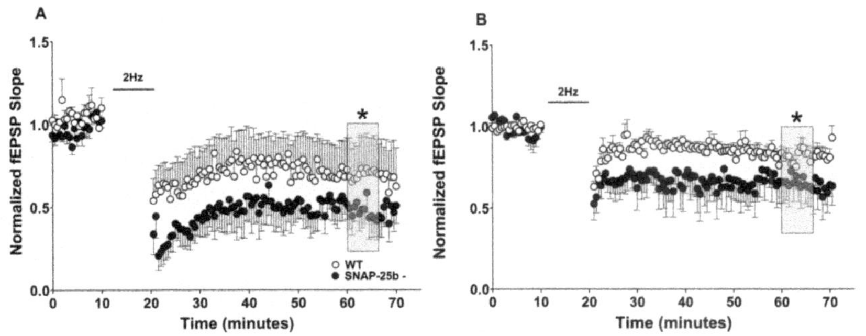

Figure 6. LTD is enhanced in young and old SNAP-25b-deficient mice of both sexes by hippocampal-dependent training. (**A**) One month old SNAP-25b-deficient mice (filled circles, $n = 7$) exhibited significantly larger sLTD compared to WT littermates (open circles, $n = 4$). (*, $p < 0.01$; Student's *t*-test). Bar indicates 2Hz/10mins LTD stimulus application. Each point is normalized to the averaged baseline and is mean ± SEM of n slices; (**B**) Four month old SNAP-25b-deficient mice (filled circles, $n = 6$) also exhibited greater sLTD than WT controls (open circles, $n = 13$, $p < 0.05$; Student's t-test). Bar indicates 2 Hz/10 min LTD LFS stimulus train. Each point is normalized to the averaged baseline and is mean ± SEM of n slices.

2.5. mGluRII and NMDA Receptor-Dependent LTD of Synaptic Transmission

We investigated if changes in activation of glutamate receptors might be responsible for differences observed in synaptic plasticity resulting from a lack of SNAP-25b. Multiple reports indicate an essential role for SNAP-25 in transient presynaptic suppression of transmitter release by the 5-HT serotonin receptor at lamprey synapses [32], and in mGluRII-dependent chemical- and stimulus-evoked LTD at Schaffer collateral-CA1 hippocampal synapses. Evidence points to an interaction between the C-terminus of SNAP-25 with the G-protein Gβγ being necessary for suppression of neurotransmitter release at all of these synapses. C-terminus cleavage of SNAP-25 by botulinum toxin A prevents Gβγ and SNAP-25 interaction necessary for depression of transmitter release [32,33], as does infusion of a C-terminal fragment of SNAP-25 that binds to Gβγ [34]. Given the role of these glutamate receptors in the induction of LTD of presynaptic transmitter release, we tested whether there were differences in the expression of mGluRII or NMDAR chemical LTD (cLTD) produced by a lack of SNAP-25b. To evaluate the involvement of each form of presynaptic cLTD and how they interface with SNAP-25 to affect presynaptic vesicular release, mGluRII- and NMDAR-dependent LTD were selectively induced in hippocampal slices from one and four month old mice lacking SNAP-25b, by bath application of either NMDA (20μM) or the mGluRII agonist DCG-IV (25 μM).

In mGluRII cLTD, lack of SNAP-25b did not alter the amplitude of mGluRII cLTD in either one (Figure 7A) or four month old (Figure 7B) SNAP-25b-deficient mice compared to WT littermate controls, indicating that molecular mechanisms downstream of synaptic stimulation underlie the induction of mGluRII-dependent presynaptic LTD. Moreover, they are not differentially regulated by the two isoforms of SNAP-25.

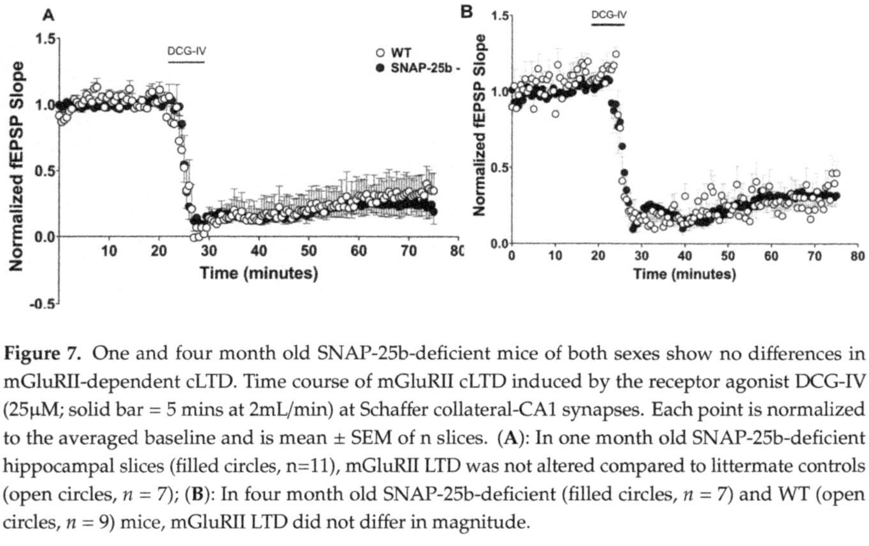

Figure 7. One and four month old SNAP-25b-deficient mice of both sexes show no differences in mGluRII-dependent cLTD. Time course of mGluRII cLTD induced by the receptor agonist DCG-IV (25μM; solid bar = 5 mins at 2mL/min) at Schaffer collateral-CA1 synapses. Each point is normalized to the averaged baseline and is mean ± SEM of n slices. (**A**): In one month old SNAP-25b-deficient hippocampal slices (filled circles, n=11), mGluRII LTD was not altered compared to littermate controls (open circles, $n = 7$); (**B**): In four month old SNAP-25b-deficient (filled circles, $n = 7$) and WT (open circles, $n = 9$) mice, mGluRII LTD did not differ in magnitude.

NMDAR cLTD also exhibited the same amplitude of fEPSP in one month old SNAP-25b-deficient mice as in control mice (Figure 8). If components of the NMDAR cascade interact with SNAP-25, this interaction does not appear to be influenced by a lack of SNAP-25b in adolescent mice. Further, we found that NMDAR-dependent LTD could not be elicited in either mutant or control mice by bath application of NMDA at four months of age, indicating that synaptic stimulation is an essential component for the induction of NMDAR-dependent LTD in adult mice, since the majority of stimulus-evoked LTD is blocked by NMDAR antagonists. Taken together, these findings suggest that the 'a' and 'b' isoforms of SNAP-25 differentially regulate the induction of LTD through regulation of the magnitude and patterns of synaptic stimulation and through its effects on presynaptic terminals that require synaptic activity, rather than by altering the downstream activation of glutamate receptors, either post- or presynaptically.

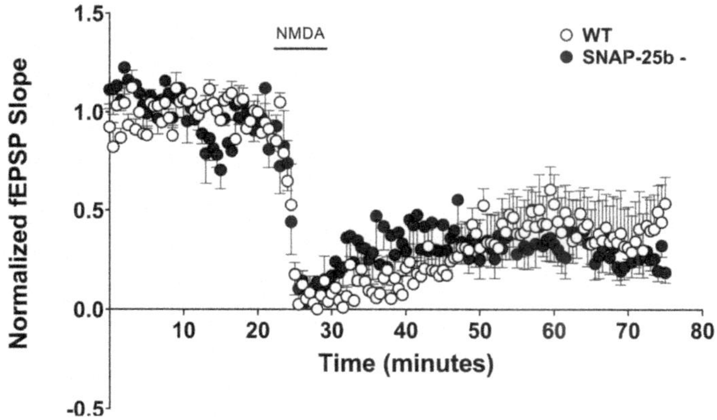

Figure 8. Differences observed in synaptic plasticity in one month old SNAP-25b-deficient mice of both sexes were not due to alterations in NMDAR response to glutamate. The presence of SNAP-25a in mutant mice (filled circles, n = 5) did not alter the extracellular amplitude of NMDA-induced cLTD (20 μM; solid bar = 5 mins at 2mL/min) compared to WT littermates (open circles, n = 7) at Schaffer collateral-CA1 synapses. Each point is normalized to averaged baseline and is mean ± SEM of n slices.

3. Discussion

SNAP-25 is an essential SNARE protein and a regulatory target for the expression of short- and long-term plasticity of presynaptic transmitter release. The individual contributions of each isoform differ in regulation of synaptic transmission and activity-dependent long-term synaptic plasticity [29]. Given the importance of LTP and LTD in activity-dependent networks that underlie learning and memory [35,36], we assessed alterations in LTP, LTD and behavioral learning in SNAP-25b-deficient mice, compared to WT mice that exhibited normal developmental regulation of the 'a' and 'b' isoform. Although relative expression levels of SNAP-25 isoforms throughout different brain regions are not yet fully evaluated, the switch from SNAP-25a to SNAP-25b is well described in the hippocampus. We used Schaffer collateral-CA1 synapses in hippocampal slices to assess LTP and LTD, and a hippocampus-dependent behavioral task, to correlate the observed changes with expression of SNAP-25a or SNAP-25b as the dominant isoforms at that time in development.

The reduced level of LTP and enhanced magnitude of LTD expressed in one month old SNAP-25b-deficient mice relative to littermate controls suggests that SNAP-25a favors the induction of LTD over LTP, and is consistent with our previous study showing that both male and female SNAP-25b-deficient mice exhibited significantly smaller LTP [29]. By preventing a down regulation of LTD during development, the absence of SNAP-25b could prevent the shift in synaptic plasticity favoring larger LTP that normally occurs around three weeks of age. During the first four weeks of development, WT mice display a dramatic increase in SNAP-25b mRNA expression, and this corresponds to a developmental window for complete maturation of synaptic connections. Therefore, the loss of SNAP-25b could alter synaptic maturation by modifying activity-dependent synaptic plasticity. In this study, SNAP-25b-deficient mice showed a delayed progression in normal expression levels of LTP and LTD. It is probable that, without the dramatic increase in SNAP-25b, which markedly decreases the ratio of SNAP-25a to SNAP-25b, one month old SNAP-25b-deficient mice were more like younger WT mice. In the latter mice the ratio of SNAP-25a/SNAP25b is still high, and LTD is favored over LTP. This observation is in line with previous reports that SNAP-25a is less efficient at coordinating bursts of vesicular release [30]. The results of our study support the conclusion that the switch from SNAP-25a to SNAP-25b may be a key mediator of the developmental switch in relative magnitudes of LTD and LTP seen in the hippocampus during development.

By four months of age, SNAP-25b-deficient mice appeared to have over-compensated for the lack of SNAP-25b. In the process of up-regulating the strength of LTP to match adult WT LTP levels, hippocampal Schaffer collateral-CA1 synapses in SNAP-25b-deficient brain slices exhibited enhanced LTP relative to WT controls. It is intriguing that these synapses also showed larger PTP immediately after HFS. The increase in PTP may be due to increased residual calcium in the presynaptic terminal after HFS, suggesting the possibility of alterations in presynaptic calcium regulation that increase vesicular release immediately after HFS. An enhancement of both PTP and LTP could also signify an upregulation of NMDAR-gated postsynaptic currents. In particular, NR2B subunit containing NMDARs can be upregulated, resulting in enhanced LTP [37,38].

In contrast, levels of LTD in these older SNAP-25b-deficient mice were similar to controls, suggesting that, if NR2B NMDAR subtypes are lesser participants in the induction of LTD, presynaptic sites of regulation may account for compensatory shifts in expression of LTD in older mice. With regard to the amplitude of LTD, SNAP-25b-deficient mice were able to accurately compensate for the enhancement seen in younger mice, bringing LTD back to levels observed in littermate controls. Taken together, findings at one and four months of age confirm the hypothesis that the developmental shift from SNAP-25a to SNAP-25b is an important factor in regulating the relative magnitude of sLTP and sLTD in the developing and adult brain. Findings from previous studies and the present study (highlighted in grey) of these SNAP-25b mice at 1–4 months of age are summarized in Table 1.

It is unclear whether compensatory mechanisms underlying shifts in LTP and LTD result from a lack of SNAP-25b or over-expression of SNAP-25a. Although the adolescent 'a' isoform is less effective at coordinating vesicle fusion, it can rescue SNAP-25 null neurons in culture and coordinate synchronous stimulus evoked release [39]. The absence of SNAP-25b may promote enhanced functionality or increased recruitment of SNAP-25a or altered association of SNAP-25a with accessory proteins as mice age. Munc18-1, an accessory SNARE protein which promotes SNARE complex formation, also interacts with SNAP-25 and binds to SNAP-25b more easily than SNAP-25a [23]. Differential interaction of SNAP-25 isoforms with Munc18-1 could potentially alter the rate of SNAP-25a incorporation into SNARE complexes during vesicle priming. As mice age and more SNAP-25a is incorporated into SNARE complexes, more vesicles may be accessible for synaptic transmission and account for the compensatory effects seen in older SNAP-25b-deficient mice. To test this possibility, it would be helpful to investigate changes in the function and size of the readily-releasable vesicle pool (RRP) in SNAP-25b-deficient mice at one and four months, using imaging of FM1-43 release and electron microscopy. Alternatively, these compensatory mechanisms may be postsynaptic in origin, involving changes such as up-regulation of NR2B-NMDAR expression to compensate for continued impairments in presynaptic function.

At the same time, SNARE proteins can also play significant roles in postsynaptic vesicle-mediated trafficking, including trafficking of NMDAR [40]. Hussain and colleagues have recently shown that SNAP-25 is also present in the postsynaptic density, although its location was not altered one hour after induction of LTP [41]. At this time, an additional postsynaptic role for SNAP-25 in developmental regulation of long-term synaptic plasticity cannot be ruled out.

Table 1. Developmental Effects of SNAP-25b Deficiency.

		Developmental Effects of SNAP-25b Deficiency			
Postnatal age	LTP	LTD	Basal synaptic transmission (I/O, PPF, Pr)	Active Avoidance Learning	Metabolic and Expression Effects
1 month old	Reduced LTP (1,2)	• Larger LTD (2) • No diff in mGluRII-dependent LTD (2) • No diff in NMDAR-dependent LTD (2)	• Higher ceiling fEPSP baseline amplitude (1) • Reduced PPF and increased Pr in males, but not females (1) • Faster presynaptic vesicular release (2)	• Impaired (1) • Enhanced LTP, but did not alter LTD (remained larger than controls)	• WT females higher SNAP-25a expression levels than WT males, suggesting SNAP-25b switch may occur later in females
2 month old	Reduced LTP (1)		• Reduced PPF and increased Pr in males, but not females • (1)		• Decreased interaction of SNAP-25 with MUNC-18 and Gβ_2 subnits (3) • Increased insulin secretion from β–islet cells (4) • Slowed/prolonged glucose-stimulated Ca^{2+} increases and desynchronized oscillations in β–islet cells (4) • Metabolic disease (obesity, hyperglycemia, liver steatosis, adipocyte hypertrophy) and enhanced response to a high-fat diet (5)
4 month old	• Increased LTP (2)	• No diff in LTD (2) • No diff in mGluRII-dependent LTD (2)		• Rescued LTP to control levels and LTD larger than controls (2)	

CITATIONS: (1) Irfan et al., Scientific Reports, 9:6403,2019; (2) Gopaul et al., this study; (3) Daraio et al., Neurosci Lett, 674: 75–80; (4) Daraio et al., Scientific Reports, 7:7744, 2017; (5) Valladolid-Acebes et al., PNAS, 112: E4326–4335. Abbreviations: LTP (long-trem potentiation), LTD (long-term depression), mGluR1 (metabotropic glutamate receptor 1), NMDAR (N-methyl-D-aspartate receptor), PPF (paired-pulse facilitation), I/O (input/output relations), Pr (release probability), WT (wild type).

To assess whether hippocampal-dependent learning and memory is affected by lack of the adult SNAP-25b isoform throughout development, hippocampal-dependent spatial learning was evaluated in four month old SNAP-25b-deficient mice using an active avoidance assay. Four month old SNAP-25b-deficient mice initially showed no differences in acquisition of learned place avoidance, but they did show enhanced conflict learning immediately after extinction, i.e., a more rapid shift away from a previous shock zone to learn a new zone. On the second day of conflict reversal learning, there was only a nonsignificant trend towards fewer shock zone entries for the mice lacking SNAP-25b. The possibility exists that, since older SNAP-25b-deficient mice did not experience the initial delay in learning we observed in younger mice, their underlying memory deficit had largely disappeared as compensatory mechanisms were expressed during development. Given that, at one month of age, this same mouse line showed impaired cognitive learning, we hypothesize that enhanced LTD in these adolescent animals accounted for their performance during the relearning phase of the test, and both LTD and learning had returned to WT levels in adulthood. The hypothesis of compensatory changes in older SNAP-25b-deficient mice is also supported by our earlier study, in which SNAP-25b-deficient mice exhibited higher escape latencies in the hippocampal-dependent Morris water maze behavioral task, although anxiety levels were significantly higher in the elevated plus maze test [29].

To address any underlying memory deficit not detected by the active avoidance assay, additional behavioral learning and memory paradigms will be needed to test hippocampus-dependent and -independent forms of retention and relearning, to determine if behavioral flexibility is altered by the subtle differences in presynaptic function afforded by the isoforms of SNAP-25. For example, it is possible to increase the complexity of the spatial avoidance learning task by adding a rotating shock segment with the stationary segment during the initial learning phase. Mice expressing only SNAP-25a exhibited a higher rate of neurogenesis, as shown by a higher number of doublecortin-positive precursor cells [14]. Increased neurogenesis may lead to improved performance on cognitive flexibility tasks and could serve as an alternative explanation for the results observed here.

Prior to training in the active avoidance task, mice were monitored for anxiety. The time spent in the open arm of the elevated plus maze relative to total time in the apparatus was used as an indicator of anxiety. At four months of age, SNAP-25b-deficient animals were no different from controls and showed comparable levels of locomotion, indicating that our spatial learning findings cannot be attributed to alterations in motivation or motor function.

Some of the same mice that went through active avoidance behavioral testing, referred to as trained mice, were then monitored for changes in sLTP and sLTD. Trained one month old SNAP-25b-deficient mice expressed enhanced, instead of reduced, LTP, but still showed the larger LTD relative to littermate controls that we observed in untrained mice. The relative switch in the strength of LTP in trained versus untrained mice lacking SNAP-25b suggests that the act of training functionally improved synaptic plasticity, or that SNAP-25b-deficient mice demonstrate less LTP elicited by active avoidance training itself, leaving more room below ceiling levels for LTP to be elicited by HFS.

In contrast to adolescent mice, adult trained mice exhibited LTP similar in magnitude to controls, but LTD was still significantly enhanced in magnitude. In untrained SNAP-25b-deficient mice at one month of age, LTP was lower than controls. However, by four months of age, untrained SNAP-25b-deficient mice showed enhanced LTP. Since adolescent, untrained SNAP-25b-deficient mice exhibited smaller LTP than controls, compensatory mechanisms may have been activated. This may have resulted in overshooting the normal homeostatic set point in naive mice, leading to larger LTP in untrained SNAP-25b-deficient mice. Conceivably, the training paradigm could have activated cascades that reset synapses to normal levels of LTP, and a secondary consequence of this compensation to adjust LTP may have caused an enhancement of LTD. Moreover, the difference seen in SNAP-25b-deficient mice post training could have been due to a shift in the ability to regulate the sliding threshold of the synapse. If the sliding threshold scale is being altered in SNAP-25b-deficient mice as a compensatory mechanism to accommodate the deficits in synaptic plasticity, it could imply that SNAP-25 function contributes to a synapse's ability to regulate metaplasticity or the threshold for changing synaptic strength.

To evaluate the effects of lack of SNAP-25b on baseline transmitter release, we measured presynaptic neurotransmitter release with FM1-43. The destaining kinetics of this vesicle-specific dye indicated a faster rate of neurotransmitter release. This result can either be interpreted as i) higher baseline release probability in mice lacking SNAP-25b or ii) a smaller pool of RRP, as described by [30].

Impaired transmitter release mechanisms may result in memories that are not as persistent and stable as in WT mice, making them more susceptible to reversal in conflict learning assays. Different transmitter release probabilities are suggested to impair the induction of LTP, resulting in weaker synaptic strengthening and memory formation. Therefore, extinction in adult SNAP-25b-deficient mice may proceed faster than in control mice that express adult SNAP-25b.

We have previously shown that mutations in the C-terminus of SNAP-25 in a gene targeted mouse model enhances the magnitude of LTP at Schaffer collateral-CA1 synapses [42]. This is because of the reduced interaction between SNAP-25 and the inhibitory $G_{i/o}$ proteins [43]. The absence of SNAP-25b did not alter the amplitude of mGluRII-dependent LTD measured at Schaffer collateral-CA1 synapses when DCG-IV was bath applied, or NMDAR-dependent LTD when NMDA was bath applied to hippocampal slices from such mice compared to four month old littermates. This suggests that SNAP-25a and Gβγ can interact to induce cLTD in adult SNAP-25b-deficient mice with similar affinity as in WT mice during both adolescence and adulthood. These results are not consistent with recent immunoprecipitation data showing that SNAP-25a-containing SNARE complexes associate with Gβγ 50% less than SNAP-25b-containing complexes [23]. Since a difference in chemically induced NMDA or mGLuRII-dependent LTD was not detected, even if Gβγ is less likely to interact with SNAP-25a, it appears that lower levels of Gβγ and SNAP-25 interaction are still sufficient to permit expression of normal magnitude of NMDAR and mGLuRII cLTD that is downstream of, and does not depend upon, glutamate release. These results also imply that the altered synaptic plasticity induced by stimulus in young and old transgenic mice versus WT mice may have been in response to altered presynaptic activity, since we observed no difference in fEPSPs when mGluRII-LTD was expressed in SNAP-25b-deficient mice. Our previous work in rats has shown that presynaptic infusion of Gβγ scavenging peptides, ct-SNAP-25 and structurally distinct mSIRK, can each occlude mGluRII presynaptic LTD [44]. Therefore, Gβγ binding to SNAP-25a in mice expressing only this isoform may show larger sLTD compared to littermate controls.

In summary, one role of the switch from expression of the SNAP-25a isoform to SNAP-25b isoform seems to be to down-regulate the predominance of LTD of synaptic strength in early activity-dependent long-term plasticity, favoring an up-regulation of LTP that promotes faster learning acquisition and retention, but reducing behavioral flexibility when learning contingencies change.

4. Materials and Methods

4.1. Animals

SNAP-25b deficient on C57BL/6NCrl background and age-matched wild type littermate mice of both sexes were generated as described previously [14] in the laboratory of Dr. Christina Bark at the Karolinska Institute, Stockholm, Sweden, and genotyped at NYMC by PCR using published methods [14]. A colony of these mice was bred and maintained in the AALAC-accredited animal facility of New York Medical College in accordance with AALAC standards and applicable guidelines. All breeding, genotyping, electrophysiological and behavioral experiments were performed at New York Medical College according to AAALAC International standards and guidelines and approved by the Institutional Animal Care and Use Committee of New York Medical College, Valhalla, NY, USA (IACUC Ethical Permit # 25-12-0312, approved March 20, 2013 and # 91-2-1012, approved October 16, 2012).

4.2. Hippocampal Slice Preparation and Recording

Coronal brain slices were used to analyze dorsal hippocampal Schaffer collateral-CA1 synapses. Brains were removed from male and female mice decapitated under deep isoflurane anesthesia, and immersed in ice-cold (2–4 °C) oxygenated sucrose-containing artificial cerebrospinal fluid (ACSF) in mM: NaCl 87, $NaHCO_3$ 25, NaH_2PO_4 1.25, KCl 2.5, $CaCl_2$ 0.5, $MgCl_2$ 7, D-glucose 25 and sucrose 75, (saturated with 95% O_2/5% CO_2, pH 7.4). The cerebellum and frontal lobes were removed, brains were hemisected, and each half was mounted on a metal stage with cyanoacrylate glue and again submerged in ice-cold sucrose-containing ACSF. Coronal slices 350–400 μm thick were cut with a vibratome (VT1200S, Leica Microsystems, Buffalo Grove, IL, USA), placed in an interface holding chamber at 32 ± 2 °C for at least 30 min, and then transferred to room-temperature ACSF in mM: NaCl 126, $NaHCO_3$ 26, NaH_2PO_4 1.25, KCl 3, $CaCl_2$ 2.5, $MgCl_2$ 2 and D-glucose 10 (constantly bubbled with 95% O_2/5% CO_2, pH 7.4) for an additional 30 min before start of recording. Slices were maintained at room temperature until being moved to an interface recording chamber and maintained thereafter at 32 ± 2 °C for the remainder of the experiment.

Slices were continuously perfused at a rate of 2 mL/min with oxygenated normal ACSF. Field excitatory postsynaptic potentials (fEPSPs) were evoked in the CA1 area by activation of Schaffer collateral axons using a stainless steel bipolar stimulating electrode placed in the stratum radiatum. A thin-walled glass microelectrode was pulled (1–2 MΩ; Flaming/Brown micropipette puller, Model P-97, Sutter Instruments, Novato, CA, USA) and filled with normal ACSF. The borosilicate micropipette was placed approximately 200–300 μm from the stimulating electrode in the similar stratum radiatum area touching distal dendritic arbors to record fEPSPs. A constant current stimulation was applied by an ISO-Flex isolator driven by a Master-8 programmable pulse generator (A.M.P.I, Jerusalem, Israel) and recorded at half maximal fEPSP amplitude once every 30 s. Recordings were amplified by Multiclamp 700B Axon Instruments (Molecular Devices, San Jose, CA, USA), digitized by an AD board (National Semiconductor, Santa Clara, CA, USA), and analyzed using SciWorks software (DataWave Technologies, Loveland, CO, USA) by calculating the maximum slope within 20–80% of the maximum fEPSP initial negative slope. At least 15-min stable baselines were recorded at 0.033 Hz prior to application of either high-frequency theta burst stimulation (TBS; 4 trains of 10 bursts of 5 pulses each at 100 Hz with a 200 msec interburst interval, applied at 3 min intervals) to induce LTP, or a single train of low-frequency stimuli (LFS; 2Hz/10min) to induce LTD.

4.3. Two-Photon Laser Scanning Microscopy of FM1-43 Vesicular Release from Schaffer Collateral Presynaptic Terminals

Fluorescence was visualized using a customized two-photon laser-scanning Olympus BX61WI microscope with a 60x/0.90W water immersion infrared objective lens and an Olympus multispectral confocal laser scan unit. The light source was a Mai-Tai™ laser (Solid-State Laser Co., Mountain View, CA, USA), tuned to 820 nm for exciting FM1-43. Epifluorescence was detected with photomultiplier tubes of the confocal laser scan head with pinhole maximally opened and emission spectral window optimized for signal over background. In the transfluorescent pathway, a 565 nm dichroic mirror was used to separate green and red fluorescence to eliminate transmitted or reflected excitation light (Chroma Technology, Rockingham, VT, USA). After confirming the presence of Schaffer collateral-evoked fEPSPs >1 mV in amplitude in CA1 stratum radiatum, 10 μM 6-cyano-7-nitroquinoxaline-2,3-dione (CNQX) was bath-applied throughout the rest of the experiment to prevent synaptically driven action potentials in the pyramidal neurons and prevent the accelerated dye release. Presynaptic boutons were loaded by bath-applying 5 μM FM1-43 (Invitrogen Molecular Probes, Eugene, OR, USA) in hypertonic ACSF supplemented with sucrose to 800 mOsm for 25 s to selectively load the readily releasable pool (RRP), then returned to normal ACSF. Stimulus-induced destaining was measured after 30 min perfusion with dye-free ACSF, by 0.1 Hz bipolar stimuli (150 μs DC pulses).

The technique of loading presynaptic vesicles with a lipophilic styryl dye permits analysis of vesicle fusion dynamics in presynaptic terminals [45], since FM1-43 preferentially loads into presynaptic vesicles. Once the dye is loaded into the vesicles, it can only be released when vesicles fuse with the membrane and discharge their contents. Two-photon analysis of FM1-43 makes it possible to directly image presynaptic rates of vesicular release [46].

4.4. Elevated Plus Maze

The elevated plus maze (EPM) is used as a general indicator of anxiety. The apparatus consists of two open arms (50 × 10 cm) and two closed arms (50 × 10 × 40 cm), connected by a center platform (10 × 10 cm) made of opaque dark grey plexiglass (Stoelting Co., Wood Dale, IL, USA). The arms of the EPM are elevated 50 cm above the floor. Animals (n = 12–17) were placed in the center platform of the EPM, facing an open arm and allowed to explore the maze for 5 min. The middle point of the animal was used as the reference point to determine the position of a mouse and recorded using AnyMaze behavioral analysis software (Stoelting Co., Wood Dale, IL, USA). Percent of time spent in the open arms was calculated, where a decrease in percent of time spentg in the open arms indicates an anxiety phenotype. The software also measured total distance travelled during EPM assessment.

4.5. Active Place Avoidance Learning Task

The active place avoidance task used was previously described by Burghardt et al. [31]. In this paradigm, mice are placed on a circular rotating platform that continuously turns clockwise at a speed of 1 rpm. Over several days and multiple trials (Figure 3a), mice (n = 12–17) learn to identify the 60° shock zone guided by spatial markers on the walls surrounding the apparatus. Entrance into the shock zone triggered a brief foot-shock (500 ms, 60 Hz, 0.2 mA) with an inter-shock interval of 1.5s that would cease upon leaving the shock zone. The middle point of the animal was used as the reference point to determine the position of a mouse and recorded using AnyMaze behavioral analysis software (Stoelting Co., Wood Dale, IL, USA) Using the same software, the number of shock-zone entries was measured, where a decrease in shock-zone entries indicates learning. During the initial pre-training trial (10 min) when the shock was turned off, mice were allowed to habituate to the apparatus and showed no preference for any area of the platform. Subsequently, the shock was turned on, and the mice completed three training sessions (10 min each) per day for three days, followed by an extinction trial (10 min) on the next day when the shock was turned off and the animals were allowed to ambulate freely into the zone previously associated with the shock. After extinction, a conflict variant task was performed in order to test cognitive flexibility. The shock zone was moved 180° from where the original shock zone was placed and three conflict-training sessions (10 min each) were conducted per day for two days. Mice had to avoid the new shock zone which requires cognitive flexibility and is represented in this task by the simultaneous suppression of the learned condition response of avoiding the original shock zone and learning the new association of a foot-shock with a different zone.

4.6. Statistical Analysis

In LTP and LTD experiments, comparison of fEPSP slopes normalized to baseline slopes from slices from wildtype and SNAP-25b-deficient mice were analyzed by two-tailed Student's t-test for unpaired data. The null hypothesis was that true mean normalized slope between wildtype and SNAP-25b-deficient slices was equal to zero, the alternative hypothesis was that the true mean difference was not equal to zero. One-way ANOVA with repeated measures was used to determine differences between WT and SNAP-25b-deficient mice for all behavioral testing, followed by Bonferroni correction for multiple comparisons (Prism, GraphPad Software, San Diego, CA, USA). All data points are mean ± SEM of n slices, recording from 1–2 slices per mouse. Significance level was pre-set to $p < 0.05$.

5. Conclusions

The isoform SNAP-25a, which is expressed first developmentally in rodents, developmentally regulates the expression of both long-term depression (LTD) and long-term potentiation (LTP) of synaptic strength at Schaffer collateral-CA1 synapses in the hippocampus. In one month old mice lacking the adolescent/adult isoform SNAP-25b, Schaffer collateral-CA1 synapses showed faster release kinetics, decreased stimulus-evoked LTP and enhanced LTD, while chemically induced mGluR-dependent and NMDAR-dependent LTD not requiring synaptic stimulation were not affected. By four months of age, SNAP-25b-deficient mice had compensated for the lack of adult SNAP-25b, exhibiting larger LTP and normal LTD compared to wild type littermates. These compensatory changes restored normal learning and memory in adult mice, as well as enhancing reversal learning. These data indicate a connection between the regulation of long-term synaptic plasticity and development of cognitive learning capacity. The recovery of plasticity and learning in adulthood indicates mechanisms that can compensate for a lack of the adult SNAP-25b isoform.

Author Contributions: Conceptualization, K.R.G., M.I., T.H., C.B., P.K.S.; methodology, C.B., P.K.S.; software validation, K.R.G., M.I., L.R.V., A.M., G.S., C.B., P.K.S.; formal analysis, K.R.G., M.I.; investigation, K.R.G., M.I., O.M., L.R.V., A.M., G.S.; resources, C.B., P.K.S.; data curation, K.R.G., M.I., C.B., P.K.S.; writing—original draft preparation, K.R.G., M.I.; writing—review and editing, K.R.G., M.I., O.M., L.R.V., A.M., T.H., C.B., P.K.S.; visualization, K.R.G., M.I.; supervision, T.H., C.B., P.K.S.; project administration, T.H., C.B., P.K.S.; funding acquisition, T.H., C.B., P.K.S.; All authors have read and agreed to the published version of the manuscript.

Funding: This work was supported by grants from the Sven Mattssons Foundation, Swedish Dementia Foundation, Swedish Brain Foundation, Gamla Tjänarinnor Foundation, the Åhlén-foundation (CB), Swedish Research Council (TH), NIH NRSA Fellowship (KRG) and grant NS044421 (PKS).

Conflicts of Interest: The authors declare no conflicts of interest.

Abbreviations

EPM	Elevated plus maze
LTD	Long-term depression
LTP	Long-term potentiation
mGluR	Metabotropic glutamate receptor
NMDAR	N-methyl-D-aspartate glutamate receptor
RRP	Readily releasable vesicle pool
SNAP-25	Synaptosomal-associated protein of 25 kD
SNARE	Soluble N-ethylmaleimide–sensitive factor Attachment protein REceptor
VAMP2	Vesicle-associated membrane protein 2

References

1. Oyler, G.A.; Higgins, G.A.; Hart, R.A.; Battenberg, E.; Billingsley, M.; Bloom, F.E.; Wilson, M.C. The identification of a novel synaptosomal-associated protein, SNAP-25, differentially expressed by neuronal subpopulations. *J. Cell Biol.* **1989**, *109*, 3039–3052. [CrossRef] [PubMed]
2. Sollner, T.; Bennett, M.K.; Whiteheart, S.W.; Scheller, R.H.; Rothman, J.E. A protein assembly-disassembly pathway in vitro that may correspond to sequential steps of synaptic vesicle docking, activation, and fusion. *Cell* **1993**, *75*, 409–418. [CrossRef]
3. Li, L.; Chin, L.S. The molecular machinery of synaptic vesicle exocytosis. *Cell Mol. Life Sci.* **2003**, *60*, 942–960. [CrossRef] [PubMed]
4. Sudhof, T.C. The synaptic vesicle cycle. *Annu. Rev. Neurosci.* **2004**, *27*, 509–547. [CrossRef] [PubMed]
5. Gonzalo, S.; Greentree, W.K.; Linder, M.E. SNAP-25 is targeted to the plasma membrane through a novel membrane-binding domain. *J. Biol. Chem.* **1999**, *274*, 21313–21318. [CrossRef]
6. Lane, S.R.; Liu, Y. Characterization of the palmitoylation domain of SNAP-25. *J. Neurochem.* **1997**, *69*, 1864–1869. [CrossRef]
7. Veit, M.; Sollner, T.H.; Rothman, J.E. Multiple palmitoylation of synaptotagmin and the t-SNARE SNAP-25. *Febs. Lett.* **1996**, *385*, 119–123. [CrossRef]
8. Jahn, R.; Scheller, R.H. SNAREs–engines for membrane fusion. *Nat. Rev. Mol. Cell Biol.* **2006**, *7*, 631–643. [CrossRef]
9. Grosse, G.; Grosse, J.; Tapp, R.; Kuchinke, J.; Gorsleben, M.; Fetter, I.; Hohne-Zell, B.; Gratzl, M.; Bergmann, M. SNAP-25 requirement for dendritic growth of hippocampal neurons. *J. Neurosci. Res.* **1999**, *56*, 539–546. [CrossRef]

10. Osen-Sand, A.; Catsicas, M.; Staple, J.K.; Jones, K.A.; Ayala, G.; Knowles, J.; Grenningloh, G.; Catsicas, S. Inhibition of axonal growth by SNAP-25 antisense oligonucleotides in vitro and in vivo. *Nature* **1993**, *364*, 445–448. [CrossRef]
11. Baralle, F.E.; Giudice, J. Alternative splicing as a regulator of development and tissue identity. *Nat. Rev Mol. Cell Biol.* **2017**, *18*, 437–451. [CrossRef] [PubMed]
12. Pan, Q.; Shai, O.; Lee, L.J.; Frey, B.J.; Blencowe, B.J. Deep surveying of alternative splicing complexity in the human transcriptome by high-throughput sequencing. *Nat. Genet.* **2008**, *40*, 1413–1415. [CrossRef] [PubMed]
13. Wang, E.T.; Sandberg, R.; Luo, S.; Khrebtukova, I.; Zhang, L.; Mayr, C.; Kingsmore, S.F.; Schroth, G.P.; Burge, C.B. Alternative isoform regulation in human tissue transcriptomes. *Nature* **2008**, *456*, 470–476. [CrossRef] [PubMed]
14. Johansson, J.U.; Ericsson, J.; Janson, J.; Beraki, S.; Stanic, D.; Mandic, S.A.; Wikstrom, M.A.; Hokfelt, T.; Ogren, S.O.; Rozell, B.; et al. An ancient duplication of exon 5 in the Snap25 gene is required for complex neuronal development/function. *PLoS Genet.* **2008**, *4*, e1000278. [CrossRef]
15. Valladolid-Acebes, I.; Daraio, T.; Brismar, K.; Harkany, T.; Ogren, S.O.; Hokfelt, T.G.; Bark, C. Replacing SNAP-25b with SNAP-25a expression results in metabolic disease. *Proc. Natl. Acad. Sci. USA* **2015**, *112*, E4326–E4335. [CrossRef]
16. Bark, I.C. Structure of the chicken gene for SNAP-25 reveals duplicated exon encoding distinct isoforms of the protein. *J. Mol. Biol.* **1993**, *233*, 67–76. [CrossRef]
17. Bark, I.C.; Wilson, M.C. Human cDNA clones encoding two different isoforms of the nerve terminal protein SNAP-25. *Gene* **1994**, *139*, 291–292. [CrossRef]
18. Bark, I.C.; Hahn, K.M.; Ryabinin, A.E.; Wilson, M.C. Differential expression of SNAP-25 protein isoforms during divergent vesicle fusion events of neural development. *Proc. Natl. Acad. Sci. USA* **1995**, *92*, 1510–1514. [CrossRef]
19. Nagy, G.; Milosevic, I.; Fasshauer, D.; Muller, E.M.; de Groot, B.L.; Lang, T.; Wilson, M.C.; Sorensen, J.B. Alternative splicing of SNAP-25 regulates secretion through nonconservative substitutions in the SNARE domain. *Mol. Biol. Cell* **2005**, *16*, 5675–5685. [CrossRef]
20. Gonelle-Gispert, C.; Halban, P.A.; Niemann, H.; Palmer, M.; Catsicas, S.; Sadoul, K. SNAP-25a and -25b isoforms are both expressed in insulin-secreting cells and can function in insulin secretion. *Biochem. J.* **1999**, *339*, 159–165. [CrossRef]
21. Grant, N.J.; Hepp, R.; Krause, W.; Aunis, D.; Oehme, P.; Langley, K. Differential expression of SNAP-25 isoforms and SNAP-23 in the adrenal gland. *J. Neurochem.* **1999**, *72*, 363–372. [CrossRef] [PubMed]
22. Irfan, M.; Daraio, T.; Bark, C. SNAP-25 Puts SNAREs at Center Stage in Metabolic Disease. *Neuroscience* **2019**, *420*, 86–96. [CrossRef] [PubMed]
23. Daraio, T.; Valladolid-Acebes, I.; Brismar, K.; Bark, C. SNAP-25a and SNAP-25b differently mediate interactions with Munc18-1 and Gbetagamma subunits. *Neurosci. Lett.* **2018**, *674*, 75–80. [CrossRef] [PubMed]
24. Bark, C.; Bellinger, F.P.; Kaushal, A.; Mathews, J.R.; Partridge, L.D.; Wilson, M.C. Developmentally regulated switch in alternatively spliced SNAP-25 isoforms alters facilitation of synaptic transmission. *J. Neurosci.* **2004**, *24*, 8796–8805. [CrossRef] [PubMed]
25. Collingridge, G.L.; Peineau, S.; Howland, J.G.; Wang, Y.T. Long-term depression in the CNS. *Nat Rev. Neurosci.* **2010**, *11*, 459–473. [CrossRef]
26. Dudek, S.M.; Bear, M.F. Bidirectional long-term modification of synaptic effectiveness in the adult and immature hippocampus. *J. Neurosci.* **1993**, *13*, 2910–2918. [CrossRef]
27. Kemp, N.; Bashir, Z.I. Long-term depression: A cascade of induction and expression mechanisms. *Prog. Neurobiol.* **2001**, *65*, 339–365. [CrossRef]
28. Lante, F.; Cavalier, M.; Cohen-Solal, C.; Guiramand, J.; Vignes, M. Developmental switch from LTD to LTP in low frequency-induced plasticity. *Hippocampus* **2006**, *16*, 981–989. [CrossRef]
29. Irfan, M.; Gopaul, K.R.; Miry, O.; Hokfelt, T.; Stanton, P.K.; Bark, C. SNAP-25 isoforms differentially regulate synaptic transmission and long-term synaptic plasticity at central synapses. *Sci. Rep.* **2019**, *9*, 6403. [CrossRef]
30. Sorensen, J.B.; Nagy, G.; Varoqueaux, F.; Nehring, R.B.; Brose, N.; Wilson, M.C.; Neher, E. Differential control of the releasable vesicle pools by SNAP-25 splice variants and SNAP-23. *Cell* **2003**, *114*, 75–86. [CrossRef]
31. Burghardt, N.S.; Park, E.H.; Hen, R.; Fenton, A.A. Adult-born hippocampal neurons promote cognitive flexibility in mice. *Hippocampus* **2012**, *22*, 1795–1808. [CrossRef] [PubMed]

32. Gerachshenko, T.; Blackmer, T.; Yoon, E.J.; Bartleson, C.; Hamm, H.E.; Alford, S. Gbetagamma acts at the C terminus of SNAP-25 to mediate presynaptic inhibition. *Nat. Neurosci.* **2005**, *8*, 597–605. [CrossRef] [PubMed]
33. Blackmer, T.; Larsen, E.C.; Bartleson, C.; Kowalchyk, J.A.; Yoon, E.J.; Preininger, A.M.; Alford, S.; Hamm, H.E.; Martin, T.F. G protein betagamma directly regulates SNARE protein fusion machinery for secretory granule exocytosis. *Nat. Neurosci.* **2005**, *8*, 421–425. [CrossRef] [PubMed]
34. Zurawski, Z.; Rodriguez, S.; Hyde, K.; Alford, S.; Hamm, H.E. Gbetagamma Binds to the Extreme C Terminus of SNAP25 to Mediate the Action of Gi/o-Coupled G Protein-Coupled Receptors. *Mol. Pharm.* **2016**, *89*, 75–83. [CrossRef] [PubMed]
35. Hebb, D.O. *The Organization of Behavior: A Neuropsychological Theory*; Wiley: Oxford, UK, 1949; p. 335.
36. Martin, S.J.; Grimwood, P.D.; Morris, R.G. Synaptic plasticity and memory: An evaluation of the hypothesis. *Annu. Rev. Neurosci.* **2000**, *23*, 649–711. [CrossRef] [PubMed]
37. Ku, H.Y.; Huang, Y.F.; Chao, P.H.; Huang, C.C.; Hsu, K.S. Neonatal isolation delays the developmental decline of long-term depression in the CA1 region of rat hippocampus. *Neuropsychopharmacology* **2008**, *33*, 2847–2859. [CrossRef]
38. Volianskis, A.; France, G.; Jensen, M.S.; Bortolotto, Z.A.; Jane, D.E.; Collingridge, G.L. Long-term potentiation and the role of N-methyl-D-aspartate receptors. *Brain Res.* **2015**, *1621*, 5–16. [CrossRef]
39. Delgado-Martinez, I.; Nehring, R.B.; Sorensen, J.B. Differential abilities of SNAP-25 homologs to support neuronal function. *J. Neurosci.* **2007**, *27*, 9380–9391. [CrossRef]
40. Lau, C.G.; Takayasu, Y.; Rodenas-Ruano, A.; Paternain, A.V.; Lerma, J.; Bennett, M.V.; Zukin, R.S. SNAP-25 is a target of protein kinase C phosphorylation critical to NMDA receptor trafficking. *J. Neurosci.* **2010**, *30*, 242–254. [CrossRef]
41. Hussain, S.; Ringsevjen, H.; Schupp, M.; Hvalby, O.; Sorensen, J.B.; Jensen, V.; Davanger, S. A possible postsynaptic role for SNAP-25 in hippocampal synapses. *Brain Struct. Funct.* **2019**, *224*, 521–532. [CrossRef]
42. Irfan, M.; Zurawski, Z.; Hamm, H.E.; Bark, C.; Stanton, P.K. Disabling Gbetagamma-SNAP-25 interaction in gene-targeted mice results in enhancement of long-term potentiation at Schaffer collateral-CA1 synapses in the hippocampus. *Neuroreport* **2019**, *30*, 695–699. [CrossRef]
43. Zurawski, Z.; Thompson Gray, A.D.; Brady, L.J.; Page, B.; Church, E.; Harris, N.A.; Dohn, M.R.; Yim, Y.Y.; Hyde, K.; Mortlock, D.P.; et al. Disabling the Gbetagamma-SNARE interaction disrupts GPCR-mediated presynaptic inhibition, leading to physiological and behavioral phenotypes. *Sci. Signal.* **2019**, *12*, 569. [CrossRef] [PubMed]
44. Zhang, X.L.; Upreti, C.; Stanton, P.K. Gbetagamma and the C terminus of SNAP-25 are necessary for long-term depression of transmitter release. *PLoS ONE* **2011**, *6*, e20500.
45. Stanton, P.K.; Winterer, J.; Zhang, X.L.; Muller, W. Imaging LTP of presynaptic release of FM1-43 from the rapidly recycling vesicle pool of Schaffer collateral-CA1 synapses in rat hippocampal slices. *Eur. J. Neurosci.* **2005**, *22*, 2451–2461. [CrossRef] [PubMed]
46. Winterer, J.; Stanton, P.K.; Muller, W. Direct monitoring of vesicular release and uptake in brain slices by multiphoton excitation of the styryl FM 1-43. *Biotechniques* **2006**, *40*, 343–351. [CrossRef] [PubMed]

International Journal of
Molecular Sciences

MDPI

Article

CB_1 Activity Drives the Selection of Navigational Strategies: A Behavioral and c-Fos Immunoreactivity Study

Daniela Laricchiuta *, Francesca Balsamo, Carlo Fabrizio, Anna Panuccio, Andrea Termine and Laura Petrosini

Laboratory of Experimental and Behavioral Neurophysiology, Fondazione Santa Lucia, 00143 Rome, Italy; francesca.balsamo93@gmail.com (F.B.); carlo.fabrizio217@gmail.com (C.F.); anna.panuccio@gmail.com (A.P.); andreatermine1@hotmail.it (A.T.); laura.petrosini@uniroma1.it (L.P.)

* Correspondence: daniela.laricchiuta@gmail.com

Received: 9 December 2019; Accepted: 31 January 2020; Published: 6 February 2020

Abstract: To promote efficient explorative behaviors, subjects adaptively select spatial navigational strategies based on landmarks or a cognitive map. The hippocampus works alone or in conjunction with the dorsal striatum, both representing the neuronal underpinnings of the navigational strategies organized on the basis of different systems of spatial coordinate integration. The high expression of cannabinoid type 1 (CB_1) receptors in structures related to spatial learning—such as the hippocampus, dorsal striatum and amygdala—renders the endocannabinoid system a critical target to study the balance between landmark- and cognitive map-based navigational strategies. In the present study, mice treated with the CB_1-inverse agonist/antagonist AM251 or vehicle were trained on a Circular Hole Board, a task that could be solved through either navigational strategy. At the end of the behavioral testing, c-Fos immunoreactivity was evaluated in specific nuclei of the hippocampus, dorsal striatum and amygdala. AM251 treatment impaired spatial learning and modified the pattern of the performed navigational strategies as well as the c-Fos immunoreactivity in the hippocampus, dorsal striatum and amygdala. The present findings shed light on the involvement of CB_1 receptors as part of the selection system of the navigational strategies implemented to efficiently solve the spatial problem.

Keywords: endocannabinoid system; spatial learning; hippocampus; dorsal striatum; amygdala; Circular Hole Board; AM251; mice

1. Introduction

Spatial navigation is a complex ability based on the integration of sensory-motor, perceptual, cognitive and emotional processes [1,2]. Subjects show a great individual variability in navigational strategies they carried out, because of external factors, such as landmark differentiation, visual access to environmental cues, complexity of spatial layout, as well as of internal factors, such as biological substrates, individual differences and spatial cognitive style [3]. The navigational strategies are organized on the basis of different systems of spatial coordinate integration. Namely, the navigational strategies related to landmark are based on representations encompassing visual information or connecting landmarks and paths through an egocentric (body-centered) system of reference [4,5]. Thus, the navigational strategies based on landmark are supported by associative learning involving responses to visually salient cues, without providing any spatial information [6]. These strategies are habitual and almost automatic [7,8], cognitively less demanding and scarcely flexible [9].

Conversely, the navigational strategies based on a global representation of the environment mainly involve an allocentric (world-centered) system of reference. Such a system allows moving into the space to reach a goal by referring to an internalized cognitive map, in which the location of an object is represented in spatial relationships with the other objects of the environment [10]. The allocentric navigational strategies involve learning events and environmental relations, integrating them in a spatial cognitive map, remembering the map and then deploying it to plan trajectories and paths [11]. Thus, they are cognitive tasks highly demanding and flexible. Notably, to promote efficient explorative behaviors the subject uses Landmark-based Navigational Strategy (L-NS) or Cognitive Map-based Navigational Strategy (CM-NS) according to specific factors [12]. For example, emotional and stressful factors or pharmacological treatments tend to increase the implicit processes linked to habits, and to be detrimental on allocentric navigational strategies [13,14].

Convergent evidence has proposed that the processes related to formation and maintenance of the implicit habits mainly rely on the intact functioning of the striatum [15–18], while allocentric spatial learning [19,20] and the formation of cognitive maps [21] critically depend on intact functioning of the hippocampus. However, this distinction is probably an overgeneralization. In reality, the striatal and hippocampal regions appear to be not entirely segregated and both of them may contribute to navigational abilities regardless of the strategy carried out. In fact, both structures contain neurons whose firing is correlated with animal's location, directional heading, some egocentric movements and behavioral trajectories, as well as neurons sensitive to changes in reward location [22]. Amidst these similarities, however, some different representational properties may promote a differential role of hippocampus and striatum during spatial learning and bias the significance of their behavioral output. Thus, similar spatial information may be processed and represented in more than one brain structure, and in particular hippocampal and striatal neurons may be sensitive to the same environmental manipulations to contribute to the final behavioral output, that is, the navigational strategy carried out [23].

Outstanding issues on this topic are what neuro-modulatory factors and how brain structures direct the selection of a navigational strategy from a wide repertoire of strategies. The functionality of Endocannabinoid System (ECS) in structures featured by high synaptic plasticity and related to associative learning and goal-directed behavior has made ECS a critical target for studies on learning and memory [24–28]. ECS is formed by cannabinoid receptors, their endogenous lipid ligands (endocannabinoids) and the machinery for synthesis and degradation of endocannabinoids [29,30]. Most central ECS functions are mediated by cannabinoid type-1 receptors (CB_1) [31,32] that are densely expressed in neocortex, basal ganglia, amygdala, hippocampus, hypothalamus and cerebellum [30–33]. These receptors are principally found at presynaptic terminals and modulate the delivery of excitatory and inhibitory neurotransmitters, usually inhibiting their release. Demanding or stressful experiences trigger ECS activation in hippocampus [34,35], amygdala [36], striatum [37] and thalamus [38]. Systemic or intracerebral administrations of CB_1 agonists and paradoxically antagonists impair the consolidation of spatial [39–41] and fear [26,42] memories. These controversial findings indicate complex ECS effects on memory that cannot be merely defined as impairing or enhancing effects of endocannabinoid activation and deactivation.

Starting from these remarks, the present study focuses on the involvement of ECS in the selection of navigational strategies related to landmark or cognitive mapping. To this aim, adult mice were treated with AM251 or with vehicle (VHL). Although exhibiting properties of inverse agonist at CB_1 receptors, the AM251 may behave as a CB_1 receptor antagonist according to its pharmaco-kinetic and -dynamic properties, in the various brain regions. Even if it is difficult to distinguish in vivo whether the AM251 acts as inverse agonist or antagonist at the CB_1 receptors, it has to be underlined that both actions (as inverse agonist or as antagonist) are consistent in deactivating the ECS. Thus, in the present study the AM251 was used to reduce the ECS functionality acting on CB_1 receptors during a navigational task. In order to achieve this aim, the treated animals were subjected to the Circular Hole Board (CHB) (Figure 1A,B), a task that can be solved by using different navigational strategies. Mice can escape from the maze either by remembering the spatial position of the exit hole, thus using the allocentric navigational strategy (CM-NS), or by following the proximal landmark, thus using the implicit associative learning (L-NS) (Figure 1B). At the end of the behavioral testing, the brain activation was evaluated by using cellular imaging of the activity-dependent protoncogene c-Fos in specific regions of the hippocampus (CA1, CA3 and Dentate gyrus—DG), dorsal striatum (dorsolateral striatum—DLS, dorsomedial striatum—DMS) and basolateral amygdala (BLA) (Figure 1C).

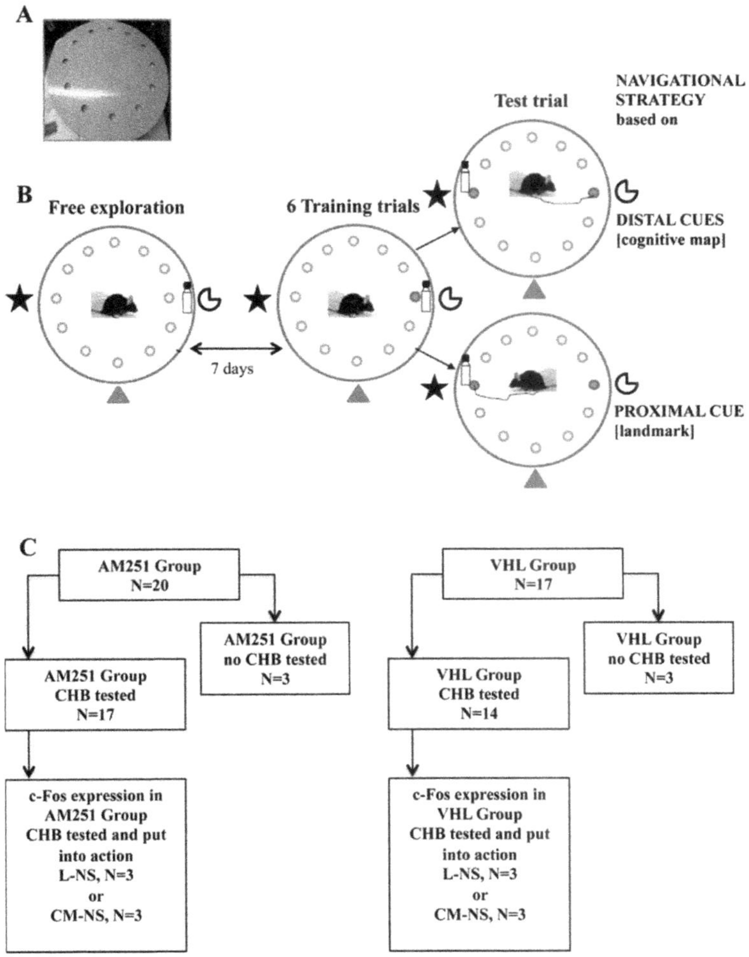

Figure 1. Circular Hole Board (CHB). (**A**) apparatus. (**B**) experimental procedures. (**C**) experimental groups: On the left, animals injected with AM251 (AM251 group, n = 20) and tested (n = 17) or not (n = 3) in the CHB. On the right, animals injected with vehicle (VHL group, n = 17) and tested (n = 3) or not (n = 3) in the CHB. To analyze c-Fos expression, we selected six animals (of which three belonging to the AM251 group and three to the VHL group) that used the Landmark-related Navigational Strategy (L-NS) in the CHB, and six animals (of which three belonging to the AM251 group and three to the VHL group) that used the Cognitive Map-related Navigational Strategy (CM-NS) in the CHB.

The present study highlights the involvement of CB_1 receptors as part of the selection system of the navigational strategies implemented to efficiently solve the spatial problem.

2. Results

2.1. CHB Behavioral Testing

2.1.1. Free Exploration Trial

During the initial free exploration trial (Figure 1B), no differences in exploratory behavior of animals before treatment were found, as revealed by one-way ANOVAs on total distances (Figure 2A) ($F_{1,29}$ = 0.20, p = 0.66), velocity (Figure 2B) ($F_{1,29}$ = 0.62, p = 0.43), visited holes (Figure 2C) ($F_{1,29}$ = 3.70, p = 0.07), rim stretched attend postures (Figure 2D) ($F_{1,29}$ = 1.66, p = 0.21), grooming ($F_{1,29}$ = 0.86, p = 0.36) and defecations ($F_{1,29}$ = 0.53, p = 0.47).

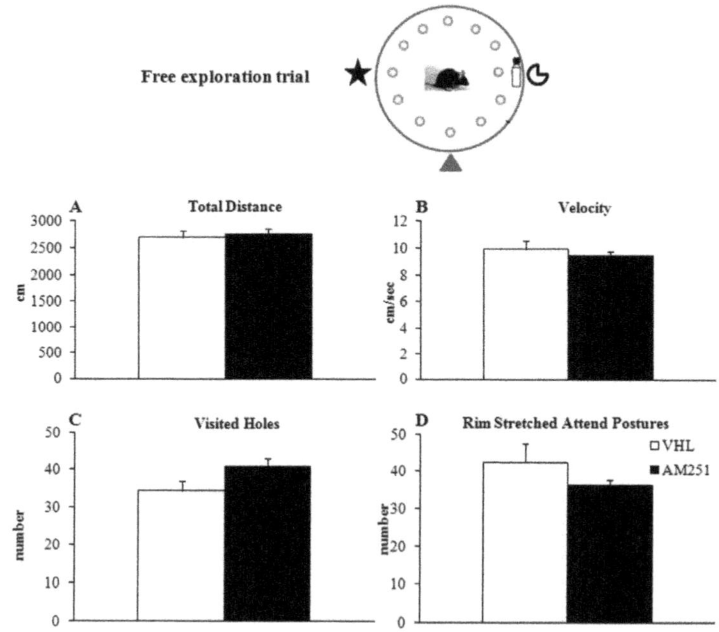

Figure 2. Behavior in the free exploration trial. Total distance (**A**), velocity (**B**), visited holes (the mouse put at least its nose in the hole) (**C**) and rim stretched attend postures (the mouse looked over the edge of the board) (**D**) exhibited by animals injected with AM251 or vehicle (VHL). The data presented as mean and standard errors were analyzed by one-way ANOVAs.

2.1.2. Training Trials

During the six training trials (Figure 1B), when the position of the exit hole was kept fixed with respect to the proximal and distal cues, AM251 group exhibited impaired spatial learning. Namely, AM251 animals travelled distances longer than VHL mice (Figure 3A) although they exhibited the same velocity (Figure 3B). Furthermore, while VHL group decreased the number of visited holes (Figure 3C) as trials went by, AM251 group maintained the same performance throughout the whole training. In comparison to VHL group, AM251 group showed longer latencies in exploring the first hole (Figure 3D) and reaching the exit hole (Figure 3E), and exhibited values not significantly changing throughout the whole training. While the number of perseverations (Figure 3F), rim stretched attend postures (Figure 3G) and grooming (Figure 3H) was similar between groups, the number of defecations of AM251 animals was higher than in VHL animals (Figure 3I). Statistical results of two-way ANOVAs on all parameters of the training trials are reported in Table 1.

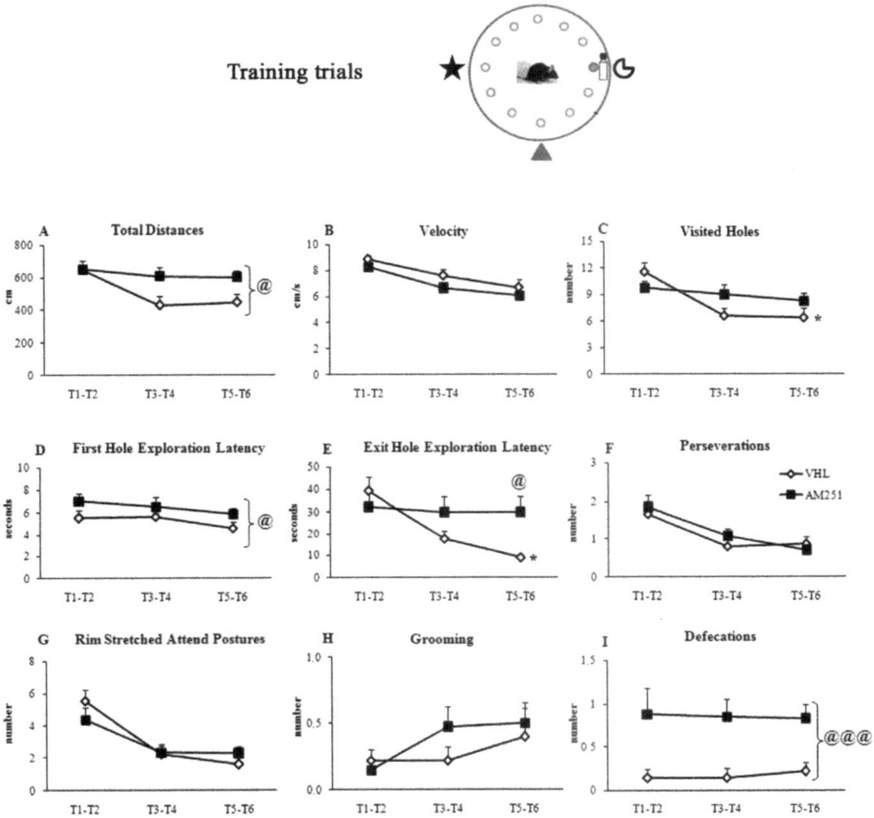

Figure 3. Behavior in the training trials. Total distances (**A**), velocity (**B**), visited holes (**C**), first hole exploration latency (**D**), exit hole exploration latency (**E**), perseverations (the mouse visited the same hole or at least two adjacent holes twice in a row) (**F**), rim stretched attend postures (**G**), grooming (**H**) and defecations (**I**) exhibited by animals injected with AM251 or vehicle (VHL). The data are presented as mean and standard errors. For each parameter, the values of trials 1-2 (T1-2), 3-4 (T3-4) and 5-6 (T5-6) were mediated and analyzed by two-way ANOVAs (group x trials). Significant *group* effect: @: $p \leq 0.05$; @@@: $p < 0.0005$; significant Interaction: * $p \leq 0.05$.

Table 1. Statistical results of two-way ANOVAs on the behavioral parameters of the training trials. In **bold** * are reported significant results.

Effect (Freedom Degrees)	Total Distances	Velocity	Visited Holes	First Hole Latency	Exit Hole Latency	Perseverations	Rim Stretched Attend Postures	Grooming	Defecations
Group F(1,29)	4.01	2.42	0.80	6.27	2.42	1.16	0.05	0.66	16.29
p	**0.05** *	0.13	0.37	**0.02** *	0.13	0.69	0.83	0.42	**0.0004** *
Trials F(2,29)	6.40	25.67	10.50	1.65	5.20	14.42	21.17	1.92	0.01
p	**0.003** *	**≤0.0001** *	0.0001 *	0.20	**0.008** *	**≤0.0001** *	**≤0.0001** *	0.15	0.99
Interaction F(2,58)	2.63	0.26	4.20	0.11	3.70	0.71	1.64	0.70	0.07
p	0.08	0.77	**0.02** *	0.89	**0.03** *	0.49	0.20	0.50	0.93

2.1.3. Test Trial

During the test trial (Figure 1B), the distal cues remained fixed with respect to the position of the exit hole, but the landmark (bottle) was relocated from hole 3 to hole 9. One-way ANOVAs on total distances (Figure 4A) ($F_{1,29} = 1.92$, $p = 0.18$), velocity (Figure 4B) ($F_{1,29} = 0.23$, $p = 0.64$), visited holes (Figure 4C) ($F_{1,29} = 0.81$, $p = 0.37$), first hole exploration latency (Figure 4D) ($F_{1,29} = 0.45$, $p = 0.51$) and exit hole exploration latency (Figure 4E) ($F_{1,29} = 0.01$, $p = 0.90$) failed to reveal any significant effect. More interestingly, significant association ($\chi^2 = 5.80$, $p = 0.02$) was found between AM251 treatment and navigational strategy put into action. In fact, in AM251 group the number of animals exhibiting the L-NS ($n = 11$) was not significantly different from the number of animals exhibiting the CM-NS ($n = 6$) ($\chi^2 = 1.47$, $p = 0.23$). Conversely, in VHL group the number of animals exhibiting the CM-NS ($n = 11$) was significantly higher than the number of animals exhibiting the L-NS ($n = 3$) ($\chi^2 = 4.57$, $p = 0.03$) (Figure 4F).

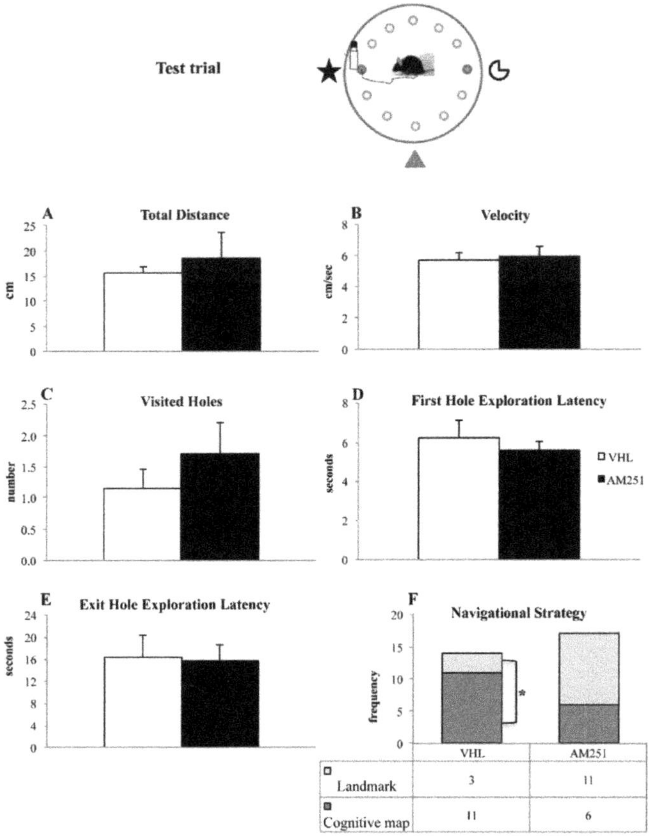

Figure 4. Behavior in the test trial. Total distances (**A**), velocity (**B**), visited holes (**C**), first hole exploration latency (**D**), exit hole exploration latency (**E**) and navigational strategy (**F**) exhibited by animals injected with AM251 or vehicle (VHL). For parameters **A–E**, the data were analyzed by one-way ANOVAs, while the frequencies of the parameter Navigational Strategy were compared by χ^2 test. $^* p \leq 0.05$.

2.2. c-Fos Immunohistochemistry

As regards the VHL group, significant differences in c-Fos expression were found among animals that put into action either navigational strategy and animals not behaviorally tested in all regions analyzed (CA1: $H = 6.49$, $p = 0.03$; CA3: $H = 7.20$, $p = 0.03$; DG: $H = 5.95$, $p = 0.05$; BLA, DLS and DMS: always $H = 7.20$, $p = 0.02$) (Figure 5). Interestingly, VHL animals which carried out the CM-NS showed a number of c-Fos$^+$ cells significantly (always Z = 1.96, p = 0.05) higher than the VHL animals which carried out the L-NS in all regions of interest but in CA1 (H = 1.52, p = 0.13). Furthermore, VHL animals which carried out the CM-NS showed a number of c-Fos$^+$ cells significantly (always Z = 1.96, p = 0.05) higher than the not tested VHL animals in all regions analyzed. In comparison to not tested VHL animals, the VHL animals which carried out the L-NS showed a number of c-Fos$^+$ cells significantly (always Z = 1.96, p = 0.05) higher in CA1 and DMS, lower in CA3, BLA and DLS, and not statistically different in DG (H = 1.09, p = 0.27).

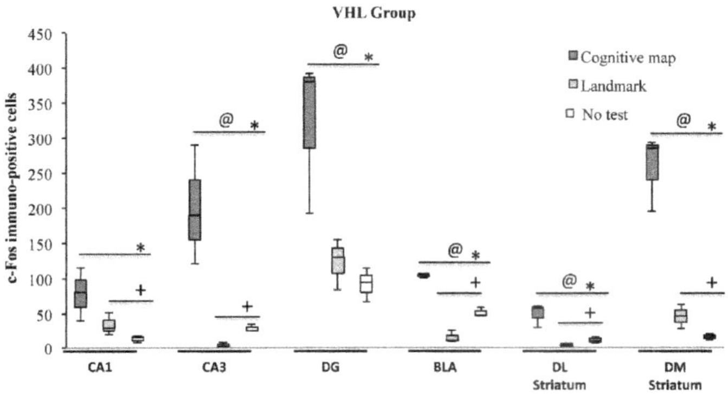

Figure 5. c-Fos activation in hippocampal CA1 and CA3 and dentate gyrus (DG), basolateral amygdala (BLA), dorsolateral (DL) and dorsomedial (DM) striatum exhibited by animals injected with vehicle (VHL) according to the landmark- or cognitive map-related navigational strategy carried out in CHB and in No test condition. Data were analyzed by Kruskal-Wallis test followed by Mann-Whitney U when appropriate. VHL animals which carried out the cognitive map-related navigational strategy showed a number of c-Fos$^+$ cells significantly (@, p = 0.05) higher than the VHL animals which carried out landmark-related navigational strategy in all regions of interest but in CA1. Furthermore, VHL animals that carried out the cognitive map-related navigational strategy showed a number of c-Fos$^+$ cells significantly (*, p = 0.05) higher than the not tested VHL animals in all regions analyzed. In comparison to not tested VHL animals, the VHL animals that carried out the landmark-related navigational strategy showed a number of c-Fos$^+$ cells significantly (+, p = 0.05) higher in CA1 and DM Striatum, lower in CA3, BLA and DL Striatum and not statistically different in DG.

As regards the AM251 group, no significant differences in c-Fos expression were found in CA1 (H = 3.47, p = 0.18), CA3 (H = 5.60, p = 0.06), DG (H = 3.28, p = 0.19), BLA (H = 2.75, p = 0.25) and DLS (H = 5.60, p = 0.06), while a significant difference was found in DMS (H = 5.95, p = 0.05) (Figure 6). In this region, AM251 animals which carried out the CM-NS expressed a number of c-Fos$^+$ cells not significantly different (Z = 1.09, p = 0.27) in comparison to AM251 animals which carried out the L-NS, but significantly (Z = 1.96, p = 0.05) higher than the not tested AM251 animals. Furthermore, AM251 animals which carried out the L-NS expressed a number of c-Fos$^+$ cells significantly (Z = 1.96, p = 0.05) higher than the not tested AM251 animals.

Additional analyses were carried out by computing the normalized c-Fos$^+$ cell count per group. Significant differences were found in CA3 and DLS (always H = 9.05, p = 0.03) as well as in DMS (H = 7.51, p= 0.05), but not in CA1 (H = 4.18, p = 0.24), DG (H = 3.92, p = 0.27) and BLA (H = 6.69, p = 0.08) between animals exhibiting either navigational strategy (Figure 7). Namely, in CA3 and DLS the VHL or AM251 animals which carried out the CM-NS expressed a number of c-Fos$^+$ cells significantly (Z = 1.96, p = 0.05) higher than the VHL or AM251 animals that carried out the L-NS. The results found in DMS revealed very interesting differences between groups: in fact, VHL (but not AM251) animals which carried out the CM-NS expressed a number of c-Fos$^+$ cells significantly (always

Z = 1.96, p = 0.05) higher than the VHL animals that carried out the L-NS, and higher than the AM251 animals using the CM-NS.

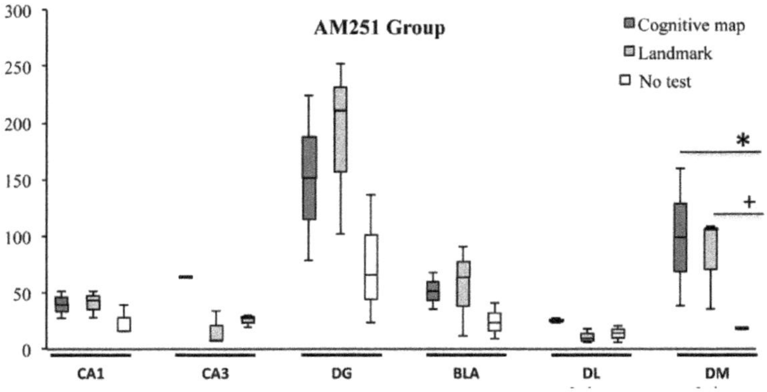

Figure 6. c-Fos activation in hippocampal CA1, CA3 and dentate gyrus (DG), basolateral amygdala (BLA) and dorsolateral (DL) and dorsomedial (DM) striatum exhibited by animals injected with AM251 according to the landmark- or cognitive map-related navigational strategy carried out in CHB and in No test condition. Data were analyzed by Kruskal-Wallis test followed by Mann-Whitney U when appropriate. In DM Striatum, AM251 animals that carried out the cognitive map-related navigational strategy expressed a number of c-Fos$^+$ cells significantly (*, p = 0.05) higher than the not tested AM251 animals, while the AM251 animals that carried out the landmark-related navigational strategy expressed a number of c-Fos$^+$ cells significantly (+, p = 0.05) higher than the not tested AM251 animals.

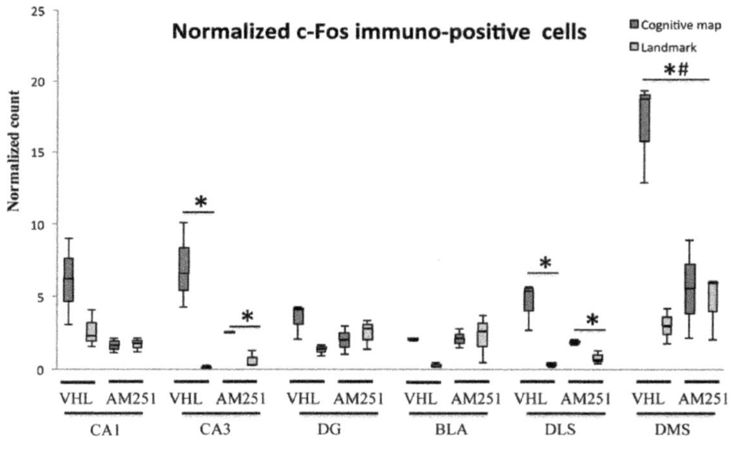

Figure 7. Comparison of normalized c-Fos activations in hippocampal CA1 and CA3 and dentate gyrus (DG), basolateral amygdala (BLA) and dorsolateral (DLS) and dorsomedial (DMS) striatum exhibited by animals injected with vehicle (VHL) or AM251 according to the landmark- or cognitive map-related navigational strategy carried out in CHB. Data were analyzed by Kruskal-Wallis test followed by Mann-Whitney U when appropriate. In CA3 and DLS, the VHL or AM251 animals that carried out the cognitive map-related navigational strategy expressed a normalized number of c-Fos$^+$ cells significantly (*, p = 0.05) higher than the VHL or AM251 animals that carried out the landmark-related navigational strategy. In DMS, VHL (but not AM251) animals that carried out the cognitive map-related navigational strategy expressed a number of c-Fos$^+$ cells significantly (*, p = 0.05) higher than the VHL animals that carried out the landmark-related navigational strategy, and higher (#, p = 0.05) than the AM251 animals using the cognitive map-related navigational strategy.

Summing up: in VHL animals, the navigational strategy based on cognitive mapping determined a significant activation in all brain regions taken into account—the navigational strategy based on landmark determined a significant activation in CA1 and DMS, a significant de-activation in CA3, BLA and DLS, while no effect was found in DG in comparison to the no-test condition; AM251 treatment blocked the activation in DMS that characterized the navigational strategy based on cognitive mapping.

3. Discussion

The present study investigated the role of ECS in selecting navigational strategies carried out by mice treated with AM251 or vehicle. We also analyzed the neuronal activation of hippocampus, dorsal striatum and amygdala according to the navigational strategy carried out.

In the first CHB trial, which assessed animals' free explorative behavior before CB_1 pharmacological blockade, no significant differences were found on any variable between groups to be treated with AM251 or VHL.

Conversely, during the six training trials AM251 modulated spatial learning and emotional reactivity. Namely, when compared to VHL animals, AM251 animals showed longer distances, longer latencies in visiting the first hole and greater number of defecation boluses. Moreover, while the VHL group significantly decreased number of visited holes and exploration latencies to find the exit hole throughout all training trials, the AM251 group continued to show high values in these parameters, without displaying any learning effect.

Various studies highlighted ECS involvement in spatial learning and memory [43–46]. Namely, it has been described that CB_1 receptor knock-out mice and wild-type mice exhibited identical rate of acquisition of platform position in the Morris Water Maze (MWM) when the hidden platform was kept fixed; vice versa, in a reversal task in which the location of the hidden platform was moved in a different position, the mutant mice continued to visit the original platform location. Furthermore, CB_1 functionality affects spatial working memory and facilitates extinction and/or forgetting processes as indicated by (endo)cannabinoid agonists administration [26,47]. It has been reported that the CB_1 antagonist Rimonabant, when systemically administered, impaired MWM spatial learning, by increasing swimming speed and thigmotaxis, while when intrahippocampally infused, it facilitated learning ability, by shortening path lengths to locate the platform, without affecting long-term consolidation of spatial memory [48]. The authors have interpreted the different effects between administration regimes in terms of CB_1 blockade in non-memory- and memory-related brain regions, advancing a memory enhancing effect linked to the selective inactivation of hippocampal CB_1 receptors [48]. However, since the effects of endocannabinoids on memory are strongly dependent on more or less aversive context and on stress levels during training [49], in evaluating the effects of CB_1 blockade on spatial learning the emotional load of the task has to be taken into account. A sign of enhanced emotional reactivity of the AM251 animals could be represented by the significantly higher number of defecation boluses in comparison to VHL animals. However, it should be noted that the ECS controls several aspects of gastrointestinal functions, independently from emotional load. Namely, CB_1 antagonists activate gastrointestinal motility, peristalsis, defecation and secretions in a dose-dependent manner, while CB_1 agonists inhibit these effects [50].

During the test phase, in which the exit hole position remained constant with respect to the distal cues but shifted with respect to the landmark, the two experimental groups differently put the two spatial strategies into action. Namely, while most VHL animals revealed as cognitive map-learners, AM251 animals carried out either navigational strategy without any preference. Notably, VHL group that exhibited an efficient learning during the training phase displayed a dominance of cognitive map-related strategy at the expense of landmark-related strategy. This prominence was not present in AM251 group that not by chance exhibited an impaired learning during the training phase.

Interestingly, the navigational strategy each animal used was closely associated to the profile of c-Fos expression in specific brain areas. In VHL animals, the predominant strategy based on cognitive mapping was associated with the neuronal activation in the entire hippocampus (CA1, CA3 and DG), amygdala (BLA) and dorsal striatum (DLS and DMS), while the less frequently used strategy based on landmark was associated with the activation in CA1 and DMS and de-activation in CA3, BLA and DLS. Furthermore, in comparison to L-NS c-Fos expression profile also evidenced the distinct contribution of hippocampal DG as well as the elevated activation of DMS in CM-NS. Thus, on one hand the two navigational strategies share the parallel activation of some hippocampal and dorso-striatal areas, and on the other hand they are characterized by distinct neuronal patterns. The present data are in line with those reported by Fouquet et al. [51] as well as with the previously advanced conceptualization of the organization of the memory systems. This vision assumes that many, but not all, hippocampal and striatal regions are continuously engaged in similar, though not identical, context-dependent tasks [22]. Specifically, the striatum and hippocampus co-operate to implement the appropriate behavioral output in front of new environmental requests, as after a change in visual context, regardless of whether the task is solved through a flexible cognitive mapping or a landmark-directed strategy. This latter strategy differs from the bare egocentric strategies that represent the object location in space relative to the body axes, as well as from the sequential or serial strategies that require a temporal order memory of successive choice points [51] or the unidirectional exploration of adjacent points [41], respectively. In fact, the landmark-based strategy is centered on a learned association between a well-defined, current external stimulus and a behavior to achieve the goal, without exploiting any spatial competence. As a result, such a strategy is relatively inflexible and slow to acquire because its processing does not take advantage of the more rapid learning feasible when outcomes are driven by the spatial cognitive mapping.

Interestingly, we found that the strength of influence of the neuronal structures on behavioral output was modulated by endocannabinoid functioning. To study such a modulation, here, we administrated AM251. As previously described, this drug works as an inverse agonist or antagonist at CB_1 receptors. The inverse agonism may be explained considering the two-state model of receptors, which proposes that they can be constitutively active ("on" state: they are coupled to their effector mechanisms even in the absence of agonists) or inactive ("off" state: they are not spontaneously coupled to their effector mechanisms). Agonists increase the proportion of receptors in the "on" state, inverse agonists increase it in the "off" state, and antagonists leave the number of receptors in each state unaffected. In detail, while the agonists are featured by both affinity (binding to the target receptor) and intrinsic efficacy (changing receptor activity to produce a response) and the antagonists are featured by affinity but no intrinsic efficacy (binding to the target receptor without producing any response), the inverse agonists have negative intrinsic efficacy (decreasing the activity of a receptor) [52]. When there the constitutive receptor activity and the action of an endogenous agonist are present, as occurs in the case of ECS, an antagonist will reduce the component of the response due to the endogenous agonist, and an inverse agonist will reduce both the endogenous agonist component and constitutive receptor activity. Therefore, the effect of the inverse agonist will be consistent with, but greater than, that of the antagonist. Notably, it has to be taken into account that AM251 that exhibits properties of inverse agonist may behave as an antagonist in different brain regions depending on dosage, its intrinsic efficacy and the magnitude of constitutive receptor activity specific in the various cell phenotypes. In fact, there is evidence that AM251 may act as CB_1 receptor antagonist at low concentrations and as inverse agonist at high concentrations [53,54]. In the present study AM251 was used at a dose of 1 mg/kg i.p., a dose that is controversially retained either evoke an inverse agonist response or block CB_1 receptors and prevent the effects of CB_1 agonists [55–59]. In short, it is challenging to discriminate in vivo the AM251 action as inverse agonist at the CB_1 receptor or antagonist blocking endogenous endocannabinoids acting at a constitutively active CB_1 receptor.

As an additional complexity in facing ECS, it has to be taken into account that AM251 exhibits significant activity not only at CB_1 as an inverse agonist/antagonist, but also at orphan cannabinoid G-protein-coupled receptor 55 (GPR55) as an agonist [60–63]. The lack of homology between cannabinoid receptors may explain the differences among CB_1 and GPR55 in the signaling system and downstream cascade. While the CB_1 receptor is coupled to inhibitory G proteins (which inhibit the adenylyl cyclase and increase the activation of the mitogen-activated protein kinase), and regulates the activity of Ca^{2+} and K^+ channels, and consequently, inhibits neurotransmitters release, the GPR55 promotes the outflow of calcium from intracellular stores via phospholipase C activation, and accordingly, facilitates neurotransmitters release.

Studies in rodent showed that GPR55 is present in hippocampus, thalamus, forebrain, frontal cortex, hypothalamus, cerebellum and striatum [64]. There is increasing evidence of co-expression and cross-talk between cannabinoid receptors supporting the notion that ligands binding to the CB_1 receptor could influence the response to ligands acting through GPR55, and vice versa [65]. In this framework, under the action of AM251, on the one hand the activity of inhibitory G proteins coupled to CB_1 receptors is inhibited, and consequently, the neurotransmitter release is allowed; on the other hand, the function of G proteins coupled to GPR55 is activated and consequently the neurotransmitter release is facilitated. As a result, the opposite signaling between CB_1 and GPR55 leads to accordant behavioral consequences. We cannot exclude that some behavioral and immunochemical effects observed in the present research may involve GPR55 instead or additionally to the classic activity of CB_1; however, the detailed examination of the respective roles goes beyond the scope of the present work.

Recently, the functionality of GPR55 in the procedural memory has been investigated. Rats bilaterally injected into the dorsolateral striatum with a GPR55 and CB_1 agonist and simultaneously with AM251 showed the most efficient performance in a procedural memory task [66]. To further investigate the GPR55 role in spatial learning, rats that received bilateral infusions of a GPR55 agonist into the hippocampus exhibited serial strategy at the expense of spatial strategy, suggesting that integrity of GPR55 pathway could be required to establish a specific navigational strategy [67]. Accordingly, injections of palmitoylethanolamide into the ventral hippocampus affected spatial memory, probably via GPR55 [68], and hippocampal GPR55 stimulation improved synaptic plasticity [69].

Coming back to the more classic literature on CB_1 that underlines how the selection of navigational strategies is biased by manipulating CB_1 functionality, previous findings describe that during the light phase of the diurnal cycle the AM251 intra-hippocampal administration results in a decrease of the spatial strategy, while AM251 intra-striatal administration results in an increase in random strategy [41]. Not surprisingly, in the hippocampus but not in the striatum, CB_1 circadian variations were present [41]. Furthermore, systemic and local administrations of drugs acting on ECS affect the consolidation of spatial and emotional memories without altering memory and time of exploration in Novel Object Recognition test [55,70].

In the present study, the CB_1 blockade elicited by AM251 treatment markedly reduced DMS activation that characterizes the strategy based on cognitive mapping, and at same time reduced the numbers of animals putting this strategy into action. DMS is involved in processes mediating goal-directed behavior or decision-making, such as action-outcome associations, behavioral flexibility and action selection [71,72]. Namely, DMS integrates sensory, cognitive and value-based information, along with motor feedback to form the complex associations that control action selection and goal-directed behaviors; furthermore, it links the goal with more suitable spatial behavior [73]. These operations are required for the acquisition of new strategies and for the flexible re-adjustment of already present strategies when the environment changes [72]. Much of the evidence that DMS is the site of plasticity for these behaviors derives from the observations that disrupting specific inputs to DMS impairs goal-directed learning [74] and that DMS contains reward-responsive, stimulus-related and location-related neurons [75–77]. From a computational point of view, it has been formalized the DMS role in learning and expression of navigational strategies, no matter if the used internal model is allocentric or egocentric [78]. Thus, DMS acts as a transition point in which learning is translated into action and behavioral flexibility is allowed [72,79,80].

At the cellular level, DMS is composed primarily of 95% GABAergic medium spiny neurons (MSN) that integrate cortically processed information about the environment, internal body state and past experience, to select actions that maximize positive and minimize negative outcomes. The different portions of MSN dendrites receive synaptic connections from distinct brain areas: in particular, inputs to the cell body and proximal dendrites derive from GABAergic and cholinergic striatal interneurons, while inputs to distal dendrites derive from glutamatergic cortical, dopaminergic nigrostriatal and thalamic afferents [81]. Recently, the plastic structural changes that cannabinoid administration elicits on striatal MSN have been investigated: treatment with THC, a CB_1 partial agonist, selectively increased dendritic spine density in distal dendrites of MSN belonging to DMS, whereas no effects were observed in DLS neurons. This observation fits the effects of AM251 treatment on DMS found in the present research, and makes us consider DMS as the crucial structure that may provide information about the role of CB_1 in the selection between navigational strategies. In fact, the multi-pronged observations that the manipulations on CB_1 activity modified DMS spine density [82], brought about changes in navigational behaviors (present findings, [41]), modulated c-Fos activation in DMS (present findings), are related to the presence of presynaptic CB_1 receptors (which is known to inhibit the neurotransmitter release) in both cortical glutamatergic afferents on and GABAergic terminals in MSN and in aspiny interneurons of dorsal striatum [83,84]. Furthermore, the present data indicate that to select more frequently the navigational strategy related to cognitive map it is required the full functionality of DMS neurons. When AM251 blocks CB_1 activity and consequently either navigational strategy can be selected without any preference, GABA release by DMS is favored either by a direct dis-inhibition of MSN and aspiny interneurons and by an indirect facilitation of the cortical glutamatergic afferents impinging on MSN [25]. As a result, a downstream inhibition is promoted, in line with the general reduced c-Fos activation observed in the present research. The observation that rats with a unilateral cortico-DMS disconnection are still capable of acquiring goal-directed actions, while rats with bilateral disconnection are severely impaired [74] represents further evidence to support the present finding that different activation levels in a specific structure are related to different behavioral outcomes.

4. Material and Methods

4.1. Subjects

Male adult (2.5 month-old) C57BL/6JOlaHsd mice (Envigo, Udine, Italy) were used in the present research. The animals were group-housed (four mice/cage) with food (Mucedola, Milan, Italy) and water *ad libitum*, and kept under a 12-h light/dark cycle with light on at 07:00 h, controlled temperature (22–23 °C) and constant humidity (60 ± 5%). All experiments took place during the light phase. All efforts were made to minimize animal suffering and to reduce their number, in accordance with the European Directive (Directive 2010/63/EU, 22/09/2010).

Mice were randomly assigned to: AM251 group encompassing animals ($n = 17$) injected with AM251 and then tested in the CHB; VHL group encompassing animals ($n = 14$) injected with VHL and then tested in the CHB. For c-Fos quantification we selected six animals (of which three belonging to the AM251 group and three to the VHL group) that in used the L-NS in the CHB, and six animals (of which three belonging to the AM251 group and three to the VHL group) that in used the CM-NS in the CHB. Furthermore, to compare neuronal activation in the absence of any behavioral testing we performed c-Fos immunostaining in a further six mice injected with AM251 ($n = 3$) or VHL ($n = 3$) without being behaviorally tested (Figure 1C).

4.2. Drugs

The animals treated with the drug acting on ECS ($n = 20$) were intraperitoneally (i.p.) injected with the AM251 (1 mg/kg; Tocris, Bristol, UK) dissolved in the VHL composed of saline solution with 10% DMSO and 5% Tween 80 and administered at volume of 5 mL/kg of body weight. The VHL animals ($n = 17$) received the same volume of VHL i.p.

The selection of AM251 dosage at 1 mg/kg was based on behavioral results on locomotor- [85] anxiety- [86] and reward- [87] related effects. The same dosage was used by Maione et al. [56] to investigate AM251 neurochemical properties.

4.3. CHB Testing

4.3.1. Apparatus

The CHB is a revolving plexiglas grey round plate (110 cm in diameter; situated 1 m above the floor) with 12 holes at equal distances from each other, located 10 cm from the rim of the board. The holes are 5 cm in diameter and can be closed by a lid at a depth of 5 cm (Figure 1A). Whether a hole is open or closed can only be detected by the mouse putting its head over the edge of the hole. The holes were virtually numbered in a clockwise direction, considering holes like hours on a clock.

A ladder (5 cm width, 15 cm length) leads from the open exit hole down to a cage containing the sawdust from the animal's cage. The proximal intra-maze cue was represented by a glass water bottle (50 cl) positioned near the exit hole, while the distal extra-maze cues allowing the spatial orientation on the board were provided by multiple colorful and geometric shapes on the room walls kept fixed throughout the experiment.

Behavior was digitally recorded and analyzed with Ethovision XT (Noldus, Wageningen, The Netherlands) that sampled the position of the mouse 12.5 times per second. Experimenters who performed the behavioral testing were blind to the drug treatment.

4.3.2. Procedures

Each trial started by placing the mouse in a cylinder (Plexiglas; 25 cm height; 10 cm diameter) located at CHB center. After 5 s, the cylinder was lifted and the mouse could explore the board and exit through the open tunnel. If the mouse did not find the exit hole within 120 s, it was gently guided to the exit using a grid. After each trial the board was cleaned with 30% alcohol to dissipate odor cues.

Five days before behavioral experiments started, the mice were trained to descend the ladder to reach the home cage for three consecutive days (10 min/day).

4.3.3. Free Exploration Trial

To evaluate the general locomotor activity before the pharmacological treatment in the first trial (free exploration trial) (Figure 1B) the mice were allowed exploring the CHB for 5 min. In this first trial, all holes were closed, the intra-maze landmark (proximal cue: glass bottle, 13 cm height) was positioned near hole 3 that would become the exit hole in the subsequent trials, but which was still closed; multiple extra-maze distal cues were located on room walls.

The following parameters were calculated: total distances (cm), velocity (cm/s), visited holes, when the mouse put at least its nose in the hole, rim stretched attend postures, when the mouse looked over the edge of the board, grooming and defecations. At the end of the 5 min of exploration, hole 3 was opened and the mouse was guided there by the experimenter to get the ladder down for three consecutive times.

4.3.4. Training Trials

One week after the free exploration trial, mice were pharmacologically treated with AM251 or VHL. After 30 min, mice were given six successive 120 s-training trials with inter-trial interval of 15 min. The proximal landmark was positioned near the exit hole (hole 3) that was the only hole opened while the multiple distal cues were kept fixed on room walls. Thus, the location of the exit hole was always fixed relatively to intra- and extra-maze cues, allowing the animal to learn its location by taking into account the spatial relationships among distal extra-maze spatial cues as well as the association between proximal intra-maze cue and exit hole (Figure 1B). It should be noted that the proximal intra-maze cue was the preferential stimulus to build the L-NS (implicit associative learning), while the distal extra-maze cues were the appropriate stimuli to develop the CM-NS (spatial allocentric learning).

The following parameters were calculated: total distances (cm), velocity (cm/s), visited holes, first hole exploration latency (s), exit hole exploration latency (s), perseverations, when the mouse visited the same hole or at least two adjacent holes twice in a row, rim stretched attend postures, grooming and defecations. For each parameter, the values of the trials 1–2, 3–4 and 5–6 were mediated.

4.3.5. Test Trial

After 15 min, the test trial started (Figure 1B), in which distal extra-maze cues were kept fixed and the proximal intra-maze cue was relocated in the position opposite to its location during the training trials (from hole 3 to hole 9). Thus, hole 3 remained opened, and the "new exit" through hole 9 was filled with sawdust from the animal's cage. Test trial allowed analyzing whether the navigational strategy used to solve the spatial task was based mainly on cognitive map or landmark.

The following parameters were calculated: total distances (cm), velocity (cm/s), visited holes, first hole exploration latency (s), exit hole exploration latency (s) and navigational strategy (frequency) based on cognitive map or landmark.

4.4. Biochemical Analyses

4.4.1. Tissue Preparation

Mice tested on CHB and mice not behaviorally tested were isolated for 1 h to perform c-Fos immunohistochemistry (modified from Conversi et al. [88]). Mice were deeply anesthetized and sacrificed by decapitation. The brains were removed, immersed overnight in paraformaldehyde 4% (PAF), cryoprotected with 30% sucrose solution, cut in serial 40 µm coronal sections with a freezing microtome and alternatively processed for c-Fos immunohistochemistry and Nissl staining. According to the mice stereotaxic atlas [89] by using Nissl sections as anatomical reference, the following regions of interest were identified: hippocampus (from −1.34 to −1.94 mm in relation to bregma), amygdala (from −1.34 to −1.94 mm), and dorsal striatum (from 1.18 to 0.62 mm).

4.4.2. c-Fos Immunohistochemistry

Free floating coronal sections were washed three times in Phosphate-Buffer (PB) + 0.3% Triton X-100 (PBTX), incubated for 30 min in 0.3% H_2O_2 in PBTX to prevent endogenous peroxidase activity, washed three times in PBTX, incubated for 30 min in PBTX containing Avidin blocking solution (Vectastain Elite ABC Kit, Vector Laboratories, Peterborough, UK) (two drops/5 mL), washed three times in PBTX, incubated for 30 min in PBTX containing Biotin blocking solution (Vectastain Elite ABC Kit) (two drops/5 mL), washed again three times in PBTX and incubated overnight with primary antibody (Anti-c-Fos antibody ab 190289, AbCam, Cambridge, UK) diluted 1:5000 in PBTX + 5% normal goat serum. After three washes in PBTX, the sections were incubated for 2 h in PBTX containing secondary antibody (1:200, biotinylated Goat anti-Rabbit, Vectastain Elite ABC Kit), washed again three times in PBTX, incubated for 1 h in Avidin-Biotin complex (Vectastain Elite ABC Kit) diluted 1:50 in PBTX, washed three times in PB 0.1 M and then visualized with diaminobenzidine, as chromogen (DAB, ScyTek Laboratories, Logan, UT, USA) diluted 1:30. Finally, the sections were washed three times in PB 0.1 M, dehydrated in ethanol, cleared in xylene and coverslipped.

We confirmed the specificity of the immunohistochemical pattern by omitting the primary antibody. Such a negative control resulted in the absence of c-Fos immunoreactivity in all brain regions.

4.4.3. Cell Counting

The quantitative analysis of the immunohistochemical reaction for c-Fos immuno-positive ($^+$) cells was performed by doing a cell count along the coronal sections containing the regions of interest by using an optical microscope (Zeiss AxioLab, Oberkochen, Germany) integrated with an image acquisition system (DEI-750 Camera, Optronics, Goleta, CA, USA). For each structure of each subject, at least three sections were identified.

Immunoreactive cells were counted in hippocampal CA1 and CA3, and DG (Figure 8), DMS and DLS, as well as BLA. The boundaries of the structures of interest were recognized with the help of the stereotaxic atlas [89] and the corresponding Nissl slide. For the cell counting, 4× images were used as TIFF files, in which the light and contrast levels were kept stable. c-Fos$^+$ cell counting was performed using the public domain software ImageJ (http://rsb.info.nih.gov/ij) and in particular using the "Image-based Tool for Counting Nuclei" plugin (ITCN) to count active cores. The photomicrographs were converted into 8-bit images. The levels of "width", "minimum distance" and "threshold" were also changed until the automated count was comparable to the same count made manually. The number of c-Fos$^+$ cells was computed and compared within the AM251 groups (animals using L-NS or CM-NS as well as no tested animals) and VHL groups (animals using L-NS or CM-NS as well as no tested animals) (Figure 1C). Furthermore, additional analyses were carried out by calculating the mean count in each structure for each animal divided by the mean count in that region of the respective not tested animals to generate a normalized count for each animal.

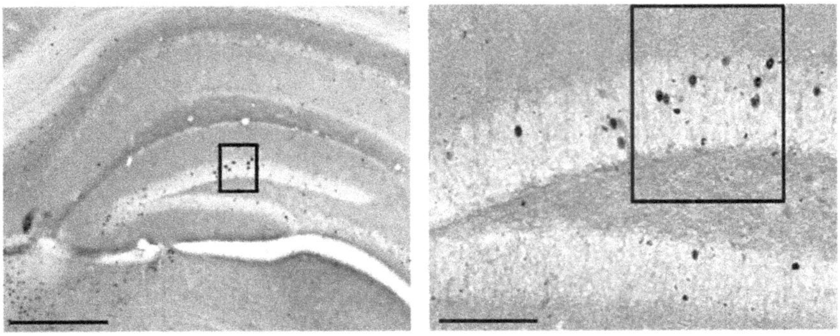

Figure 8. Representative photomicrographs of c-Fos+ cells (highlighted in the black boxes) in hippocampal CA1, CA3 and dentate gyrus (DG) of an animal treated with vehicle at lower (4×) (on the left, Scale bar: 1000 μm) and higher (20×) (on the right, Scale Bar: 200 μm) magnification.

4.5. Statistical Analysis

Behavioral data (mean ± SEM) were firstly tested for normality (Wilk-Shapiro's test) and homoscedasticity (Levene's test) and then compared by one- and two-way ANOVAs, followed by Newman-Keuls test when appropriate. Frequencies of navigational strategies were compared by χ^2 test. Since immunohistochemical data did not meet parametric assumptions, non-parametric analyses of variance (Kruskal-Wallis test, Mann-Whitney U) were used. All analyses were performed by using Statistica 7.0 for Windows (TIBCO Software Inc., Aliso Viejo, CA, USA), and differences were considered significant at $p \leq 0.05$.

5. Conclusions

Our data show that the information coding only in the hippocampus is not sufficient to bias towards strategy based on cognitive mapping, and the information coding only in the DMS is not sufficient to drive strategy based on landmark. In fact, both structures continuously cooperate to perform successfully either allocentric or implicit associative learning. In this enlarged and concerted network, the prefrontal cortex may represent the hub modulating balance between the striatal and hippocampal joint activity since no direct anatomical connectivity is known [90].

Assuming that representations that persist across context changes reflect learned information, we make the following conclusions. A parallel processing occurs within hippocampus and striatum regardless of the navigational strategy selected, although the different hippocampal and striatal sub-regions effectively compete for control of behavioral outcome. The strength of the influence of the neural systems on the behavioral output is modulated by the CB_1 receptors, prompting them to use different spatial navigational strategies to cope with the environmental demands.

Author Contributions: Conceptualization, D.L. and L.P.; Methodology, D.L. and L.P.; Software, C.F., A.T., A.P. and F.B.; Validation, all authors; Formal Analysis, C.F., A.T., A.P. and F.B.; Investigation, all authors; Resources, D.L. and L.P.; Data Curation, D.L.; Writing—Original Draft Preparation, all authors; Writing—Review and Editing, D.L. and L.P.; Visualization, all authors; Supervision, D.L. and L.P.; Project Administration, L.P.; Funding Acquisition, L.P. All authors have read and agreed to the published version of the manuscript.

Funding: This work was supported by the Italian Ministry of Health (Grant number RF 2018-12365391 to L.P.).

Conflicts of Interest: The authors declare no conflict of interest.

References

1. Hinman, J.R.; Dannenberg, H.; Alexander, A.S.; Hasselmo, M.E. Neural mechanisms of navigation involving interactions of cortical and subcortical structures. *J. Neurophysiol.* **2018**, *119*, 2007–2029. [CrossRef] [PubMed]
2. Qiu, Y.; Wu, Y.; Liu, R.; Wang, J.; Huang, H.; Huang, R. Representation of human spatial navigation responding to input spatial information and output navigational strategies: An ALE meta-analysis. *Neurosci. Biobehav. Rev.* **2019**, *103*, 60–72. [CrossRef] [PubMed]
3. Bocchi, A.; Palmiero, M.; Nori, R.; Verde, P.; Piccardi, L. Does spatial cognitive style affect how navigational strategy is planned? *Exp. Brain Res.* **2019**, *237*, 2523–2533. [CrossRef] [PubMed]
4. Ekstrom, A.D.; Isham, E.A. Human spatial navigation: Representations across dimensions and scales. *Curr. Opin. Behav. Sci.* **2017**, *17*, 84–89. [CrossRef]
5. Olson, C.R. Brain representation of object-centered space in monkeys and humans. *Annu. Rev. Neurosci.* **2003**, *26*, 331–354. [CrossRef]
6. Siegel, A.W.; White, S.H. The development of spatial representations of large-scale environments. *Adv. Child. Dev. Behav.* **1975**, *10*, 9–55.
7. Dezfouli, A.; Lingawi, N.W.; Balleine, B.W. Habits as action sequences: Hierarchical action control and changes in outcome value. *Philos. Trans. R. Soc. B Biol. Sci.* **2014**, *369*, 20130482. [CrossRef]
8. Graybiel, A.M. Habits, Rituals, and the Evaluative Brain. *Annu. Rev. Neurosci.* **2008**, *31*, 359–387. [CrossRef]
9. Bonicalzi, S.; Haggard, P. From Freedom from to Freedom to: New Perspectives on Intentional Action. *Front. Psychol.* **2019**, *10*, 1193. [CrossRef]
10. Ruotolo, F.; Iachini, T.; Ruggiero, G.; van der Ham, I.J.M.; Postma, A. Frames of reference and categorical/coordinate spatial relations in a "what was where" task. *Exp. Brain Res.* **2016**, *234*, 2687–2696. [CrossRef]
11. Eichenbaum, H. Time (and space) in the hippocampus. *Curr. Opin. Behav. Sci.* **2017**, *17*, 65–70. [CrossRef] [PubMed]
12. Dolan, R.J.; Dayan, P. Goals and Habits in the Brain. *Neuron* **2013**, *80*, 312–325. [CrossRef] [PubMed]
13. Elliott, A.E.; Packard, M.G. Intra-amygdala anxiogenic drug infusion prior to retrieval biases rats towards the use of habit memory. *Neurobiol. Learn. Mem.* **2008**, *90*, 616–623. [CrossRef] [PubMed]
14. Schwabe, L.; Wolf, O.T. Stress Modulates the Engagement of Multiple Memory Systems in Classification Learning. *J. Neurosci.* **2012**, *32*, 11042–11049. [CrossRef] [PubMed]
15. Malvaez, M.; Wassum, K.M. Regulation of habit formation in the dorsal striatum. *Curr. Opin. Behav. Sci.* **2018**, *20*, 67–74. [CrossRef]
16. Packard, M.G.; Knowlton, B.J. Learning and Memory Functions of the Basal Ganglia. *Annu. Rev. Neurosci.* **2002**, *25*, 563–593. [CrossRef]
17. Yin, H.H.; Knowlton, B.J.; Balleine, B.W. Lesions of dorsolateral striatum preserve outcome expectancy but disrupt habit formation in instrumental learning. *Eur. J. Neurosci.* **2004**, *19*, 181–189. [CrossRef] [PubMed]
18. White, N.M.; Packard, M.G.; McDonald, R.J. Dissociation of memory systems: The story unfolds. *Behav. Neurosci.* **2013**, *127*, 813–834. [CrossRef] [PubMed]
19. O'Keefe, J.; Nadel, L. *The Hippocampus AS a Cognitive Map*; Clarendon Press: Oxford, UK, 1978.
20. Morris, R.G.M.; Garrud, P.; Rawlins, J.N.P.; O'Keefe, J. Place navigation impaired in rats with hippocampal lesions. *Nature* **1982**, *297*, 681–683. [CrossRef] [PubMed]
21. Rudy, J.W.; Sutherland, R.J. Configural association theory and the hippocampal formation: An appraisal and reconfiguration. *Hippocampus* **1995**, *5*, 375–389. [CrossRef]
22. Mizumori, S.J.Y.; Yeshenko, O.; Gill, K.M.; Davis, D.M. Parallel processing across neural systems: Implications for a multiple memory system hypothesis. *Neurobiol. Learn. Mem.* **2004**, *82*, 278–298. [CrossRef]

23. Yeshenko, O.; Guazzelli, A.; Mizumori, S.J.Y. Context-dependent reorganization of spatial and movement representations by simultaneously recorded hippocampal and striatal neurons during performance of allocentric and egocentric tasks. *Behav. Neurosci.* **2004**, *118*, 751–769. [CrossRef]
24. Deoliveiraalvares, L.; Deoliveira, L.; Camboim, C.; Diehl, F.; Genro, B.; Lanziotti, V.; Quillfeldt, J. Amnestic effect of intrahippocampal AM251, a CB1-selective blocker, in the inhibitory avoidance, but not in the open field habituation task, in rats. *Neurobiol. Learn. Mem.* **2005**, *83*, 119–124. [CrossRef] [PubMed]
25. Katona, I.; Rancz, E.A.; Acsady, L.; Ledent, C.; Mackie, K.; Hajos, N.; Freund, T.F. Distribution of CB1 cannabinoid receptors in the amygdala and their role in the control of GABAergic transmission. *J. Neurosci. Off. J. Soc. Neurosci.* **2001**, *21*, 9506–9518. [CrossRef]
26. Laricchiuta, D.; Centonze, D.; Petrosini, L. Effects of endocannabinoid and endovanilloid systems on aversive memory extinction. *Behav. Brain Res.* **2013**, *256*, 101–107. [CrossRef]
27. Laricchiuta, D.; Saba, L.; De Bartolo, P.; Caioli, S.; Zona, C.; Petrosini, L. Maintenance of aversive memories shown by fear extinction-impaired phenotypes is associated with increased activity in the amygdaloid-prefrontal circuit. *Sci. Rep.* **2016**, *6*, 21205. [CrossRef] [PubMed]
28. Peterfi, Z.; Urban, G.M.; Papp, O.I.; Nemeth, B.; Monyer, H.; Szabo, G.; Erdelyi, F.; Mackie, K.; Freund, T.F.; Hajos, N.; et al. Endocannabinoid-Mediated Long-Term Depression of Afferent Excitatory Synapses in Hippocampal Pyramidal Cells and GABAergic Interneurons. *J. Neurosci.* **2012**, *32*, 14448–14463. [CrossRef] [PubMed]
29. Katona, I.; Freund, T.F. Multiple functions of endocannabinoid signaling in the brain. *Annu. Rev. Neurosci.* **2012**, *35*, 529–558. [CrossRef] [PubMed]
30. Marsicano, G.; Lutz, B. Expression of the cannabinoid receptor CB1 in distinct neuronal subpopulations in the adult mouse forebrain: CB1 expression in murine forebrain. *Eur. J. Neurosci.* **1999**, *11*, 4213–4225. [CrossRef] [PubMed]
31. Katona, I. Endocannabinoid receptors: CNS localization of the CB_1 cannabinoid receptor. *Curr. Top. Behav. Neurosci.* **2009**, *1*, 65–86.
32. Piomelli, D. The molecular logic of endocannabinoid signalling. *Nat. Rev. Neurosci.* **2003**, *4*, 873–884. [CrossRef] [PubMed]
33. Di Marzo, V.; Fontana, A.; Cadas, H.; Schinelli, S.; Cimino, G.; Schwartz, J.-C.; Piomelli, D. Formation and inactivation of endogenous cannabinoid anandamide in central neurons. *Nature* **1994**, *372*, 686–691. [CrossRef] [PubMed]
34. Campos, A.C.; Ferreira, F.R.; Guimarães, F.S.; Lemos, J.I. Facilitation of endocannabinoid effects in the ventral hippocampus modulates anxiety-like behaviors depending on previous stress experience. *Neuroscience* **2010**, *167*, 238–246. [CrossRef] [PubMed]
35. Hill, M.N.; Patel, S.; Carrier, E.J.; Rademacher, D.J.; Ormerod, B.K.; Hillard, C.J.; Gorzalka, B.B. Downregulation of Endocannabinoid Signaling in the Hippocampus Following Chronic Unpredictable Stress. *Neuropsychopharmacology* **2005**, *30*, 508–515. [CrossRef] [PubMed]
36. Pietrzak, R.H.; Huang, Y.; Corsi-Travali, S.; Zheng, M.-Q.; Lin, S.; Henry, S.; Potenza, M.N.; Piomelli, D.; Carson, R.E.; Neumeister, A. Cannabinoid Type 1 Receptor Availability in the Amygdala Mediates Threat Processing in Trauma Survivors. *Neuropsychopharmacology* **2014**, *39*, 2519–2528. [CrossRef] [PubMed]
37. De Chiara, V.; Angelucci, F.; Rossi, S.; Musella, A.; Cavasinni, F.; Cantarella, C.; Mataluni, G.; Sacchetti, L.; Napolitano, F.; Castelli, M.; et al. Brain-Derived Neurotrophic Factor Controls Cannabinoid CB1 Receptor Function in the Striatum. *J. Neurosci.* **2010**, *30*, 8127–8137. [CrossRef] [PubMed]
38. Sciolino, N.R.; Bortolato, M.; Eisenstein, S.A.; Fu, J.; Oveisi, F.; Hohmann, A.G.; Piomelli, D. Social isolation and chronic handling alter endocannabinoid signaling and behavioral reactivity to context in adult rats. *Neuroscience* **2010**, *168*, 371–386. [CrossRef]
39. Abush, H.; Akirav, I. Cannabinoids modulate hippocampal memory and plasticity. *Hippocampus* **2010**, *20*, 1126–1138. [CrossRef]
40. Galanopoulos, A.; Polissidis, A.; Georgiadou, G.; Papadopoulou-Daifoti, Z.; Nomikos, G.G.; Pitsikas, N.; Antoniou, K. WIN55,212-2 impairs non-associative recognition and spatial memory in rats via CB1 receptor stimulation. *Pharmacol. Biochem. Behav.* **2014**, *124*, 58–66. [CrossRef]
41. Rueda-Orozco, P.E.; Soria-Gomez, E.; Montes-Rodriguez, C.J.; Martínez-Vargas, M.; Galicia, O.; Navarro, L.; Prospero-García, O. A potential function of endocannabinoids in the selection of a navigation strategy by rats. *Psychopharmacology (Berl.)* **2008**, *198*, 565–576. [CrossRef]

42. Kruk-Slomka, M.; Biala, G. CB1 receptors in the formation of the different phases of memory-related processes in the inhibitory avoidance test in mice. *Behav. Brain Res.* **2016**, *301*, 84–95. [CrossRef] [PubMed]
43. Lichtman, A.H. SR 141716A enhances spatial memory as assessed in a radial-arm maze task in rats. *Eur. J. Pharmacol.* **2000**, *404*, 175–179. [CrossRef]
44. Wise, L.E.; Shelton, C.C.; Cravatt, B.F.; Martin, B.R.; Lichtman, A.H. Assessment of anandamide's pharmacological effects in mice deficient of both fatty acid amide hydrolase and cannabinoid CB1 receptors. *Eur. J. Pharmacol.* **2007**, *557*, 44–48. [CrossRef] [PubMed]
45. Wolff, M.C.; Leander, J.D. SR141716A, a cannabinoid CB1 receptor antagonist, improves memory in a delayed radial maze task. *Eur. J. Pharmacol.* **2003**, *477*, 213–217. [CrossRef] [PubMed]
46. Assini, F.L.; Nakamura, C.A.; Piermartiri, T.C.; Tasca, C.I.; Takahashi, R.N. Coadministration of cannabinoid CB1-receptor and adenosine A1-receptor antagonists improves the acquisition of spatial memory in mice: Participation of glutamatergic neurotransmission. *Behav. Pharmacol.* **2012**, *23*, 292–301. [CrossRef] [PubMed]
47. Varvel, S.A.; Lichtman, A.H. Evaluation of CB_1 Receptor Knockout Mice in the Morris Water Maze. *J. Pharmacol. Exp. Ther.* **2002**, *301*, 915–924. [CrossRef]
48. Robinson, L.; McKillop-Smith, S.; Ross, N.L.; Pertwee, R.G.; Hampson, R.E.; Platt, B.; Riedel, G. Hippocampal endocannabinoids inhibit spatial learning and limit spatial memory in rats. *Psychopharmacology (Berl.)* **2008**, *198*, 551–563. [CrossRef]
49. Morena, M.; Campolongo, P. The endocannabinoid system: An emotional buffer in the modulation of memory function. *Neurobiol. Learn. Mem.* **2014**, *112*, 30–43. [CrossRef]
50. Darmani, N.A. Methods Evaluating Cannabinoid and Endocannabinoid Effects on Gastrointestinal Functions. In *Marijuana and Cannabinoid Research*; Humana Press: Totowa, NJ, USA, 2005; Volume 123, pp. 169–189. ISBN 978-1-59259-999-8.
51. Fouquet, C.; Babayan, B.M.; Watilliaux, A.; Bontempi, B.; Tobin, C.; Rondi-Reig, L. Complementary Roles of the Hippocampus and the Dorsomedial Striatum during Spatial and Sequence-Based Navigation Behavior. *PLoS ONE* **2013**, *8*, e67232. [CrossRef]
52. Berg, K.A.; Clarke, W.P. Making Sense of Pharmacology: Inverse Agonism and Functional Selectivity. *Int. J. Neuropsychopharmacol.* **2018**, *21*, 962–977. [CrossRef]
53. Pertwee, R.G. Inverse agonism and neutral antagonism at cannabinoid CB1 receptors. *Life Sci.* **2005**, *76*, 1307–1324. [CrossRef] [PubMed]
54. Sim-Selley, L.J.; Brunk, L.K.; Selley, D.E. Inhibitory effects of SR141716A on G-protein activation in rat brain. *Eur. J. Pharmacol.* **2001**, *414*, 135–143. [CrossRef]
55. Barbieri, M.; Ossato, A.; Canazza, I.; Trapella, C.; Borelli, A.C.; Beggiato, S.; Rimondo, C.; Serpelloni, G.; Ferraro, L.; Marti, M. Synthetic cannabinoid JWH-018 and its halogenated derivatives JWH-018-Cl and JWH-018-Br impair Novel Object Recognition in mice: Behavioral, electrophysiological and neurochemical evidence. *Neuropharmacology* **2016**, *109*, 254–269. [CrossRef] [PubMed]
56. Maione, S.; Costa, B.; Piscitelli, F.; Morera, E.; De Chiaro, M.; Comelli, F.; Boccella, S.; Guida, F.; Verde, R.; Ortar, G.; et al. Piperazinyl carbamate fatty acid amide hydrolase inhibitors and transient receptor potential channel modulators as "dual-target" analgesics. *Pharmacol. Res.* **2013**, *76*, 98–105. [CrossRef] [PubMed]
57. Newsom, R.J.; Garcia, R.J.; Stafford, J.; Osterlund, C.; O'Neill, C.E.; Day, H.E.W.; Campeau, S. Remote CB1 receptor antagonist administration reveals multiple sites of tonic and phasic endocannabinoid neuroendocrine regulation. *Psychoneuroendocrinology* **2020**, *113*, 104549. [CrossRef] [PubMed]
58. McLaughlin, P.J.; Winston, K.M.; Swezey, L.A.; Vemuri, V.K.; Makriyannis, A.; Salamone, J.D. Detailed analysis of food-reinforced operant lever pressing distinguishes effects of a cannabinoid CB1 inverse agonist and dopamine D1 and D2 antagonists. *Pharmacol. Biochem. Behav.* **2010**, *96*, 75–81. [CrossRef] [PubMed]
59. Hill, M.N.; McLaughlin, R.J.; Bingham, B.; Shrestha, L.; Lee, T.T.Y.; Gray, J.M.; Hillard, C.J.; Gorzalka, B.B.; Viau, V. Endogenous cannabinoid signaling is essential for stress adaptation. *Proc. Natl. Acad. Sci. USA* **2010**, *107*, 9406–9411. [CrossRef]
60. Henstridge, C.M.; Balenga, N.A.B.; Ford, L.A.; Ross, R.A.; Waldhoer, M.; Irving, A.J. The GPR55 ligand L-α-lysophosphatidylinositol promotes RhoA-dependent Ca^{2+} signaling and NFAT activation. *FASEB J.* **2009**, *23*, 183–193. [CrossRef]
61. Ruz-Maldonado, I.; Liu, B.; Atanes, P.; Pingitore, A.; Huang, G.C.; Choudhary, P.; Persaud, S.J. The cannabinoid ligands SR141716A and AM251 enhance human and mouse islet function via GPR55-independent signalling. *Cell. Mol. Life Sci.* **2020**. [CrossRef]

62. Ryberg, E.; Larsson, N.; Sjögren, S.; Hjorth, S.; Hermansson, N.-O.; Leonova, J.; Elebring, T.; Nilsson, K.; Drmota, T.; Greasley, P.J. The orphan receptor GPR55 is a novel cannabinoid receptor. *Br. J. Pharmacol* **2007**, *152*, 1092–1101. [CrossRef]
63. Walsh, S.K.; Hepburn, C.Y.; Keown, O.; Åstrand, A.; Lindblom, A.; Ryberg, E.; Hjorth, S.; Leslie, S.J.; Greasley, P.J.; Wainwright, C.L. Pharmacological profiling of the hemodynamic effects of cannabinoid ligands: A combined in vitro and in vivo approach. *Pharmacol. Res. Perspect.* **2015**, *3*, e00143. [CrossRef] [PubMed]
64. Tudurí, E.; Imbernon, M.; Hernández-Bautista, R.J.; Tojo, M.; Fernø, J.; Diéguez, C.; Nogueiras, R. GPR55: A new promising target for metabolism? *J. Mol. Endocrinol.* **2017**, *58*, R191–R202. [CrossRef] [PubMed]
65. Pertwee, R.G.; Howlett, A.C.; Abood, M.E.; Alexander, S.P.H.; Di Marzo, V.; Elphick, M.R.; Greasley, P.J.; Hansen, H.S.; Kunos, G.; Mackie, K.; et al. International Union of Basic and Clinical Pharmacology. LXXIX. Cannabinoid Receptors and Their Ligands: Beyond CB_1 and CB_2. *Pharmacol. Rev.* **2010**, *62*, 588–631. [CrossRef] [PubMed]
66. Marichal-Cancino, B.A.; Sánchez-Fuentes, A.; Méndez-Díaz, M.; Ruiz-Contreras, A.E.; Prospéro-García, O. Blockade of GPR55 in the dorsolateral striatum impairs performance of rats in a T-maze paradigm. *Behav. Pharmacol.* **2016**, *27*, 393–396. [CrossRef]
67. Marichal-Cancino, B.A.; Fajardo-Valdez, A.; Ruiz-Contreras, A.E.; Méndez-Díaz, M.; Prospéro-García, O. Possible role of hippocampal GPR55 in spatial learning and memory in rats. *Acta Neurobiol. Exp.* **2018**, *78*, 41–50. [CrossRef]
68. Kramar, C.; Loureiro, M.; Renard, J.; Laviolette, S.R. Palmitoylethanolamide Modulates GPR55 Receptor Signaling in the Ventral Hippocampus to Regulate Mesolimbic Dopamine Activity, Social Interaction, and Memory Processing. *Cannabis Cannabinoid Res.* **2017**, *2*, 8–20. [CrossRef]
69. Hurst, K.; Badgley, C.; Ellsworth, T.; Bell, S.; Friend, L.; Prince, B.; Welch, J.; Cowan, Z.; Williamson, R.; Lyon, C.; et al. A putative lysophosphatidylinositol receptor GPR55 modulates hippocampal synaptic plasticity. *Hippocampus* **2017**, *27*, 985–998. [CrossRef]
70. Rueda-Orozco, P.E.; Montes-Rodriguez, C.J.; Ruiz-Contreras, A.E.; Mendez-Diaz, M.; Prospero-Garcia, O. The effects of anandamide and oleamide on cognition depend on diurnal variations. *Brain Res.* **2017**, *1572*, 129–136. [CrossRef]
71. Ragozzino, M.E. The Contribution of the Medial Prefrontal Cortex, Orbitofrontal Cortex, and Dorsomedial Striatum to Behavioral Flexibility. *Ann. N. Y. Acad. Sci.* **2007**, *1121*, 355–375. [CrossRef]
72. Peak, J.; Hart, G.; Balleine, B.W. From learning to action: The integration of dorsal striatal input and output pathways in instrumental conditioning. *Eur. J. Neurosci.* **2019**, *49*, 658–671. [CrossRef]
73. Moussa, R.; Poucet, B.; Amalric, M.; Sargolini, F. Contributions of dorsal striatal subregions to spatial alternation behavior. *Learn. Mem.* **2011**, *18*, 444–451. [CrossRef] [PubMed]
74. Hart, G.; Bradfield, L.A.; Balleine, B.W. Prefrontal Corticostriatal Disconnection Blocks the Acquisition of Goal-Directed Action. *J. Neurosci.* **2018**, *38*, 1311–1322. [CrossRef] [PubMed]
75. Wiener, S.I. Spatial and behavioral correlates of striatal neurons in rats performing a self-initiated navigation task. *J. Neurosci. Off. J. Soc. Neurosci.* **1993**, *13*, 3802–3817. [CrossRef]
76. Wiener, S.I. Spatial, behavioral and sensory correlates of hippocampal CA1 complex spike cell activity: Implications for information processing functions. *Prog. Neurobiol.* **1996**, *49*, 335–361. [CrossRef]
77. Schmitzer-Torbert, N.; Redish, A.D. Neuronal Activity in the Rodent Dorsal Striatum in Sequential Navigation: Separation of Spatial and Reward Responses on the Multiple T Task. *J. Neurophysiol.* **2004**, *91*, 2259–2272. [CrossRef]
78. Khamassi, M.; Humphries, M.D. Integrating cortico-limbic-basal ganglia architectures for learning model-based and model-free navigation strategies. *Front. Behav. Neurosci.* **2012**, *6*, 79. [CrossRef]
79. Hunnicutt, B.J.; Jongbloets, B.C.; Birdsong, W.T.; Gertz, K.J.; Zhong, H.; Mao, T. A comprehensive excitatory input map of the striatum reveals novel functional organization. *eLife* **2016**, *5*. [CrossRef]
80. Hintiryan, H.; Foster, N.N.; Bowman, I.; Bay, M.; Song, M.Y.; Gou, L.; Yamashita, S.; Bienkowski, M.S.; Zingg, B.; Zhu, M.; et al. The mouse cortico-striatal projectome. *Nat. Neurosci.* **2016**, *19*, 1100–1114. [CrossRef]
81. Steiner, H.; Tseng, K.-Y. *Handbook of Basal Ganglia Structure and Function*, 2nd ed.; Academic Press: Cambridge, MA, USA, 2016.
82. Fernández-Cabrera, M.R.; Higuera-Matas, A.; Fernaud-Espinosa, I.; DeFelipe, J.; Ambrosio, E.; Miguéns, M. Selective effects of Δ9-tetrahydrocannabinol on medium spiny neurons in the striatum. *PLoS ONE* **2018**, *13*, e0200950. [CrossRef]

83. Huang, C.C.; Lo, S.W.; Hsu, K.S. Presynaptic mechanisms underlying cannabinoid inhibition of excitatory synaptic transmission in rat striatal neurons. *J. Physiol.* **2001**, *532*, 731–748. [CrossRef]
84. Köfalvi, A.; Rodrigues, R.J.; Ledent, C.; Mackie, K.; Vizi, E.S.; Cunha, R.A.; Sperlágh, B. Involvement of cannabinoid receptors in the regulation of neurotransmitter release in the rodent striatum: A combined immunochemical and pharmacological analysis. *J. Neurosci. Off. J. Soc. Neurosci.* **2005**, *25*, 2874–2884. [CrossRef] [PubMed]
85. Eisenstein, S.A.; Clapper, J.R.; Holmes, P.V.; Piomelli, D.; Hohmann, A.G. A role for 2-arachidonoylglycerol and endocannabinoid signaling in the locomotor response to novelty induced by olfactory bulbectomy. *Pharmacol. Res.* **2010**, *61*, 419–429. [CrossRef] [PubMed]
86. Umathe, S.N.; Manna, S.S.S.; Utturwar, K.S.; Jain, N.S. Endocannabinoids mediate anxiolytic-like effect of acetaminophen via CB1 receptors. *Prog. Neuropsychopharmacol. Biol. Psychiatry* **2009**, *33*, 1191–1199. [CrossRef] [PubMed]
87. Laricchiuta, D.; Musella, A.; Rossi, S.; Centonze, D. Behavioral and electrophysiological effects of endocannabinoid and dopaminergic systems on salient stimuli. *Front. Behav. Neurosci.* **2014**, *8*, 183. [CrossRef]
88. Conversi, D.; Orsini, C.; Cabib, S. Distinct patterns of Fos expression induced by systemic amphetamine in the striatal complex of C57BL/6JICo and DBA/2JICo inbred strains of mice. *Brain Res.* **2004**, *1025*, 59–66. [CrossRef]
89. Paxinos, G.; Franklin, K.B.J. *Paxinos and Franklin's the Mouse Brain in Stereotaxic Coordinates*, 5th ed.; Academic Press: Cambridge, MA, USA, 2019.
90. Albouy, G.; King, B.R.; Maquet, P.; Doyon, J. Hippocampus and striatum: Dynamics and interaction during acquisition and sleep-related motor sequence memory consolidation. *Hippocampus* **2013**, *23*, 985–1004. [CrossRef]

International Journal of
Molecular Sciences

MDPI

Article

Dystonia: Sparse Synapses for D2 Receptors in Striatum of a DYT1 Knock-out Mouse Model

Vincenza D'Angelo [1], Emanuela Paldino [2], Silvia Cardarelli [3], Roberto Sorge [1], Francesca Romana Fusco [2], Stefano Biagioni [3], Nicola Biagio Mercuri [1,2], Mauro Giorgi [3,*] and Giuseppe Sancesario [1,*]

1 Department of Systems Medicine, Tor Vergata University of Rome, via Montpellier 1, 00133 Rome, Italy; dangelo@med.uniroma2.it (V.D.); sorge@uniroma2.it (R.S.); mercurin@med.uniroma2.it (N.B.M.)

2 Santa Lucia Foundation, via del Fosso di Fiorano 64, 00143 Rome, Italy; e.paldino@hsantalucia.it (E.P.); f.fusco@hsantalucia.it (F.R.F.)

3 Department of Biology and Biotechnology "Charles Darwin", Sapienza University of Rome, Piazzale A. Moro 5, 00185 Rome, Italy; silvia.cardarelli@uniroma1.it (S.C.); stefano.biagioni@uniroma1.it (S.B.)

* Correspondence: mauro.giorgi@uniroma1.it (M.G.); sancesario@med.uniroma2.it (G.S.)

Received: 27 December 2019; Accepted: 1 February 2020; Published: 6 February 2020

Abstract: Dystonia pathophysiology has been partly linked to downregulation and dysfunction of dopamine D2 receptors in striatum. We aimed to investigate the possible morpho-structural correlates of D2 receptor downregulation in the striatum of a DYT1 Tor1a mouse model. Adult control Tor1a+/+ and mutant Tor1a+/− mice were used. The brains were perfused and free-floating sections of basal ganglia were incubated with polyclonal anti-D2 antibody, followed by secondary immune-fluorescent antibody. Confocal microscopy was used to detect immune-fluorescent signals. The same primary antibody was used to evaluate D2 receptor expression by western blot. The D2 receptor immune-fluorescence appeared circumscribed in small disks (~0.3–0.5 µm diameter), likely representing D2 synapse aggregates, densely distributed in the striatum of Tor1a+/+ mice. In the Tor1a+/− mice the D2 aggregates were significantly smaller (μm^2 2.4 ± SE 0.16, compared to μm^2 6.73 ± SE 3.41 in Tor1a+/+) and sparse, with ~30% less number per microscopic field, value correspondent to the amount of reduced D2 expression in western blotting analysis. In DYT1 mutant mice the sparse and small D2 synapses in the striatum may be insufficient to "gate" the amount of presynaptic dopamine release diffusing in peri-synaptic space, and this consequently may result in a timing and spatially larger nonselective sphere of influence of dopamine action.

Keywords: dystonia; striatum; D2 receptors; synapses; dopamine volume transmission

1. Introduction

Dystonia is a disorder of movement characterized by disturbed agonist-antagonist muscle activation [1], with particular difficulty switching between sequential muscles involved in a complex task, either voluntary or involuntary [2]. Symptomatic dystonia can be observed in various neurological disorders, including cerebrovascular diseases, Parkinson's disease, and Wilson's disease [3]. Unlike symptomatic dystonia, no pathologic correlate is still known for many dystonic disorders categorized as idiopathic dystonia, and further divided into generalized, focal, and segmental dystonia [4].

In the past, various dystonic forms were often interpreted in psychological or psychiatric terms until the late 19th century [3,5], when the descriptions of familial forms of generalized primary torsion dystonia suggested that it is an organic disease [6,7]. The identification of genetic mutations in some families in the late 20th century established an organic framework for idiopathic dystonia (see 3).

The most common and severe form of genetic dystonia is caused by a 3 bp deletion (GAG) in the coding region of the TOR1A (DYT1) gene, which results in a defective protein called torsinA [8].

Although hereditary, DYT1 dystonia has a childhood or adolescent clinical onset [4]. However, significant neurodegeneration could not be documented at the histological level in any brain areas of patients with DYT1 dystonia [9,10], suggesting that dystonia pathophysiology is determined by functional rather than structural abnormalities. Several neurophysiological and neuroimaging studies, as well as new genetic insights, have been so far performed in DYT1 dystonia, contributing to define functional abnormalities in the basal ganglia [11–15]. Clinical neuroimaging studies have revealed decreased caudate-putamen dopamine D2 receptor availability in DYT1 patients compared to controls [16,17]. Moreover, reduced striatal D2 receptor binding and protein level have also been reported in several different DYT1 experimental models [18–20]. Aside from D2 receptor downregulation, multiple lines of evidence demonstrated reduced coupling between the D2 receptor and its cognate G proteins, as well as severely altered D2 receptor electrophysiological plasticity in the striatum but not in the *substantia nigra* [18,21–25]. Comparative studies on the functions of D1, D2, an A2A receptors, as well as of different neurotransmitters (dopamine, GABA, glutamate, acetylcholine) have been performed by Pisani et al. in mouse models of dystonia, demonstrating a selective downregulation and dysfunction of D2 receptors [18,21,23]. In addition, a recent paper has clarified the mechanisms of D2 receptor downregulation in the striatum, mediated by increased lysosomal degradation, associated in turn with lower levels of striatal RGS9-2 and spinophiling, opening a new approach to the therapy [26]. Therefore, it is generally assumed that abnormal striatal synaptic plasticity, and D2 receptor-dependent striatal outflow abnormalities have a leading role in determining basal ganglia pathophysiology in DYT1 dystonia [27,28]. The developmental profile of the aberrant D2 receptor function has been studied in DYT1 mutant mice, recording in cholinergic neurons an abnormal excitatory response to the D2 receptor agonist quinpirole since postnatal day 14, which persisted at three and nine months in hMT mice [22].

We aimed to investigate possible morpho-structural correlates of D2 receptor downregulation in striatum of an adult DYT1 knock-out mouse model.

2. Results

We first quantified the levels of D2 receptors on proteins extracted from the striatum. In line with a previous study [26] western blotting analysis revealed a significant ~ 30% reduction ($p < 0.05$) of D2 receptor levels in the striatum of mutant Tor1a+/− compared to control Tor1a+/+ mice (Figure 1).

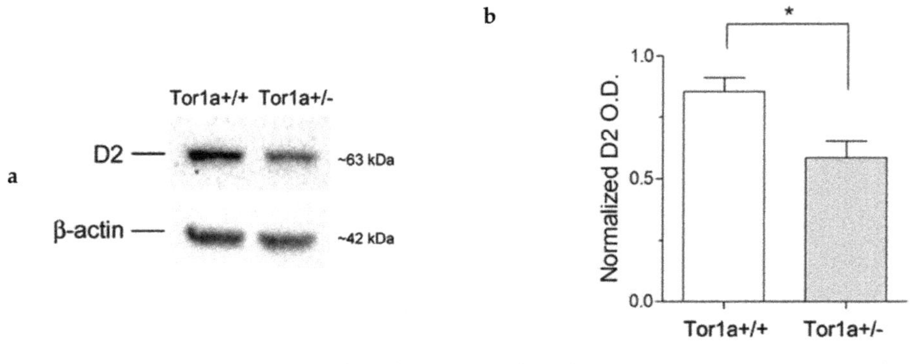

Figure 1. (**a**) Comparative immunoblots of D2 receptors from the striatum of control Tor1a+/+, and mutant Tor1a+/− mice. β-actin content was detected as internal reference standard. (**b**) Densitometric analysis of relative optical density (OD) on D2 receptors immune-stained bands. Results were expressed as the mean ± SEM of the values obtained for each separate hemisphere from four mice per group. * $p < 0.05$.

Light microscopy immune-histochemistry demonstrated an intense D2 receptor brown peroxidase reaction product reactivity in the striatum (Figure 2A). In control Tor1a+/+, the striatum displayed D2 positive neuronal perikarya, peripherally outlined by an intense reaction product, and surrounded by a diffuse lighter neuropil staining. These data document a large distribution of D2 receptors on neuronal bodies, and neuropil of striatal neurons. In Tor1a+/− the D2 peroxidase reaction product appeared less intense around neuronal bodies, as well as in the neuropil of the striatum (Figure 2B), confirming the western blot analysis. However, the diffuse brown reaction product detectable by the D2 light microscopy immune-histochemistry can give just a rough idea of the D2 densitometric changes around neuronal bodies and neuropil, but does not allow a precise definition of the morpho-structural characteristics of the D2 receptor aggregates in control and mutant mice.

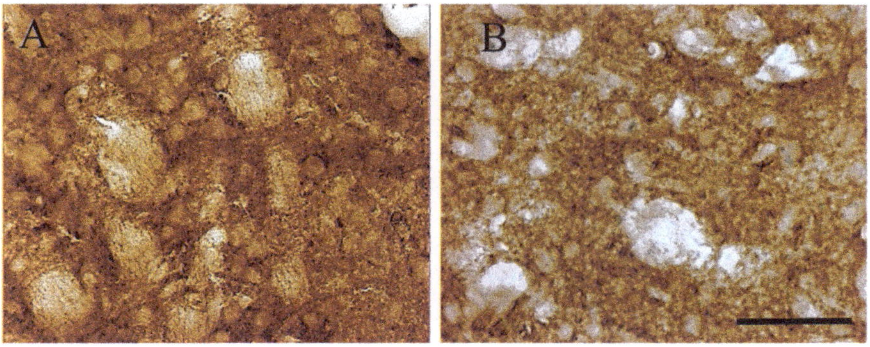

Figure 2. Representative microphotographs of D2 receptor immune-histochemistry in control Tor1a+/+ (**A**), and mutant Tor1a+/− knock-out (**B**) mice. The brown reaction product is clustered around neuronal bodies and diffused in the neuropil. Scale bar in **B** = 100 µm.

Immune-fluorescence images were acquired with a LSM700 Zeiss confocal laser scanning microscope (Zeiss, Germany): 5× and 20× objectives were used to define areas of interest in the dorsolateral striatum; distribution of D2 receptors was first acquired using 63× oil immersion lens (1.4 numerical aperture), and thereafter with an additional digital zoom factor (1×–1.5×–2×). Images of D2 immune-fluorescence acquired with a 63× oil immersion lens at first look appeared as a shiny galaxy in a starkly sky, with clusters of extremely small grains covering diffusely the neuronal compartments of the striatum, without apparent difference between perikarya and neuropil, whereas grains were rare and almost absent on the cell nuclei and in striatal axonal bundles (Figure 3). The density of D2 positive fluorescent grains was clearly different between the striatum of Tor1a+/+ and Tor1a+/− mice. In Tor1a+/+ the D2 positive grain were contiguous and even superimposed each other, whereas in the striatum of Tor1a+/− mice the D2 positive grains were close but separated from each other (Figure 3).

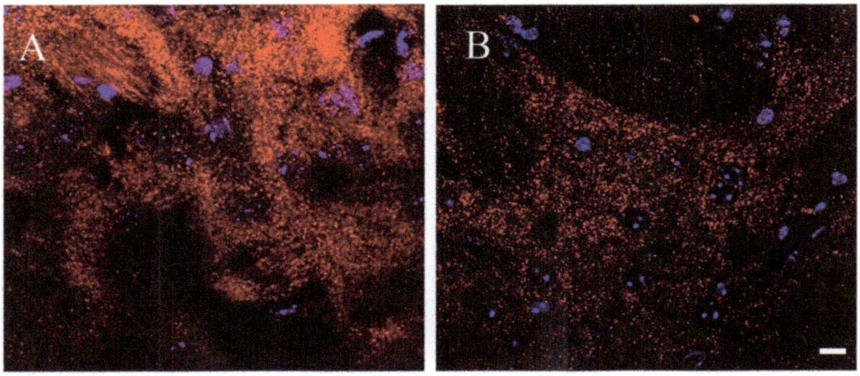

Figure 3. Representative immune-fluorescence microphotographs of confocal laser scanning microscopy (63×), double-labeled for D2 receptors and for nuclei in the striatum of control Tor1a+/+ (**A**), and of mutant Tor1a+/− knock-out (**B**) mice. D2 receptor immune-labeling is visualized in red-Cy3 fluorescence, while nuclei are visualized by DAPI fluorescence in blue. Bar in **B** = 10 μm.

A better understanding of D2 receptors subcellular distribution came out in images acquired using a 63× oil immersion lens (1.4 numerical aperture) with an additional digital zoom factor (1×–1.5×–2×). The immune-fluorescent signal appeared extremely specific without background staining so that the D2 receptor localization appeared circumscribed in very small disks, roundish or elliptical in shape, of size variable with a diameter between ~0.3–0.5 μm in the striatum of Tor1a+/+ mice (Figure 4A).

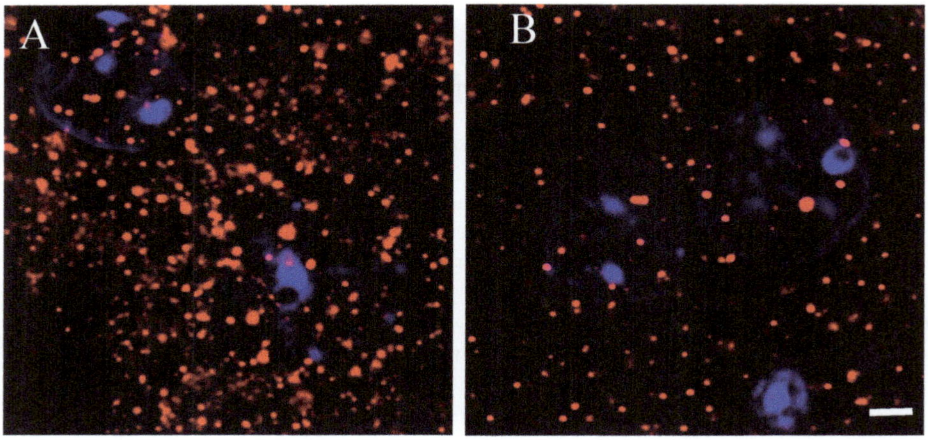

Figure 4. Representative immune-fluorescence microphotographs of high magnification (63× plus zoom factor 1×–1.5×–2×) confocal laser scanning microscopy, double-labeled for D2 receptors and for nuclei in the striatum of control Tor1a+/+ (**A**), and of mutant Tor1a+/− knock-out (**B**) mice. D2 receptor immune-labeling is visualized in red-Cy3 fluorescence, while nuclei are visualized by DAPI fluorescence in blue. Bar in **B** = 2 μm.

In the striatum of Tor1a+/− mice, the D2 positive disks appeared sparse (Figure 4B), and the number and size of D2 positive disks and their total area per microscopic field was significantly lower in the striatum of Tor1a+/− than in corresponding areas of Tor1a+/+ mice (Figure 5A–C). Moreover, the mean distance between D2 positive disks was <1 μm for most of the disks in the striatum of Tor1a+/+ but > 2 μm for most of the disks in corresponding areas of Tor1a+/− mice (as can be easily seen in Figure 4).

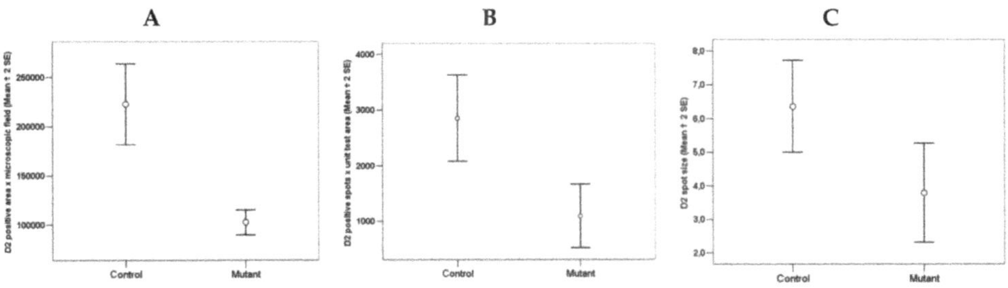

Figure 5. Comparative densitometric analysis of the D2 immune-fluorescence in the striatum of control Torla+/+ and mutant Torla+/− knock-out transgenic mice. (**A**) Total D2 positive area per microscopic field through 63× oil immersion objective. df = 1, F = 31.202; $p < 0.0001$. (**B**) Number of D2 positive disks per microscopic field through 63× oil immersion objective (1.4 numerical aperture) with an additional digital zoom factor (1×–1.5×–2×). df = 1, F = 13.406; $p < 0.001$. (**C**) Size of D2 disk (μm^2) through 63× oil immersion objective (1.4 numerical aperture) with an additional digital zoom factor (1×–1.5×–2×). df = 1, F = 6.543; $p < 0.04$. Results were expressed as the mean + SE of average values detected within the randomly selected microscopic areas of the dorsolateral caudate-putamen of both hemispheres from four mice in each group. Comparison among groups was performed with the ANOVA one-way.

However, quantitative analysis of the relative immune-fluorescence intensity per single D2 positive disk in the striatum was slightly less intense in Torla+/− than in Torla+/+ mice (Figure 6, Table 1), suggesting that the density of D2 receptors in large and small aggregates is substantially similar in control and mutant animals.

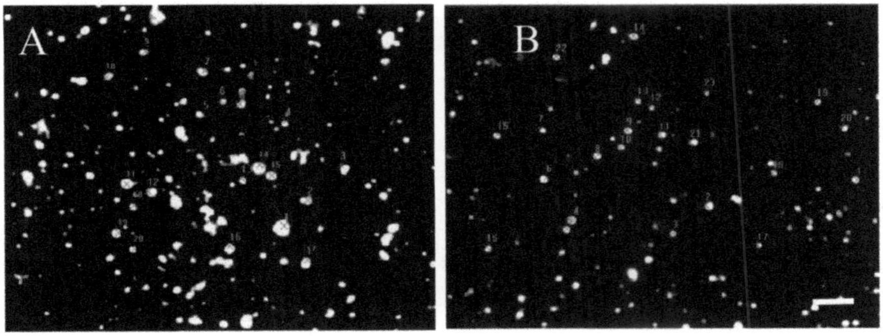

Figure 6. Table 1—Comparative fluorescence intensity of D2 receptor spots evaluated by Java image processing and ImageJ in microphotographs from confocal laser scanning microscopy. D2 positive spots were randomly selected and marked in red in control Torla+/+ (**A**), and mutant Torla+/− knock-out (**B**) mice. Bar in **B** = 2 μm. Table 1—Results were expressed as the mean + SE of the area and the average grey values (0–255) within the selected spots of the dorsolateral caudate-putamen of both hemispheres from three mice in each group. Marked decrease of the area (see analysis in Figure 5), but not of the grey level in D2 positive spots was observed in mutant Torla+/− knock-out mice, compared with controls.

Table 1. Control Mutant.

Spots	Area μm^2	Grey level	Area μm^2	Grey Level
Mean	**6.73**	**252.00**	**2.4**	**216.18**
± SE	3.41	0.30	0.16	13.63

3. Discussion

The main finding of our work is that the reduced expression of D2 receptors in the Tor1a+/−striatum is associated with marked reduction in the number and size, but not in fluorescence intensity of the D2 positive residual aggregates. These results have however, some limitations in providing a morpho-structural correlate of the downregulation of D2 receptors in dystonia.

In our study, high-resolution confocal microscopy can give well-defined immune-fluorescent images of every D2 receptors' aggregate in the microscopic field, but without configuring at the same time the mosaic of the surrounding cell structures, unlike electron microscopy immunocytochemistry. Indeed, immune-fluorescent D2 receptors appear arranged in homogeneous roundish shapes, variable in size with a diameter between ~0.3–0.5 μm, likely approaching dimensions of dopaminergic synapses on small dendritic spines (with shaft diameter ~0.04–1 μm and length ~0.1–2 μm) of striatal medium spiny neurons [29]. The D2 receptor positive disks likely represent the subcellular structures formed by D2 molecular aggregates, but we can just hypothesize the subcellular sites where the D2 aggregates are inserted in Tor1a+/+ and in Tor1a+/− mice.

Instead, at the electron microscopy level, in the striatum the D2 receptors were extensively localized to 38% of dendrites and 48% of spines of medium spiny neurons, and quite frequently on axon terminals [30]. Moreover, synapses formed by D2 immune-reactive terminals were not always easy to be identified in the striatum due to a lack of pronounced pre- or postsynaptic densities, suggesting a significant D2 extra-synaptic localization [30]. Indeed, in the striatum 60%–70% of dopaminergic axon terminals, supposed to be dopamine release sites, do not make synaptic junctional complex [31]. On the other hand, while photomicrographs obtained by transmission electron microscopy can allow detailed morphological characteristics of synapses, they can just give a rough estimate of synaptic sizes, shapes [32], and of spatial distribution frequency, due to the high magnification and limited electron microscopic fields.

By confocal microscopy in our study, we can obtain a definite spatial arrangement of the D2 signals with an estimate of sizes and shapes of the D2 receptor aggregates, and of their distribution density through a relatively large microscopic field. D2 receptor aggregates have generally a disk-shape and a size in the range of dimension of dendritic spines, and so D2 receptor aggregates have morphological similarity with synapses which can be categorized as macular synapses among the three main types of synaptic junctions (macular, perforated, and horseshoe-shaped synapses) [33]. Therefore, we would try to approximately interpret the reduction in number and size of the D2 aggregates in the striatum of Tor1a+/− mice as a global reduction of D2 synapses, but without differentiating between the synaptic and extra-synaptic sites. Reduction in synapses' number and size is a well-known mechanism reducing efficiency of neuronal connectivity. However, the residual D2 synapses in Tor1a+/− mice display immune-fluorescence intensity and macular shape similar to those seen in Tor1a+/+ mice, suggesting that the synaptic arrangement and intensity of D2 receptors can be preserved in mutant animals.

A reduced topographic density of D2 synapses would be particularly relevant for dopamine transmission in the striatum of Tor1a+/− mice. Indeed, dopaminergic transmission is predominantly characterized as nonsynaptic or peri-synaptic volume transmission, since most dopaminergic receptors are extra-synaptic, i.e., rather distant from directly opposing dopaminergic terminal varicosities [29]. Once released by tyrosine hydroxylase (TH) positive terminals, dopamine is normally not limited in a synaptic cleft but diffuses far away from release sites, traveling through extracellular spaces usually along an effective radius of 7 μm in three dimensions, until dopamine molecules bind to synaptic and extra-synaptic D1 and D2 receptors or are cleared up by presynaptic dopamine transporters [29,34,35].

In Tor1a+/− mice, sparse D2 synapses in the striatum have to face a normal density of dopamine axon terminals, since the TH-positive axonal networks in mutant animals do not show morphologically obvious abnormalities in the striatum [36]. Moreover, there was no significant difference in striatal dopamine content between mutant and control mice, while its metabolite homovanillic acid (HVA) was higher suggesting that dopamine turnover was significantly increased in the mutant mice [36,37], according to evidence for increased dopamine turnover observed in the striatum of DYT1 patients [38].

Therefore, in DYT1 mutant mice the dopamine peri-synaptic landscape appears specifically changed, since D2 synapses are sparse in the striatum, but the sites of presynaptic dopamine release appear to be preserved [36,37]. In this scenario, dopamine molecules diffusing away from release sites can have a minor probability of binding in contiguous D2 synapses [39], so that they may have a large radius of diffusion, until they are bound by D1 and D2 receptors and/or are cleared up by presynaptic dopamine transporters. On the basis of the peri-synaptic dopamine volume efflux model, we hypothesize that in DYT1 mutant mice the sparse D2 synapses may be insufficient to "gate" a relatively abundant dopamine efflux, and this consequently may result in a timing and spatially larger nonselective sphere of influence of dopamine action.

The model of sparse D2 synapses can help interpret some aspects of DYT1 physiopathology in basal ganglia. At first glance, it seems inexplicable that the baseline and amphetamine-stimulated dopamine release detected by microdialysis is lower in mutant DYT1 mice compared with controls, while at the same time the caudate-putamen tissue dopamine levels were normal [36]. We can interpret the reduced dopamine release in mutant animals to be compensatory to D2 sparse synapses. Moreover, sparse D2 synapses can explain why DYT1 dystonia individuals do not respond to medications that alter dopamine transmission [36], since D2 synapses are sparse, there would be already a relative ceiling action of endogenous dopamine quantal release from dopamine preserved synaptic terminals. Finally, sparse D2 synapses can clearly differentiate the morpho-functional correlates of basal ganglia physiopathology in dystonia and Parkinson's disease, the one characterized as post-synaptic pathology with D2 sparse synapses insufficient to gate dopamine efflux from preserved dopaminergic terminals, the other as a presynaptic pathology with loss of striatal dopamine release terminals but conserved D1 and D2 receptors in experimental parkinsonian monkey [40]. Moreover, in tissues from parkinsonian patients who died without receiving L-dopa in the last weeks of life, the binding of a radioactive ligand to Dl or D2 was elevated 20%–60% above the neurological control tissue [41]. Accordingly, dopamine receptor agonists are useful medications even regarded as first choice to delay the starting of L-dopa therapy in Parkinson's disease [42], but ineffective in DYT1 dystonia [36].

Although the sparse D2 synapses may have a relevant role in DYT1 physio-pathology, it cannot be inferred that the observed D2 alteration is primary and specific, since the D2 receptors distribution in the striatum was not compared with other receptors in our study. In a previous work, we have demonstrated that in DYT1 mice carrying human mutant torsinA (hMT), the enkephalinergic neurons express a higher cellular content of the neuropeptide enkephalin and of PDE10A, a key enzyme in the catabolism of the second messenger nucleotides [43]. Therefore, we can infer that the downregulation of D2 receptors, predominantly expressed by the medium spiny enkephalinergic neurons, seems to be a primary process not secondary to reduced cellular capacity of protein synthesis in DYT1 dystonia.

However, it is not known whether the downregulation of the D2 receptors is selective and specific to the physiopathology of dystonia, or it is part of a larger process involving altered expressions of other receptor types, and in particular of the adenosine 2A (A2A) receptor subtype. It is worth noting that in the enkephalin positive striatal medium spiny neurons there is evidence for colocalization of the D2 receptors with the A2A receptors, which together form intramembrane cell complexes called heteromers, functionally interacting in a reciprocal antagonistic manner [44]. Likely, sparse D2 synapses may also result in an impaired D2/A2A interaction in DYT1 dystonia.

Further studies should investigate the synaptic and extra-synaptic distribution of the sparse D2 aggregates and of D2/A2A heteromers, evaluating their relationship with dopamine volume transmission in the physiopathology of basal ganglia microcircuits in DYT1 mutant mice. Moreover, clarifying the developmental morpho-structural characteristics of the D2 receptor aggregates and of D2/A2A etheromers in the DYT1 mouse model would be helpful to develop preventative treatment before clinical manifestation of disease.

4. Materials and Methods

4.1. Animals and Methods

C57BL/6 Tor1a+/− knock-out transgenic mice, that mimic the loss of function of the *DYT1* dystonia *TOR1A* mutation [45], were bred at the Santa Lucia Foundation Animal Facility; mice were sacrificed at 5–6-months-old. DNA was isolated and amplified from 1- to 2-mm tail fragments with the Extract-N-Amp Tissue polymerase chain reaction (PCR) kit (XNAT2 kit; Sigma-Aldrich, Milan, Italy, and genotyping performed as reported by Bonsi et al. [26]. All the efforts were made to minimize the number of animals utilized and their suffering. Treatment and handling of mice were carried out in compliance with both the EC and Italian guidelines (2010/63EU, D.lgs. 26/2014; 86/609/EEC; D.Lvo 116/1992), according to experimental protocols approved by the Animal Ethics Committee of the University of Rome "Tor Vergata" (D.M. 153/2001-A and 43/2002-A), and by the Santa Lucia Foundation Animal Care and Use Committee (D.M. 9/2006-A), and authorized by the Italian Ministry of Health (authorization # 223/2017-PR).

For biochemical studies, the animals (n = 5 Tor1a+/+, n = 5 Tor1a+/− knock-out) were killed by cervical dislocation, and their brains were removed rapidly and placed on an ice-cold plate. Thick brain sections were cut with an Oxford vibratome, and the caudate-putamen was dissected out rapidly from both hemispheres, and promptly frozen in liquid nitrogen and stored at −80 °C [46].

For morphological studies, the animals were deeply anesthetized with tiletamine/zolazepam (80 mg/Kg) and xylazine (10 mg/Kg), and perfused trans-cardially with 1% heparin in a 50 mL 0.1 M sodium phosphate buffer, and with 250 mL 4% paraformaldehyde in a 0.1 M sodium phosphate buffer (pH 7.4). The brains were removed immediately and post-fixed in the same fixative solution overnight at 4 °C. Coronal brain sections 40-µm-thick were cut with an Oxford vibratome across the entire rostro-caudal extent of the basal ganglia, and collected and stored at 4 °C in a 0.1 M phosphate buffer that contained 0.02% sodium azide, as previously described [43].

4.2. Quantitative Analysis of D2 Protein

The quantitative analysis of D2 receptor expressions in the striatum was assessed by western blotting. Tissues were lysated in 10 mM Tris–HCl pH 8, 150 mM NaCl, 1% *v*/*v* Triton X-100, 0.1% *w*/*v* sodium deoxycholate, 0.1% *w*/*v* sodium dodecyl sulfate, 1 mM EDTA, 5 mM β-mercaptoethanol, 1 mM PMSF, and protease inhibitor cocktail (Sigma-Aldrich). Thirty µg of proteins were loaded on a 9% SDS polyacrylamide gel and subjected to electrophoresis under reducing condition. The proteins were then transferred to a nitrocellulose membrane (Bio-Rad). The blots were incubated overnight at 4 °C with a rabbit polyclonal anti-D2 receptor antibody (1:1000, AB5084P; Millipore); or mouse anti-β-actin (1:10,000; Sigma-Aldrich) as a reference standard. Immune-reactive bands were revealed by horseradish peroxidase-conjugated secondary antibodies (1:10,000, Jackson Immunoresearch), incubated in a lumi-light-enhanced chemiluminescence substrate (Bio-Rad) and exposed to chemidoc (Bio-Rad). Densitometric analysis of scanned blots was performed using the NIH ImageJ version 1.29 program (NIH, Bethesda, MD, USA).

4.3. Immune-Histochemistry

Coronal brain sections including the sensorimotor cerebral cortex, and the caudate-putamen were processed for identification of D2 positive receptors. Briefly, free floating sections were washed three times with Tris-buffered saline, pH 7.4, and endogenous peroxidase activity was inactivated by incubation for 5 min in Tris-buffered saline that contained 2% H2O2. Sections were rinsed with a Tris-buffered saline that contained 0.1% Triton X-100, and incubated with 2% normal goat serum, followed by overnight incubation at 4 °C with the primary polyclonal antibody (1:500) rabbit anti-D2 receptors (AB5084P, Millipore). The primary antibody was detected using a biotinylated secondary antibody (Vectastain ABC kit; Vector Laboratories, Burlingame, CA, USA), and an avidin horseradish peroxidase/diaminobenzidine/H2O2 chromogen system (Sigma-Aldrich, Milan, Italy). The specificity of the immunocytochemical reaction was confirmed by the absence of staining after omission of the primary antibody. After the diaminobenzidine reaction, the sections were rinsed with Tris-buffered saline and mounted on gelatine-coated slides, dehydrated, and cover-slipped with Permount for light microscopy examination. The sections were observed and photographed with a light microscope (Olympus BX51, Tokyo, Japan) equipped with an automatic micro camera (LeicaDC 300 F, Q550 IW Soft, Wetzlar, Germany).

4.4. Immune-Fluorescence Techniques and Confocal Microscopy

To evaluate the D2 receptor immune-fluorescence in striatal neurons, slices processing and confocal image acquisition were performed according to Bonsi et al. [26]. Mice were deeply anesthetized and perfused with cold 4% paraformaldehyde in a 0.12 M phosphate buffer (pH 7.4). The brain was post-fixed for at least 3 h at 4 °C and equilibrated with 30% sucrose overnight. Coronal brain sections (40 μm thick) were cut with a freezing microtome. Slices were dehydrated with serial alcohol dilutions (50%–70%–50%) and then incubated 1 h at room temperature in a 10% donkey serum solution in PBS 0.25%-Triton X-100 (PBS-Tx). The primary rabbit anti-D2 antibody (AB5084P, Millipore) was utilized (1:500, three days at 4 °C). The following secondary antibodies were used (1:200, room temperature, 2 h): Alexa 488 and Alexa 647 (Invitrogen), and cyanine 3 (cy3)-conjugated secondary antibodies (Jackson ImmunoResearch, Cambridge House, UK). Cell nuclei were detected with a blue-fluorescent DNA stain by 4′,6-diamidino-2-phenylindole (DAPI) (D9542, Sigma-Aldrich, Milan, Italy). After washout, slices were mounted on plus polarized glass slides with a Vectashield mounting medium (Super Frost Plus; Thermo Fisher Scientific) and coverslipped.

Images were acquired with a LSM800 Zeiss confocal laser scanning microscope (Zeiss, Germany), with a 5×, 20× objective, or 63× oil immersion lens (1.4 numerical aperture) with an additional digital zoom factor (1×–1.5×–2×) under no saturation conditions. Single-section images (1024 × 1024) or *z*-stack projections in the *z*-dimension (*z*-spacing, 1 μm) were collected. *Z*-stack images were acquired to analyze the whole neuronal soma, which spans multiple confocal planes. The confocal pinhole was kept at 1, the gain and the offset were adjusted to prevent saturation of the brightest signal and sequential scanning for each channel was performed. The confocal settings, including laser power, photomultiplier gain, and offset, were kept constant for each marker. For quantitative immune-fluorescence analysis, images were collected from at least 3–4 slices processed simultaneously from each striatum ($n \geq 4$ mice/genotype) and exported for analysis with the ImageJ software (NIH). Software background subtraction was utilized to reduce noise.

The intensity of D2 fluorescence per D2 positive spots was evaluated by Java image processing and ImageJ. Briefly, the D2 positive spots in immune-fluorescence microphotographs were randomly selected using a drawing selection circle closely around, and the red color image of D2 fluorescence intensity was analyzed by converting each pixel in greyscale (0–255). The minimum and maximum grey values within the selection were detected, and the mean grey value (the sum of the grey values of all the pixels in the selection area divided by the number of pixels) per spot was calculated.

4.5. Statistical Analysis

Data analysis was performed using the Statistical Package for the Social Sciences Windows, version 15.0 (SPSS). Descriptive statistics consisted of the mean ± SE for parameters with Gaussian distributions (after confirmation with histograms and the Kolgomorov–Smirnov test). The homogeneity of the variance was evaluated by Levene's test. Comparison among groups was performed with the ANOVA one-way. *P*-value < 0.05 was considered statistically significant.

5. Conclusions

The reduced expression of D2 receptors in the striatum of Tor1a+/− dystonic mice is associated with marked reduction in number and size of the D2 positive synapses. D2 sparse and small synapses may be insufficient to "gate" a relatively abundant dopamine release diffusing extrasynaptically through the interstitial space among striatal neurons. This may result in a longer duration and larger sphere of influence of dopamine transmission, making its action non-selective in striatal micro-circuitries. Our study suggests a relevant role for dopamine volume transmission in DYT1 dystonia physiopathology.

Author Contributions: V.D., M.G., and G.S. conceived and designed the study. V.D., E.P., and S.C. performed the experiments and acquired the data. R.S. performed data analysis. G.S. evaluated the results and wrote the manuscript. V.D., F.R.F., S.B., and N.B.M. contributed to critical reviewing and editing the manuscript. All authors approved the manuscript. All authors have read and agreed to the published version of the manuscript.

Funding: This research received no external funding.

Acknowledgments: We wish to thank Paola Bonsi and Antonio Pisani for kindly providing the control Tor1a+/+ and the mutant Tor1a+/− mice. We are also grateful to Paola Bonsi, and to Antonio Pisani for valuable technical suggestions and critical reading of the manuscript. We kindly acknowledge James Linch for style reviewing.

Conflicts of Interest: The authors declare no conflict of interest.

References

1. Marsden, C.D.; Sheehy, M.P. Writer's cramp. *Trends Neurosci.* **1990**, *13*, 148–153. [CrossRef]
2. Berardelli, A.; Rothwell, J.C.; Hallett, M.; Thompson, P.D.; Manfredi, M.; Marsden, C.D. The pathophysiology of primary dystonia. *Brain* **1998**, *121*, 1195–1212. [CrossRef] [PubMed]
3. Lanska, D.J. Chapter 33: The History of Movement Disorders. In *Handbook of Clinical Neurology. The History of Neurology*; Elsevier: Amsterdam, The Netherlands, 2010; Volume 95, pp. 501–546.
4. Albanese, A.; Bhatia, K.; Bressman, S.B.; Delong, M.R.; Fahn, S.; Fung, V.S.; Hallett, M.; Jankovic, J.; Jinnah, H.A.; Klein, C.; et al. Phenomenology and classification of dystonia: A consensus update. *Mov. Disord.* **2013**, *28*, 863–873. [CrossRef] [PubMed]
5. Munts, A.G.; Koehler, P.J. How psychogenic is dystonia? Views from past to present. *Brain* **2010**, *133 Pt 5*, 1552–1564. [CrossRef]
6. Schwalbe, W. *Eine Eigentumliche Tonische Krampfform mit hysterischen Symptomen*; Inaugural-Dissertation-UniversitätsBuch-druckerei, G. Schade:: Berlin, Germany, 1908.
7. Oppenheim, H. Uber eine eigenartige Krampfkrankheit des kindlichen und jugendlichen Alters (Dysbasia lordotica progressiva, Dystonia musculorum deformans). *Neurol. Centrabl.* **1911**, *30*, 1090–1107.
8. Ozelius, L.J.; Hewett, J.W.; Page, C.E.; Bressman, S.B.; Kramer, P.L.; Shalish, C.; de Leon, D.; Brin, M.F.; Raymond, D.; Corey, D.P.; et al. The early-onset torsion dystonia gene (DYT1) encodes an ATP-binding protein. *Nat. Genet.* **1997**, *17*, 40–48. [CrossRef]
9. Fahn, S.; Bressman, S.B.; Marsden, C.D. Classification of dystonia. *Adv. Neurol.* **1998**, *78*, 1–10. [CrossRef]
10. Standaert, D.G. Update on the pathology of dystonia. *Neurobiol. Dis.* **2011**, *42*, 148–151. [CrossRef]
11. Rondot, P. The shadow of movement. *J. Neurol.* **1991**, *238*, 411–419. [CrossRef]
12. Kanowsky, P. Dystonia: A disorder of motor programming or motor execution? *Mov. Disord.* **2002**, *17*, 1143–1147. [CrossRef]
13. DeFazio, G.; Berardelli, A.; Hallett, M. Do primary adult-onset focal dystonias share aetiological factors? *Brain* **2007**, *130*, 1183–1193. [CrossRef] [PubMed]
14. Ulug, A.M.; Vo, A.; Argyelan, M.; Tanabe, L.; Schiffer, W.K.; Dewey, S.; Dauer, W.T.; Eidelberg, D. Cerebellothalamocortical pathway abnormalities in torsinA DYT1 knock-in mice. *Proc. Natl. Acad. Sci. USA* **2011**, *108*, 6638–6643. [CrossRef] [PubMed]
15. Goodchild, R.E.; Grundmann, K.; Pisani, A. New genetic insights highlight 'old' ideas on motor dysfunction in dystonia. *Trends Neurosci.* **2013**, *36*, 717–725. [CrossRef] [PubMed]
16. Asanuma, K.; Ma, Y.; Okulski, J.; Dhawan, V.; Chaly, T.; Carbon, M.; Bressman, S.B.; Eidelberg, D. Decreased striatal D2 receptor binding in non-manifesting carriers of the DYT1 dystonia mutation. *Neurology* **2005**, *64*, 347–349. [CrossRef] [PubMed]
17. Carbon, M.; Niethammer, M.; Peng, S.; Raymond, D.; Dhawan, V.; Chaly, T.; Ma, Y.; Bressman, S.; Eidelberg, D. Abnormal striatal and thalamic dopamine neurotransmission: Genotype-related features of dystonia. *Neurology* **2009**, *72*, 2097–2103. [CrossRef]

18. Napolitano, F.; Pasqualetti, M.; Usiello, A.; Santini, E.; Pacini, G.; Sciamanna, G.; Errico, F.; Tassone, A.; Di Dato, V.; Martella, G.; et al. Dopamine D2 receptor dysfunction is rescued by adenosine A2A receptor antagonism in a model of DYT1 dystonia. *Neurobiol. Dis.* **2010**, *38*, 434–445. [CrossRef]
19. Yokoi, F.; Dang, M.T.; Li, J.; Standaert, D.G.; Li, Y. Motor deficits and decreased striatal dopamine receptor 2 binding activity in the striatum-specific Dyt1 conditional knock-out mice. *PLoS ONE* **2011**, *6*, e24539. [CrossRef]
20. Dang, M.T.; Yokoi, F.; Cheetham, C.C.; Lu, J.; Vo, V.; Lovinger, D.M.; Li, Y. An anticholinergic reverses motor control and corticostriatal LTD deficits in Dyt1 DGAG knock-in mice. *Behav. Brain Res.* **2012**, *226*, 465–472. [CrossRef]
21. Pisani, A.; Martella, G.; Tscherter, A.; Bonsi, P.; Sharma, N.; Bernardi, G.; Standaert, D. Altered responses to dopaminergic D2 receptor activation and N-type calcium currents in striatal cholinergic interneurons in a mouse model of DYT1 dystonia. *Neurobiol. Dis.* **2006**, *24*, 318–325. [CrossRef]
22. Sciamanna, G.; Tassone, A.; Martella, G.; Mandolesi, G.; Puglisi, F.; Cuomo, D.; Madeo, G.; Ponterio, G.; Standaert, D.G.; Bonsi, P.; et al. Developmental profile of the aberrant dopamine D2 receptor response in striatal cholinergic interneurons in DYT1 dystonia. *PLoS ONE* **2011**, *6*, e24261. [CrossRef]
23. Sciamanna, G.; Hollis, R.; Ball, C.; Martella, G.; Tassone, A.; Marshall, A.; Parsons, D.; Li, X.; Yokoi, F.; Zhang, L.; et al. Cholinergic dysregulation produced by selective inactivation of the dystonia-associated protein torsinA. *Neurobiol. Dis.* **2012**, *47*, 416–427. [CrossRef] [PubMed]
24. Martella, G.; Maltese, M.; Nistico, R.; Schirinzi, T.; Madeo, G.; Sciamanna, G.; Ponterio, G.; Tassone, A.; Mandolesi, G.; Vanni, V.; et al. Regional specificity of synaptic plasticity deficits in a knock-in mouse model of DYT1 dystonia. *Neurobiol. Dis.* **2014**, *65*, 124–132. [CrossRef] [PubMed]
25. Scarduzio, M.; Zimmerman, C.N.; Jaunarajs, K.L.; Wang, Q.; Standaert, D.G.; McMahon, L.L. Strength of cholinergic tone dictates the polarity of dopamine D2 receptor modulation of striatal cholinergic interneuron excitability in DYT1 dystonia. *Exp. Neurol.* **2017**, *295*, 162–175. [CrossRef] [PubMed]
26. Bonsi, P.; Ponterio, G.; Vanni, V.; Tassone, A.; Sciamanna, G.; Migliarini, S.; Martella, G.; Meringolo, M.; Dehay, B.; Doudnikoff, E.; et al. RGS 9-2 rescues dopamine D2 receptor levels and signaling in DYT1 dystonia mouse models. *EMBO Mol. Med.* **2018**, *11*, e9283.
27. Peterson, D.A.; Sejnowski, T.J.; Poizner, H. Convergent evidence for abnormal striatal synaptic plasticity in dystonia. *Neurobiol. Dis.* **2010**, *37*, 558–573. [CrossRef] [PubMed]
28. Quartarone, A.; Pisani, A. Abnormal plasticity in dystonia: Disruption of synaptic homeostasis. *Neurobiol. Dis.* **2011**, *42*, 162–170. [CrossRef]
29. Yao, W.-D.; Spealman, R.D.; Zhang, J. Dopaminergic signaling in dendritic spines. *Biochem. Pharmacol.* **2008**, *75*, 2055–2069. [CrossRef]
30. Hersch, S.M.; Ciliax, B.J.; Gutekunst, C.A.; Rees, H.D.; Heilman, C.J.; Yung, K.K.L.; Bolam, J.P.; Ince, E.; Yi, H.; Levey, A.I. Electron microscopic analysis of D1 and D2 dopamine receptor proteins in the dorsal striatum and their synaptic relationships with motor corticostriatal afferents. *J. Neurosci.* **1995**, *75*, 5222–5237. [CrossRef]
31. Descarries, L.; Watkins, K.C.; Garcia, S.; Bosler, O.; Doucet, G. Dual character, asynaptic and synaptic, of the dopamine innervation in adult rat neostriatum: A quantitative autoradiographic and immunocytochemical analysis. *J. Comp. Neurol.* **1996**, *375*, 167–186. [CrossRef]
32. DeFelipe, J.; Marco, P.; Busturia, I.; Merchán-Pérez, A. Estimation of the number of synapses in the cerebral cortex: Methodological considerations. *Cereb. Cortex* **1999**, *9*, 722–732. [CrossRef]
33. Santuy, A.; Rodríguez, J.-R.; De Felipe, J.; Merchán-Pérez, A. Study of the size and shape of synapses in the juvenile rat somatosensory cortex with 3D electron microscopy. *eNeuro* **2018**, *5*. [CrossRef] [PubMed]
34. Arbuthnott, G.W.; Wickens, J. Space, time and dopamine. *Trends Neurosci.* **2007**, *30*, 62–69. [CrossRef] [PubMed]
35. Rice, M.E.; Patel, J.C.; Cragg, S.J. Dopamine release in the basal ganglia. *Neuroscience* **2011**, *198*, 112–137. [CrossRef] [PubMed]
36. Song, C.-H.; Fan, X.; Exeter, C.J.; Hess, E.J.; Jinnah, H.A. Functional analysis of dopaminergic systems in a DYT1 knock-in mouse model of dystonia. *Neurobiol. Dis.* **2012**, *48*, 66–78. [CrossRef] [PubMed]
37. Zhao, Y.; De Cuypere, M.; LeDoux, M.S. Abnormal motor function and dopamine neurotransmission in DYT1 ΔGAG transgenic mice. *Exp. Neurol.* **2008**, *210*, 719–730. [CrossRef] [PubMed]

38. Augood, S.J.; Hollingsworth, Z.; Albers, D.S.; Yang, L.; Leung, J.-C.; Muller, B.; Klein, C.; Breakefield, X.O.; Standaert, D.G. Dopamine transmission in DYT1 dystonia: A biochemical and autoradiographical study. *Neurology* **2002**, *59*, 445–448. [CrossRef]
39. Simonyan, K.; Berman, B.D.; Herscovitch, P.; Hallett, M. Abnormal striatal dopaminergic neurotransmission during rest and task production in spasmodic dysphonia. *J. Neurosci.* **2013**, *33*, 14705–14714. [CrossRef]
40. Hadipour-Niktarash, A.; Rommelfanger, K.S.; Masilamoni, G.J.; Smith, Y.; Wichmann, T. Extrastriatal D2-like receptors modulate basal ganglia pathways in normal and Parkinsonian monkeys. *J. Neurophysiol.* **2012**, *107*, 1500–1512. [CrossRef]
41. Seeman, P.; Niznik, H.B. Dopamine receptors and transporters in Parkinson's disease and schizophrenia. *FASEB J.* **1990**, *4*, 2737–2744. [CrossRef]
42. Hisahara, S.; Shimohama, S. Dopamine receptors and Parkinson's disease. *Int. J. Med. Chem.* **2011**, *2011*, 1–16. [CrossRef]
43. D'Angelo, V.; Castelli, V.; Giorgi, M.; Cardarelli, S.; Saverioni, I.; Palumbo, F.; Bonsi, P.; Pisani, A.; Giampà, C.; Sorge, R.; et al. Phosphodiesterase-10A inverse changes in striatopallidal and striatoentopeduncular pathways of a transgenic mouse model of DYT1 dystonia. *J. Neurosci.* **2017**, *37*, 2112–2124. [CrossRef] [PubMed]
44. Ferré, S.; Quiroz, C.; Woods, A.S.; Cunha, R.; Popoli, P.; Ciruela, F.; Lluis, C.; Franco, R.; Azdad, K.; Schiffmann, S.N. An update on adenosine A2A-Dopamine D2 receptor interactions. Implications for the function of G Protein-coupled receptors. *Curr. Pharm. Des.* **2008**, *14*, 1468–1474. [CrossRef] [PubMed]
45. Goodchild, R.E.; Kim, C.E.; Dauer, W.T. Loss of the dystonia-associated protein torsinA selectively disrupts the neuronal nuclear envelope. *Neuron* **2005**, *48*, 923–932. [CrossRef]
46. Giorgi, M.; D'Angelo, V.; Esposito, Z.; Nuccetelli, V.; Sorge, R.; Martorana, A.; Stefani, A.; Bernardi, G.; Sancesario, G. Lowered cAMP and cGMP signalling in the brain during levodopa-induced dyskinesias in hemiparkinsonian rats: New aspects in the pathogenetic mechanisms. *Eur. J. Neurosci.* **2008**, *28*, 941–950. [CrossRef]

International Journal of
Molecular Sciences

MDPI

Review

Role of the Serotonin Receptor 7 in Brain Plasticity: From Development to Disease

Marianna Crispino [1], Floriana Volpicelli [2,*] and Carla Perrone-Capano [2,3]

1 Department of Biology, University of Naples Federico II, 80126 Naples, Italy; crispino@unina.it
2 Department of Pharmacy, University of Naples Federico II, 80131 Naples, Italy; perrone@unina.it
3 Institute of Genetics and Biophysics "Adriano Buzzati Traverso", National Research Council, (CNR), 80131 Naples, Italy
* Correspondence: floriana.volpicelli@unina.it; Tel.: +39-081-678453

Received: 6 December 2019; Accepted: 10 January 2020; Published: 13 January 2020

Abstract: Our knowledge on the plastic functions of the serotonin (5-HT) receptor subtype 7 (5-HT7R) in the brain physiology and pathology have advanced considerably in recent years. A wealth of data show that 5-HT7R is a key player in the establishment and remodeling of neuronal cytoarchitecture during development and in the mature brain, and its dysfunction is linked to neuropsychiatric and neurodevelopmental diseases. The involvement of this receptor in synaptic plasticity is further demonstrated by data showing that its activation allows the rescue of long-term potentiation (LTP) and long-term depression (LTD) deficits in various animal models of neurodevelopmental diseases. In addition, it is becoming clear that the 5-HT7R is involved in inflammatory intestinal diseases, modulates the function of immune cells, and is likely to play a role in the gut-brain axis. In this review, we will mainly focus on recent findings on this receptor's role in the structural and synaptic plasticity of the mammalian brain, although we will also illustrate novel aspects highlighted in gastrointestinal (GI) tract and immune system.

Keywords: brain connectivity; brain development; gut-brain axis; neurodevelopmental diseases; neuronal cytoarchitecture; neuroplasticity; regulatory T cells; serotonin (5-HT)

1. Serotonin Overview

1.1. Serotonin Metabolism

Brain 5-HT is a neurotransmitter playing a key role in modulating neuronal circuit development and activities. The serotonergic neurons, through their extensive axonal network, are able to reach and influence nearly all the Central Nervous System (CNS) areas. As a consequence, 5-HT regulates a plethora of functions such as sleep and circadian rhythms, mood, memory and reward, emotional behavior, nociception and sensory processing, autonomic responses, and motor activity [1].

Our current understanding of the development, evolution, and function of 5-HT neurotransmission is derived from different model organisms, spanning from invertebrates to vertebrates [2]. It is noteworthy that in all species, the serotonergic network is highly plastic, showing changes in its anatomical organization all through the life of the organisms.

5-HT metabolic pathways, reuptake, and degradation are broadly conserved among multicellular organisms [2]. 5-HT is synthesized from the amino acid tryptophan, which is an essential dietary supplement. Tryptophan is hydroxylated to 5-hydroxytryptophan (5-HTP) by the tryptophan-hydroxylase (TPH)—the rate limiting enzyme for 5-HT biosynthesis. 5-HTP, in turn, is converted in 5-HT by the aromatic L-amino acid decarboxylase. The enzyme TPH has two distinct isoforms encoded by two genes: the *Tph1* is expressed in peripheral tissues and pineal gland, while the *Tph2* is selectively expressed in the CNS and in the enteric neurons of the gut [3]. Studies on TPH -knockout (KO) mice confirmed that the synthesis of 5-HT in the brain is driven by TPH2, whereas the synthesis of 5-HT in peripheral organs is driven by TPH1 [4]. Since 5-HT is unable to cross the blood–brain barrier, at least in adult life, the central and the peripheral serotonergic systems are independently regulated. The synaptic effects of 5-HT are mainly terminated by its reuptake into 5-HT nerve terminals mediated by the 5-HT transporter.

The vast array of brain functions exerted by 5-HT neurotransmission in the CNS is made more complex by the interaction of the 5-HT system with many other classical neurotransmitter systems. Through the activation of serotonergic receptors located on cholinergic, dopaminergic, GABAergic or glutamatergic neurons, 5-HT exerts its effects modulating the neurotransmitter release of these neurons [5,6]. In addition, cotransmission—here defined as the release of more than one classical neurotransmitter by the same neuron—occurs also in 5-HT neurons. Among the cotransmitters released by 5-HT neurons, glutamate [7], and possibly other amino acids [8] were identified. The regulation and functional effects of this neuronal cotransmission are still poorly understood and are the object of intense investigation [9].

1.2. Role of Serotonin in Morphological Remodeling of CNS Circuits

In the mammalian brain, 5-HT neurons are among the earliest neurons to be specified during development [10]. They are located in the hindbrain and are grouped in nine raphe nuclei, designated as B1–B9 [11]. Although they are relatively few (about 30,000 in the mouse and 300,000 in humans), they give rise to extensive rostral and caudal axonal projections to the entire CNS, representing the most widely distributed neuronal network in the brain [12].

In addition to its well-established role as a neurotransmitter, 5-HT exerts morphogenic actions on the brain, influencing several neurodevelopmental processes such as neurogenesis, cell migration, axon guidance, dendritogenesis, synaptogenesis and brain wiring [13].

Besides the endogenous 5-HT, the brain of the fetus also receives it from the placenta of the mother. Thus, the placenta represents a crucial micro-environment during neurodevelopment, orchestrating a series of complex maternal-fetal interactions. The contribution of this interplay is essential for the correct development of the CNS and for long-term brain functions [14]. Therefore, maternal insults to placental microenvironment may alter embryonic brain development, resulting in prenatal priming of neurodevelopmental disorders [15]. For instance, in mice it has been shown that maternal inflammation results in an upregulation of tryptophan conversion to 5-HT within the placenta, leading to altered serotonergic axonal growth in the fetal forebrain. These results indicate that the level of 5-HT during embryogenesis is critical for proper brain circuit wiring, and open a new perspective for understanding the early origins of neurodevelopmental disorders [16–18].

The importance of a correct 5-HT level in the brain has been demonstrated by numerous studies on mice models. When the genes involved in 5-HT uptake or degradation are knocked out, the increased 5-HT levels in the brain lead to the altered topographical development of the somatosensory cortex and incorrect cortical interneuron migration [19,20]. On the other hand, the transient disruption of 5-HT signaling, during a restricted period of pre- or postnatal development, using pharmacological (selective serotonin reuptake inhibitor exposure) animal models, leads to long-term behavioral abnormalities, such as increased anxiety in adulthood [21,22]. These animals do not show gross morphological alterations in the CNS suggesting that the lack of cerebral 5-HT may only affect the fine tuning of specific serotonergic circuits. This hypothesis has been recently confirmed using a mouse model in which the enhanced green fluorescent protein is knocked into the Tph2 locus, resulting in lack of brain 5-HT, and allowing the detection of serotonergic system through enhanced fluorescence, independently of 5-HT immunoreactivity. In these mice, the serotonergic innervation was apparently normal in cortex and striatum. On the other hand, mutant adult mice showed a dramatic reduction of serotonergic axon terminal arborization in the diencephalic areas, and a marked serotonergic hyperinnervation in the nucleus accumbens and in the hippocampus [23]. These results demonstrate that brain 5-HT plays a key role in regulating the wiring of the serotonergic system during brain development. Interestingly, the transient silencing of 5-HT transporter expression in neonatal thalamic neurons affects somatosensory barrel architecture through the selective alteration of dendritic structure and trajectory of late postnatal interneuron development in the mouse cortex [24]. Altogether, these findings indicate that perturbing 5-HT levels during critical periods of early development influences later neuronal development through alteration of CNS connectivity that may persist into the adulthood [17,25,26]. Interestingly, recent evidence demonstrated that changes in 5-HT homeostasis affect axonal branch complexity, not only during development but also in adult life [27]. In adult TPH2-conditional KO mice it was shown that the administration of the serotonin precursor 5-hydroxytryptophan was able to re-establish the 5-HT signaling and to rescue defects in serotonergic system organization [27].

Interestingly, in recent elegant experiments that combined chemogenetics and fMRI, it was demonstrated that, in adult mice, the endogenous stimulation of 5-HT-producing neurons does not affect global brain activity but selectively activates specific cortical and subcortical areas. By contrast, the pharmacological increase of 5-HT levels determined widespread fMRI deactivation, possibly reflecting the mixed contribution of central and perivascular constrictive effects [28].

On the whole, findings from genetic mouse models confirm that the level of 5-HT during brain ontogeny is critical for proper CNS circuit wiring, and suggest that alterations in 5-HT signaling during brain development have profound implications for behavior and mental health across the life span. Indeed, a plethora of genetic and pharmacological studies have linked defects of brain 5-HT signaling with psychiatric and neurodevelopmental disorders, such as major depression, anxiety, schizophrenia, obsessive compulsive disorder and Autism Spectrum Disorders (ASD) [17,29,30]. In addition, it is becoming increasingly clear that 5-HT has a crucial role also in the maintenance of mature neuronal circuitry in the brain, opening novel perspectives in rescuing defects of CNS connectivity in the adult. For instance, the potential of 5-HT neurons to remodel their morphology during the entire life is indicated by the well-known capability of 5-HT axons of the adult to regenerate and sprout after lesions [26,31].

However, understanding the cellular and molecular mechanisms underlying the effects of 5-HT during brain development, maintenance and dysfunction is challenging, in part due to the existence of at least 14 subtypes of receptors (5-HTRs) grouped in seven distinct classes (from 5-HT1R to 5-HT7R). All 5-HT receptors are broadly distributed in the brain where they display a highly dynamic developmental and region-selective expression pattern and trigger different signaling pathways. The 5-HT receptors are typical G-protein-coupled-receptors with seven transmembrane domains, with the exception of the 5-HT3 receptor, which is a ligand-gated ion channel [32].

2. Role of the 5-HT7R in Shaping Neuronal Circuits

2.1. The 5-HT7R

The 5-HT7R, the last discovered member of the 5-HTR family [33,34], has always been the subject of intense investigation, due to its high expression in functionally relevant regions of the brain [35,36]. Accordingly, several recent data have elucidated its role in a wide range of physiological functions in the mammalian CNS and also in peripheral organs [37]. Interestingly, emerging findings indicate that 5-HT7R is involved in brain plasticity, being one of the players contributing not only to shape brain networks during development but also to remodel neuronal wiring in the mature brain, thus controlling higher cognitive functions (see Sections 2.2 and 2.3). Therefore, this receptor is currently considered as potential target for the treatment of several neuropsychiatric and neurodevelopmental disorders, (as discussed in Section 3), also in view of the fact that its ligands have a wide range of neuropharmacological effects [38,39].

In the mammalian CNS, the 5-HT7R is mainly expressed in the spinal cord, thalamus, hypothalamus, hippocampus, prefrontal cortex, striatal complex, amygdala and in the Purkinje neurons of the cerebellum [40,41]. This wide distribution reflects the numerous functions in which the receptor is involved, such as circadian rhythms, sleep-wake cycle, thermoregulation, learning and memory processing, and nociception [37].

In mammals, this receptor exhibits a number of functional splice variants due to the presence of introns in the 5-HT7R gene and to alternative splicing. The splice variants of the receptor, named 5-HT7(a), (b), (c) in rodents, and 5-HT7(a), (b), (d) in humans [36,42,43], do not show significant differences in localization, ligand binding affinities, and activation of adenylate cyclase [36]. To date, the only functional difference between the splice variants is that the human 5-HT7(d) isoform displays a different pattern of receptor internalization compared to the other isoforms [44].

The 5-HT7R is a G protein-coupled receptor, that activates at least two different signaling pathways. The classical pathway relies on the activation of $G\alpha s$ and the consequent stimulation of adenylate cyclase, leading to an increase in cyclic adenosine monophosphate (cAMP). The latter activates protein kinase A (PKA), that in turn phosphorylates various proteins such as the mitogen-activated protein kinase and extracellular signal-regulated kinases (ERK) [39].

Another 5-HT7R pathway depends on the activation of $G\alpha_{12}$, that in turn triggers stimulation of Rho GTPases, Cdc42 and RhoA; these intracellular signaling proteins, critical for the regulation of cytoskeleton organization, lead to morphological modifications of fibroblasts and neurons [45].

5-HT7R signaling also involves changes in intracellular Ca^{2+} concentration and Ca^{2+}/calmodulin pathways [46,47], as well as PKA independent mechanisms which include exchange protein directly activated by cAMP (EPAC) signaling [48].

5-HT receptor signaling has been recently shown to also depend on their oligomerization. In particular the 5-HT7R can form homodimers, as well as heterodimers with 5-HT1AR [49]. The latter, when is in a monomeric conformation, causes a decrease in cAMP concentration through activation of the Gi. Heterodimerization with 5-HT7R inhibits the 5-HT1AR cAMP signaling pathway, while homodimerization of both receptors do not influence the respective cAMP pathways. These findings suggest that oligomerization of G-protein-coupled-receptors may have profound functional consequences on their downstream signaling, thus triggering cellular and developmental-specific regulatory effects.

2.2. Role of the 5-HT7R in Shaping Neuronal Circuits during Development

The influence of the 5-HT7R on neuronal morphology has stimulated interest in studying its potential role in the establishment and maintenance of brain connectivity and in synaptic plasticity. The availability of selective agonists and antagonists, as well as that of genetically modified mice lacking the 5-HT7R, has shed light on the physio-pathological role of this receptor [39,50,51]. By using rodents' primary cultures of hippocampal neurons and various 5-HT7R agonists in combination with selective antagonists, it was consistently shown that the pharmacological stimulation of the endogenous 5-HT7R promotes a pronounced extension of neurite length [48,52,53]. The morphogenic effects of 5-HT7R stimulation have also been demonstrated in cultured neurons from additional embryonic forebrain areas, such as the striatum and the cortex [54,55] (Figure 1). Neurite elongation was shown to rely on *de novo* protein synthesis and multiple signaling systems, such as ERK, Cdk5, the RhoGTPase Cdc42 and mTOR. These pathways converge to promote the reorganization of the neuronal cytoskeleton through qualitative and quantitative changes of selected proteins, such as microtubule-associated proteins and cofilin [54,56]. In hippocampal neurons, it has been demonstrated that 5-HT7R finely modulates the NMDA receptors activity [57,58]. Furthermore, 5-HT7R activation increases phosphorylation of the GluA1 AMPA receptor subunit and AMPA receptor-mediated neurotransmission in the hippocampus [59,60]. Consistent with these findings, 5-HT7R-KO mice display reduced LTP in the hippocampus [61].

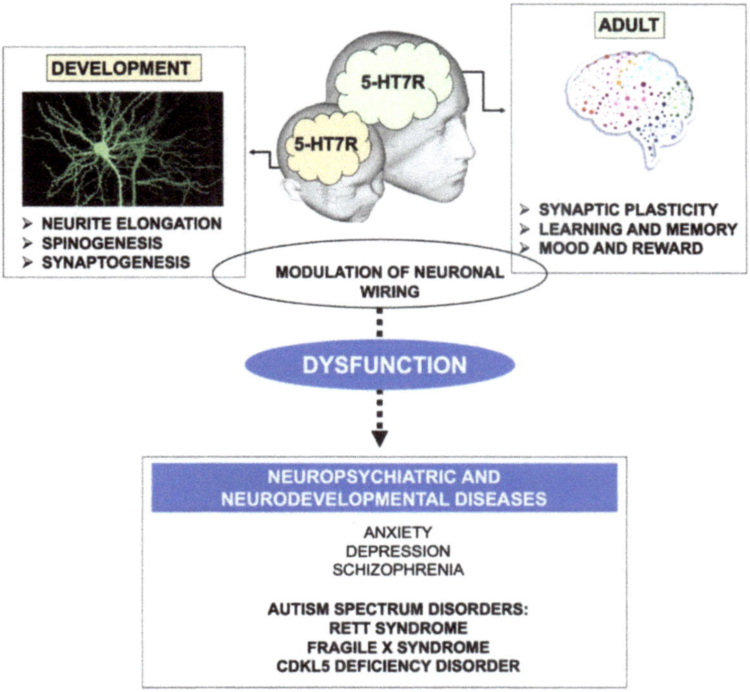

Figure 1. Schematic drawing illustrating the role of the 5-HT7R in brain plasticity and connectivity. During development, the 5-HT7R contributes to proper neuronal wiring through the stimulation of neurite elongation, growth and maturation of dendritic spines, and synaptogenesis. During adulthood, the 5-HT7R signaling stimulates synaptic plasticity (LTP, LTD and structural remodeling of neuronal connections), which in turn affects many physiological functions, such as learning, memory, mood and reward. Dysregulated 5-HT7R signaling was demonstrated in neuropsychiatric and neurodevelopmental diseases characterized by altered brain connectivity. Notably, 5-HT7R stimulation exerts a widespread beneficial effect on behavioral and molecular alterations in various mouse models of Autism Spectrum Disorders (highlighted in bold).

Chronic stimulation of the 5-HT7R/Gα_{12} signaling pathway promotes dendritic spine formation, enhances basal neuronal excitability, and modulates LTP in organotypic slices preparation from the hippocampus of juvenile mice. Interestingly, 5-HT7R stimulation does not affect neuronal morphology, synaptogenesis, and synaptic plasticity in hippocampal slices from adult animals, probably due to decreased hippocampal expression of the 5-HT7R during later postnatal stages [62]. It has been recently hypothesized that this decline could be due to the simultaneous upregulation of the microRNA (miR)-29a in the developing hippocampus. Indeed 5-HT7R mRNA is downregulated by the miR-29a in cultured hippocampal neurons, and miR-29a overexpression impairs the 5-HT7R-dependent neurite elongation [63].

Neuronal remodeling is highly influenced by the extracellular matrix. Accordingly, it has been shown that the physical interaction between the 5-HT7R and the hyaluronan receptor CD44, a main component of the extracellular matrix, plays a crucial role in synaptic remodeling. Briefly, stimulation of the 5-HT7R increases the activity of the metalloproteinase MMP-9, which, in turn, cleaves the extracellular domain of CD44. This signaling cascade promotes detachment from the extracellular matrix, thus triggering dendritic spine elongation in the hippocampal neurons of the mice [64].

In accordance with the influence of the 5-HT7R signaling pathways in remodeling developing forebrain neuron morphology, it was shown that prolonged stimulation of this receptor and the downstream activation of Cdk5 and Cdc42 increased the density of filopodia-like dendritic spines and synaptogenesis in cultured striatal and cortical neurons [65]. The crucial role of 5-HT7R in shaping developing synapses (Figure 1) was confirmed by the pharmacological inactivation of the receptor as well as through the analysis of early postnatal neurons isolated from 5-HT7R-deficient mice. It is noteworthy that, when 5-HT7R was blocked pharmacologically, and in 5-HT7R-KO neurons, the number of dendritic spines decreased, suggesting that constitutive receptor activity is critically involved in dendritic spinogenesis. From this point of view, a detailed analysis of dendritic spine shape and density in the brain of 5-HT7R-KO mice at various ages would be crucial to assess the physiological effects of this receptor on neuronal cytoarchitecture.

The involvement of 5-HT7R in spinogenesis and synaptogenesis—together with the demonstration that its activation is able to stimulate protein synthesis-dependent neurite elongation, as well as axonal elongation [54,56]—suggests the intriguing possibility that the activation of this receptor may be linked to the axonal and synaptic system of protein synthesis. The local system of protein synthesis has been demonstrated to play a crucial role in synaptic plasticity—although its regulatory mechanisms are only partially understood [66–68]—and 5-HT7R and its related pathways are good candidates to be part of this system.

2.3. Role of the 5-HT7R in Remodeling Neuronal Circuits in Adults

Neuronal circuits remain able to reorganize in response to experience well into adulthood, continuing to exhibit robust plasticity along the entire life [69]. Consistently, the action of 5-HT7R on the modulation of neuronal plasticity is not restricted to embryonic and early postnatal development, but can also occur in later developmental stages and in adulthood (Figure 1).

Interestingly, it was shown that selective pharmacological stimulation of 5-HT7R during adolescence determines its persistent upregulation in adult rat forebrain areas [70]. Likewise, it has been hypothesized that 5-HT7R may underlie the persistent structural rearrangements of the brain reward pathways occurring during postnatal development, following exposure to methylphenidate, the elective drug for the treatment of Attention Deficit Hyperactivity Disorder [71]. Accordingly, stimulation of the 5-HT7R in adolescent rats leads to increased dendritic arborization in the nucleus accumbens—a limbic area involved in reward—as well as increased functional connectivity in different forebrain networks likely to be involved in anxiety-related behavior [72]. Changes in dendritic spine formation, turnover and shape occur during the entire life span in response to stimuli that trigger long-term alterations in synaptic efficacy, such as LTP and LTD [73–75]. Consistently, it has been shown that the activation of 5-HT7R in hippocampal slices from wild type mice (as well as in Fragile X Syndrome mice, see next paragraph) reverses LTD mediated by metabotropic glutamate receptors (mGluR-LTD), a form of plasticity playing a crucial role in cognition and in behavioral flexibility [59]. Moreover, the acute in vivo administration of a selective 5-HTR7 agonist improved cognitive performance in mice [76]. These results are consistent with the hypothesis that long-term changes of synaptic plasticity, which are a substrate of learning and memory formation, lead to neural network rewiring (Figure 1). Accordingly, the 5-HT7R-KO mice exhibit reduced hippocampal LTP, and specific impairments in contextual learning, seeking behavior and allocentric spatial memory [61,77].

Interestingly, the expression level of 5-HT7R in the hippocampal CA3 region, an area of the brain involved in allocentric navigation, decreases with age [78], suggesting that the spatial memory deficits associated with aging could be attributed to decreased 5-HT7R activity in this region of the brain. Conversely, another group reported that hippocampal expression of 5-HT7R does not change with age, but exhibits 24 h rhythms [79]. This observation should be taken into account in the interpretation of previous findings, as well as in planning future experiments. Several other studies have produced contradictory results related to the involvement of 5-HT7R in memory and attention-related processes [80,81], probably due to experimental differences (animal strain, behavioral tests, compounds and doses, route of administration, etc.). In conclusion, although the role of this receptor on cognitive functions needs to be fully elucidated, it is clear that it modulates various aspects of learning and memory processes.

Interestingly, the 5-HT7R is also involved in bidirectional modulation of cerebellar synaptic plasticity, since its activation induces LTD at the parallel fiber-Purkinje cell synapse, whereas it blocks LTP induced by parallel fiber stimulation [41]. These results suggest that the receptor might be involved in motor learning, a cognitive function depending on the activity of cerebellar circuits [82].

Altogether, these findings strongly suggest that the 5-HT7R plays a role in modulating synaptic plasticity and neuronal connectivity in both developing and mature brain circuits, although the molecular and cellular mechanisms underlying this modulation are only partially understood (Figure 1).

3. The 5-HT7R and Neurological Diseases

Numerous brain disorders, such as ASD, cognitive and mood dysfunctions, schizophrenia, depression, anxiety, impulsivity, epilepsy, migraine and neuropathic pain show altered 5-HT7R-mediated signaling [38]. The potential involvement of 5-HT7R in most of these diseases was discovered studying the effects of a broad range of antidepressant and antipsychotic drugs that interact with the receptor, displaying high affinity [39]. Recently, the importance of 5-HT7R modulation was brought to the attention of psychiatric and pharmacological communities, since a novel very effective and atypical mood-stabilizing antipsychotic drug, lurasidone, predominantly blocks this receptor. This drug also acts as agonist of the 5-HT1AR. Experiments on animal models indicate that chronic treatment with lurasidone enhances 5-HT transmission in dorsal raphe nuclei by coordinated 5-HT1AR agonism and 5-HT7R antagonism through modulation of GABAergic and glutamatergic pathways, thus contributing to the augmentation of the drug's antidepressive effects [83,84].

In line with the possible involvement of 5-HT7R in the mechanisms of action of antidepressants, genome-wide association studies in humans have suggested a relationship between 5-HTR7 genetic polymorphisms and schizophrenia [85]. Likewise, a very recent work showed that one single nucleotide polymorphism located in the promoter region of the 5-HT7R gene is associated with a better response to two antidepressants, paroxetine and fluoxetine, that are selective 5-HT reuptake inhibitors. These data provide novel pharmacogenomic evidence to support the role of 5-HT7R in antidepressant response [86].

However, the pharmacological and genetic manipulation of 5-HT7R in animal models of depression, anxiety and schizophrenia has often given inconsistent or conflicting results. Experimental differences (for instance animals' strain, behavioral tests, drugs and their doses, route of administration), as well as the use of non-selective drugs targeting other receptors in addition to the 5-HT7R, might account for these mixed results. In addition, the interpretation of behavioral data from 5-HT7R-KO mice is complicated by the indirect effects of the missing gene, such as changes in developmental processes or dysregulation of compensatory genes and pathways [87]. Very recently, mixed outcomes on various behavioral assays for anxiety, depression and psychosis, performed on mice treated with two selective 5-HT7R antagonists [88], and on 5-HT7R-KO mice [89], raised doubts on the role played by the receptor in these neuropsychiatric diseases. Ultimately, the available data suggest that additional research will be required to further evaluate and dissect the contribution of this receptor in anxiety/depression and schizophrenia, and its potential involvement for the treatment of these neuropsychiatric diseases.

Conversely, compelling evidence strongly suggests that the 5-HT7R is involved in CNS disorders characterized by intellectual disabilities and cognitive impairment, such as Rett syndrome (RTT) and Fragile X syndrome (FXS; Figure 1). These diseases belong to ASD, a heterogeneous group of neurodevelopmental disorders characterized by impaired social interaction and communication, repetitive and stereotyped behaviors, often accompanied by cognitive defects [90]. Growing evidence indicates that the brain 5-HT neurotransmission system is altered in ASD patients, and in various animal models of the disease [91,92]. For instance, mice lacking brain 5-HT, in addition to several abnormal phenotypes (growth retardation, high aggressive behavior, maternal neglect), show selective deficits resembling ASD's symptoms, including impairment in social interactions and repetitive behavior [3,4]. Various pharmacological studies are providing evidence that targeting 5-HTRs has the potential to treat the core symptoms of ASD and associated intellectual disabilities [93,94]. Recent evidence in animal models suggests that, among other subtypes, the 5-HT7R might be one of the players involved in ASD (see below). In line with this hypothesis, the only two drugs that to date are approved for the treatment of behavioral manifestations of ASD, risperidone and aripiprazole, are 5-HT7R antagonists [95], although their efficacy may be attributed also to their interactions with other receptors. Indeed, none of the approved ASD drugs are highly selective for 5-HT7R, hampering our understanding of its potential as a target for pharmacological treatment of ASD in humans. Nevertheless, brain-permeant and selective agonists of the 5-HT7R have been successfully employed to rescue ASD dysfunctions in animal models of FXS, RTT and CDKL5 Deficiency (CDD).

FXS mice exhibit cognitive impairment and stereotyped behavior, accompanied by altered morphology and density of dendritic spines in the forebrain, alongside synapse malfunctioning in the hippocampus, with the abnormal enhancement of mGluR-LTD. The activation of 5-HT7R by a selective brain-permeant agonist in hippocampal slices from FXS mice is able to correct excessive mGluR-LTD through activation of cAMP/PKA pathway, bringing it back to its physiological level and thereby restoring its synaptic plasticity. Noteworthy, acute in vivo administration of the agonist rescues learning and autistic-like behavior in 3/4 months-old FXS mice [76].

Beneficial effects of the same agonist, chronically administered, were also observed in adult mouse models of RTT. This syndrome is a severe X-linked neurological disorder characterized by deficits in autonomic, cognitive, motor functions and autistic features. In vivo systemic repeated stimulation of 5-HT7R with a selective brain-permeant agonist was able to improve cognitive and motor coordination deficits, as well as spatial memory and synaptic plasticity in RTT mice. 5-HT7R stimulation also restored the normal level of key molecules regulating actin cytoskeleton dynamics, such as Rho GTPases and mTOR signaling pathways that showed altered expression levels in the hippocampus of RTT mice [96,97]. The 5-HT7R-mediated neurobehavioral and molecular changes were still present 2 months after the last injection, suggesting long-lasting, beneficial effects on RTT-related impairments. Subsequent studies uncovered functional alterations of brain mitochondria in RTT mouse models, that were rescued by the chronic pharmacological stimulation of the 5-HT7R [98,99]. Similar promising preclinical results have been recently obtained in a mouse model of CDD, a rare neurodevelopmental syndrome characterized by severe neurobehavioral and motor deficits and stereotyped movements [100].

Altogether, the above findings provide compelling evidence that 5-HT7R stimulation exerts a widespread beneficial effect on behavioral and molecular symptomatology in various mouse models of neurodevelopmental disorders, in particular those belonging to ASD (Figure 1). Moreover, these results have important therapeutic implications, indicating that it is possible to reverse severe behavioral and molecular deficits in the animal models by pharmacological treatment at adult age. Intriguingly, all these diseases are accompanied by the alteration of dendritic spines in forebrain areas involved in higher cognitive functions, suggesting altered connectivity. Although data on the 5-HT7R-dependent remodeling of dendritic spines in the ASD animal models are still missing, it is possible to hypothesize that the activation of 5-HT7R may also promote structural rearrangements of neural circuits in the adult brain, that in turn might underlie the rescue of long-term synaptic plasticity.

4. The 5-HT7R in the Gut and in the Immune System

Despite the vast repertoire of neurodevelopmental, behavioral and cognitive processes modulated by the brain 5-HT, only ~5% of the total body content of 5-HT is located in the CNS, while the remaining part is synthesized and stored in peripheral tissues.

Outside the CNS, the vast majority of 5-HT is found in the gastrointestinal (GI) epithelium, where it is mainly produced by enterochromaffin (EC) cells of the gut mucosa and only in small quantity by neurons of the Enteric Nervous System and by the resident gut microbiota. 5-HT released by EC cells is actively taken up and stored by blood platelets, and released upon their activation, modifying vascular smooth muscle tone and a variety of other functions controlled by peripheral organs [101]. In the gut, in physiological conditions, 5-HT made by EC cells and enteric neurons act synergistically to regulate the intestinal functions, such as motility, sensation, and secretion, whereas alteration in the 5-HT metabolism is associated with various diseases of the GI tract (see below).

Peripheral 5-HT is also a potent immune system modulator and can affect various immune cells through its receptors. In addition, 5-HT is synthesized and released by some cells of the immune system (T limphocytes and mast cells), expanding the range of tissues involved in its signaling [102]. Peripheral 5-HT7R expression roughly mirrors peripheral 5-HT distribution, since it has been observed in the GI tract, as well as in the peripheral organs (kidney, liver, pancreas, spleen, and stomach), and in cells of the immune system [103].

Here, we briefly review recent studies showing that 5-HT7R plays a crucial role in generation/ perpetuation of intestinal inflammation, and in immune cell activation. The immune system is known to play an important intermediary role in the dynamic equilibrium between the CNS and the GI tract [104]. Therefore, it is intriguing to hypothesize the involvement of 5-HT7R in bidirectional communication between the brain and gut, possibly mediated by the immune system (Figure 2). As a key element of this axis, 5-HT signaling may link emotional and cognitive areas of the brain with peripheral gut activity. Interestingly, recent findings suggest that the alteration of this two-way serotonergic system of communication between brain and gut may play a role in the pathogenesis of various diseases, including ASD [105]. Moreover, it is becoming increasingly evident that the resident gut microbiota, that produce tryptophan and 5-HT, is a critical component of the gut-brain communication, modulating brain development and behavioral responses [106,107]. The 5-HT effects in this complex microbiota-gut-brain communication are mediated by 5-HT receptors and, among other subtypes, the 5-HT7R is a very interesting candidate, being expressed both in the gut and in the brain.

Various findings suggest that 5-HT7R may have a crucial role in the pathogenesis of inflammatory disorders affecting the GI tract, such as Inflammatory Bowel Disease (IBD), which includes ulcerative colitis and Crohn's Disease. IBD is characterized by activation of the immune cells accompanied by their infiltration in the gut and inflammation of the GI tract, leading to profound alteration of the GI function and dysfunctions of 5-HT signaling [108].

Interestingly, genetic or pharmacological silencing of 5-HT7R in mouse models of ulcerative colitis reduced the severity of intestinal inflammation and decreased the production of inflammatory markers by GI dendritic cells. These antigen-presenting cells initiate adaptive immune responses upon inflammation [109]. These results, indicating that 5-HT7R inhibition reduces inflammation symptoms in gut inflammatory disorders, are in contrast with other findings. Indeed, Guseva et al. [110] reported that pharmacological blockade or genetic ablation of 5-HT7R resulted in increased severity of symptoms in both acute and chronic mouse models of colitis, whereas receptor stimulation produced an anti-inflammatory effect. In addition, expression of 5-HT7R significantly increased after induction of colitis in mice and in inflamed intestinal dendritic cells of patients with Crohn's disease.

Novel epigenetic mechanisms regulating 5-HT7R expression have been recently highlighted by studies on animal models and patients with Irritable Bowel Syndrome (IBS), a functional GI disorder often associated to visceral hyperalgesia without inflammatory processes. It was shown that miR-29a modulates visceral hypersensitivity in a mouse model of IBS by directly targeting the 5-HT7R and downregulating its expression. The authors found that intestinal tissues from mice and patients with IBS displayed increased levels of miR-29a and reduced levels of 5-HT7R. Consistently, in mice with IBS, when miR-29a was knocked-out, 5-HT7R was overexpressed and intestinal hyperalgesia was attenuated [111]. These findings suggest that colon hypersensitivity may be mediated by the endogenous interaction between miRNA-29a and 5-HT7R, offering a potential promising therapeutic approach for reversing abdominal pain in IBS patients.

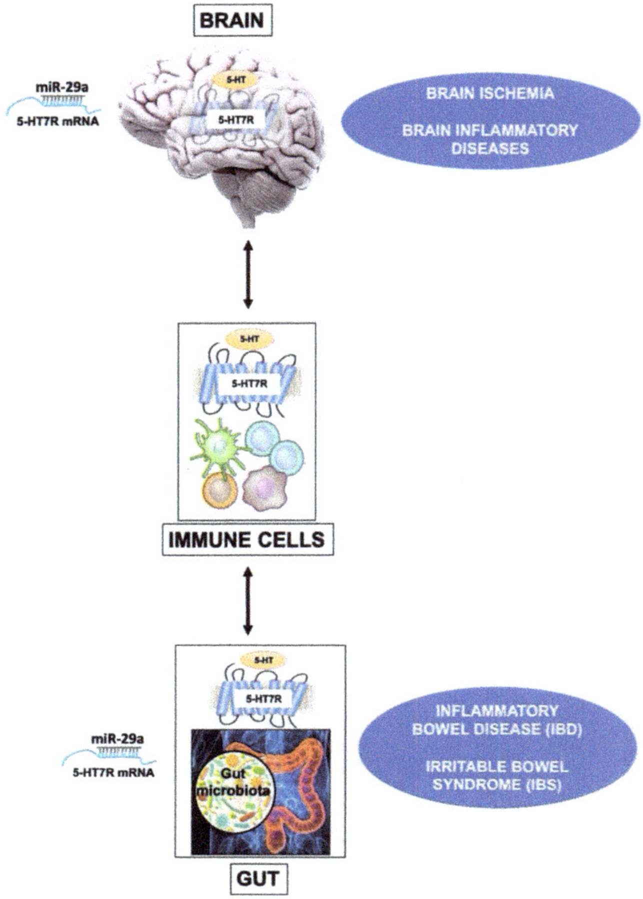

Figure 2. The drawing outlines the influence of 5-HT7R in brain, gut and immune cells. In the CNS, the 5-HT7R expressed by brain Treg cells displays specialized functions that contribute to their proliferation after ischemic stroke, promoting neurological recovery. A similar mechanism is likely to occur also in other neuroinflammatory diseases. In immune cells, the basal activity of 5-HT7R is likely to play a key role in the maintenance of homeostasis. In the gut, alteration of the 5-HT metabolism is associated with various diseases such as Inflammatory Bowel Disease (IBD), and Irritable Bowel Syndrome (IBS). In these dysfunctions the activation of 5-HT7R on dendritic cells modulates their immune response, resulting in either beneficial or detrimental effects, depending on experimental models. It is worth to underline that the expression level of 5-HT7R is epigenetically modulated by the microRNA-29a (miR-29a) in the brain as well as in the gut.

The discrepancies on the role of 5-HT7R in gut disorders may depend on the notable differences in animal models and experimental design. Moreover, it is possible that immune cells are differentially recruited depending on the experimental model of induced intestinal damage and human gut diseases, and that their production of pro-inflammatory and anti-inflammatory cytokines is modulated by the level of 5-HT and by 5-HT7R expression. Indeed, it has been demonstrated that the 5-HT7R–Cdc42-mediated signaling regulates dendritic cell morphology and enhances chemotactic motility [112]. Likewise, a prominent role of 5-HT7R in regulating endothelial cell migration has been identified, suggesting that this receptor is a potential modulator of physiological and pathophysiological processes involving cell migration, adhesion [113] and inflammatory fibrotic infiltration [103,114].

Thus, 5-HT7R-dependent activation and the migration of dendritic cells might be significantly different in various intestinal inflammatory diseases, accounting—at least in part—for conflicting results on the role of 5-HT7R in the pathogenesis of gut inflammatory disorders. As a conclusion, although it is clear that 5-HT7R can influence gut inflammation, additional studies are required to precisely understand 5-HT7R function and dysfunction in the intestine.

Nevertheless, these findings highlight the involvement of the 5-HT7R in the physiology and pathology of the immune system. Indeed, it is well known that 5-HT plays a key role in inflammation, immunity and immunomodulatory diseases and that almost all immune cells express at least one 5-HT receptor, both in rodents and humans. 5-HT7R expression has been detected in lymphoid progenitor cells, mast cells, monocytes, macrophages, dendritic cells and T lymphocytes [103]. This receptor is also expressed by microglia, the brain resident macrophages, and its stimulation in human microglial cell lines leads to increased IL-6 expression—a proinflammatory cytochine [115]. The stimulation of 5-HT7R by 5-HT enhances the proliferation and activation of mouse naive T cells through ERK signaling [116]. Likewise, in human macrophages, the anti-inflammatory and pro-fibrotic activity of 5-HT is primarily mediated by 5-HT7R-PKA pathway [114]. Notably, it has been recently shown that brain regulatory T (Treg) cells are distinct from those of other tissues, since they express unique genes related to the Nervous System, including the 5-HT7R. The specific features and functions of brain Treg cells are poorly understood, because their number is very low in the brain under normal conditions. Conversely, a large number of Treg cells infiltrate the mouse brain during the chronic phase of ischemic stroke, suppressing astrogliosis and potentiating neurological recovery [117]. Brain Treg cells, but not splenic Treg cells, respond to 5-HT by increased proliferation and this response was blocked by a selective antagonist of the 5-HT7R. Notably, 5-HT7R-deficient Treg cells do not expand correctly into the brain and do not promote neurological recovery after ischemic stroke. These findings demonstrated that 5-HT7R play a specialized role in Treg cells and suggest that the 5-HT signaling mediated by Treg cells might represent one of the mechanisms that contribute to the cross talk between immune system and brain inflammation.

Altogether, the reported findings on the 5-HT7R signaling in the CNS, GI tract and immune system suggest the involvement of this receptor in inflammatory and immune-mediated disorders affecting the gut and brain (Figure 2). For instance, as mentioned above, two very recent studies showed that 5-HT7R is a direct target of miR-29a in the brain, as well as in the intestine [63,111], suggesting that this miRNA might modulate 5-HT7R expression in both tissues in a coordinated way.

5. Conclusions and Future Perspectives

The modulation of 5-HTR7 expression using pharmacological and genetic tools, coupled to cellular, molecular, electrophysiological and behavioral approaches, has greatly increased our knowledge on the functions of this receptor in the brain, as well as in other organs.

The results highlighted here indicate that 5-HT7R is an important player involved in the modulation of synaptic and structural plasticity in both developing and mature brain circuits. However, the detailed molecular mechanisms and signaling pathways underlying 5-HT7R morphogenic effects are still objects of intense investigation. We believe that conditional KO mouse models are able to silence/overexpress the 5-HT7R gene during selected age windows, and—in specific cell types—would help to more accurately define the contribution of the 5-HT7R to brain physiology.

Altered 5-HT7R-mediated signaling is involved in numerous brain diseases. In particular, compelling evidence indicates that 5-HT7R stimulation reverts behavioral, molecular and functional deficits in various animal models of neurodevelopmental disorders. These 5-HT7R beneficial effects, at least in RTT, might operate through the rescue of the mitochondrial dysfunctions associated with the disease. Thus, it would be of great interest to deepen our understanding on the mechanisms underlying the regulatory effects of 5-HT7R on mitochondrial function.

Finally, we briefly discussed recent findings highlighting a crucial role of the 5-HT7R in intestinal inflammation and immune cell activation, and suggested its possible involvement in the complex interaction between the brain, immune cells and gut.

Importantly, the findings described here open new avenues in the development of selective drugs targeting 5-HT7R as novel potential therapeutic agents in many diseases so far considered incurable.

Author Contributions: Writing and editing, M.C.; original draft preparation and editing, F.V.; original draft preparation, writing and review, C.P.-C. All authors have read and agreed to the published version of the manuscript.

Funding: This work was funded by "Finanziamento Ricerca di Ateneo" from University of Naples Federico II, and by "POR Campania FESR 2014/2020" from Regione Campania (Project N. B61G18000470007).

Acknowledgments: We would like to thank Umberto di Porzio for critical reading of the manuscript.

Conflicts of Interest: The authors declare no conflict of interest.

Abbreviations

5-HT	Serotonin
5-HT7R	Serotonin Receptor subtype 7
5-HTP	5-hydroxytryptophan
ASD	Autism Spectrum Disorders
CDD	CDKL5 Deficiency
CNS	Central Nervous System
EC	enterochromaffin
FXS	Fragile X Syndrome
GI	GastroIntestinal
IBD	Inflammatory Bowel Disease
IBS	Irritable Bowel Syndrome
KO	Knockout
LTP	Long-Term Potentiation
LTD	Long-Term Depression
miR	microRNA
PKA	Protein Kinase A
RTT	Rett syndrome
TPH	Tryptophan-5-hydroxylase
Treg	regulatory T cells

References

1. Pilowsky, P.M. *Serotonin The Mediator that Spans Evolution*, 1st ed.; Elsevier: Philadelphia, PA, USA, 2019; pp. 1–420.
2. Lillesaar, C.; Gaspar, P. Serotonergic Neurons in Vertebrate and Invertebrate Model Organisms (Rodents, Zebrafish, Drosophila melanogaster, Aplysiacalifornica, Caenorhabditis elegans). In *Serotonin The Mediator that Spans Evolution*, 1st ed.; Pilowsky, P.M., Ed.; Elsevier: Philadelphia, PA, USA, 2019; pp. 49–80.
3. Pratelli, M.; Pasqualetti, M. Serotonergic neurotransmission manipulation for the understanding of brain development and function: Learning from Tph2 genetic models. *Biochimie* **2019**, *161*, 3–14. [CrossRef]
4. Mosienko, V.; Alenina, N. Life Without Brain Serotonin: Phenotypes of Animals Deficient in Central Serotonin Synthesis. In *Serotonin The Mediator that Spans Evolution*, 1st ed.; Pilowsky, P.M., Ed.; Elsevier: Philadelphia, PA, USA, 2019; pp. 405–420.
5. Seyedabadi, M.; Fakhfouri, G.; Ramezani, V.; Mehr, S.E.; Rahimian, R. The role of serotonin in memory: Interactions with neurotransmitters and downstream signaling. *Exp. Brain Res.* **2014**, *232*, 723–738. [CrossRef]
6. De Deurwaerdère, P.; Di Giovanni, G. Serotonergic modulation of the activity of mesencephalic dopaminergic systems: Therapeutic implications. *Prog. Neurobiol.* **2017**, *151*, 175–236. [CrossRef]
7. Sengupta, A.; Bocchio, M.; Bannerman, D.M.; Sharp, T.; Capogna, M. Control of Amygdala Circuits by 5-HT Neurons via 5-HT and Glutamate Cotransmission. *J. Neurosci.* **2017**, *37*, 1785–1796. [CrossRef]

8. Fernandez, S.P.; Cauli, B.; Cabezas, C.; Muzerelle, A.; Poncer, J.C.; Gaspar, P. Multiscale single-cell analysis reveals unique phenotypes of raphe 5-HT neurons projecting to the forebrain. *Brain Struct. Funct.* **2016**, *221*, 4007–4025. [CrossRef]
9. Svensson, E.; Apergis-Schoute, J.; Burnstock, G.; Nusbaum, M.P.; Parker, D.; Schiöth, H.B. General Principles of Neuronal Co-transmission: Insights From Multiple Model Systems. *Front. Neural Circuits* **2018**, *12*, 117. [CrossRef]
10. Levitt, P.; Rakic, P. The time of genesis, embryonic origin and differentiation of the brain stem monoamine neurons in the rhesus monkey. *Brain Res.* **1982**, *256*, 35–57. [CrossRef]
11. Jacobs, B.L.; Azmitia, E.C. Structure and function of the brain serotonin system. *Physiol. Rev.* **1992**, *72*, 165–229. [CrossRef]
12. Gagnon, D.; Parent, M. Distribution of VGLUT3 in highly collateralized axons from the rat dorsal raphe nucleus as revealed by single-neuron reconstructions. *PLoS ONE* **2014**, *9*, e87709. [CrossRef]
13. Wirth, A.; Holst, K.; Ponimaskin, E. How serotonin receptors regulate morphogenic signalling in neurons. *Prog. Neurobiol.* **2017**, *151*, 35–56. [CrossRef]
14. Bonnin, A.; Goeden, N.; Chen, K.; Wilson, M.L.; King, J.; Shih, J.C.; Blakely, R.D.; Deneris, E.S.; Levitt, P. A transient placental source of serotonin for the fetal forebrain. *Nature* **2011**, *472*, 347–350. [CrossRef]
15. Shallie, P.D.; Naicker, T. The placenta as a window to the brain: A review on the role of placental markers in prenatal programming of neurodevelopment. *Int. J. Dev. Neurosci.* **2019**, *73*, 41–49. [CrossRef]
16. Goeden, N.; Velasquez, J.; Arnold, K.A.; Chan, Y.; Lund, B.T.; Anderson, G.M.; Bonnin, A. Maternal Inflammation Disrupts Fetal Neurodevelopment via Increased Placental Output of Serotonin to the Fetal Brain. *J. Neurosci.* **2016**, *36*, 6041–6049. [CrossRef]
17. Shah, R.; Courtiol, E.; Castellanos, F.X.; Teixeira, C.M. Abnormal Serotonin Levels During Perinatal Development Lead to Behavioral Deficits in Adulthood. *Front. Behav. Neurosci.* **2018**, *12*, 114. [CrossRef]
18. Brummelte, S.; Mc Glanaghy, E.; Bonnin, A.; Oberlander, T.F. Developmental changes in serotonin signaling: Implications for early brain function, behavior and adaptation. *Neuroscience* **2017**, *342*, 212–231. [CrossRef]
19. Cases, O.; Vitalis, T.; Seif, I.; De Maeyer, E.; Sotelo, C.; Gaspar, P. Lack of barrels in the somatosensory cortex of monoamine oxidase A-deficient mice: Role of a serotonin excess during the critical period. *Neuron* **1996**, *16*, 297–307. [CrossRef]
20. Riccio, O.; Potter, G.; Walzer, C.; Vallet, P.; Szabó, G.; Vutskits, L.; Kiss, J.Z.; Dayer, A.G. Excess of serotonin affects embryonic interneuron migration through activation of the serotonin receptor 6. *Mol. Psychiatry* **2009**, *14*, 280–290. [CrossRef]
21. Ansorge, M.S.; Morelli, E.; Gingrich, J.A. Inhibition of serotonin but not norepinephrine transport during development produces delayed, persistent perturbations of emotional behaviors in mice. *J. Neurosci.* **2008**, *28*, 199–207. [CrossRef]
22. Oberlander, T.F.; Gingrich, J.A.; Ansorge, M.S. Sustained neurobehavioral effects of exposure to SSRI antidepressants during development: Molecular to clinical evidence. *Clin. Pharmacol. Ther.* **2009**, *86*, 672–677. [CrossRef] [PubMed]
23. Migliarini, S.; Pacini, G.; Pelosi, B.; Lunardi, G.; Pasqualetti, M. Lack of brain serotonin affects postnatal development and serotonergic neuronal circuitry formation. *Mol. Psychiatry* **2013**, *18*, 1106–1118. [CrossRef]
24. De Gregorio, R.; Chen, X.; Petit, E.I.; Dobrenis, K.; Sze, J.Y. Disruption of Transient SERT Expression in Thalamic Glutamatergic Neurons Alters Trajectory of Postnatal Interneuron Development in the Mouse Cortex. *Cereb. Cortex* **2019**. [CrossRef] [PubMed]
25. Marín, O. Developmental timing and critical windows for the treatment of psychiatric disorders. *Nat. Med.* **2016**, *22*, 1229–1238. [CrossRef] [PubMed]
26. Teissier, A.; Soiza-Reilly, M.; Gaspar, P. Refining the Role of 5-HT in Postnatal Development of Brain Circuits. *Front. Cell. Neurosci.* **2017**, *11*, 139. [CrossRef] [PubMed]
27. Pratelli, M.; Migliarini, S.; Pelosi, B.; Napolitano, F.; Usiello, A.; Pasqualetti, M. Perturbation of Serotonin Homeostasis during Adulthood Affects Serotonergic Neuronal Circuitry. *eNeuro* **2017**, *4*. [CrossRef] [PubMed]
28. Giorgi, A.; Migliarini, S.; Galbusera, A.; Maddaloni, G.; Mereu, M.; Margiani, G.; Gritti, M.; Landi, S.; Trovato, F.; Bertozzi, S.M.; et al. Brain-wide Mapping of Endogenous Serotonergic Transmission via Chemogenetic fMRI. *Cell Rep.* **2017**, *21*, 910–918. [CrossRef] [PubMed]
29. Lesch, K.P.; Waider, J. Serotonin in the modulation of neural plasticity and networks: Implications for neurodevelopmental disorders. *Neuron* **2012**, *76*, 175–191. [CrossRef]

30. Dayer, A. Serotonin-related pathways and developmental plasticity: Relevance for psychiatric disorders. *Dialogues Clin. Neurosci.* **2014**, *16*, 29–41.
31. Deneris, E.; Gaspar, P. Serotonin neuron development: Shaping molecular and structural identities. *Wiley Interdiscip. Rev. Dev. Biol* **2018**, *7*, e301. [CrossRef]
32. McCorvy, J.D.; Roth, B.L. Structure and function of serotonin G protein-coupled receptors. *Pharmacol. Ther.* **2015**, *150*, 129–142. [CrossRef]
33. Bard, J.A.; Zgombick, J.; Adham, N.; Vaysse, P.; Branchek, T.A.; Weinshank, R.L. Cloning of a novel human serotonin receptor (5-HT7) positively linked to adenylate cyclase. *J. Biol. Chem.* **1993**, *268*, 23422–23426.
34. Ruat, M.; Traiffort, E.; Leurs, R.; Tardivel-Lacombe, J.; Diaz, J.; Arrang, J.M.; Schwartz, J.C. Molecular cloning, characterization, and localization of a high-affinity serotonin receptor (5-HT7) activating cAMP formation. *Proc. Natl. Acad. Sci. USA* **1993**, *90*, 8547–8551. [CrossRef] [PubMed]
35. Matthys, A.; Haegeman, G.; Van Craenenbroeck, K.; Vanhoenacker, P. Role of the 5-HT7 receptor in the central nervous system: From current status to future perspectives. *Mol. Neurobiol.* **2011**, *43*, 228–253. [CrossRef] [PubMed]
36. Gellynck, E.; Heyninck, K.; Andressen, K.W.; Haegeman, G.; Levy, F.O.; Vanhoenacker, P.; Van Craenenbroeck, K. The serotonin 5-HT7 receptors: Two decades of research. *Exp. Brain Res.* **2013**, *230*, 555–568. [CrossRef] [PubMed]
37. Blattner, K.M.; Canney, D.J.; Pippin, D.A.; Blass, B.E. Pharmacology and Therapeutic Potential of the 5-HT. *ACS Chem. Neurosci.* **2019**, *10*, 89–119. [CrossRef] [PubMed]
38. Nikiforuk, A.; Hołuj, M.; Potasiewicz, A.; Popik, P. Effects of the selective 5-HT7 receptor antagonist SB-269970 on premature responding in the five-choice serial reaction time test in rats. *Behav. Brain Res.* **2015**, *289*, 149–156. [CrossRef]
39. Leopoldo, M.; Lacivita, E.; Berardi, F.; Perrone, R.; Hedlund, P.B. Serotonin 5-HT7 receptor agents: Structure-activity relationships and potential therapeutic applications in central nervous system disorders. *Pharmacol. Ther.* **2011**, *129*, 120–148. [CrossRef]
40. Volpicelli, F.; Speranza, L.; di Porzio, U.; Crispino, M.; Perrone-Capano, C. The serotonin receptor 7 and the structural plasticity of brain circuits. *Front. Behav. Neurosci.* **2014**, *8*, 318. [CrossRef]
41. Lippiello, P.; Hoxha, E.; Speranza, L.; Volpicelli, F.; Ferraro, A.; Leopoldo, M.; Lacivita, E.; Perrone-Capano, C.; Tempia, F.; Miniaci, M.C. The 5-HT7 receptor triggers cerebellar long-term synaptic depression via PKC-MAPK. *Neuropharmacology* **2016**, *101*, 426–438. [CrossRef]
42. Heidmann, D.E.; Szot, P.; Kohen, R.; Hamblin, M.W. Function and distribution of three rat 5-hydroxytryptamine7 (5-HT7) receptor isoforms produced by alternative splicing. *Neuropharmacology* **1998**, *37*, 1621–1632. [CrossRef]
43. Heidmann, D.E.; Metcalf, M.A.; Kohen, R.; Hamblin, M.W. Four 5-hydroxytryptamine7 (5-HT7) receptor isoforms in human and rat produced by alternative splicing: Species differences due to altered intron-exon organization. *J. Neurochem.* **1997**, *68*, 1372–1381. [CrossRef]
44. Guthrie, C.R.; Murray, A.T.; Franklin, A.A.; Hamblin, M.W. Differential agonist-mediated internalization of the human 5-hydroxytryptamine 7 receptor isoforms. *J. Pharmacol. Exp. Ther.* **2005**, *313*, 1003–1010. [CrossRef] [PubMed]
45. Kvachnina, E.; Liu, G.; Dityatev, A.; Renner, U.; Dumuis, A.; Richter, D.W.; Dityateva, G.; Schachner, M.; Voyno-Yasenetskaya, T.A.; Ponimaskin, E.G. 5-HT7 receptor is coupled to G alpha subunits of heterotrimeric G12-protein to regulate gene transcription and neuronal morphology. *J. Neurosci.* **2005**, *25*, 7821–7830. [CrossRef] [PubMed]
46. Lenglet, S.; Louiset, E.; Delarue, C.; Vaudry, H.; Contesse, V. Activation of 5-HT(7) receptor in rat glomerulosa cells is associated with an increase in adenylyl cyclase activity and calcium influx through T-type calcium channels. *Endocrinology* **2002**, *143*, 1748–1760. [CrossRef] [PubMed]
47. Johnson-Farley, N.N.; Kertesy, S.B.; Dubyak, G.R.; Cowen, D.S. Enhanced activation of Akt and extracellular-regulated kinase pathways by simultaneous occupancy of Gq-coupled 5-HT2A receptors and Gs-coupled 5-HT7A receptors in PC12 cells. *J. Neurochem.* **2005**, *92*, 72–82. [CrossRef] [PubMed]
48. Fields, D.P.; Springborn, S.R.; Mitchell, G.S. Spinal 5-HT7 receptors induce phrenic motor facilitation via EPAC-mTORC1 signaling. *J. Neurophysiol.* **2015**, *114*, 2015–2022. [CrossRef] [PubMed]
49. Prasad, S.; Ponimaskin, E.; Zeug, A. Serotonin receptor oligomerization regulates cAMP-based signaling. *J. Cell Sci.* **2019**, *132*, jcs230334. [CrossRef]

50. Hedlund, P.B.; Danielson, P.E.; Thomas, E.A.; Slanina, K.; Carson, M.J.; Sutcliffe, J.G. No hypothermic response to serotonin in 5-HT7 receptor knockout mice. *Proc. Natl. Acad. Sci. USA* **2003**, *100*, 1375–1380. [CrossRef]
51. Di Pilato, P.; Niso, M.; Adriani, W.; Romano, E.; Travaglini, D.; Berardi, F.; Colabufo, N.A.; Perrone, R.; Laviola, G.; Lacivita, E.; et al. Selective agonists for serotonin 7 (5-HT7) receptor and their applications in preclinical models: An overview. *Rev. Neurosci.* **2014**, *25*, 401–415. [CrossRef]
52. Tajiri, M.; Hayata-Takano, A.; Seiriki, K.; Ogata, K.; Hazama, K.; Shintani, N.; Baba, A.; Hashimoto, H. Serotonin 5-HT(7) receptor blockade reverses behavioral abnormalities in PACAP-deficient mice and receptor activation promotes neurite extension in primary embryonic hippocampal neurons: Therapeutic implications for psychiatric disorders. *J. Mol. Neurosci.* **2012**, *48*, 473–481. [CrossRef]
53. Rojas, P.S.; Neira, D.; Muñoz, M.; Lavandero, S.; Fiedler, J.L. Serotonin (5-HT) regulates neurite outgrowth through 5-HT1A and 5-HT7 receptors in cultured hippocampal neurons. *J. Neurosci. Res.* **2014**, *92*, 1000–1009. [CrossRef]
54. Speranza, L.; Chambery, A.; Di Domenico, M.; Crispino, M.; Severino, V.; Volpicelli, F.; Leopoldo, M.; Bellenchi, G.C.; di Porzio, U.; Perrone-Capano, C. The serotonin receptor 7 promotes neurite outgrowth via ERK and Cdk5 signaling pathways. *Neuropharmacology* **2013**, *67*, 155–167. [CrossRef] [PubMed]
55. Lacivita, E.; Podlewska, S.; Speranza, L.; Niso, M.; Satała, G.; Perrone, R.; Perrone-Capano, C.; Bojarski, A.J.; Leopoldo, M. Structural modifications of the serotonin 5-HT7 receptor agonist N-(4-cyanophenylmethyl)-4-(2-biphenyl)-1-piperazinehexanamide (LP-211) to improve in vitro microsomal stability: A case study. *Eur. J. Med. Chem.* **2016**, *120*, 363–379. [CrossRef] [PubMed]
56. Speranza, L.; Giuliano, T.; Volpicelli, F.; De Stefano, M.E.; Lombardi, L.; Chambery, A.; Lacivita, E.; Leopoldo, M.; Bellenchi, G.C.; di Porzio, U.; et al. Activation of 5-HT7 receptor stimulates neurite elongation through mTOR, Cdc42 and actin filaments dynamics. *Front. Behav. Neurosci.* **2015**, *9*, 62. [CrossRef] [PubMed]
57. Vasefi, M.S.; Yang, K.; Li, J.; Kruk, J.S.; Heikkila, J.J.; Jackson, M.F.; MacDonald, J.F.; Beazely, M.A. Acute 5-HT7 receptor activation increases NMDA-evoked currents and differentially alters NMDA receptor subunit phosphorylation and trafficking in hippocampal neurons. *Mol. Brain* **2013**, *6*, 24. [CrossRef]
58. Vasefi, M.S.; Kruk, J.S.; Heikkila, J.J.; Beazely, M.A. 5-Hydroxytryptamine type 7 receptor neuroprotection against NMDA-induced excitotoxicity is PDGFβ receptor dependent. *J. Neurochem.* **2013**, *125*, 26–36. [CrossRef]
59. Costa, L.; Spatuzza, M.; D'Antoni, S.; Bonaccorso, C.M.; Trovato, C.; Musumeci, S.A.; Leopoldo, M.; Lacivita, E.; Catania, M.V.; Ciranna, L. Activation of 5-HT7 serotonin receptors reverses metabotropic glutamate receptor-mediated synaptic plasticity in wild-type and Fmr1 knockout mice, a model of Fragile X syndrome. *Biol. Psychiatry* **2012**, *72*, 924–933. [CrossRef]
60. Andreetta, F.; Carboni, L.; Grafton, G.; Jeggo, R.; Whyment, A.D.; van den Top, M.; Hoyer, D.; Spanswick, D.; Barnes, N.M. Hippocampal 5-HT7 receptors signal phosphorylation of the GluA1 subunit to facilitate AMPA receptor mediated-neurotransmission in vitro and in vivo. *Br. J. Pharmacol.* **2016**, *173*, 1438–1451. [CrossRef]
61. Roberts, A.J.; Krucker, T.; Levy, C.L.; Slanina, K.A.; Sutcliffe, J.G.; Hedlund, P.B. Mice lacking 5-HT receptors show specific impairments in contextual learning. *Eur. J. Neurosci.* **2004**, *19*, 1913–1922. [CrossRef]
62. Kobe, F.; Guseva, D.; Jensen, T.P.; Wirth, A.; Renner, U.; Hess, D.; Müller, M.; Medrihan, L.; Zhang, W.; Zhang, M.; et al. 5-HT7R/G12 signaling regulates neuronal morphology and function in an age-dependent manner. *J. Neurosci.* **2012**, *32*, 2915–2930. [CrossRef]
63. Volpicelli, F.; Speranza, L.; Pulcrano, S.; De Gregorio, R.; Crispino, M.; De Sanctis, C.; Leopoldo, M.; Lacivita, E.; di Porzio, U.; Bellenchi, G.C.; et al. The microRNA-29a Modulates Serotonin 5-HT7 Receptor Expression and Its Effects on Hippocampal Neuronal Morphology. *Mol. Neurobiol.* **2019**, *56*, 8617–8627. [CrossRef]
64. Bijata, M.; Labus, J.; Guseva, D.; Stawarski, M.; Butzlaff, M.; Dzwonek, J.; Schneeberg, J.; Böhm, K.; Michaluk, P.; Rusakov, D.A.; et al. Synaptic Remodeling Depends on Signaling between Serotonin Receptors and the Extracellular Matrix. *Cell Rep.* **2017**, *19*, 1767–1782. [CrossRef] [PubMed]
65. Speranza, L.; Labus, J.; Volpicelli, F.; Guseva, D.; Lacivita, E.; Leopoldo, M.; Bellenchi, G.C.; di Porzio, U.; Bijata, M.; Perrone-Capano, C.; et al. Serotonin 5-HT7 receptor increases the density of dendritic spines and facilitates synaptogenesis in forebrain neurons. *J. Neurochem.* **2017**, *141*, 647–661. [CrossRef] [PubMed]
66. Crispino, M.; Chun, J.T.; Cefaliello, C.; Perrone Capano, C.; Giuditta, A. Local gene expression in nerve endings. *Dev. Neurobiol.* **2014**, *74*, 279–291. [CrossRef] [PubMed]

67. Crispino, M.; Cefaliello, C.; Kaplan, B.; Giuditta, A. Protein synthesis in nerve terminals and the glia-neuron unit. *Results Probl. Cell Differ.* **2009**, *48*, 243–267.
68. Holt, C.E.; Martin, K.C.; Schuman, E.M. Local translation in neurons: Visualization and function. *Nat. Struct. Mol. Biol.* **2019**, *26*, 557–566. [CrossRef]
69. Hübener, M.; Bonhoeffer, T. Neuronal plasticity: Beyond the critical period. *Cell* **2014**, *159*, 727–737. [CrossRef]
70. Nativio, P.; Zoratto, F.; Romano, E.; Lacivita, E.; Leopoldo, M.; Pascale, E.; Passarelli, F.; Laviola, G.; Adriani, W. Stimulation of 5-HT7 receptor during adolescence determines its persistent upregulation in adult rat forebrain areas. *Synapse* **2015**, *69*, 533–542. [CrossRef]
71. Leo, D.; Adriani, W.; Cavaliere, C.; Cirillo, G.; Marco, E.M.; Romano, E.; di Porzio, U.; Papa, M.; Perrone-Capano, C.; Laviola, G. Methylphenidate to adolescent rats drives enduring changes of accumbal Htr7 expression: Implications for impulsive behavior and neuronal morphology. *Genes Brain Behav.* **2009**, *8*, 356–368. [CrossRef]
72. Canese, R.; Zoratto, F.; Altabella, L.; Porcari, P.; Mercurio, L.; de Pasquale, F.; Butti, E.; Martino, G.; Lacivita, E.; Leopoldo, M.; et al. Persistent modification of forebrain networks and metabolism in rats following adolescent exposure to a 5-HT7 receptor agonist. *Psychopharmacology* **2015**, *232*, 75–89. [CrossRef]
73. Bosch, M.; Hayashi, Y. Structural plasticity of dendritic spines. *Curr. Opin. Neurobiol.* **2012**, *22*, 383–388. [CrossRef]
74. Chang, J.Y.; Parra-Bueno, P.; Laviv, T.; Szatmari, E.M.; Lee, S.R.; Yasuda, R. CaMKII Autophosphorylation Is Necessary for Optimal Integration of Ca. *Neuron* **2017**, *94*, 800–808.e4. [CrossRef] [PubMed]
75. Lai, K.O.; Ip, N.Y. Structural plasticity of dendritic spines: The underlying mechanisms and its dysregulation in brain disorders. *Biochim. Biophys. Acta* **2013**, *1832*, 2257–2263. [CrossRef] [PubMed]
76. Costa, L.; Sardone, L.M.; Bonaccorso, C.M.; D'Antoni, S.; Spatuzza, M.; Gulisano, W.; Tropea, M.R.; Puzzo, D.; Leopoldo, M.; Lacivita, E.; et al. Activation of Serotonin 5-HT. *Front. Mol. Neurosci.* **2018**, *11*, 353. [CrossRef]
77. Beaudet, G.; Jozet-Alves, C.; Asselot, R.; Schumann-Bard, P.; Freret, T.; Boulouard, M.; Paizanis, E. Deletion of the serotonin receptor type 7 disrupts the acquisition of allocentric but not egocentric navigation strategies in mice. *Behav. Brain Res.* **2017**, *320*, 179–185. [CrossRef]
78. Beaudet, G.; Bouet, V.; Jozet-Alves, C.; Schumann-Bard, P.; Dauphin, F.; Paizanis, E.; Boulouard, M.; Freret, T. Spatial memory deficit across aging: Current insights of the role of 5-HT7 receptors. *Front. Behav. Neurosci.* **2014**, *8*, 448. [CrossRef] [PubMed]
79. Duncan, M.J.; Smith, J.T.; Franklin, K.M. Time of day but not aging regulates 5-HT. *Neurosci. Lett.* **2018**, *662*, 306–311. [CrossRef] [PubMed]
80. Freret, T.; Paizanis, E.; Beaudet, G.; Gusmao-Montaigne, A.; Nee, G.; Dauphin, F.; Bouet, V.; Boulouard, M. Modulation of 5-HT7 receptor: Effect on object recognition performances in mice. *Psychopharmacology* **2014**, *231*, 393–400. [CrossRef]
81. Zareifopoulos, N.; Papatheodoropoulos, C. Effects of 5-HT-7 receptor ligands on memory and cognition. *Neurobiol. Learn. Mem.* **2016**, *136*, 204–209. [CrossRef]
82. D'Angelo, E. Physiology of the cerebellum. *Handb. Clin. Neurol.* **2018**, *154*, 85–108.
83. Okada, M.; Fukuyama, K.; Nakano, T.; Ueda, Y. Pharmacological Discrimination of Effects of MK801 on Thalamocortical, Mesothalamic, and Mesocortical Transmissions. *Biomolecules* **2019**, *9*, 746. [CrossRef]
84. Okada, M.; Fukuyama, K.; Okubo, R.; Shiroyama, T.; Ueda, Y. Lurasidone Sub-Chronically Activates Serotonergic Transmission via Desensitization of 5-HT1A and 5-HT7 Receptors in Dorsal Raphe Nucleus. *Pharmaceuticals* **2019**, *12*, 149. [CrossRef] [PubMed]
85. Ikeda, M.; Iwata, N.; Kitajima, T.; Suzuki, T.; Yamanouchi, Y.; Kinoshita, Y.; Ozaki, N. Positive association of the serotonin 5-HT7 receptor gene with schizophrenia in a Japanese population. *Neuropsychopharmacology* **2006**, *31*, 866–871. [CrossRef] [PubMed]
86. Wei, Y.B.; McCarthy, M.; Ren, H.; Carrillo-Roa, T.; Shekhtman, T.; DeModena, A.; Liu, J.J.; Leckband, S.G.; Mors, O.; Rietschel, M.; et al. A functional variant in the serotonin receptor 7 gene (HTR7), rs7905446, is associated with good response to SSRIs in bipolar and unipolar depression. *Mol. Psychiatry* **2019**. [CrossRef]
87. Nelson, R.J.; Young, K.A. Behavior in mice with targeted disruption of single genes. *Neurosci. Biobehav. Rev.* **1998**, *22*, 453–462. [CrossRef]
88. Maxwell, J.; Gleason, S.D.; Falcone, J.; Svensson, K.; Balcer, O.M.; Li, X.; Witkin, J.M. Effects of 5-HT. *Behav. Brain Res.* **2019**, *359*, 467–473. [CrossRef] [PubMed]

89. Balcer, O.M.; Seager, M.A.; Gleason, S.D.; Li, X.; Rasmussen, K.; Maxwell, J.K.; Nomikos, G.; Degroot, A.; Witkin, J.M. Evaluation of 5-HT. *Behav. Brain Res.* **2019**, *360*, 270–278. [CrossRef] [PubMed]
90. Sztainberg, Y.; Zoghbi, H.Y. Lessons learned from studying syndromic autism spectrum disorders. *Nat. Neurosci.* **2016**, *19*, 1408–1417. [CrossRef]
91. Muller, C.L.; Anacker, A.M.J.; Veenstra-VanderWeele, J. The serotonin system in autism spectrum disorder: From biomarker to animal models. *Neuroscience* **2016**, *321*, 24–41. [CrossRef]
92. Garbarino, V.R.; Gilman, T.L.; Daws, L.C.; Gould, G.G. Extreme enhancement or depletion of serotonin transporter function and serotonin availability in autism spectrum disorder. *Pharmacol. Res.* **2019**, *140*, 85–99. [CrossRef]
93. Chugani, D.C.; Chugani, H.T.; Wiznitzer, M.; Parikh, S.; Evans, P.A.; Hansen, R.L.; Nass, R.; Janisse, J.J.; Dixon-Thomas, P.; Behen, M.; et al. Efficacy of Low-Dose Buspirone for Restricted and Repetitive Behavior in Young Children with Autism Spectrum Disorder: A Randomized Trial. *J. Pediatr.* **2016**, *170*, 45–53. [CrossRef]
94. De Bruin, E.I.; Graham, J.H.; Louwerse, A.; Huizink, A.C. Mild dermatoglyphic deviations in adolescents with autism spectrum disorders and average intellectual abilities as compared to typically developing boys. *Autism Res. Treat.* **2014**, *2014*, 968134. [CrossRef] [PubMed]
95. Ghanizadeh, A.; Sahraeizadeh, A.; Berk, M. A head-to-head comparison of aripiprazole and risperidone for safety and treating autistic disorders, a randomized double blind clinical trial. *Child Psychiatry Hum. Dev.* **2014**, *45*, 185–192. [CrossRef] [PubMed]
96. De Filippis, B.; Nativio, P.; Fabbri, A.; Ricceri, L.; Adriani, W.; Lacivita, E.; Leopoldo, M.; Passarelli, F.; Fuso, A.; Laviola, G. Pharmacological stimulation of the brain serotonin receptor 7 as a novel therapeutic approach for Rett syndrome. *Neuropsychopharmacology* **2014**, *39*, 2506–2518. [CrossRef]
97. De Filippis, B.; Chiodi, V.; Adriani, W.; Lacivita, E.; Mallozzi, C.; Leopoldo, M.; Domenici, M.R.; Fuso, A.; Laviola, G. Long-lasting beneficial effects of central serotonin receptor 7 stimulation in female mice modeling Rett syndrome. *Front. Behav. Neurosci.* **2015**, *9*, 86. [CrossRef]
98. De Filippis, B.; Valenti, D.; Chiodi, V.; Ferrante, A.; de Bari, L.; Fiorentini, C.; Domenici, M.R.; Ricceri, L.; Vacca, R.A.; Fabbri, A.; et al. Modulation of Rho GTPases rescues brain mitochondrial dysfunction, cognitive deficits and aberrant synaptic plasticity in female mice modeling Rett syndrome. *Eur. Neuropsychopharmacol.* **2015**, *25*, 889–901. [CrossRef] [PubMed]
99. Valenti, D.; de Bari, L.; Vigli, D.; Lacivita, E.; Leopoldo, M.; Laviola, G.; Vacca, R.A.; De Filippis, B. Stimulation of the brain serotonin receptor 7 rescues mitochondrial dysfunction in female mice from two models of Rett syndrome. *Neuropharmacology* **2017**, *121*, 79–88. [CrossRef] [PubMed]
100. Vigli, D.; Rusconi, L.; Valenti, D.; La Montanara, P.; Cosentino, L.; Lacivita, E.; Leopoldo, M.; Amendola, E.; Gross, C.; Landsberger, N.; et al. Rescue of prepulse inhibition deficit and brain mitochondrial dysfunction by pharmacological stimulation of the central serotonin receptor 7 in a mouse model of CDKL5 Deficiency Disorder. *Neuropharmacology* **2019**, *144*, 104–114. [CrossRef]
101. Martin, A.M.; Sun, E.W.; Keating, D. Mechanisms controlling hormone secretion in human gut and its relevance to metabolism. *J. Endocrinol.* **2019**, *244*, R1–R15. [CrossRef]
102. Roumier, A.; Bechade, C.; Maroteaux, L. Serotonin and the Immune System. In *Serotonin The Mediator that Spans Evolution*, 1st ed.; Pilowsky, P.M., Ed.; Elsevier: Philadelphia, PA, USA, 2019; pp. 181–196.
103. Polat, B.; Halici, Z.; Cadirci, E.; Karakus, E.; Bayir, Y.; Albayrak, A.; Unal, D. Liver 5-HT7 receptors: A novel regulator target of fibrosis and inflammation-induced chronic liver injury in vivo and in vitro. *Int. Immunopharmacol.* **2017**, *43*, 227–235. [CrossRef]
104. Bengmark, S. Gut microbiota, immune development and function. *Pharmacol. Res.* **2013**, *69*, 87–113. [CrossRef]
105. Jenkins, T.A.; Nguyen, J.C.; Polglaze, K.E.; Bertrand, P.P. Influence of Tryptophan and Serotonin on Mood and Cognition with a Possible Role of the Gut-Brain Axis. *Nutrients* **2016**, *8*, 56. [CrossRef] [PubMed]
106. Israelyan, N.; Margolis, K.G. Serotonin as a link between the gut-brain-microbiome axis in autism spectrum disorders. *Pharmacol. Res.* **2018**, *132*, 1–6. [CrossRef] [PubMed]
107. Banskota, S.; Ghia, J.E.; Khan, W.I. Serotonin in the gut: Blessing or a curse. *Biochimie* **2019**, *161*, 56–64. [CrossRef] [PubMed]
108. Shajib, M.S.; Baranov, A.; Khan, W.I. Diverse Effects of Gut-Derived Serotonin in Intestinal Inflammation. *ACS Chem. Neurosci.* **2017**, *8*, 920–931. [CrossRef] [PubMed]

109. Kim, J.J.; Bridle, B.W.; Ghia, J.E.; Wang, H.; Syed, S.N.; Manocha, M.M.; Rengasamy, P.; Shajib, M.S.; Wan, Y.; Hedlund, P.B.; et al. Targeted inhibition of serotonin type 7 (5-HT7) receptor function modulates immune responses and reduces the severity of intestinal inflammation. *J. Immunol.* **2013**, *190*, 4795–4804. [CrossRef] [PubMed]
110. Guseva, D.; Holst, K.; Kaune, B.; Meier, M.; Keubler, L.; Glage, S.; Buettner, M.; Bleich, A.; Pabst, O.; Bachmann, O.; et al. Serotonin 5-HT7 receptor is critically involved in acute and chronic inflammation of the gastrointestinal tract. *Inflamm. Bowel Dis.* **2014**, *20*, 1516–1529. [CrossRef] [PubMed]
111. Zhu, H.; Xiao, X.; Chai, Y.; Li, D.; Yan, X.; Tang, H. MiRNA-29a modulates visceral hyperalgesia in irritable bowel syndrome by targeting HTR7. *Biochem. Biophys. Res. Commun.* **2019**, *511*, 671–678. [CrossRef]
112. Holst, K.; Guseva, D.; Schindler, S.; Sixt, M.; Braun, A.; Chopra, H.; Pabst, O.; Ponimaskin, E. The serotonin receptor 5-HT$_7$R regulates the morphology and migratory properties of dendritic cells. *J. Cell Sci.* **2015**, *128*, 2866–2880. [CrossRef]
113. Profirovic, J.; Strekalova, E.; Urao, N.; Krbanjevic, A.; Andreeva, A.V.; Varadarajan, S.; Fukai, T.; Hen, R.; Ushio-Fukai, M.; Voyno-Yasenetskaya, T.A. A novel regulator of angiogenesis in endothelial cells: 5-hydroxytriptamine 4 receptor. *Angiogenesis* **2013**, *16*, 15–28. [CrossRef]
114. Domínguez-Soto, Á.; Usategui, A.; Casas-Engel, M.L.; Simón-Fuentes, M.; Nieto, C.; Cuevas, V.D.; Vega, M.A.; Luis Pablos, J.; Corbí, Á. Serotonin drives the acquisition of a profibrotic and anti-inflammatory gene profile through the 5-HT7R-PKA signaling axis. *Sci. Rep.* **2017**, *7*, 14761. [CrossRef]
115. Mahé, C.; Loetscher, E.; Dev, K.K.; Bobirnac, I.; Otten, U.; Schoeffter, P. Serotonin 5-HT7 receptors coupled to induction of interleukin-6 in human microglial MC-3 cells. *Neuropharmacology* **2005**, *49*, 40–47. [CrossRef] [PubMed]
116. León-Ponte, M.; Ahern, G.P.; O'Connell, P.J. Serotonin provides an accessory signal to enhance T-cell activation by signaling through the 5-HT7 receptor. *Blood* **2007**, *109*, 3139–3146. [CrossRef] [PubMed]
117. Ito, M.; Komai, K.; Mise-Omata, S.; Iizuka-Koga, M.; Noguchi, Y.; Kondo, T.; Sakai, R.; Matsuo, K.; Nakayama, T.; Yoshie, O.; et al. Brain regulatory T cells suppress astrogliosis and potentiate neurological recovery. *Nature* **2019**, *565*, 246–250. [CrossRef] [PubMed]

International Journal of
Molecular Sciences

MDPI

Review

On the Modulatory Roles of Neuregulins/ErbB Signaling on Synaptic Plasticity

Ada Ledonne [1,*] and Nicola B. Mercuri [1,2]

[1] Department of Experimental Neuroscience, Santa Lucia Foundation, Via del Fosso di Fiorano, no 64, 00143 Rome, Italy; mercurin@med.uniroma2.it

[2] Department of Systems Medicine, University of Rome "Tor Vergata", Via Montpellier no 1, 00133 Rome, Italy

* Correspondence: adaledonne@gmail.com; Tel.: +3906-501703160; Fax: +3906-501703307

Received: 9 December 2019; Accepted: 29 December 2019; Published: 31 December 2019

Abstract: Neuregulins (NRGs) are a family of epidermal growth factor-related proteins, acting on tyrosine kinase receptors of the ErbB family. NRGs play an essential role in the development of the nervous system, since they orchestrate vital functions such as cell differentiation, axonal growth, myelination, and synapse formation. They are also crucially involved in the functioning of adult brain, by directly modulating neuronal excitability, neurotransmission, and synaptic plasticity. Here, we provide a review of the literature documenting the roles of NRGs/ErbB signaling in the modulation of synaptic plasticity, focusing on evidence reported in the hippocampus and midbrain dopamine (DA) nuclei. The emerging picture shows multifaceted roles of NRGs/ErbB receptors, which critically modulate different forms of synaptic plasticity (LTP, LTD, and depotentiation) affecting glutamatergic, GABAergic, and DAergic synapses, by various mechanisms. Further, we discuss the relevance of NRGs/ErbB-dependent synaptic plasticity in the control of brain processes, like learning and memory and the known involvement of NRGs/ErbB signaling in the modulation of synaptic plasticity in brain's pathological conditions. Current evidence points to a central role of NRGs/ErbB receptors in controlling glutamatergic LTP/LTD and GABAergic LTD at hippocampal CA3–CA1 synapses, as well as glutamatergic LTD in midbrain DA neurons, thus supporting that NRGs/ErbB signaling is essential for proper brain functions, cognitive processes, and complex behaviors. This suggests that dysregulated NRGs/ErbB-dependent synaptic plasticity might contribute to mechanisms underlying different neurological and psychiatric disorders.

Keywords: neuregulins; ErbB receptors; synaptic plasticity; LTP; LTD; hippocampus; midbrain dopamine neurons; dopamine

1. Introduction

Neuregulins (NRGs) are a family of neurotrophic factors, which are essential for proper development of the peripheral and central nervous system, as well as adult brain homeostasis. NRGs were discovered more than 25 years ago, independently by four groups that identified a protein, now recognized as the first member of the NRGs family, neuregulin 1 (NRG1), as a factor able to activate ErbB2 tyrosine kinase receptors (called heregulin or neu differentiation factor (NDF)) [1–3], to induce proliferation of Schwann cells (called glial growth factor (GGF)) [4–8], or to stimulate the synthesis of acetylcholine receptors at developing neuromuscular junctions (NMJ) (called acetylcholine receptor inducing activity (ARIA)) [9]. Such a diverse array of names and corresponding functions of NRG1 are indicative of the versatility and importance of NRGs in human brain. Actually, far from their first identification, now it is recognized that NRGs, by acting on ErbB tyrosine kinases (ErbB2, ErbB3, and ErbB4), are crucial developmental factors, mediating neural differentiation, neuronal guidance, myelination, synapse formation, and NMJ development, as well as they represent important modulators of neuronal excitability, neurotransmission, and synaptic plasticity in adult brain. By these multifaceted roles, NRGs/ErbB signaling controls key physiological neural functions, affecting learning and memory processes and complex behaviors, and its dysfunction is emerging as a pathological feature in different neurological and psychiatric disorders, including schizophrenia, bipolar disorder, autism spectrum disorders, Alzheimer's disease, major depressive disorder, addiction, Parkinson's disease, and peripheral and central nervous system injury diseases.

In this review, we will provide an overview of NRGs/ErbB roles in the regulation of synaptic transmission, by focusing on its contribution to the control of synaptic plasticity in the hippocampus and midbrain dopamine (DA) nuclei. Available evidence supports diverse mechanisms engaged by NRGs/ErbB receptors in the regulation of long term modifications of synaptic transmission in these brain areas, depicting a complex scenario involving either presynaptic or postsynaptic NRGs-mediated mechanisms, which affect glutamatergic, GABAergic, and DAergic transmission.

2. Neuregulins and ErbB Receptors: Subtypes and Signaling Pathways

2.1. Neuregulins

Neuregulins (NRGs) are a family of epidermal growth factor (EGF)-related proteins encoded by six genes (*Nrg1–Nrg6*). Each gene, by specific controlled transcription and splicing mechanisms, typically produces many mRNA and protein isoforms all expressing a shared domain, the EGF-like domain, required for signaling activation. Neuregulin 1 (NRG1) is the most extensively studied and well characterized member of the NRGs family. The *Nrg1* gene produces six different types (NRG1 I–VI) and 33 spliced isoforms, due to specific uses of six different transcriptional initiation sites and by alternative splicing [10–14]. NRG1 types (I–VI) are identified based on differences in the N terminal domain, besides the presence of immunoglobulin (Ig)-like domains, and/or a cysteine-rich domain (CRD) (Figure 1A). NRG1 type I, II, IV, and V isoforms have the Ig domain, which contributes to distinct interactions with extracellular matrix components (e.g., heparan-sulfate proteoglycans (HSPGs)) and defines the distance and concentration over which these growth factors act [15]. NRG1 type III is the only presenting a CRD, which serves as a secondary transmembrane domain, and thus is a membrane-anchored isoform, acting in an autocrine manner.

NRG1 type I is also known as heregulin, Neu differentiation factor (NDF), or acetylcholine receptor inducing activity (ARIA) [1,2], whereas types II and III have been identified as glial growth factor (GGF) [6], and sensory and motor neuron derived factor (SMDF), respectively [16], based on the first function/cellular population for which they were firstly identified.

NRG1 types are widely and diffusely expressed in both central and peripheral nervous system, as well as in heart, liver, stomach, lung, kidney, spleen, and skin. Human brain areas showing higher NRG1 expression are prefrontal and cingulate cortex, hippocampus, habenula, amygdala, substantia nigra, dorsal and ventral striatum (caudate, putamen, and nucleus accumbens), hypothalamus, spinal cord, and cerebellum [14,17]. All six types of NRG1 are detectable in brain, but the abundance of each form varies significantly, being also related to the developmental period and neuronal activity [14]. In general, NRG1 type III appears to be the predominant brain type of NRG1, while type I and II NRG1 are less expressed, similar to type IV, which, however, is brain specific.

The *Nrg2* gene produces two types of NRG2s, with different EGF-like domains, termed NRG2α and NRG2β, respectively, and at least 10 isoforms due to alternative splicing [18,19]. NRG2 is expressed in the developing nervous system and is also found in the embryonic heart, lung, and bladder [19,20]. In adult brain, NRG2's highest expression is in hippocampal dentate gyrus (granule cells), cerebellum, and olfactory bulbs [18–20], whereas weaker expression has been reported in the neocortex, hippocampal CA1–CA3 neurons, and striatum [21].

NRG3 is present in different splicing forms, up to 15 [22–25], and is mainly diffused in the brain, either in the embryonic or adult stage. NRG3 has been detected in spinal cord and numerous brain regions, including anterior olfactory nucleus, cerebral cortex, piriform cortex, vestibular nuclei, medial habenula, hypothalamus, thalamus, deep cerebellar nuclei, and hippocampus [26].

NRG4 exists in five isoforms [27,28]. Different from the other NRGs, NRG4 expression appears more confined to peripheral organs, being expressed at high levels in the pancreas, in the skeletal muscle, and in the brown adipose tissue, with its expression in adult brain considered negligible [27,29]. Recently, however, NRG4 expression has been reported in the developing brain, in various brain areas like cortex, hippocampus, cerebellum, olfactory bulb, midbrain, and brain stem [30].

NRG5 is also known as tomoregulin or transmembrane protein with EGF-like and two follistatin-like domains 1 (TMEFF1) [31,32]. NRG5 has five spliced isoforms [27,28], and it is highly expressed in several brain areas, olfactory bulb, amygdala, different regions of cortex (entorhinal, cingulate, motor, somatosensory cortex), various areas of hippocampus (CA3, CA1, and subiculum), locus coeruleus, substantia nigra pars compacta (SNpc), hypothalamic nuclei (ventromedial and paraventricular), and cerebellum [33].

NRG6 is also called neuroglycan C (NGC), or chondroitin sulfate proteoglycan 5 (CSFG5), or chicken acidic leucine rich EGF-like domain containing brain protein (CALEB) [32,34]. In adult human brain, the strongest NRG6 expression has been reported in the striatum (caudate and putamen nuclei), hippocampus, amygdala, and cerebral cortex, whereas weaker expression has been reported in substantia nigra, thalamus, pons, medulla oblongata, and cerebellum [35].

2.2. NRGs Processing

NRGs are synthesized as precursors, called pro-NRGs, being the active forms produced following a proteolytic shedding. Different proteases have been associated with the NRGs' proteolysis, mainly belonging to the "A disintegrin and metalloprotease" (ADAM) subfamily of the matrix metalloproteinases (MMPs), like ADAM9, ADAM10, ADAM12, ADAM15 (also called metargidin), ADAM19 (also called meltrin-β), and ADAM17 (also called tumor necrosis factor-α converting enzyme (TACE-1)). In addition, NRGs are processed by the β-amyloid converting enzyme (BACE-1), a γ-secretase that is better known for its involvement in the synthesis of β-amyloid peptides. Different NRGs can have distinct mechanisms regulating their maturation and release. More detailed information is available for NRG1. Almost all types of NRG1 are secreted as soluble factors, released after the conversion from pro-NRG1 to mature NRG1 forms, thus acting in a paracrine manner (Figure 1B). The diffusion and specificity of action on distinct cellular populations might be also allowed by type specific domains (Ig-like or additional spacer domain, as well as differences in N terminus). NRG1 type III, also after its conversion to the mature form, remains attached to the cell membrane due to its CRD, and thus, it acts in an autocrine manner, signaling only to immediately neighboring cells [13,36] (Figure 1B). Regarding peculiar mechanisms that might control the synthesis of specific NRGs, it has been reported that NRG1 types I and II or NRG2 accumulate as pro-forms on cell bodies and are released by MMPs, in an activity dependent manner (e.g., following NMDAR activation), whereas NRG1 type III and NRG3 seem to be constitutively processed by BACE and accumulate on axons where they interact with ErbB in juxtacrine mode [37].

Production of mature NRGs, by ectodomain shedding, occurs in response to diverse stimuli, and possibly in a type specific and area related way. NRG1 levels are increased by an enhancement of neuronal activity, as the one prompted during epileptic seizures in a rat model of epilepsy induced by kainic acid administration, as well as by milder neuronal activities, as induced following forced locomotor activity [38]. Interestingly, neuronal activity seems to shape in a different way the expression of distinct NRG1 types/isoforms, since epileptic seizures appear to not alter NRG1 type II and III levels, while strongly increasing NRG1 type I expression [14]. Increased expression of NRG1 type II after seizure insurgence appears instead in non-neuronal cells. In this regard, a depolarizing treatment increases NRG1 type I expression also in astrocytes from cellular cultures, thus suggesting that brain activity might differently regulate the levels of distinct NRG1 isoforms in diverse cellular populations/brain regions, thus contributing to compartmentalized NRGs related actions.

Moreover, production of soluble NRGs, by ectodomain shedding, occurs in response to other stimuli, like the activation of G protein coupled receptors (GPCR) or ionotropic receptors, through a process that ultimately leads to ErbB transactivation. Such mechanisms involve the activation of MMPs and/or ADAM-dependent shedding of pro-NRGs as a consequence of GPCRs activation, such as angiotensin II receptors, protease-activated receptor 1 (PAR-1), as well as NMDAR stimulation [37,39]. Interestingly, the interplay between NRGs and proteases is bidirectional. Indeed, it has been seen that NRG1 increases the expression of MMP-9, a metalloproteinase critically involved in the regulation of extracellular matrix regulation [40].

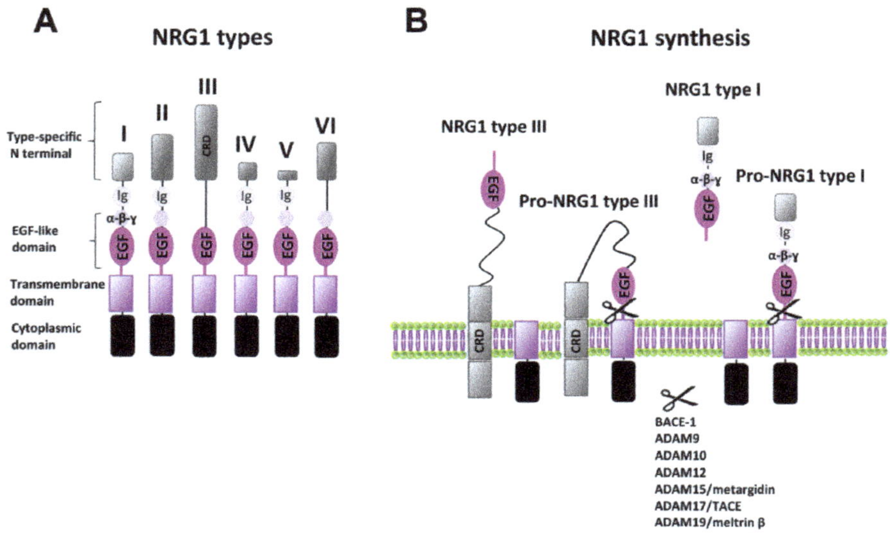

Figure 1. NRG1 types and synthesis. (**A**) Diagram of structural differences of various NRG1 types. (**B**) Synthesis of NRGs' mature forms is induced by different ADAMs and MMPs, with the release of soluble forms (as for example for NRG1 type I) or the exposure of the membrane-anchored EGF-like domain (as for NRG1 type III).

2.3. *ErbB Receptors*

NRGs signal by activating tyrosine kinases receptors of the ErbB family. Four ErbB subtypes (ErbB1–4) have been identified, of which ErbB2, ErbB3, and ErbB4 mediate NRGs signaling, whereas ErbB1, also known as the epidermal growth factor (EGF) receptor, is activated by EGF. Upon NRGs binding, ErbB receptors undergo a structural/conformational change in the juxta-membrane region, which potentiates the affinity for other ErbB subunits, thus promoting the formation of homo- and/or hetero-dimers of functional ErbB receptors.

Each ErbB subunit displays a peculiar profile, regarding ligand binding properties and affinities or catalytic activity, which confer them distinct abilities in the possibility to form homodimers or heterodimers. ErbB4 represents the only autonomous subunit, since it either expresses NRGs binding sites or has active kinase domains, thus possibly forming both homodimers and heterodimers of ErbB receptors. ErbB3 has ligand binding sites, but does not express an active kinase domain [41], thus, it cannot form homodimers, nor directly phosphorylate other ErbB receptors, but can associate with other ErbB subunits (ErbB2 and ErbB4) in the formation of heterodimers. ErbB2, instead, has an active kinase domain, but does not bind NRGs or other identified ligands (being for this still considered an orphan receptor). Despite this, ErbB2 represents the preferred dimerization partner among all ErbB receptors [42], and dimerization with ErbB2 strongly increases NRGs' binding affinity for ErbB3 and ErbB4 [43,44]. Hence, NRGs actions are mediated by heterodimers ErbB2/ErbB3, ErbB2/ErbB4, and ErbB3/ErbB4, as well as ErbB4/ErbB4 homodimers.

Different NRGs display specific profiles of affinity for different ErbB subunits. NRG1 and NRG2 bind both ErbB3 and ErbB4, thus possibly activating all ErbB dimers (ErbB2/ErbB3, ErbB2/ErbB4, ErbB3/ErbB4, ErbB4/ErbB4), whereas NRG3 and NRG4 only bind to ErbB4; thus, they signal through ErbB4/ErbB4 and/or ErbB2/ErbB4 dimers. Notably, NRG3 has a much higher affinity for ErbB4 than any other NRGs [45]. Despite NRGs are not ligands for ErbB1, it can form heterodimers with ErbB4 [46–48], thus implying that NRGs signaling can also be mediated by ErbB1/ErbB4. However, evidence suggests that the activation of ErbB1/ErbB4 causes significantly different cellular effects with respect to ErbB1/ErbB1 homodimers (which mediate specific EGF effects), hence still implying segregated functions of EGF and NRGs in neurons [13,46,49–51].

2.4. NRG/ErbB Signaling

In the canonical ErbB signaling (Figure 2A), NRGs binding induces a conformational change of ErbB subunits fostering trans-phosphorylation and the consequent recruitment of proteins containing phosphotyrosine binding or Src homology-2 (Shc-2) domains, which act as adaptor/effector molecules, triggering the activation of multiple signaling pathways. Intracellular pathways commonly activated by NRGs, downstream of ErbB receptors, are PI3K-Akt-mTOR-S6K and Ras-Raf-MEK-ERK [13]. Other NRGs/ErbB-activated pathways include the PLC-PKC pathway, as well as kinases like c-Abl, JNK, CDK5, Kyn, and Pyk2 [52–54]. NRGs-induced effects that involve protein synthesis and cause neuronal growth and survival are mainly mediated by the stimulation of PI3K-Akt-mTOR and glycogen synthase 3 kinase (GS3K), activated downstream of Akt. Phosphatydil-inositole-3 kinase (PI3K) catalyzes the phosphorylation of phosphatidylinositol 4,5-diphosphate (PIP2) to the second messenger phosphatidylinositol 3,4,5-triphosphate (PIP3), which fosters the activation of the protein serine/threonine kinase Akt, then activating the mammalian target of rapamycin (mTOR) (Figure 2A). Among ErbB subunits, ErbB3 expresses the highest prevalence of docking sites for PI3K [55]; thus, the activation of the PI3K-Akt-mTOR pathway might be preferentially due to the stimulation of ErbB3-containing receptors, like ErbB2/ErbB3 and/or ErbB3/ErbB4.

Ras-Raf-MEK-ERK is another pathway commonly stimulated by NRGs/ErbB receptors. NRGs-induced activation of Ras-Raf-MEK-ERK is allowed by the recruitment of the adaptor protein, the growth factor receptor/bound protein 2 (GRB2) to the phosphotyrosine residues of activated ErbB subunits. GRB2 binding with ErbB can be either direct or intermediated by the interaction with the Src homolog and collagen homolog (SHC) adaptor protein, which directly binds tyrosine phosphorylated ErbB sites, thus being itself phosphorylated and then binding GRB2. ErbB-GRB2 then recruits and activates Son of Sevenless (SOS), a guanine nucleotide exchange factor, which fosters GTP availability for binding to Ras, thus activating this kinase. Active Ras starts the kinase cascade pathway including c-Raf, MEK1/2, and ERK1/2. Phosphorylated ERK1/2 translocates to the nucleus, where it activates transcriptional factors (like Elk1), inducing the transcription of genes promoting cell growth and survival. A residual pool of active cytoplasmic ERK1/2 also phosphorylates cytoskeletal proteins such as actin, which promotes cell motility, regulators of cell division and cytokines, vesicle, and organelle movement, and mitochondrial targets such as Bcl2 that render cells resistant to apoptosis (Figure 2A). Other NRGs/ErbB-activated pathways, like the PLC-PKC pathway or c-Abl, JNK, CDK5, Kyn, and Pyk2 kinases [52–54], are mainly involved in the regulation of gene expression, by controlling the activity of transcriptional factors such as c-Fos, Elk1, STAT, c-Jun, and c-Myc.

Besides canonical ErbB activation, NRGs/ErbB-dependent effects can be mediated by additional signaling modalities, i.e., "non-canonical forward ErbB signaling" and "NRG1 backward signaling" (Figure 2B). Non-canonical ErbB signaling is started by proteolysis of ErbB4, that can be subjected to cleavage, by γ-secretase, in the membrane-bound fragment with the release of the ErbB4 intracellular domain (ErbB4-ICD) [56,57]. Such ErbB4-ICD can translocate to the nucleus, thus regulating gene transcription [58] (Figure 2B). ErbB4 can be also cleaved by TACE in the extracellular membrane region with the release of a soluble peptide (called ecto-Erb4), which contains the NRG1 binding site and acts as a ligand for membrane-anchored NRGs (i.e., immature pro-NRGs forms or the membrane-bound form of NRG1 type III). Ecto-ErbB4 binding to presynaptic NRGs has dual effects, since either it neutralizes NRGs' canonical actions by preventing NRGs/ErbB forward signaling or can trigger the so-called "NRG1 backward signaling" [59] (Figure 2B). Such a modality of non-canonical NRG1 signaling is elicited by a proteolytic cleavage of the intracellular domain of pro-NRG1, by γ-secretase, with the release of the NRG1 intracellular domain (NRG1-ICD). After its translocation to nucleus NRG1-ICD, through the interaction with the transcription factor Eos, regulates the transcription of different genes, including postsynaptic density (PSD) protein PSD95 [13,60]. Some evidence also documented that NRG3, like NRG1, can participate in back signaling by the C-terminal domain [59].

Overall, the emerging scenario demonstrates that NRGs can activate a complex network of signaling pathways, since NRGs/ErbB signals can be transmitted either in forward (canonical and non-canonical) and backward modalities (Figure 2A,B).

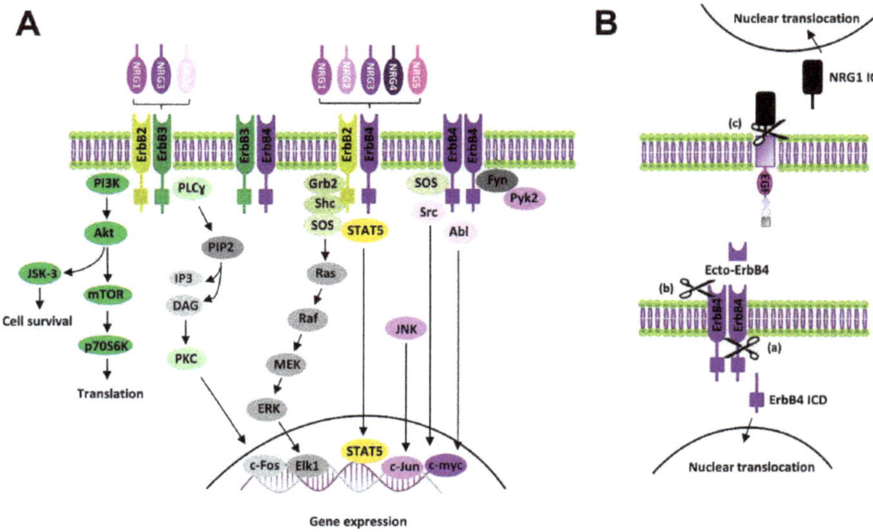

Figure 2. NRGs/ErbB signaling. (**A**) Diagram showing canonical ErbB signaling pathways activated by NRGs. (**B**) Scheme of non-canonical ErbB signaling modalities: (a) Non-canonical forward ErbB4 signaling: proteolytic cleavage of ErbB4 causes the release of its intracellular domain (ErbB4 ICD), which translocates to the nucleus, modulating gene expression; (b) Ecto-ErbB4 domain, containing the binding site of NRGs, is released by proteolytic cleavage of ErbB4 extracellular domain. Ecto-ErbB4 can bind pro-NRGs, thus interfering with NRGs canonical signaling or triggering NRG1 backward signaling; (c) NRG1 backward signaling: proteolytic cleavage of NRG1 (pro-NRG1s or membrane-anchored NRG1 type III) in the intracellular domain induces the release of NRG1 ICD, which triggers backward signaling by nuclear translocation and modulation of gene transcription.

3. NRGs/ErbB Roles in the Modulation of Synaptic Plasticity

3.1. NRGs/ErbB-dependent Regulation of Synaptic Plasticity in the Hippocampus

Synaptic plasticity, the ability to finely adjust the strength of synaptic transmission, is a critical feature of the CNS and is supposed to underlie essential physiological processes, such as learning and memory, as well as complex behaviors, including goal-oriented behaviors. Hippocampal synapses operate over a dynamic range of efficacy and are subject to both short- and long-term forms of plasticity, including long-term potentiation (LTP) and long-term depression (LTD), leading candidates as synaptic mechanisms for memory [61–64]. While the description of different types of synaptic plasticity in the hippocampus is out of the scope of this review, here we will focus on those that are known to be modulated by NRGs/ErbB signaling.

3.1.1. Glutamatergic LTP at CA3–CA1 Synapses

LTP at CA3–CA1 synapses of the hippocampus is the most extensively studied form of activity-dependent synaptic plasticity in the vertebrate nervous system. LTP induction can be triggered by different stimulation protocols, resembling those occurring physiologically, and it mainly requires NMDARs activation during a strong postsynaptic depolarization. The increase in postsynaptic calcium concentration leads to the consequent activation of biochemical events necessary to sustain LTP expression and maintenance, which is ultimately due to increased AMPAR-mediated excitatory postsynaptic currents (EPSCs) [61].

Stimulating protocols able to induce hippocampal LTP include high-frequency (tetanic) stimulation (HFS, 100 Hz) or "pairing protocols", during which delivery of low-frequency synaptic activation is paired with a direct depolarization of the postsynaptic cell. LTP induction is also achieved by a stimulation that generates a synaptic response within a discrete time window prior to the firing of the postsynaptic cell, producing the so called "spike-time dependent plasticity" (STDP) [65,66], as well as through the delivery of theta-burst stimulation (TBS), which consists of a pattern of neuronal firing (complex spikes) applied at the frequency of the hippocampal theta rhythm, which spontaneously occurs during behavior [67]. Depotentiation of potentiated synapses, mimicking an LTP reversal, is an additional mechanism preserving synaptic homeostasis at hippocampal CA3–CA1 synapses. Such depotentiation can be induced in acute hippocampal slices and in freely moving animals by brief TBS, delivered shortly after LTP induction [68].

The first demonstration that NRGs/ErbB signaling plays a central role in the modulation of synaptic plasticity in adult brain is represented by the evidence that NRG1 application to hippocampal rat slices prevents the induction of LTP at Schaffer collaterals-CA1 synapses [69]. In particular, a treatment with NRG1 suppresses the induction of HFS-induced LTP of field excitatory postsynaptic potentials (fEPSPs) from the CA1 dendritic region, without affecting basal glutamatergic synaptic transmission [69]. NRG1's effects on tetanic-induced LTP are concentration dependent and considered to be reliant on postsynaptic mechanisms, since paired-pulse facilitation of evoked glutamatergic synaptic responses, representing a measure of presynaptic function, was not altered by NRG1 treatment [69]. The involvement of NRGs/ErbB signaling in the modulation of hippocampal synaptic plasticity has been further confirmed by several other pieces of evidence, also providing insights into the cellular mechanisms engaged by NRGs in the maintenance of synaptic efficacy and neuronal connectivity in hippocampal formation. NRG1 acutely reverses TBS-induced LTP of fEPSPs in the hippocampal CA1 area at pre-stimulation values [70]. Such NRG1-dependent depotentiation depends on the activation of ErbB receptors, since it is counteracted by ErbB inhibitors. Notably, besides preventing NRG1-dependent synaptic depotentiation, ErbB inhibition per se increases the magnitude of TBS-induced LTP, thus supporting a role of endogenous ErbB signaling in synaptic strength regulation. NRG1-induced synaptic downscaling appears to be mediated by a specific internalization of AMPARs, whereas NMDARs surface expression and function are not altered [70].

Although the precise cellular mechanisms by which NRG1 prevents LTP induction [69] or depresses its expression [70] at hippocampal CA3–CA1 synapses are not completely elucidated, several hypotheses have been postulated, mainly based on evidence from studies analyzing synaptic plasticity at CA3–CA1 synapses following conditional ErbB genetic ablation or pharmacological inhibition. GABAergic transmission plays an important role in the NRG1-dependent suppression of LTP at CA3–CA1 synapses in the hippocampus, as proven by the evidence that NRG1's effects on LTP are impaired in mice harboring a conditional genetic ablation of ErbB4 in parvalbumin (PV) expressing GABAergic interneurons [71]. NRG1 has a facilitatory effect on GABAergic transmission, since it increases $GABA_A$-mediated currents in CA1 pyramidal cells, as a consequence of a direct stimulation of ErbB4 located on PV+ GABAergic interneurons. Thus, by boosting GABA-mediated inhibition in CA1 pyramidal neurons, NRG1 could counteract LTP induction/expression at CA3–CA1 synapses.

Interestingly, since the conditional deletion of ErbB4 in CAMKII positive neurons (i.e., pyramidal neurons in the forebrain) does not alter the NRG1-induced suppression of LTP, it has been suggested that ErbB4 in CA1 pyramidal cells has a minor contribution to the regulation of glutamatergic LTP [32,71]. Thus, the proposed scheme accounting for this evidence points to a direct stimulation, by NRG1, of ErbB4 receptors expressed on PV+ GABAergic neurons, which enhances GABA release on CA1 pyramidal cells, thus increasing their inhibition and shifting up the threshold for LTP induction at CA3–CA1 synapses (Figure 3A). Notably, NRG1/ErbB-dependent LTP impairment has been observed also in the presence of $GABA_A$ antagonists [70,72], thus questioning NRG1/ErbB-induced GABA release acting on the $GABA_A$ receptor as specific mechanism underlying LTP impairment. Thus, additional mechanisms, independent on $GABA_A$ stimulation, might be engaged by NRGs/ErbB signaling in LTP depression [32].

Besides GABAergic transmission, also dopamine (DA) transmission takes part in NRG1-dependent modulation of LTP at hippocampal CA3–CA1 synapses. NRG1 injection in the dorsal CA1 area of the hippocampus of rats acutely stimulates DA release, with the consequent activation of DAergic D4 receptors, which directly mediate the inhibition of glutamatergic synaptic plasticity at CA3–CA1 synapses [73]. Indeed, NRG1-dependent reversal of LTP is selectively blocked in hippocampal slices treated with a D4 antagonist, L-745,870 [73]. The functional role of D4 in shaping hippocampal synaptic plasticity at CA3–CA1 synapses, as well as its engagement in NRG1-dependent regulation of LTP have been further confirmed by experiments showing that D4 pharmacological activation mimics NRG1's effects on LTP, which are conversely abolished in D4 knockout (KO) mice [73]. However, while this evidence suggests involvement of DA transmission in NRGs/ErbB-dependent LTP regulation at hippocampal CA3–CA1 synapses, the specific mechanisms underlying NRG1-induced DA release in the CA1 dorsal hippocampus are not completely clarified yet. In this respect, it has been previously postulated that NRG1-induced ErbB activation in GABAergic PV+ interneurons might be instrumental to NRG1-induced DA release [32,73]. This hypothesis is mostly based on the preferential localization of ErbB4 on GABAergic PV+ interneurons in the hippocampus and the evidence that ErbB4 activation is involved in LTP modulation. Nevertheless, a functional link by which ErbB4 stimulation of GABAergic interneurons could lead to increased DA extracellular levels is lacking. More recently, such a contribution of GABAergic interneurons has been questioned by the evidence that NRG1-dependent regulation of DA outflow in CA1 dorsal hippocampus is dependent on ErbB4 expressed on DA neuronal terminals [74]. Indeed, in mice harboring a conditional genetic ablation of ErbB4 in tyrosine hydroxylase (TH) positive neurons, NRG1 injection in CA1 dorsal hippocampus fails to potentiate DA release [74]. Regarding the underlying functional mechanisms, NRG1-dependent regulation of DA levels appears to be mediated by an interplay between ErbB4 and the dopamine transporter (DAT) expressed on DAergic terminals. Indeed, based on the evidence that ErbB4 stimulation inhibits DAT activity in neuronal cultures, it has been proposed that NRG1-induced potentiation of DA transmission is consequent to an ErbB4-induced DAT inhibition [74]. Hence, according to this hypothesis, hypofunctional DA reuptake results in increased DA extracellular levels, which fosters D4 activation in CA1 pyramidal neurons, depressing LTP at CA3–CA1 synapses (Figure 3B). Regarding the specific involvement of D4 receptors downstream of NRG1-induced DA release, however, there is recent evidence demonstrating that, although either NRG1/ErbB signaling or D4 stimulation is able to mediate LTP reversal/suppression, they act in an independent way in the modulation of glutamatergic synaptic plasticity in the CA1 dorsal hippocampus [75].

NRG1-dependent LTP depression might also be induced by additional postsynaptic mechanisms, directly occurring in hippocampal CA1 pyramidal neurons. It has been reported that NRG1/ErbB4 signaling interferes with the Src-dependent regulation of NMDAR in CA1 pyramidal neurons [72]. Specifically, NMDAR-mediated transmission in CA1 pyramidal neurons is enhanced by activating Src kinases, as demonstrated by the injection of Src activating peptides in single pyramidal neurons [72]. Such Src-dependent potentiation of NMDAR-mediated transmission is impaired by NRG1/ErbB4 signaling, which directly inhibits Src kinase activity. Hence, a model accounting for this evidence depicts ErbB4 activation in CA1 pyramidal cells as the triggering event to reduce, in a cell autonomous manner, NMDAR-mediated transmission and consequently impairing LTP expression (Figure 3C). Notably, ErbB4 localization in the hippocampus is more restricted to GABAergic interneurons [76–78], while expression in CA1 pyramidal cells appears low [79–82] or absent [76–78]. This questions the postsynaptic/cell autonomous function of ErbB4 in the NRGs-dependent LTP impairment at CA3–CA1 synapses. Notwithstanding, in spite of undetected expression, ErbB4 functions in hippocampal and cortical pyramidal neurons have been reported [30,72,83–86]. Thus, the factual role of ErbB4 in hippocampal CA1 pyramidal cells in NRGs-dependent LTP regulation remains to be better elucidated.

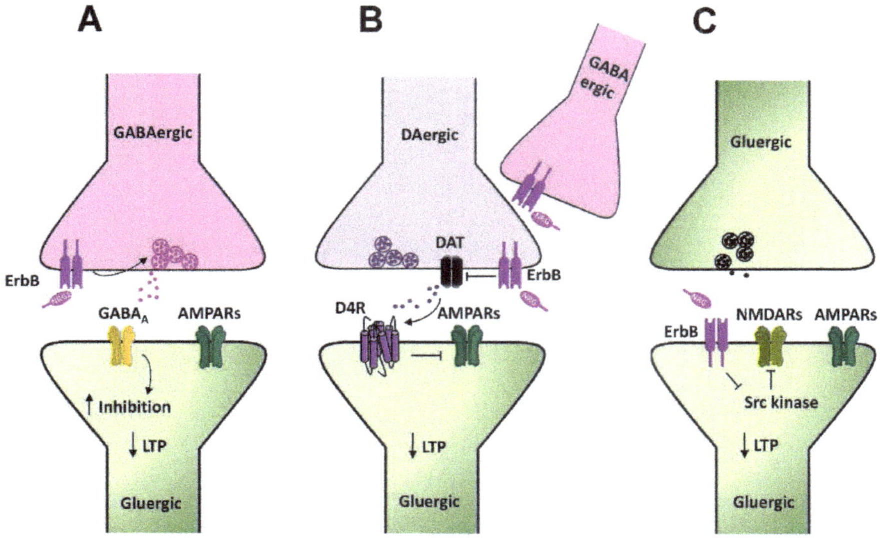

Figure 3. NRGs/ErbB-dependent regulation of LTP in the hippocampal CA1 area. (**A–C**) Diagrams illustrating the cellular mechanisms underlying NRGs/ErbB-dependent impairment of LTP at hippocampal CA3–CA1 synapses. (**A**) NRG1-dependent activation of ErbB4 in GABAergic interneurons, by triggering GABA release, fosters the activation of GABA$_A$ receptors expressed in CA1 pyramidal cells, thus reducing neuronal excitability and dampening LTP induction. (**B**) NRG1-induced ErbB4 activation on DAergic terminals inhibits DAT activity, thus increasing DA extracellular levels. The consequent activation of D4 receptors localized on CA1 pyramidal neurons impairs LTP induction/expression, via internalization of AMPARs. (**C**) NRG1-dependent ErbB4 activation in CA1 pyramidal neurons inhibits Src kinase and consequently NMDAR activity, thus reducing LTP expression.

3.1.2. mGluRI-dependent Glutamatergic LTD

Long term depression (LTD) of glutamatergic synaptic transmission at CA3–CA1 synapses represents a well characterized and largely studied form of synaptic plasticity in the hippocampus. Group 1 metabotropic glutamate receptors (mGluRI), encompassing the mGluR1 and mGluR5 subtypes, are involved in the induction of this form of synaptic plasticity either in the hippocampus or in different other areas, including dorsal and ventral striatum, medial prefrontal cortex, cerebellum, and midbrain DAergic nuclei [87]. mGluRI-LTD at hippocampal CA3–CA1 synapses ultimately relies on mGluRI-induced AMPARs internalization [61,88,89]. Intracellular mechanisms, downstream of mGluRI activation, encompass the activation of several kinases pathways, including ERK1/2, PI3K-Akt-mTOR, and mitogen activated protein kinases (MAPKs) [90–92], which fosters the synthesis of "LTD proteins", instrumental to LTD expression and maintenance [61,87,93]. mGluRI-LTD can be easily achieved through mGluRI activation with the agonist DHPG or by endogenous glutamate, synaptically released through specific electrical protocols, like paired pulse (PP) low frequency stimulation (PP-LFS 1 Hz, 15 min) [94–96]. We recently reported that NRG1/ErbB signaling is a critical modulatory pathway of mGluRI-dependent LTD at CA3–CA1 synapses [86]. Actually, NRG1 fosters the induction of mGluRI-LTD in CA1 pyramidal neurons from hippocampal mice slices. Such NRG-dependent facilitation of LTD is especially unmasked in conditions of minimal stimulation of mGluRI, which per se induces only a short term depression of AMPAR-mediated transmission, but causes a stable LTD following NRG1 treatment. More interestingly, preventing endogenous ErbB activation, with ErbB inhibitors, impairs mGluRI-LTD expression. Indeed, the intracellular injection of pan-ErbB inhibitors or selective ErbB2 inhibitors in single CA1 pyramidal cells depresses mGluRI-induced LTD, thus supporting an NRG1/ErbB-dependent cell autonomous mechanism of regulation of mGluRI-mediated synaptic plasticity, involving ErbB2-containing receptors [86] (Figure 4A). Such a postsynaptic model is also supported by the evidence that NRG1/ErbB-dependent effects are observed in the presence of a $GABA_A$ antagonist, thus not comprising the interplay of GABAergic transmission, but rather a direct mGluRI regulation in CA1 pyramidal neurons, as similarly occurs in midbrain DA cells (see the section below).

Overall, evidence on the modulatory roles of NRGs/ErbB signaling in glutamatergic synaptic plasticity in the hippocampus indicate that NRG1-dependent ErbB activation dampens glutamatergic LTP and favors mGluRI-dependent glutamatergic LTD at hippocampal CA3–CA1 synapses, thus representing a central mechanism balancing the LTP/LTD equilibrium, which shapes the strength of the excitatory transmission at CA3–CA1 synapses.

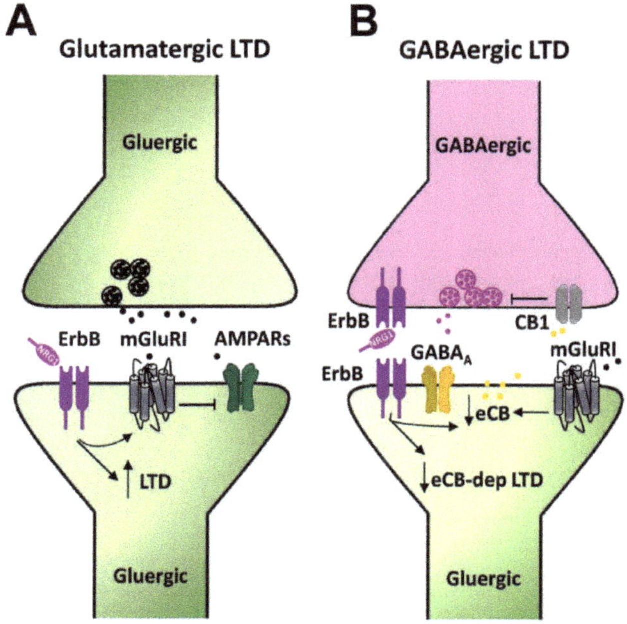

Figure 4. Mechanisms underlying NRGs/ErbB-dependent regulation of mGluRI-LTD in the hippocampus. (**A**) NRG1-dependent activation of ErbB receptors in CA1 pyramidal neurons, by regulating mGluRI function, controls the LTD magnitude of AMPAR-mediated transmission. (**B**) NRG1, by activating ErbB receptors, reduces endocannabinoids (eCBs) levels, thus impairing the eCB-dependent LTD of GABAergic transmission, which is physiologically triggered by the eCB-mediated activation of CB1 located on presynaptic GABAergic terminals, controlling GABA release.

3.1.3. mGluRI-dependent GABAergic LTD

While it is largely recognized that NRGs/ErbB signaling plays a prominent role in the regulation of the GABAergic system, affecting synapses development, as well as neurotransmitter release and GABAergic receptors expression/function, evidence regarding long term modifications of GABAergic transmission, by NRGs/ErbB-dependent mechanisms, are scarce.

The activation of mGluRI in hippocampal CA1 pyramidal neurons induces an LTD of GABAergic transmission, which is reliant on mGluRI-induced production of endocannabinoids (eCBs), which act retrogradely on their presynaptic receptors, cannabinoid receptors 1 (CB1Rs), thus inhibiting GABA release [97,98]. Such a form of mGluRI-dependent eCBs-mediated LTD of GABAergic transmission, which is inducible upon mGluRI stimulation, is modulated by NRG1/ErbB signaling [99]. Indeed, a chronic treatment with NRG1 (8–11 days on organotypic hippocampal slices) impairs the expression of mGluRI-dependent GABAergic LTD at CA3–CA1 synapses, via an eCB-dependent mechanism, which is reliant on a reduced tone of eCBs, as a consequence of their increased catabolism [99] (Figure 4B). The intracellular events, by which NRG1-induced ErbB activation potentiates the expression of catabolic enzymes of eCBs, as well as the synaptic compartments where they occur, are still uncharacterized [99]. Notwithstanding, this evidence points to an important role of NRGs/ErbB signaling in the regulation of GABAergic LTD at hippocampal CA3–CA1 synapses, thus adding further to the picture describing the relevance of NRGs/ErbB signaling in the regulation of synaptic plasticity in the hippocampus.

3.2. NRGs/ErbB-dependent Regulation of Synaptic Plasticity in Midbrain DA Neurons

Long lasting modifications of synaptic transmission have been reported in midbrain DA nuclei. The evidence is mainly related to the ventral tegmental area (VTA) DA cells, where different forms of synaptic plasticity (LTP and LTD) can be induced by either electrical stimulation protocols or pharmacological activation of mGluRI [100–105]. In VTA DA cells, glutamatergic LTP is inducible by HFS protocols paired with neuronal depolarization [100], as well as by a spike-timing-dependent stimulation protocol [106], by mechanisms involving NMDAR activation and increase in intracellular Ca^{2+}.

LTD of glutamatergic transmission can be observed in VTA and SNpc DA neurons, upon delivery of LFS (1 Hz, 10 min) paired with neuronal depolarization during stimulation. Such LFS-induced LTD is reliant on the activation of voltage-dependent Ca^{2+} channels, but does not require the activation of NMDAR or glutamatergic metabotropic receptors [101,102]. As occurs in other brain areas, mGluR1 activation triggers LTD of AMPAR-mediated transmission either in VTA [103] or SNpc DA cells [107]. In nigral DA cells, mGluR1-LTD is easily inducible by mGluR1 activation either achieved with an agonist or by synaptic glutamate, endogenously released during electrical LFS [107]. In VTA DA cells from mice, mGluRI-LTD is especially unmasked in already potentiated synapses (e.g., following exposure to psychostimulants) more than in naive synapses [108] and is due to a modification of AMPARs subunit composition, which decreases AMPARs ion conductances [103].

NRGs/ErbB signaling critically regulates glutamatergic synaptic plasticity in SNpc DA neurons [107]. Indeed, ErbB signaling is required to preserve mGluRI-dependent LTD expression in nigral DAergic neurons, the mGluRI-LTD magnitude being reduced by preventing endogenous ErbB stimulation, with ErbB inhibitors that preferentially act on ErbB2/ErbB4 subunits [107]. Furthermore, exogenous NRG1 application fosters mGluRI-dependent LTD in DA neurons, allowing its induction also in the presence of a minimal mGluRI stimulation. Hence, NRG1/ErbB signaling controls the strength of glutamatergic synaptic transmission in midbrain DA neurons, fine tuning the magnitude of LTD of AMPAR-mediated transmission caused by mGluR1 activation [107].

Regarding the cellular mechanisms underlying NRG1-dependent regulation of mGluRI-LTD, we previously found that NRG1/ErbB signaling bidirectionally controlled mGluR1 expression levels on the surface membrane of midbrain DA neurons [109]. Indeed, NRG1-induced ErbB activation rapidly stimulates mGluR1 synthesis and membrane trafficking. Exogenous NRG1 increases mGluR1 protein levels in SNpc/VTA homogenates and mGluR1 immunolabeling in TH+ DA neurons, in addition to potentiating mGluR1-mediated currents in SNpc DA cells [109]. The mechanisms underlying NRG1-induced increase in mGluR1 expression/function in midbrain DA neurons involved ErbB receptors, possibly as ErbB2/ErbB4 dimers, and downstream activation of PI3K-Akt-mTOR pathways [109]. Importantly, endogenous ErbB signaling controls mGluR1 docking on the membrane surface of midbrain DA neurons, since pharmacological ErbB inhibition causes mGluR1 internalization, which occurs through dynamin-dependent endocytosis. Hence, our evidence suggests that NRG1/ErbB-dependent regulation of mGluRI-LTD in nigral DA neurons is reliant on a direct ErbB-dependent regulation of mGluR1 expression levels, with ErbB receptor activation orchestrating the synthesis, trafficking, and membrane docking of mGluR1 in midbrain DA neurons (Figure 5). Notably, in such a way, NRG1/ErbB signaling controls several mGluR1-dependent functions in SNpc DA neurons, besides mGluRI-LTD, including mGluR1-dependent activation of the nigrostriatal DA pathway in vivo [109], supporting a central role of NRG1/ErbB signaling in the regulation of the midbrain DA system.

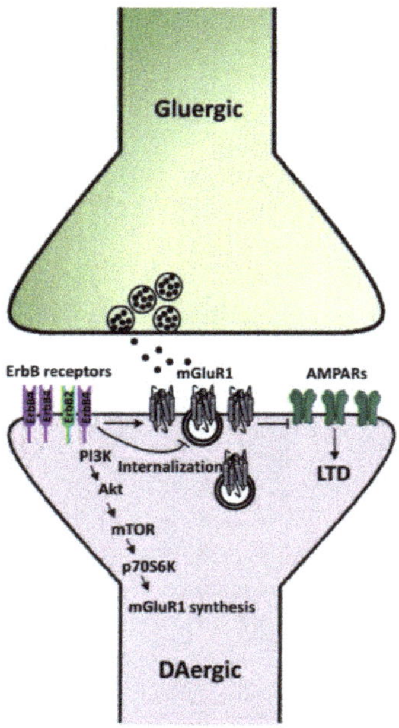

Figure 5. Mechanisms underlying NRG1/ErbB-dependent regulation of mGluRI-LTD in midbrain DA neurons. NRG1-dependent activation of ErbB receptors in SNpc DA neurons induces the synthesis of mGluR1 receptors, thus fostering mGluR1-LTD of AMPARs-mediated transmission, whereas inhibition of endogenous ErbB signaling causes mGluR1 internalization and consequent impairment of mGluR1-LTD.

4. Implications of NRG/ErbB-dependent Regulation of Synaptic Plasticity

What is the physiopathological relevance of NRGs/ErbB-induced modulation of synaptic plasticity? NRGs/ErbB-dependent regulation of the hippocampal synaptic plasticity has obvious implications in learning and memory processes, in which enduring changes of synaptic strength represent the underlying cellular mechanisms. Hippocampal LTP at CA3–CA1 synapses is involved in different learning processes, including contextual fear conditioning. Notably, NRG1/ErbB signaling, besides affecting hippocampal LTP, regulates contextual fear memory, mainly acting on PV+ GABAergic interneurons, as demonstrated by the evidence that a conditional genetic ablation of ErbB4 in this neuronal population impairs this learning process [71]. Contextual fear memory is similarly affected by intra-hippocampal injection of an ErbB inhibitor [110].

mGluRI-dependent LTD at hippocampal CA3–CA1 synapses has been associated with novelty detection and related learning processes, like object recognition memory [87,111]. We recently demonstrated that NRGs/ErbB signaling in the hippocampal CA1 area regulated either mGluRI-dependent LTD or object recognition memory in mice [86]. Indeed, intra-hippocampal injection of an ErbB inhibitor in the CA1 area prevented the induction of mGluRI-LTD and impaired the acquisition of object configurations, affecting the recognition memory [86].

NRG1/ErbB-dependent modulation of synaptic plasticity in the hippocampus has been reported also in pathological contexts, as in animal models of neurological and psychiatric disorders, like Angelman's syndrome (AS) and Alzheimer's disease (AD). AS is characterized by autism, mental retardation, motor abnormalities, and epilepsy and is mostly caused by the deletion of portions of the maternal copy of chromosome 15, which includes the UBE3A gene. The AS mouse model, produced by UBE3A deletion, displays hippocampal LTP impairment and deficits in hippocampal-dependent learning processes, like contextual fear memory [112,113], as well as an alteration of NRGs/ErbB signaling, consisting of enhanced levels of NRG1 and ErbB4 phosphorylation in the hippocampus. Interestingly, the injection of an ErbB inhibitor, PD158780, in the dorsal hippocampus, rescues synaptic and behavioral deficits, allowing proper LTP expression at CA3–CA1 synapses and rescuing contextual fear memory [110].

NRGs/ErbB-dependent regulation of hippocampal synaptic plasticity in AD animal models has been also reported. Indeed, NRG1 counteracts amyloid β-induced impairment of LTP in the hippocampal CA1 area and dentate gyrus in mouse slices, through a mechanism involving ErbB4 stimulation [114] and downstream PI3K activation [115]. Moreover, the activation of NRGs/ErbB signaling ameliorates LTP deficits, as well as cognitive abnormalities in adult Tg2576 mice, an AD animal model [116,117]. Overall, this evidence supports a relevant function of NRG1/ErbB-dependent regulation of synaptic plasticity in hippocampal-dependent learning/memory processes and suggests that a modulation of ErbB signaling might represent a rescue strategy for neuropsychiatric disorders, like autism spectrum disorders, intellectual disabilities, and AD, which display abnormalities in hippocampal-related forms of synaptic plasticity.

Conversely, the factual relevance of NRGs/ErbB-dependent regulation of glutamatergic synaptic plasticity in nigral DA neurons is still completely mysterious, mainly because the physiological functions of specific forms of synaptic plasticity in this DA nucleus are yet to be deciphered. However, since it is well recognized that the nigrostriatal DA pathway plays an important role in the establishment of goal-oriented behaviors, like feeding and locomotion, as well as in different cognitive functions, including reward/aversion-based learning, mental flexibility, and habit formation [118–124], it could be speculated that synaptic plasticity-related mechanisms within SNpc DA cells might contribute and/or underlie these brain processes. In such a perspective, NRGs/ErbB-dependent modulation of mGluRI-dependent LTD might have several implications in the regulation of various cognitive processes and complex behaviors dependent on the midbrain DA system. Regarding a potential link between NRG1/ErbB signaling in midbrain DA neurons and learning processes dependent on mGluR1-dependent synaptic plasticity in SNpc, it should be mentioned that ErbB4 deletion in midbrain DA neurons specifically impairs spatial/working memory [74], which is similarly affected by either systemic administration of mGluR1 antagonists or by a neurotoxin-induced lesion of SNpc [118,125–129]. Therefore, despite a direct connection between nigral ErbB, mGluR1, and working memory is lacking, an interplay between mGluR1-dependent synaptic plasticity and ErbB signaling in this cognitive process related to the nigrostriatal pathway could be conceived. Moreover, mGluR1-LTD in nigral DA neurons might be instrumental to motor learning, in which nigral mGluR1 has been also involved [87,130,131].

Notably, increasing evidence supports the contribution of mGluR1-dependent mechanisms in the pathogenesis of neurological and psychiatric disorders, such as schizophrenia, Parkinson's disease (PD), addiction, and autism [87,132–134], which are either characterized by alterations in midbrain DA transmission and also linked to NRG1/ErbB dysfunctions [13,46,135–137]. Whether or not a dysfunction in NRGs/ErbB-dependent regulation of synaptic plasticity in midbrain DAergic neurons represents a factual contributing factor in these pathological conditions is still to be ascertained. Nevertheless, genetic evidence suggests an association between altered NRG1/ErbB signaling and drug of abuse dependence [135–138] or autism [136,139], and indeed, dysregulated mGluRI-LTD in VTA DA neurons is instrumental to the establishment of addiction-related behaviors [87] and has been reported in an animal model of autism [140]. Future studies might unveil if altered NRGs/ErbB-dependent regulation of mGluR1-LTD in midbrain DA neurons contributes to the aberrant DA transmission occurring in schizophrenia, which is the first neuropsychiatric disorder in which the alteration of NRGs/ErbB signaling has been overtly shown [13].

5. Conclusions and Outstanding Issues

Growing evidence indicates that NRGs/ErbB signaling is essential for proper brain development and functioning. While NRGs have multifaceted roles in neurodevelopment, controlling synapse formation and myelination, the modulation of synaptic plasticity in adult brain probably represents the prominent mechanism by which NRGs can affect cognitive functions and complex behaviors in adulthood. The mechanisms engaged by NRGs in the regulation of synaptic plasticity are various. They occur at either presynaptic or postsynaptic compartments and involve modifications of neurotransmitters release or direct modulation of neurotransmitter receptors, which are chief players in the induction and/or expression of different forms of synaptic plasticity, like NMDAR, AMPAR, and mGluRI.

In the hippocampus, NRGs-dependent ErbB activation impairs LTP induction and causes depotentiation at CA3–CA1 synapses. The mechanisms underlying such LTP impairment involve NRGs-dependent enhancement of GABAergic transmission, regulation of DA extracellular levels, and postsynaptic NMDARs inhibition. At hippocampal CA3–CA1 synapses, NRGs/ErbB signaling also fosters mGluRI-dependent LTD, by preserving mGluRI function, thus shifting LTP/LTD equilibrium toward inhibition of glutamatergic transmission. Moreover, NRGs-induced ErbB activation impairs mGluRI/eCB-dependent GABAergic LTD at CA3–CA1 synapses, implying that NRGs have multifaceted roles in the long term regulation of excitatory/inhibitory balance in the hippocampus.

In midbrain DA neurons, NRGs/ErbB receptors are essential for the induction of mGluRI-dependent glutamatergic LTD. Indeed, NRGs/ErbB signaling favors mGluR1 synthesis, trafficking, and membrane docking in DA neurons, thus affecting mGluRI-LTD magnitude and midbrain DA system activation.

Although, herein, the description of NRGs/ErbB-dependent roles in the modulation of synaptic plasticity has been focused on the hippocampus and midbrain DA nuclei, it should be considered that emerging evidence also documented NRGs-induced regulation of synaptic plasticity in other brain areas, and several reports mainly obtained in transgenic mice with altered NRGs/ErbB signaling support that the proper NRGs/ErbB tone (neither too much nor too little) is essential for various cognitive functions and behaviors. Since dysfunctions in NRGs/ErbB signaling have been linked to different neurological and psychiatric disorders, including schizophrenia, bipolar disorder, autism spectrum disorders, genetic intellectual disabilities, AD, major depressive disorder, PD, and addiction [23,46,136–139,141–159], intense research efforts should be aimed at deciphering NRGs-dependent mechanisms causing pathological defects, as this might have noticeable implications on the understanding and treatment of different serious brain diseases.

Author Contributions: A.L.: conceived and designed the manuscript, created the figures, and wrote the manuscript. N.B.M.: wrote the manuscript. All authors have read and agreed to the published version of the manuscript.

Funding: This work was supported by "The Fulcro Foundation Onlus" (to N.B.M. and A.L.)".

Conflicts of Interest: The authors declare no conflict of interest.

References

1. Holmes, W.E.; Sliwkowski, M.X.; Akita, R.W.; Henzel, W.J.; Lee, J.; Park, J.W.; Yansura, D.; Abadi, N.; Raab, H.; Lewis, G.D.; et al. Identification of Heregulin, a Specific Activator of p185erbB2. *Science* **1992**, *256*, 1205–1210. [CrossRef] [PubMed]
2. Peles, E.; Bacus, S.S.; Koski, R.A.; Lu, H.S.; Wen, D.; Ogden, S.G.; Levy, R.B.; Yarden, Y. Isolation of the neu/HER-2 stimulatory ligand: A 44 kd glycoprotein that induces differentiation of mammary tumor cells. *Cell* **1992**, *69*, 205–216. [CrossRef]
3. Wen, D.; Peles, E.; Cupples, R.; Suggs, S.V.; Bacus, S.S.; Luo, Y.; Trail, G.; Hu, S.; Silbiger, S.M.; Levy, R.B.; et al. Neu differentiation factor: A transmembrane glycoprotein containing an EGF domain and an immunoglobulin homology unit. *Cell* **1992**, *69*, 559–572. [CrossRef]
4. Raff, M.C.; Abney, E.; Brockes, J.P.; Hornby-Smith, A. Schwann cell growth factors. *Cell* **1978**, *15*, 813–822. [CrossRef]
5. Brockes, J.P.; Lemke, G.E.; Balzer, D.R., Jr. Purification and preliminary characterization of a glial growth factor from the bovine pituitary. *J. Biol. Chem.* **1980**, *255*, 8374–8377.
6. Lemke, G.; Brockes, J.P. Identification and purification of glial growth factor. *J. Neurosci.* **1984**, *4*, 75–83. [CrossRef]
7. Goodearl, A.D.; Davis, J.B.; Mistry, K.; Minghetti, L.; Otsu, M.; Waterfield, M.D.; Stroobant, P. Purification of multiple forms of glial growth factor. *J. Biol. Chem.* **1993**, *268*, 18095–18102.
8. Marchionni, M.A.; Goodearl, A.D.; Chen, M.S.; Bermingham-McDonogh, O.; Kirk, C.; Hendricks, M.; Danehy, F.; Misumi, D.; Sudhalter, J.; Kobayashi, K.; et al. Glial growth factors are alternatively spliced erbB2 ligands expressed in the nervous system. *Nature* **1993**, *362*, 312–318. [CrossRef]
9. Falls, D.L.; Rosen, K.M.; Corfas, G.; Lane, W.S.; Fischbach, G.D. ARIA, a protein that stimulates acetylcholine receptor synthesis, is a member of the neu ligand family. *Cell* **1993**, *72*, 801–815. [CrossRef]
10. Brown, N.J.; Holguin, B.; Lin, B. Expression of Neuregulin 1, a Member of the Epidermal Growth Factor Family, Is Expressed as Multiple Splice Variants in the Adult Human Cornea. *Investig. Opthalmol. Vis. Sci.* **2004**, *45*, 3021–3029. [CrossRef]
11. Steinthorsdottir, V.; Stefansson, H.; Ghosh, S.; Birgisdottir, B.; Bjornsdottir, S.; Fasquel, A.C.; Olafsson, O.; Stefansson, K.; Gulcher, J.R. Multiple novel transcription initiation sites for NRG1. *Gene* **2004**, *342*, 97–105. [CrossRef] [PubMed]
12. Tan, W.; Wang, Y.; Gold, B.; Chen, J.; Dean, M.; Harrison, P.J.; Weinberger, D.R.; Law, A.J. Molecular Cloning of a Brain-specific, Developmentally Regulated Neuregulin 1 (NRG1) Isoform and Identification of a Functional Promoter Variant Associated with Schizophrenia. *J. Biol. Chem.* **2007**, *282*, 24343–24351. [CrossRef] [PubMed]
13. Mei, L.; Xiong, W.C. Neuregulin 1 in neural development, synaptic plasticity and schizophrenia. *Nat. Rev. Neurosci.* **2008**, *9*, 437–452. [CrossRef] [PubMed]
14. Liu, X.; Bates, R.; Yin, D.M.; Shen, C.; Wang, F.; Su, N.; Kirov, S.A.; Luo, Y.; Wang, J.Z.; Xiong, W.C.; et al. Specific Regulation of NRG1 Isoform Expression by Neuronal Activity. *J. Neurosci.* **2011**, *31*, 8491–8501. [CrossRef] [PubMed]
15. Li, Q.; Loeb, J.A. Neuregulin-heparan-sulfate proteoglycan interactions produce sustained erbB receptor activation required for the induction of acetylcholine receptors in muscle. *J. Biol. Chem.* **2001**, *276*, 38068–38075.
16. Ho, W.-H.; Armanini, M.P.; Nuijens, A.; Phillips, H.S.; Osheroff, P.L. Sensory and Motor Neuron-derived Factor.A novel heregulin variant highly expressed in sensory and motor neurons. *J. Biol. Chem.* **1995**, *270*, 14523–14532. [CrossRef]
17. Law, A.J.; Shannon Weickert, C.; Hyde, T.M.; Kleinman, J.E.; Harrison, P.J. Neuregulin-1 (NRG-1) mRNA and protein in the adult human brain. *Neuroscience* **2004**, *127*, 125–136. [CrossRef]
18. Chang, H.; Riese, D.J.; Gilbert, W.; Stern, D.F.; McMahan, U.J. Ligands for ErbB-family receptors encoded by a neuregulin-like gene. *Nature* **1997**, *387*, 509–512. [CrossRef]
19. Carraway, K.L.; Weber, J.L.; Unger, M.J.; Ledesma, J.; Yu, N.; Gassmann, M.; Lai, C. Neuregulin-2, a new ligand of ErbB3/ErbB4-receptor tyrosine kinases. *Nature* **1997**, *387*, 512–516. [CrossRef]
20. Busfield, S.J.; Michnick, D.A.; Chickering, T.W.; Revett, T.L.; Ma, J.; Woolf, E.A.; Comrack, C.A.; Dussault, J.; Woolf, J.; Goodearl, A.D.; et al. Characterization of a neuregulin-related gene, Don-1, that is highly expressed in restricted regions of the cerebellum and hippocampus. *Mol. Cell. Biol.* **1997**, *17*, 4007–4014. [CrossRef]

21. Yan, L.; Shamir, A.; Skirzewski, M.; Leiva-Salcedo, E.; Kwon, O.B.; Karavanova, I.; Paredes, D.; Malkesman, O.; Bailey, K.R.; Vullhorst, D.; et al. Neuregulin-2 ablation results in dopamine dysregulation and severe behavioral phenotypes relevant to psychiatric disorders. *Mol. Psychiatry* **2018**, *23*, 1233–1243. [CrossRef] [PubMed]
22. Carteron, C.; Ferrer-Montiel, A.; Cabedo, H. Characterization of a neural-specific splicing form of the human neuregulin 3 gene involved in oligodendrocyte survival. *J. Cell. Sci.* **2006**, *119*, 898–909. [CrossRef] [PubMed]
23. Kao, W.T.; Wang, Y.; Kleinman, J.E.; Lipska, B.K.; Hyde, T.M.; Weinberger, D.R.; Law, A.J. Common genetic variation in Neuregulin 3 (NRG3) influences risk for schizophrenia and impacts NRG3 expression in human brain. *Proc. Natl. Acad. Sci. USA* **2010**, *107*, 15619–15624. [CrossRef] [PubMed]
24. Paterson, C.; Wang, Y.; Hyde, T.M.; Weinberger, D.R.; Kleinman, J.E.; Law, A.J. Temporal, Diagnostic, and Tissue-Specific Regulation of NRG3 Isoform Expression in Human Brain Development and Affective Disorders. *Am. J. Psychiatry* **2017**, *174*, 256–265. [CrossRef] [PubMed]
25. Zeledón, M.; Eckart, N.; Taub, M.; Vernon, H.; Szymanksi, M.; Wang, R.; Chen, P.L.; Nestadt, G.; McGrath, J.A.; Sawa, A.; et al. Identification and functional studies of regulatory variants responsible for the association of *NRG3* with a delusion phenotype in schizophrenia. *Mol. Neuropsychiatry* **2015**, *1*, 36–46. [CrossRef] [PubMed]
26. Zhang, D.; Sliwkowski, M.X.; Mark, M.; Frantz, G.; Akita, R.; Sun, Y.; Hillan, K.; Crowley, C.; Brush, J.; Godowski, P.J. Neuregulin-3 (NRG3): A novel neural tissue-enriched protein that binds and activates ErbB4. *Proc. Natl. Acad. Sci. USA* **1997**, *94*, 9562–9567. [CrossRef]
27. Harari, D.; Tzahar, E.; Romano, J.; Shelly, M.; Pierce, J.H.; Andrews, G.C.; Yarden, Y. Neuregulin-4: A novel growth factor that acts through the ErbB-4 receptor tyrosine kinase. *Oncogene* **1999**, *18*, 2681–2689. [CrossRef]
28. Hayes, N.V.; Gullick, W.J. The neuregulin family of genes and their multiple splice variants in breast cancer. *J. Mammary Gland Biol. Neoplasia* **2008**, *13*, 205–214. [CrossRef]
29. Rosell, M.; Kaforou, M.; Frontini, A.; Okolo, A.; Chan, Y.W.; Nikolopoulou, E.; Millership, S.; Fenech, M.E.; MacIntyre, D.; Turner, J.O.; et al. Brown and white adipose tissues: Intrinsic differences in gene expression and response to cold exposure in mice. *Am. J. Physiol. Endocrinol. Metab.* **2014**, *306*, E945–E964. [CrossRef]
30. Paramo, B.; Wyatt, S.; Davies, A.M. An essential role for neuregulin-4 in the growth and elaboration of developing neocortical pyramidal dendrites. *Exp. Neurol.* **2018**, *302*, 85–92. [CrossRef]
31. Uchida, T.; Wada, K.; Akamatsu, T.; Yonezawa, M.; Noguchi, H.; Mizoguchi, A.; Kasuga, M.; Sakamoto, C. A novel epidermal growth factor-like molecule containing two follistatin modules stimulates tyrosine phosphorylation of erbB-4 in MKN28 gastric cancer cells. *Biochem. Biophys. Res. Commun.* **1999**, *266*, 593–602. [CrossRef] [PubMed]
32. Mei, L.; Nave, K.A. Neuregulin-ERBB signaling in the nervous system and neuropsychiatric diseases. *Neuron* **2014**, *83*, 27–49. [CrossRef] [PubMed]
33. Kanemoto, N.; Horie, M.; Omori, K.; Nishino, N.; Kondo, M.; Noguchi, K.; Tanigami, A. Expression of TMEFF1 mRNA in the mouse central nervous system: Precise examination and comparative studies of TMEFF1 and TMEFF2. *Brain Res. Mol. Brain Res.* **2001**, *86*, 48–55. [CrossRef]
34. Kinugasa, Y.; Ishiguro, H.; Tokita, Y.; Oohira, A.; Ohmoto, H.; Higashiyama, S. Neuroglycan C, a novel member of the neuregulin family. *Biochem. Biophys. Res. Commun.* **2004**, *321*, 1045–1049. [CrossRef]
35. Aono, S.; Tokita, Y.; Yasuda, Y.; Hirano, K.; Yamauchi, S.; Shuo, T.; Matsui, F.; Keino, H.; Kashiwai, A.; Kawamura, N.; et al. Expression and identification of a new splice variant of neuroglycan C, a transmembrane chondroitin sulfate proteoglycan, in the human brain. *J. Neurosci. Res.* **2006**, *83*, 110–118. [CrossRef]
36. Falls, D.L. Neuregulins: Functions, forms, and signaling strategies. *Exp. Cell. Res.* **2003**, *284*, 14–30. [CrossRef]
37. Vullhorst, D.; Ahmad, T.; Karavanova, I.; Keating, C.; Buonanno, A. Structural Similarities between Neuregulin 1-3 Isoforms Determine Their Subcellular Distribution and Signaling Mode in Central Neurons. *J. Neurosci.* **2017**, *37*, 5232–5249. [CrossRef]
38. Eilam, R.; Pinkas-Kramarski, R.; Ratzkin, B.J.; Segal, M.; Yarden, Y. Activity dependent regulation of Neu differentiation factor/neuregulin expression in rat brain. *Proc. Natl. Acad. Sci. USA* **1998**, *95*, 1888–1893. [CrossRef]
39. Arora, P.; Cuevas, B.D.; Russo, A.; Johnson, G.L.; Trejo, J. Persistent transactivation of EGFR and ErbB2/HER2 by protease-activated receptor-1 promotes breast carcinoma cell invasion. *Oncogene* **2008**, *27*, 4434–4445. [CrossRef]

40. Li, Q.; Zhang, R.; Ge, Y.L.; Mei, Y.W.; Guo, Y.L. Effects of neuregulin on expression of MMP-9 and NSE in brain of ischemia/reperfusion rat. *J. Mol. Neurosci.* **2009**, *38*, 207–215. [CrossRef]
41. Yarden, Y.; Sliwkowski, M.X. Untangling the ErbB signalling network. *Nat. Rev. Mol. Cell. Biol.* **2001**, *2*, 127–137. [CrossRef] [PubMed]
42. Graus-Porta, D.; Beerli, R.R.; Daly, J.M.; Hynes, N.E. ErbB-2, the preferred heterodimerization partner of all ErbB receptors, is a mediator of lateral signaling. *EMBO J.* **1997**, *16*, 1647–1655. [CrossRef] [PubMed]
43. Sliwkowski, M.X.; Schaefer, G.; Akita, R.W.; Lofgren, A.; Fitzpatrick, V.D.; Nuijens, A.; Fendly, B.M.; Cerione, R.A.; Vandlen, R.L.; Carraway, K.L. Coexpression of erbB2 and erbB3 proteins reconstitutes a high affinity receptor for heregulin. *J. Biol. Chem.* **1994**, *269*, 14661–14665. [PubMed]
44. Wang, L.M.; Kuo, A.; Alimandi, M.; Veri, M.C.; Lee, C.C.; Kapoor, V.; Ellmore, N.; Chen, X.H.; Pierce, J.H. ErbB2 expression increases the spectrum and potency of ligand mediated signal transduction through ErbB4. *Proc. Natl. Acad. Sci. USA* **1998**, *95*, 6809–6814. [CrossRef]
45. Jones, J.T.; Ballinger, M.D.; Pisacane, P.I.; Lofgren, J.A.; Fitzpatrick, V.D.; Fairbrother, W.J.; Wells, J.A.; Sliwkowski, M.X. Binding interaction of the heregulinbeta egf domain with ErbB3 and ErbB4 receptors assessed by alanine scanning mutagenesis. *J. Biol. Chem.* **1998**, *273*, 11667–11674. [CrossRef]
46. Iwakura, Y.; Nawa, H. ErbB1–4 dependent EGF/neuregulin signals and their cross talk in the central nervous system: Pathological implications in schizophrenia and Parkinson's disease. *Front. Cell. Neurosci.* **2013**, *7*, 4. [CrossRef]
47. Kim, J.H.; Saito, K.; Yokoyama, S. Chimeric receptor analyses of the interactions of the ectodomains of ErbB-1 with epidermal growth factor and of those of ErbB-4 with neuregulin. *Eur. J. Biochem.* **2002**, *269*, 2323–2329. [CrossRef]
48. Olayioye, M.A.; Graus-Porta, D.; Beerli, R.R.; Rohrer, J.; Gay, B.; Hynes, N.E. ErbB-1 and ErbB-2 acquire distinct signaling properties dependent upon their dimerization partner. *Mol. Cell. Biol.* **1998**, *18*, 5042–5051. [CrossRef]
49. Iwakura, Y.; Piao, Y.S.; Mizuno, M.; Takei, N.; Kakita, A.; Takahashi, H.; Nawa, H. Influences of dopaminergic lesion on epidermal growth factor-ErbB signals in Parkinson's disease and its model: Neurotrophic implication in nigrostriatal neurons. *J. Neurochem.* **2005**, *93*, 974–983. [CrossRef]
50. Nagano, T.; Namba, H.; Abe, Y.; Aoki, H.; Takei, N.; Nawa, H. In vivo administration of epidermal growth factor and its homologue attenuates developmental maturation of functional excitatory synapses in cortical GABAergic neurons. *Eur. J. Neurosci.* **2007**, *25*, 380–390. [CrossRef]
51. Namba, H.; Zheng, Y.; Abe, Y.; Nawa, H. Epidermal growth factor administered in the periphery influences excitatory synaptic inputs onto midbrain dopaminergic neurons in postnatal mice. *Neuroscience* **2009**, *158*, 1731–1741. [CrossRef] [PubMed]
52. Bjarnadottir, M.; Misner, D.L.; Haverfield-Gross, S.; Bruun, S.; Helgason, V.G.; Stefansson, H.; Sigmundsson, A.; Firth, D.R.; Nielsen, B.; Stefansdottir, R.; et al. Neuregulin 1 (NRG1) signaling through Fyn modulates NMDA receptor phosphorylation: Differential synaptic function in NRG1C knock-outs compared with wild-type mice. *J. Neurosci.* **2007**, *27*, 4519–4529. [CrossRef] [PubMed]
53. Fu, A.K.; Fu, W.Y.; Cheung, J.; Tsim, K.W.; Ip, F.C.; Wang, J.H.; Ip, N.Y. Cdk5 is involved in neuregulin induced AChR expression at the neuromuscular junction. *Nat. Neurosci.* **2001**, *4*, 374–381. [CrossRef] [PubMed]
54. Si, J.; Mei, L. ERK MAP kinase activation is required for acetylcholine receptor inducing activity induced increase in all five acetylcholine receptor subunit mRNAs as well as synapse-specific expression of acetylcholine receptor epsilon-transgene. *Brain Res. Mol. Brain Res.* **1999**, *67*, 18–27. [CrossRef]
55. Hellyer, N.J.; Cheng, K.; Koland, J.G. ErbB3 (HER3) interaction with the p85 regulatory subunit of phosphoinositide 3-kinase. *J. Biochem.* **1998**, *333*, 757–763. [CrossRef]
56. Lee, H.J.; Jung, K.M.; Huang, Y.Z.; Bennett, L.B.; Lee, J.S.; Mei, L.; Kim, T.W. Presenilin dependent gamma-secretase-like intramembrane cleavage of ErbB4. *J. Biol. Chem.* **2002**, *277*, 6318–6323. [CrossRef]
57. Ni, C.Y.; Murphy, M.P.; Golde, T.E.; Carpenter, G. gamma -Secretase cleavage and nuclear localization of ErbB-4 receptor tyrosine kinase. *Science* **2001**, *294*, 2179–2181. [CrossRef]
58. Sardi, S.P.; Murtie, J.; Koirala, S.; Patten, B.A.; Corfas, G. Presenilin dependent ErbB4 nuclear signaling regulates the timing of astrogenesis in the developing brain. *Cell* **2006**, *127*, 185–197. [CrossRef]
59. Bao, J.; Wolpowitz, D.; Role, L.W.; Talmage, D.A. Back signaling by the Nrg-1 intracellular domain. *J. Cell. Biol.* **2003**, *161*, 1133–1141. [CrossRef]

60. Bao, J.; Lin, H.; Ouyang, Y.; Lei, D.; Osman, A.; Kim, T.-W.; Mei, L.; Dai, P.; Ohlemiller, K.K.; Ambron, R.T. Activity-dependent transcription regulation of PSD-95 by neuregulin-1 and Eos. *Nat. Neurosci.* **2004**, *7*, 1250–1258. [CrossRef]
61. Malenka, R.C.; Bear, M.F. LTP and LTD: An embarrassment of riches. *Neuron* **2004**, *44*, 5–21. [CrossRef] [PubMed]
62. Kandel, E.R.; Dudai, Y.; Mayford, M.R. The molecular and systems biology of memory. *Cell* **2014**, *157*, 163–186. [CrossRef] [PubMed]
63. Nicoll, R.A. A Brief History of Long-Term Potentiation. *Neuron* **2017**, *93*, 281–290. [CrossRef] [PubMed]
64. Collingridge, G.L.; Peineau, S.; Howland, J.G.; Wang, Y.T. Long-term depression in the CNS. *Nat. Rev. Neurosci.* **2010**, *11*, 459–473. [CrossRef]
65. Dan, Y.; Poo, M.M. Spike timing dependent plasticity: From synapse to perception. *Physiol. Rev.* **2006**, *86*, 1033–1048. [CrossRef]
66. Markram, H.; Lübke, J.; Frotscher, M.; Sakmann, B. Regulation of synaptic efficacy by coincidence of postsynaptic APs and EPSPs. *Science* **1997**, *275*, 213–215. [CrossRef]
67. Larson, J.; Wong, D.; Lynch, G. Patterned stimulation at the theta frequency is optimal for the induction of hippocampal long-term potentiation. *Brain Res.* **1986**, *368*, 347–350. [CrossRef]
68. Hölscher, C.; Anwyl, R.; Rowan, M.J. Stimulation on the Positive Phase of Hippocampal Theta Rhythm Induces Long-Term Potentiation That Can Be Depotentiated by Stimulation on the Negative Phase in Area CA1 In Vivo. *J. Neurosci.* **1997**, *17*, 6470–6477. [CrossRef]
69. Huang, Y.Z.; Won, S.; Ali, D.W.; Wang, Q.; Tanowitz, M.; Du, Q.S.; Pelkey, K.A.; Yang, D.J.; Xiong, W.C.; Salter, M.W.; et al. Regulation of neuregulin signaling by PSD-95 interacting with ErbB4 at CNS synapses. *Neuron* **2000**, *26*, 443–455. [CrossRef]
70. Kwon, O.B.; Longart, M.; Vullhorst, D.; Hoffman, D.A.; Buonanno, A. Neuregulin- 1 reverses long-term potentiation at CA1 hippocampal synapses. *J. Neurosci.* **2005**, *25*, 9378–9383. [CrossRef]
71. Chen, Y.J.; Zhang, M.; Yin, D.M.; Wen, L.; Ting, A.; Wang, P.; Lu, Y.S.; Zhu, X.H.; Li, S.J.; Wu, C.Y.; et al. ErbB4 in parvalbumin-positive interneurons is critical for neuregulin 1 regulation of long-term potentiation. *Proc. Natl. Acad. Sci. USA* **2010**, *107*, 21818–21823. [CrossRef]
72. Pitcher, G.M.; Kalia, L.V.; Ng, D.; Goodfellow, N.M.; Yee, K.T.; Lambe, E.K.; Salter, M.W. Schizophrenia susceptibility pathway Neuregulin 1-ErbB4 suppresses Src upregulation of NMDA receptors. *Nat. Med.* **2011**, *17*, 470–478. [CrossRef] [PubMed]
73. Kwon, O.B.; Paredes, D.; Gonzalez, C.M.; Neddens, J.; Hernandez, L.; Vullhorst, D.; Buonanno, A. Neuregulin-1 regulates LTP at CA1 hippocampal synapses through activation of dopamine D4 receptors. *Proc. Natl. Acad. Sci. USA* **2008**, *105*, 15587–15592. [CrossRef] [PubMed]
74. Skirzewski, M.; Karavanova, I.; Shamir, A.; Erben, L.; Garcia-Olivares, J.; Shin, J.H.; Vullhorst, D.; Alvarez, V.A.; Amara, S.G.; Buonanno, A. ErbB4 signaling in dopaminergic axonal projections increases extracellular dopamine levels and regulates spatial/working memory behaviors. *Mol. Psychiatry* **2018**, *23*, 2227–2237. [CrossRef] [PubMed]
75. Izumi, Y.; Zorumski, C.F. Neuregulin and Dopamine D4 Receptors Contribute Independently to Depotentiation of Schaffer Collateral LTP by Temperoammonic Path Stimulation. *eNeuro* **2017**, *4*, 1–9. [CrossRef] [PubMed]
76. Vullhorst, D.; Neddens, J.; Karavanova, I.; Tricoire, L.; Petralia, R.S.; McBain, C.J.; Buonanno, A. Selective expression of ErbB4 in interneurons, but not pyramidal cells, of the rodent hippocampus. *J. Neurosci.* **2009**, *29*, 12255–12264. [CrossRef] [PubMed]
77. Madisen, L.; Zwingman, T.A.; Sunkin, S.M.; Oh, S.W.; Zariwala, H.A.; Gu, H.; Ng, L.L.; Palmiter, R.D.; Hawrylycz, M.J.; Jones, A.R.; et al. A robust and high-throughput Cre reporting and characterization system for the whole mouse brain. *Nat. Neurosci.* **2010**, *13*, 133–140. [CrossRef]
78. Bean, J.C.; Lin, T.W.; Sathyamurthy, A.; Liu, F.; Yin, D.M.; Xiong, W.C.; Mei, L. Genetic labeling reveals novel cellular targets of schizophrenia susceptibility gene: Distribution of GABA and non-GABA ErbB4-positive cells in adult mouse brain. *J. Neurosci.* **2014**, *34*, 13549–13566. [CrossRef]
79. Gerecke, K.M.; Wyss, J.M.; Karavanova, I.; Buonanno, A.; Carroll, S.L. ErbB transmembrane tyrosine kinase receptors are differentially expressed throughout the adult rat central nervous system. *J. Comp. Neurol.* **2001**, *433*, 86–100. [CrossRef]

80. Thompson, M.; Lauderdale, S.; Webster, M.J.; Chong, V.Z.; McClintock, B.; Saunders, R.; Weickert, C.S. Widespread expression of ErbB2, ErbB3 and ErbB4 in non-human primate brain. *Brain Res.* **2007**, *1139*, 95–109. [CrossRef]
81. Lai, C.; Lemke, G. An extended family of protein-tyrosine kinase genes differentially expressed in the vertebrate nervous system. *Neuron* **1991**, *6*, 691–704. [CrossRef]
82. Mechawar, N.; Lacoste, B.; Yu, W.F.; Srivastava, L.K.; Quirion, R. Developmental profile of neuregulin receptor ErbB4 in postnatal rat cerebral cortex and hippocampus. *Neuroscience* **2007**, *148*, 126–139. [CrossRef] [PubMed]
83. Li, B.; Woo, R.S.; Mei, L.; Malinow, R. The neuregulin-1 receptor erbB4 controls glutamatergic synapse maturation and plasticity. *Neuron* **2007**, *54*, 583–597. [CrossRef] [PubMed]
84. Barros, C.S.; Calabrese, B.; Chamero, P.; Roberts, A.J.; Korzus, E.; Lloyd, K.; Stowers, L.; Mayford, M.; Halpain, S.; Müller, U. Impaired maturation of dendritic spines without disorganization of cortical cell layers in mice lacking NRG1/ErbB signaling in the central nervous system. *Proc. Natl. Acad. Sci. USA* **2009**, *106*, 4507–4512. [CrossRef] [PubMed]
85. Cooper, M.A.; Koleske, A.J. Ablation of ErbB4 from excitatory neurons leads to reduced dendritic spine density in mouse prefrontal cortex. *J. Comp. Neurol.* **2014**, *522*, 3351–3362. [CrossRef] [PubMed]
86. Ledonne, A.; Mango, D.; Latagliata, E.C.; Chiacchierini, G.; Nobili, A.; Nisticò, R.; D'Amelio, M.; Puglisi-Allegra, S.; Mercuri, N.B. Neuregulin 1/ErbB signalling modulates hippocampal mGluRI dependent LTD and object recognition memory. *Pharmacol. Res.* **2018**, *130*, 12–24. [CrossRef]
87. Lüscher, C.; Huber, K.M. Group 1 mGluR dependent synaptic long-term depression: Mechanisms and implications for circuitry and disease. *Neuron* **2010**, *65*, 445–459. [CrossRef]
88. Snyder, E.M.; Philpot, B.D.; Huber, K.M.; Dong, X.; Fallon, J.R.; Bear, M.F. Internalization of ionotropic glutamate receptors in response to mGluR activation. *Nat. Neurosci.* **2001**, *4*, 1079–1085. [CrossRef]
89. Xiao, M.Y.; Zhou, Q.; Nicoll, R.A. Metabotropic glutamate receptor activation causes a rapid redistribution of AMPA receptors. *Neuropharmacology* **2001**, *41*, 664–671. [CrossRef]
90. Thiels, E.; Kanterewicz, B.I.; Norman, E.D.; Trzaskos, J.M.; Klann, E. Long-term depression in the adult hippocampus in vivo involves activation of extracellular signal-regulated kinase and phosphorylation of Elk-1. *J. Neurosci.* **2002**, *22*, 2054–2062. [CrossRef]
91. Gallagher, S.M.; Daly, C.A.; Bear, M.F.; Huber, K.M. Extracellular signal-regulated protein kinase activation is required for metabotropic glutamate receptor dependent long-term depression in hippocampal area CA1. *J. Neurosci.* **2004**, *24*, 4859–4864. [CrossRef] [PubMed]
92. Hou, L.; Klann, E. Activation of the phosphoinositide 3-kinase-Akt-mammalian target of rapamycin signaling pathway is required for metabotropic glutamate receptor dependent long-term depression. *J. Neurosci.* **2004**, *24*, 6352–6361. [CrossRef] [PubMed]
93. Huber, K.M.; Kayser, M.S.; Bear, M.F. Role for rapid dendritic protein synthesis in hippocampal mGluR dependent long-term depression. *Science* **2000**, *288*, 1254–1257. [CrossRef] [PubMed]
94. Kemp, N.; Bashir, Z.I. Induction of LTD in the adult hippocampus by the synaptic activation of AMPA/kainate and metabotropic glutamate receptors. *Neuropharmacology* **1999**, *38*, 495–504. [CrossRef]
95. Huber, K.M.; Roder, J.C.; Bear, M.F. Chemical induction of mGluR5- and protein synthesis- dependent long-term depression in hippocampal area CA1. *J. Neurophysiol.* **2001**, *86*, 321–325. [CrossRef] [PubMed]
96. Volk, L.J.; Daly, C.A.; Huber, K.M. Differential roles for group 1 mGluR subtypes in induction and expression of chemically induced hippocampal long-term depression. *J. Neurophysiol.* **2006**, *95*, 2427–2438. [CrossRef] [PubMed]
97. Chevaleyre, V.; Castillo, P.E. Heterosynaptic LTD of hippocampal GABAergic synapses: A novel role of endocannabinoids in regulating excitability. *Neuron* **2003**, *38*, 461–472. [CrossRef]
98. Castillo, P.E.; Younts, T.J.; Chávez, A.E.; Hashimotodani, Y. Endocannabinoid signaling and synaptic function. *Neuron* **2012**, *76*, 70–81. [CrossRef]
99. Du, H.; Kwon, I.K.; Kim, J. Neuregulin-1 impairs the long-term depression of hippocampal inhibitory synapses by facilitating the degradation of endocannabinoid 2-AG. *J. Neurosci.* **2013**, *33*, 15022–15031. [CrossRef]
100. Bonci, A.; Malenka, R.C. Properties and plasticity of excitatory synapses on dopaminergic and GABAergic cells in the ventral tegmental area. *J. Neurosci.* **1999**, *19*, 3723–3730. [CrossRef]

101. Jones, S.; Kornblum, J.L.; Kauer, J.A. Amphetamine blocks long-term synaptic depression in the ventral tegmental area. *J. Neurosci.* **2000**, *20*, 5575–5580. [CrossRef] [PubMed]
102. Thomas, M.J.; Malenka, R.C.; Bonci, A. Modulation of long-term depression by dopamine in the mesolimbic system. *J. Neurosci.* **2000**, *20*, 5581–5586. [CrossRef] [PubMed]
103. Bellone, C.; Lüscher, C. mGluRs induce a long-term depression in the ventral tegmental area that involves a switch of the subunit composition of AMPA receptors. *Eur. J. Neurosci.* **2005**, *21*, 1280–1288. [CrossRef] [PubMed]
104. Liu, Q.S.; Pu, L.; Poo, M.M. Repeated cocaine exposure in vivo facilitates LTP induction in midbrain dopamine neurons. *Nature* **2005**, *437*, 1027–1031. [CrossRef]
105. Luu, P.; Malenka, R.C. Spike timing dependent long-term potentiation in ventral tegmental area dopamine cells requires PKC. *J. Neurophysiol.* **2008**, *100*, 533–538. [CrossRef]
106. Feldman, D.E. The spike-timing dependence of plasticity. *Neuron* **2012**, *75*, 556–571. [CrossRef]
107. Ledonne, A.; Mercuri, N.B. mGluR1-Dependent Long Term Depression in Rodent Midbrain Dopamine Neurons Is Regulated by Neuregulin 1/ErbB Signaling. *Front. Mol. Neurosci.* **2018**, *11*, 346. [CrossRef]
108. Mameli, M.; Balland, B.; Luján, R.; Lüscher, C. Rapid synthesis and synaptic insertion of GluR2 for mGluR-LTD in the ventral tegmental area. *Science* **2007**, *317*, 530–533. [CrossRef]
109. Ledonne, A.; Nobili, A.; Latagliata, E.C.; Cavallucci, V.; Guatteo, E.; Puglisi-Allegra, S.; D'Amelio, M.; Mercuri, N.B. Neuregulin 1 signalling modulates mGluR1 function in mesencephalic dopaminergic neurons. *Mol. Psychiatry* **2015**, *20*, 959–973. [CrossRef]
110. Kaphzan, H.; Hernandez, P.; In Jung, J.; Cowansage, K.K.; Deinhardt, K.; Chao, M.V.; Abel, T.; Klann, E. Reversal of Impaired Hippocampal Long-Term Potentiation and Contextual Fear Memory Deficits in Angelman Syndrome Model Mice by ErbB Inhibitors. *Biol. Psychiatry* **2012**, *72*, 182–190. [CrossRef]
111. Goh, J.J.; Manahan-Vaughan, D. Endogenous hippocampal LTD that is enabled by spatial object recognition requires activation of NMDA receptors and the metabotropic glutamate receptor, mGlu5. *Hippocampus* **2013**, *23*, 129–138. [CrossRef] [PubMed]
112. Jiang, Y.H.; Armstrong, D.; Albrecht, U.; Atkins, C.M.; Noebels, J.L.; Eichele, G. Mutation of the Angelman ubiquitin ligase in mice causes increased cytoplasmic p53 and deficits of contextual learning and long term potentiation. *Neuron* **1998**, *21*, 799–811. [CrossRef]
113. Van Woerden, G.M.; Harris, K.D.; Hojjati, M.R.; Gustin, R.M.; Qiu, S.; de Avila Freire, R.; Jiang, Y.H.; Elgersma, Y.; Weeber, E.J. Rescue of neurological deficits in a mouse model for Angelman syndrome by reduction of alphaCaMKII inhibitory phosphorylation. *Nat. Neurosci.* **2007**, *10*, 280–282. [CrossRef] [PubMed]
114. Min, S.S.; An, J.; Lee, J.H.; Seol, G.H.; Im, J.H.; Kim, H.S.; Baik, T.S.; Woo, R.S. Neuregulin-1 prevents amyloid beta induced impairment of long-term potentiation in hippocampal slices via ErbB4. *Neurosci. Lett.* **2011**, *505*, 6–9. [CrossRef] [PubMed]
115. Baik, T.K.; Kim, Y.-J.; Kang, S.-M.; Song, D.-Y.; Min, S.S.; Woo, R.-S. Blocking the phosphatidylinositol 3-kinase pathway inhibits neuregulin-1 mediated rescue of neurotoxicity induced by Ab1–42. *J. Pharm. Pharmacol.* **2016**, *68*, 1021–1029. [CrossRef] [PubMed]
116. Xu, J.; De Winter, F.; Farrokhi, C.; Rockenstein, E.; Mante, M.; Adame, A.; Cook, J.; Jin, X.; Masliah, E.; Lee, K.-F. Neuregulin 1 improves cognitive deficits and neuropathology in an Alzheimer's disease model. *Sci. Rep.* **2016**, *6*, 31692. [CrossRef]
117. Ryu, J.; Hong, B.H.; Kim, Y.J.; Yang, E.-J.; Choi, M.; Kim, H.; Ahn, S.; Baik, T.-K.; Woo, R.-S.; Kim, H.-S. Neuregulin-1 attenuates cognitive function impairments in a transgenic mouse model of Alzheimer's disease. *Cell Death Dis.* **2016**, *7*, e2117. [CrossRef]
118. Da Cunha, C.; Angelucci, M.E.M.; Canteras, N.S.; Wonnacott, S.; Takahashi, R.N. The lesion of the rat substantia nigra pars compacta dopaminergic neurons as a model for Parkinson's disease memory disabilities. *Cell. Mol. Neurobiol.* **2002**, *22*, 227–237. [CrossRef]
119. Da Cunha, C.; Silva, M.H.C.; Wietzikoski, S.; Wietzikoski, E.C.; Ferro, M.M.; Kouzmine, I.; Canteras, N.S. Place learning strategy of substantia nigra pars compacta-lesioned rats. *Behav. Neurosci.* **2006**, *120*, 1279–1284. [CrossRef]
120. Palmiter, R.D. Dopamine signaling in the dorsal striatum is essential for motivated behaviors: Lessons from dopamine-deficient mice. *Ann. N. Y. Acad. Sci.* **2008**, *1129*, 35–46. [CrossRef]

121. Wise, R.A. Roles for nigrostriatal—Not just mesocorticolimbic—Dopamine in reward and addiction. *Trends Neurosci.* **2009**, *32*, 517–524. [CrossRef] [PubMed]
122. Haber, S.N. The place of dopamine in the cortico-basal ganglia circuit. *Neuroscience* **2014**, *282*, 248–257. [CrossRef]
123. Ilango, A.; Kesner, A.J.; Keller, K.L.; Stuber, G.D.; Bonci, A.; Ikemoto, S. Similar roles of substantia nigra and ventral tegmental dopamine neurons in reward and aversion. *J. Neurosci.* **2014**, *34*, 817–822. [CrossRef]
124. Ledonne, A.; Mercuri, N.B. Current Concepts on the Physiopathological Relevance of Dopaminergic Receptors. *Front. Cell. Neurosci.* **2017**, *11*, 1–9. [CrossRef]
125. Da Cunha, C.; Wietzikoski, S.; Wietzikoski, E.C.; Miyoshi, E.; Ferro, M.M.; Anselmo-Franci, J.A.; Canteras, N.S. Evidence for the substantia nigra pars compacta as an essential component of a memory system independent of the hippocampal memory system. *Neurobiol. Learn. Mem.* **2003**, *79*, 236–242. [CrossRef]
126. Miyoshi, E.; Wietzikoski, S.; Camplessei, M.; Silveira, R.; Takahashi, R.N.; Da Cunha, C. Impaired learning in a spatial working memory version and in a cued version of the water maze in rats with MPTP induced mesencephalic dopaminergic lesions. *Brain Res. Bull.* **2002**, *58*, 41–47. [CrossRef]
127. Braga, R.; Kouzmine, I.; Canteras, N.S.; Da Cunha, C. Lesion of the substantia nigra pars compacta impairs delayed alternation in a Y-maze in rats. *Exp. Neurol.* **2005**, *192*, 134–141. [CrossRef]
128. Hsieh, M.H.; Ho, S.C.; Yeh, K.Y.; Pawlak, C.R.; Chang, H.M.; Ho, Y.J.; Lai, T.-J.; Wu, F.-Y. Blockade of metabotropic glutamate receptors inhibits cognition and neurodegeneration in an MPTP induced Parkinson's disease rat model. *Pharmacol. Biochem. Behav.* **2010**, *102*, 64–71. [CrossRef]
129. Sy, H.N.; Wu, S.L.; Wang, W.F.; Chen, C.H.; Huang, Y.T.; Liou, Y.M.; Chiou, C.-S.; Pawlak, C.R.; Ho, Y.-J. MPTP induced dopaminergic degeneration and deficits in object recognition in rats are accompanied by neuroinflammation in the hippocampus. *Pharmacol. Biochem. Behav.* **2010**, *95*, 158–165. [CrossRef]
130. Conn, P.J.; Battaglia, G.; Marino, M.J.; Nicoletti, F. Metabotropic glutamate receptors in the basal ganglia motor circuit. *Nat. Rev. Neurosci.* **2005**, *6*, 787–798. [CrossRef]
131. Hodgson, R.A.; Hyde, L.A.; Guthrie, D.H.; Cohen-Williams, M.E.; Leach, P.T.; Kazdoba, T.M.; Bleickardt, C.J.; Lu, S.X.; Parker, E.M.; Varty, G.B. Characterization of the selective mGluR1 antagonist, JNJ16259685, in rodent models of movement and coordination. *Pharmacol. Biochem. Behav.* **2011**, *98*, 181–187. [CrossRef]
132. Ferraguti, F.; Crepaldi, L.; Nicoletti, F. Metabotropic glutamate 1 receptor: Current concepts and perspectives. *Pharmacol. Rev.* **2008**, *60*, 536–581. [CrossRef] [PubMed]
133. Lesage, A.; Steckler, T. Metabotropic glutamate mGlu1 receptor stimulation and blockade: Therapeutic opportunities in psychiatric illness. *Eur. J. Pharmacol.* **2010**, *639*, 2–16. [CrossRef] [PubMed]
134. Herman, E.J.; Bubser, M.; Conn, P.J.; Jones, C.K. Metabotropic glutamate receptors for new treatments in schizophrenia. *Handb. Exp. Pharmacol.* **2012**, *213*, 297–365.
135. Han, S.; Yang, B.Z.; Kranzler, H.R.; Oslin, D.; Anton, R.; Farrer, L.A.; Gelernter, J. Linkage analysis followed by association show NRG1 associated with cannabis dependence in African Americans. *Biol. Psychiatry* **2012**, *72*, 637–644. [CrossRef]
136. Yoo, H.J.; Woo, R.S.; Cho, S.C.; Kim, B.N.; Kim, J.W.; Shin, M.S.; Park, T.W.; Son, J.W.; Chung, U.S.; Park, S.; et al. Genetic association analyses of neuregulin 1 gene polymorphism with endopheontype for sociality of Korean autism spectrum disorders family. *Psychiatry Res.* **2015**, *227*, 366–368. [CrossRef]
137. Ikawa, D.; Makinodan, M.; Iwata, K.; Ohgidani, M.; Kato, T.A.; Yamashita, Y.; Yamamuro, K.; Kimoto, S.; Toritsuka, M.; Yamauchi, T.; et al. Microglia-derived neuregulin expression in psychiatric disorders. *Brain Behav. Immun.* **2017**, *61*, 375–385. [CrossRef]
138. Turner, J.R.; Ray, R.; Lee, B.; Everett, L.; Xiang, J.; Jepson, C.; Kaestner, K.H.; Lerman, C.; Blendy, J.A. Evidence from mouse and man for a role of neuregulin 3 in nicotine dependence. *Mol. Psychiatry* **2014**, *19*, 801–810. [CrossRef]
139. Pinto, D.; Pagnamenta, A.T.; Klei, L.; Anney, R.; Merico, D.; Regan, R.; Conroy, J.; Magalhaes, T.R.; Correia, C.; Abrahams, B.S. Functional impact of global rare copy number variation in autism spectrum disorders. *Nature* **2010**, *466*, 368–372. [CrossRef]
140. Bariselli, S.; Tzanoulinou, S.; Glangetas, C.; Prévost-Solié, C.; Pucci, L.; Viguié, J.; Bezzi, P.; O'Connor, E.C.; Georges, F.; Lüscher, C.; et al. SHANK3 controls maturation of social reward circuits in the VTA. *Nat. Neurosci.* **2016**, *19*, 926–934. [CrossRef]

141. Stefansson, H.; Sigurdsson, E.; Steinthorsdottir, V.; Bjornsdottir, S.; Sigmundsson, T.; Ghosh, S.; Brynjolfsson, J.; Gunnarsdottir, S.; Ívarsson, Ó.; Chou, T.T.; et al. Neuregulin 1 and Susceptibility to Schizophrenia. *Am. J. Hum. Genet.* **2002**, *71*, 877–892. [CrossRef] [PubMed]
142. Yang, J.Z.; Si, T.M.; Ruan, Y.; Ling, Y.S.; Han, Y.H.; Wang, X.L.; Zhou, M.; Zhang, H.Y.; Kong, Q.C.; Liu, C.; et al. Association study of neuregulin 1 gene with schizophrenia. *Mol. Psychiatry* **2003**, *8*, 706–709. [CrossRef] [PubMed]
143. Thomson, P.A.; Christoforou, A.; Morris, S.W.; Adie, E.; Pickard, B.S.; Porteous, D.J. Association of Neuregulin 1 with schizophrenia and bipolar disorder in a second cohort from the Scottish population. *Mol. Psychiatry* **2007**, *12*, 94–104. [CrossRef] [PubMed]
144. Van Beveren, N.J.; Krab, L.C.; Swagemakers, S.; Buitendijk, G.H.; Boot, E.; van der Spek, P.; Elgersma, Y.; van Amelsvoort, T.A. Functional gene-expression analysis shows involvement of schizophrenia-relevant pathways in patients with 22q11 deletion syndrome. *PLoS ONE* **2012**, *7*, e33473. [CrossRef]
145. Kasnauskiene, J.; Ciuladaite, Z.; Preiksaitiene, E.; Utkus, A.; Peciulyte, A.; Kučinskas, V. A new single gene deletion on 2q34: ERBB4 is associated with intellectual disability. *Am. J. Med. Genet. A* **2013**, *161*, 1487–1490. [CrossRef] [PubMed]
146. Abbasy, S.; Shahraki, F.; Haghighatfard, A.; Qazvini, M.G.; Rafiei, S.T.; Noshadirad, E.; Farhadi, M.; Rezvani, H.; Shiryazdi, A.A.; Ghamari, R.; et al. Neuregulin1 types mRNA level changes in autism spectrum disorder, and is associated with deficit in executive functions. *EBioMedicine* **2018**, *37*, 483–488. [CrossRef]
147. Esnafoglu, E. Levels of peripheral Neuregulin 1 are increased in non-medicated autism spectrum disorder patients. *J. Clin. Neurosci.* **2018**, *57*, 43–45. [CrossRef]
148. Woo, R.S.; Lee, J.H.; Yu, H.N.; Song, D.Y.; Baik, T.K. Expression of ErbB4 in the apoptotic neurons of Alzheimer's disease brain. *Anat. Cell. Biol.* **2010**, *43*, 332–339. [CrossRef]
149. Chaudhury, A.R.; Gerecke, K.M.; Wyss, J.M.; Morgan, D.G.; Gordon, M.N.; Carroll, S.L. Neuregulin-1 and erbB4 immunoreactivity is associated with neuritic plaques in Alzheimer disease brain and in a transgenic model of Alzheimer disease. *J. Neuropathol. Exp. Neurol.* **2003**, *62*, 42–54. [CrossRef]
150. Go, R.C.P.; Perry, R.T.; Wiener, H.; Bassett, S.S.; Blacker, D.; Devlin, B.; Sweet, R.A. Neuregulin-1 polymorphism in late onset Alzheimer's disease families with psychoses. *Am. J. Med Genet. Part B Neuropsychiatr. Genet.* **2005**, *139*, 28–32. [CrossRef]
151. Wang, K.S.; Xu, N.; Wang, L.; Aragon, L.; Ciubuc, R.; Arana, T.B.; Mao, C.; Petty, L.; Briones, D.; Branda, B.S.; et al. NRG3 gene is associated with the risk and age at onset of Alzheimer disease. *J. Neural Transm. (Vienna)* **2014**, *121*, 183–192. [CrossRef] [PubMed]
152. Bertram, I.; Bernstein, H.G.; Lendeckel, U.; Bukowska, A.; Dobrowolny, H.; Keilhoff, G.; Kanakis, D.; Mawrin, C.; Bielau, H.; Falkai, P.; et al. Immunohistochemical evidence for impaired neuregulin-1 signaling in the prefrontal cortex in schizophrenia and in unipolar depression. *Ann. N. Y. Acad. Sci.* **2007**, *1096*, 147–156. [CrossRef] [PubMed]
153. Milanesi, E.; Minelli, A.; Cattane, N.; Cattaneo, A.; Mora, C.; Barbon, A.; Mallei, A.; Popoli, M.; Florio, V.; Conca, A.; et al. ErbB3 mRNA leukocyte levels as a biomarker for major depressive disorder. *BMC Psychiatry* **2012**, *12*, 145. [CrossRef] [PubMed]
154. Mahar, I.; Labonte, B.; Yogendran, S.; Isingrini, E.; Perret, L.; Davoli, M.A.; Rachalski, A.; Giros, B.; Turecki, G.; Mechawar, N. Disrupted hippocampal neuregulin-1/ErbB3 signaling and dentate gyrus granule cell alterations in suicide. *Transl. Psychiatry* **2017**, *7*, e1161. [CrossRef] [PubMed]
155. Depboylu, C.; Höllerhage, M.; Schnurrbusch, S.; Brundin, P.; Oertel, W.H.; Schrattenholz, A.; Höglinger, G.U. Neuregulin-1 receptor tyrosine kinase ErbB4 is upregulated in midbrain dopaminergic neurons in Parkinson disease. *Neurosci. Lett.* **2012**, *531*, 209–214. [CrossRef]
156. Hama, Y.; Yabe, I.; Wakabayashi, K.; Kano, T.; Hirotani, M.; Iwakura, Y.; Utsumi, J.; Sasaki, H. Level of plasma neuregulin-1 SMDF is reduced in patients with idiopathic Parkinson's disease. *Neurosci. Lett.* **2015**, *587*, 17–21. [CrossRef]
157. Fisher, M.L.; Loukola, A.; Kaprio, J.; Turner, J.R. Role of the Neuregulin Signaling Pathway in Nicotine Dependence and Co-morbid Disorders. *Int. Rev. Neurobiol.* **2015**, *124*, 113–131.

158. Gupta, R.; Qaiser, B.; He, L.; Hiekkalinna, T.S.; Zheutlin, A.B.; Therman, S.; Ollikainen, M.; Ripatti, S.; Perola, M.; Salomaa, V.; et al. Neuregulin signaling pathway in smoking behavior. *Transl. Psychiatry* **2017**, *7*, e1212. [CrossRef]

159. Vaht, M.; Laas, K.; Kiive, E.; Parik, J.; Veidebaum, T.; Harro, J. A functional neuregulin-1 gene variant and stressful life events: Effect on drug use in a longitudinal population-representative cohort study. *J. Psychopharmacol.* **2017**, *31*, 54–61. [CrossRef]

International Journal of
Molecular Sciences

Review

Modeling Resilience to Damage in Multiple Sclerosis: Plasticity Meets Connectivity

Mario Stampanoni Bassi [1], Ennio Iezzi [1], Luigi Pavone [1], Georgia Mandolesi [2,3], Alessandra Musella [2,3], Antonietta Gentile [2,4], Luana Gilio [1], Diego Centonze [1,4,*] and Fabio Buttari [1]

1 Unit of Neurology & Neurorehabilitation, IRCCS Neuromed, 86077 Pozzilli (IS), Italy; m.stampanonibassi@gmail.com (M.S.B.); ennio.iezzi@neuromed.it (E.I.); gi19gi82@gmail.com (L.P.); gilio.luana@gmail.com (L.G.); fabio.buttari@gmail.com (F.B.)
2 Laboratory of Synaptic Immunopathology, IRCCS San Raffaele Pisana, 00163 Rome, Italy; georgia.mandolesi@sanraffaele.it (G.M.); alessandra.musella@sanraffaele.it (A.M.); gntnnt01@uniroma2.it (A.G.)
3 Laboratory of Synaptic Immunopathology, San Raffaele University of Rome, 00166 Rome, Italy
4 Laboratory of Synaptic Immunopathology, Department of Systems Medicine, Tor Vergata University, 00173 Rome, Italy
* Correspondence: centonze@uniroma2.it; Tel.: +39-0865-929170

Received: 31 October 2019; Accepted: 20 December 2019; Published: 24 December 2019

Abstract: Multiple sclerosis (MS) is a chronic inflammatory disease of the central nervous system (CNS) characterized by demyelinating white matter lesions and neurodegeneration, with a variable clinical course. Brain network architecture provides efficient information processing and resilience to damage. The peculiar organization characterized by a low number of highly connected nodes (hubs) confers high resistance to random damage. Anti-homeostatic synaptic plasticity, in particular long-term potentiation (LTP), represents one of the main physiological mechanisms underlying clinical recovery after brain damage. Different types of synaptic plasticity, including both anti-homeostatic and homeostatic mechanisms (synaptic scaling), contribute to shape brain networks. In MS, altered synaptic functioning induced by inflammatory mediators may represent a further cause of brain network collapse in addition to demyelination and grey matter atrophy. We propose that impaired LTP expression and pathologically enhanced upscaling may contribute to disrupting brain network topology in MS, weakening resilience to damage and negatively influencing the disease course.

Keywords: synaptic plasticity; long-term potentiation (LTP); synaptic scaling; brain networks; inflammation; connectivity; multiple sclerosis; resting state functional MRI (rs-fMRI)

1. Introduction

Multiple sclerosis (MS) is a chronic inflammatory disease of the central nervous system characterized by demyelination and neurodegeneration. The clinical course of MS is highly variable and different clinical phenotypes have been described. While the introduction of MRI significantly improved MS diagnosis [1], a discrepancy between radiological findings and clinical manifestations is frequently observed [2]. Different mechanisms contribute to this "clinico-radiological paradox" and both synaptic plasticity and brain networks architecture may play an important role, influencing resilience to damage.

Several studies have demonstrated that even in the absence of the associated visible damage, inflammation in MS negatively affects the disease course [3,4]. It has been shown indeed that in animal models (i.e., experimental autoimmune encephalomyelitis, EAE) and in MS patients, specific proinflammatory cytokines alter synaptic transmission and plasticity [5,6]. Therefore, subverted synaptic plasticity induced by inflammation may represent an independent cause of brain network dysfunction in MS [7].

In the present article, we will overview the alterations of synaptic plasticity and brain connectivity induced by inflammation in MS. We propose that impaired synaptic plasticity expression could contribute to disrupting brain network topology, critically weakening resilience to damage.

2. Clinico-Radiological Paradox, Brain Network Resilience and Plasticity

In different neurological conditions, brain networks are able to maintain appropriate functional efficiency even in the presence of a high structural damage load, delaying or minimizing the appearance of clinical symptoms [8]. A growing number of studies have demonstrated that the organization of brain networks is highly specialized and reveals specific features evolved to improve efficiency, containing the wiring cost. This architecture protects networks from random attacks and is useful for optimal reorganization after damage [8,9]. In addition, synaptic plasticity is one of the main physiological mechanisms involved in brain network remodeling and critically contributes to brain damage compensation. Experimental and clinical studies have clearly shown that the efficiency of synaptic plasticity mechanisms, particularly of long-term potentiation (LTP), influences the chances of recovery after damage [10,11].

2.1. *Brain Network Architecture Provides Resilience to Damage*

Graph theory offers a helpful approach to analyzing connectivity data deriving from neurophysiological (i.e., electroencephalography and magnetoencephalography) or imaging (i.e., functional MRI, fMRI) techniques [12–17]. Complex brain networks can be modeled as a series of nodes, indicating different brain regions, and edges, representing the connections between nodes. The edges are defined on the basis of structural (i.e., diffusion tensor imaging, DTI) or functional connectivity (i.e., temporal covariation of spontaneous activity) between different regions [17–19]. Specific measures describing network functioning have been defined. In particular, brain networks are characterized by elevated global and local efficiency, modularity and scale-free degree distribution [8]. It has been argued that some key features of brain networks directly arise from the pattern of connections between nodes. An important characteristic of brain network organization, supporting the segregation of information processing, is the presence of modules, defined as communities of highly connected nodes with relatively few connections with nodes of other modules [20]. Moreover, the presence of a small number of hubs, consisting of highly connected nodes with elevated centrality, is essential for the integration of information and to increase network efficiency [21]. Hubs are highly interconnected, between them forming a subgroup of regions, the rich club, which critically contribute to brain network functioning [15,21]. Hubs develop through a "rich get richer" rule, meaning that the preferential attachment of new nodes occurs with those that have higher degree [22]. This architecture is associated to strong resistance to random damage due to the numerical prevalence of low connected nodes [8,23]. Conversely, targeted damage to hubs critically alters network functionality as shown in different neurological conditions [24].

2.2. *Synaptic Plasticity Enables Symptom Compensation*

Synaptic plasticity is the ability of neurons to modify the strength of reciprocal connections and is a key neurophysiological mechanism involved in network development and reorganization after damage. Different forms of synaptic plasticity have been studied, including anti-homeostatic and homeostatic mechanisms [25,26]. LTP has been first explored in hippocampal neurons and consists of persistent enhancement of synaptic excitability associated with remodeling of the presynaptic and postsynaptic terminals [26,27].

LTP has been consistently associated to learning and memory [27]; in addition, LTP is critically involved in brain damage compensation [10]. In experimental models, it has been shown that synaptic rearrangement in the peri-lesional area after focal ischemic brain damage in rats is critical for clinical recovery. In particular, electrophysiological recordings have demonstrated that increased glutamatergic transmission in surviving neurons is correlated with the degree of clinical improvement [10]. The key role of LTP in the compensation of clinical manifestations has also been documented in humans. Accordingly, in patients with acute stroke, clinical recovery after six months was positively correlated with the efficiency of LTP-like plasticity mechanisms probed using transcranial magnetic stimulation (TMS), namely LTP reserve [11]. Moreover, in RR-MS patients, the magnitude of plasticity reserve measured in the acute phase was correlated with the clinical recovery that occurred three months later [28].

2.3. Synaptic Plasticity and Network Remodeling

N-methyl-D-aspartate receptors (NMDARs) are essential for inducing LTP [29,30]. Specific properties of this form of Hebbian plasticity, including cooperativity, associativity and input-specificity, are directly associated to NMDAR characteristics [31]. LTP is an anti-homeostatic phenomenon, as it is associated with increased synaptic excitability, which in turn further promotes LTP induction [25]. Another form of Hebbian, anti-homeostatic plasticity associated with NMDAR activation is long-term depression (LTD), which induces lasting reduction in synaptic excitability [25,31]. The characteristics of Hebbian plasticity, in particular of LTP, suggest that this mechanism might be involved in the generation of specific network features. In particular, the formation of hubs and modules requires an anti-homeostatic mechanism, able to associate different specific inputs between them.

To dynamically keep neuronal activity levels in physiological ranges, different forms of synaptic plasticity are also required. In particular, homeostatic mechanisms (i.e., synaptic scaling) should intervene to prevent instability induced by Hebbian plasticity [32,33]. Synaptic scaling is characterized by an increase (upscaling) or decrease (downscaling) of synaptic excitability that is mediated by change in the expression of α-amino-3-hydroxy-5-methyl-4-isoxazolepropionic acid receptors (AMPARs). Unlike LTP, synaptic scaling lacks input-specificity and is a hetero-synaptic form of plasticity. Scaling has a negative feedback action and represents a compensatory mechanism devoted to stabilizing synaptic excitability in response to persistent increase or decrease of excitatory inputs to a neuron [32,34]. At a network level, homeostatic plasticity mechanisms could therefore contribute to stabilizing network activity, and might be particularly suitable to prevent excessive hyperconnectivity in the peripheral nodes of a network.

Anti-homeostatic and homeostatic plasticity cooperate to promote optimal brain network topology and are involved in symptom compensation and remodeling after brain damage [35,36].

3. Inflammation Alters Synaptic Transmission and Plasticity in MS

Several mediators released during neuroinflammation can affect synaptic functioning, influencing the disease course of MS [37]. Specific proinflammatory cytokines, including interleukin (IL)-1β and tumor necrosis factor (TNF), promote excitotoxic neuronal damage in EAE, exacerbating glutamatergic transmission and reducing GABAergic signaling [6,38]. A similar set of proinflammatory cytokines could also alter synaptic functioning in human MS, as CSF collected from MS patients in the acute phase of disease reproduces the neurophysiological alterations observed in EAE when incubated in vitro on control mouse slices [39].

3.1. Inflammation and LTP

Studies in EAE and in patients with MS pointed out that LTP and LTD expression is affected by inflammation [40–43]. In relapsing MS patients, reduced LTP-like plasticity in response to the intermittent theta burst stimulation protocol has been reported [40]. Moreover, a synaptic plasticity deficit was normalized after three-month treatment with interferon beta 1a, suggesting a direct role of acute inflammation [40]. Specific proinflammatory molecules, including IL-1β and IL-6, influence LTP induction in vitro both in physiological and pathological conditions [44–49]. IL-1β has been consistently involved in the pathogenesis of synaptic plasticity alterations in EAE [39,41–43]; however, contrasting data exist on how this cytokine affects LTP and LTD expression. While it has been reported that IL-1β impairs LTP induction in mice hippocampal slices [42], enhanced LTP expression has also been observed in response to IL-1β [43]. Accordingly, in RR-MS patients the CSF levels of IL-1β have been correlated with a paradoxical LTP-like effect in response to an LTD-inducing TMS protocol [41], confirming that this cytokine also produces a profound subversion of synaptic plasticity in human MS. Similarly, IL-6 has been associated to impaired LTP induction in vitro and in MS [49]. In RR-MS patients, the CSF levels of this cytokine were correlated to impaired LTP-like plasticity explored with TMS [49]. Notably, in the same study, higher CSF levels of IL-6 were also associated with enhanced clinical expression of new brain lesions, suggesting that CSF inflammation impairs resilience to damage.

It has been shown that inflammation in MS is associated with altered Amyloid-β (Aβ) homeostasis. One study evidenced in a group of RR-MS patients a negative correlation between the number of Gd+ lesions at MRI and Aβ1-42 CSF levels [50]. Furthermore, a negative correlation has been found between the CSF concentrations of Aβ1-42 and LTP-like responses to TMS [50]. In addition, it has been reported that the CSF levels of different proinflammatory molecules negatively correlate with Aβ1-42 concentrations [49,51]. Overall, these data suggest that in MS Aβ homeostasis is as a key factor bridging inflammation, plasticity and neurodegeneration.

3.2. *Inflammation and Upscaling*

Experimental studies have shown that TNF is involved in inducing and maintaining synaptic upscaling in different brain areas [52–55]. The specific role of TNF in promoting pathologically enhanced synaptic upscaling has been elucidated in EAE mice. In particular, TNF has been associated with increased expression of AMPARs and exacerbated glutamatergic excitatory post-synaptic currents [5]. These alterations preceded the onset of clinical deficits [5] and were prevented by pre-incubation with TNF inhibitors [56]. The pathogenic role of TNF has been associated particularly with progressive MS [57,58]. It has been shown that the CSF collected from patients with progressive disease course is able to induce TNF-mediated upscaling and excitotoxic neuronal damage in vitro, thus suggesting a link between CNS inflammation and neurodegeneration [57].

These data show that inflammation in MS is associated with a profound alteration of synaptic plasticity mainly characterized by impaired LTP and overexpressed synaptic upscaling. This evidence could be crucial for shedding light on the connectivity alterations and brain network reorganization found in MS patients.

4. Inflammation and Brain Network Organization in MS

4.1. *Inflammation Alters Brain Connectivity in MS*

A number of MRI studies have described alterations of structural and functional connectivity (SC and FC) in MS patients [15,59–66]. Altered connectivity has already been observed in the early phases of MS [65,67–69], and has been correlated with clinical and cognitive deficits [15,65,70], suggesting that this approach could be more sensitive than conventional MRI in investigating early alterations [71].

The nature of connectivity changes in MS is still debated [72]. In MS patients, reduced SC and FC in different networks has been related to both white matter lesion load and cortical atrophy [73–76], though connectivity alterations have also been reported independently of structural brain damage [64,68,69]. In patients with early RR-MS, Faivre and colleagues [68] have found increased FC in different resting state networks not correlated with lesion load. Furthermore, one study exploring FC in RR-MS and CIS patients has shown reduced efficiency in the left Rolandic operculum, insula and superior temporal gyrus bilaterally, not correlated to structural damage [69]. The role of other causes of disconnection should also be considered, including the presence of diffuse microstructural damage occurring in normal appearing white matter [77,78]. Rs-fMRI studies in patients with clinically isolated syndrome (CIS) have consistently shown that altered connectivity can be found independently of demyelinating lesions or neuronal atrophy, in fact widespread functional connectivity changes have been reported in both isolated optic neuritis and CIS without brain lesions [64,79]. In particular, acute optic neuritis has been associated with reduced FC in the visual system and altered connectivity between visual and non-visual networks, indicating rapid connectivity changes in response to focal inflammation [79]; in addition, both reduced and increased FC has been reported in CIS patients without brain lesions [64], suggesting that CNS inflammation may represent a prominent cause of connectivity dysfunction in MS [7].

In line with these findings, it has been reported that not only CNS-confined inflammation, but also systemic inflammation may alter brain activity and network reorganization [80,81]. A study using positron-emission tomography has revealed that administration of endotoxin in healthy subjects was associated with increased serum levels of inflammatory molecules and enhanced fatigue and depression [80]. These findings were associated with increased and reduced glucose metabolism in the insula and anterior cingulate cortex, respectively [80]. Similarly, it has been reported that interferon-alpha administration produced mood alterations associated with a rapid decrease in overall brain network efficiency and reduced connectivity of the nucleus accumbens, thalamus and inferior temporal cortex [81]. Moreover, experimental inflammation induced by lipopolysaccharide injection has been associated with connectivity alterations in healthy subjects [82,83] and the specific role of peripheral inflammatory cytokines has been evidenced in task-based and rs-fMRI studies [80,84–86]. Accordingly, it has been reported that serum concentrations of IL-6, a marker of peripheral inflammation, covaried with FC in the default mode network and with connectivity in the anterior cingulate cortex and in the medial prefrontal cortex [86].

These findings suggest that central inflammation, in addition to demyelination and atrophy, could alter brain connectivity in MS.

4.2. Brain Network Reorganization in MS

The physiological meaning of changes in brain FC and SC during the course of MS is still debated, as both decreased and increased connectivity in different brain networks have been described [66,68,74,87]. Furthermore, both positive and negative correlations have been reported between connectivity changes and clinical measures [67,68]. Differences between studies, concerning MS phenotype, degree of disability, presence of clinical activity, and MRI measures, could be partly responsible for these apparently contrasting findings. In addition, the presence of brain network reorganization, either compensatory or maladaptive, represents an additional source of variability between data. As optimal brain network reorganization requires efficient plasticity, synaptic plasticity dysfunction observed in MS, in particular reduced LTP expression and pathologically increased synaptic upscaling, could specifically contribute to altering brain network architecture and impairing reorganization after damage.

Increased synchronization in different resting state networks has been described in CIS patients compared with both RR-MS patients and controls, suggesting the existence of early brain reorganization to damage, which is lost with disease progression [67]. However, as increased connectivity in early MS patients has been associated with higher disability, the implications of FC changes should be carefully considered [68]. Studies exploring SC in CIS and MS have been shown to strengthen local connectivity from the early phases of the disease [66,88]. In particular, increased clustering coefficient and modularity have been described in MS patients during the first year of disease, largely independent of clinical activity [66,89]. It has been proposed that such reorganization cannot be simply considered a result of structural damage, but rather represents an adaptive phenomenon [66]. Accordingly, modularization and increased local processing improve the network's ability to react to attacks [90,91].

An extreme redistribution of both functional and structural brain connectivity, with hub loss and formation of new hubs has been reported in the early phases of MS [63,92]. A study using rs-fMRI evidenced connectivity loss in hub regions including the precuneus, cingulate cortex and frontal areas, and newly formed hubs in the left temporal pole and cerebellum [63]. Using DTI, reduced connection strength in the rich club has been demonstrated in RR-MS patients and has been correlated with cognitive impairment [92]. In particular, altered rich club and local connectivity were not found to be related to white matter damage [92]. Notably, in CIS patients rich club connectivity was preserved and alterations mostly involved feeders connecting rich club to non-rich club nodes and local connections [92]. One rs-fMRI study explored brain network reorganization in a group of CIS patients at baseline and after a one year follow-up [93] by calculating the hub disruption index, a useful and reliable measure of large-scale networks reorganization [94,95]. With such an approach, a significant hub reorganization was detected at baseline and even more after one year, showing the occurrence of brain regions with reduced and increased connectivity at the same time [93]. At baseline, increased connectivity in the right middle temporal gyrus was observed in CIS; moreover, after a one year follow-up, both a higher degree in the hippocampi, cingulate cortex and in the left parieto-occipital sulcus, and a reduced degree in the middle occipital gyrus and left posterior segment of the lateral fissure were observed. Hub reorganization correlated with better cognitive scores; conversely, no significant correlations were reported with clinical scores and white matter lesions [93].

4.3. Altered Synaptic Plasticity Impairs Brain Network Remodeling in MS

Inconsistencies among studies exist, but taken together they suggest that a precise pattern of global brain network reorganization can be found in MS from the early phases, characterized by reduced hub connections and increased local connectivity. Some of these alterations may represent network reorganization mediated by synaptic plasticity mechanisms, whereas others may be consequence of increasing brain damage. Altered synaptic plasticity could represent an additional mechanism contributing to lessening brain network reorganization in MS. Some key properties of synaptic plasticity share analogies with specific characteristics of neuronal networks supporting this hypothesis. Input-specificity, activity-dependency, associativity and tendency to induce synaptic instability make LTP well-suited to generate highly connected nodes and could be involved in hub formation and reorganization. Hence, reduced LTP expression could specifically impair hub remodeling, contributing to hub disruption observed in MS. Furthermore, homeostatic synaptic plasticity allows adaptation to damage promoting local and generalized connectivity increases, which could be aimed at restoring network functionality, is also likely involved in limiting the connectivity of peripheral nodes, contributing to maintaining overall excitability and wiring cost around optimal levels. Altered homeostatic plasticity may therefore promote diffuse network hyperconnectivity.

Notably, progressive MS represents a useful model to test this hypothesis. Accordingly, in progressive MS, reduced rich club connectivity and a relative increase of local connectivity have been demonstrated [96]. Consistently, absent LTP induction and pathologically enhanced synaptic upscaling have been revealed in this MS phenotype [57,97,98].

5. Conclusions

The appropriate interplay between anti-homeostatic and homeostatic plasticity allows the formation of brain networks provided with peculiar characteristics associated with high resilience to damage, including scale-free degree distribution, modularity and rich club organization (Figure 1, left panel). We propose that altered synaptic plasticity in MS may promote both loss of hub connections and increased local connectivity, leading to random brain network architecture and impaired remodeling in response to damage (Figure 1, right panel). Therefore, inflammation may critically reduce brain network resilience in MS, promoting a worse disease course. This is in line with preclinical and clinical studies showing that the inflammatory CSF milieu critically influences MS disability progression and impairs clinical compensation of ongoing brain damage [49]. Conversely, it has been demonstrated that, in MS, anti-inflammatory molecules and neurotrophic factors may concur to promote a stable disease course [99]. Accordingly, the platelet-derived growth factor has been found to be absent in the CSF of patients with progressive MS, possibly contributing to explaining the impaired LTP-like plasticity expression [97]. Overall, these observations suggest that contrasting central inflammation could limit the functional disconnection and promote efficient brain network reorganization in MS.

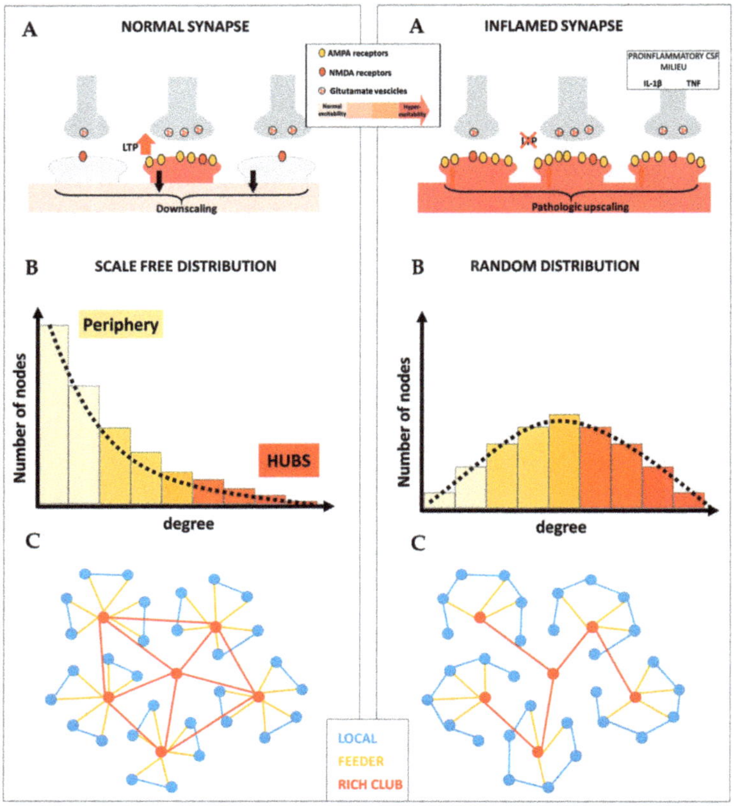

Figure 1. Different forms of synaptic plasticity cooperate to promote optimal brain network topology. **Left panel:** (**A**) In physiological conditions, the balance between anti-homeostatic and homeostatic plasticity allows the generation of potentiated synapses, associated with selective information processing, and prevents uncontrolled hyperexcitability. (**B**) Long-term potentiation (LTP) may be specifically involved in generating highly connected nodes (hubs); conversely, synaptic downscaling may be useful for maintaining low connectivity in the peripheral nodes of the network. The fine-tuning between these two forms of synaptic plasticity is required to form brain networks characterized by a scale-free degree distribution. (**C**) The resulting brain network architecture is characterized by elevated efficiency of information processing and resilience to random damage. **Right panel:** (**A**) In multiple sclerosis (MS), neuroinflammation is associated with impaired LTP and pathologically overexpressed synaptic upscaling, leading to uncontrolled neuronal hyperexcitability. (**B**) Disrupted LTP may selectively reduce hub connectivity, while overexpressed upscaling may contribute to increasing connectivity in the periphery. This is associated with loss of optimal brain network architecture as demonstrated by further random degree distribution. (**C**) Loss of LTP may selectively disrupt hub connectivity and rich club organization. Conversely, pathologic upscaling may promote increased local connectivity. The resulting brain network architecture dramatically reduces efficiency and impairs the ability to compensate for ongoing brain damage in MS.

Further studies are required to better define the relationship between CSF inflammation, synaptic plasticity and brain network structural and functional connectivity. In particular, studies combining neurophysiological investigations and fMRI measures could help to clarify the significance of brain network reorganization occurring in MS.

Author Contributions: Conceptualization, M.S.B., D.C.; writing—original draft preparation, M.S.B. and E.I.; writing—review and editing, M.S.B., E.I., L.G., L.P., G.M., A.M., A.G., F.B., D.C.; visualization, M.S.B., E.I., L.G.; supervision, D.C.; funding acquisition, D.C. All authors have read and agreed to the published version of the manuscript.

Funding: This research was funded by the Italian Ministry of Health (Ricerca Corrente), by the 5 × 1000 grant to IRCCS Neuromed, and by the Fondazione Italiana Sclerosi Multipla (FISM) COD. 2019/S/1 to DC.

Conflicts of Interest: The authors declared the following potential conflicts of interest with respect to the research, authorship, and/or publication of this article: G.M. reports grants and/or personal fees from Merck, Novartis, Biogen Idec, and Ibsa. D.C. is an Advisory Board member of Almirall, Bayer Schering, Biogen, GW Pharmaceuticals, Merck Serono, Novartis, Roche, Sanofi-Genzyme, and Teva and received honoraria for speaking or consultation fees from Almirall, Bayer Schering, Biogen, GW Pharmaceuticals, Merck Serono, Novartis, Roche, Sanofi-Genzyme, and Teva. He is also the principal investigator in clinical trials for Bayer Schering, Biogen, Merck Serono, Mitsubishi, Novartis, Roche, Sanofi-Genzyme, and Teva. His preclinical and clinical research was supported by grants from Bayer Schering, Biogen Idec, Celgene, Merck Serono, Novartis, Roche, Sanofi-Genzyme and Teva. The funders had no role in the design of the study; in the collection, analyses, or interpretation of data; in the writing of the manuscript, or in the decision to publish the results. F.B. acted as Advisory Board members of Teva and Roche and received honoraria for speaking or consultation fees from Merck Serono, Teva, Biogen Idec, Sanofi, and Novartis and non-financial support from Merck Serono, Teva, Biogen Idec, and Sanofi. M.S.B., E.I., L.P., A.M., A.G., and L.G. declare no conflict of interest.

Abbreviations

Aβ	amyloid-β
AMPA	α-amino-3-hydroxy-5-methyl-4-isoxazolepropionic acid
CNS	central nervous system
CSF	cerebrospinal fluid
CIS	clinically isolated syndrome
DTI	diffusion tensor imaging
EAE	experimental autoimmune encephalomyelitis
FC	functional connectivity
fMRI	functional MRI
IL	interleukin
K	node degree
LTD	long-term depression
LTP	long-term potentiation
MS	multiple sclerosis
NMDA	N-methyl-D-aspartate
PK	degree distribution
RR	relapsing-remitting
rs-fMRI	resting state-functional magnetic resonance imaging
SC	structural connectivity
TMS	transcranial magnetic stimulation
TNF	tumor necrosis factor

References

1. Polman, C.H.; Reingold, S.C.; Banwell, B.; Clanet, M.; Cohen, J.A.; Filippi, M.; Fujihara, K.; Havrdova, E.; Hutchinson, M.; Kappos, L.; et al. Diagnostic criteria for multiple sclerosis: 2010 Revisions to the McDonald criteria. *Ann. Neurol.* **2015**, *69*, 292–302. [CrossRef] [PubMed]
2. Barkhof, F. The clinico-radiological paradox in multiple sclerosis revisited. *Curr. Opin. Neurol.* **2002**, *15*, 239–245. [CrossRef] [PubMed]
3. Rossi, S.; Studer, V.; Motta, C.; Germani, G.; Macchiarulo, G.; Buttari, F.; Mancino, R.; Castelli, M.; De Chiara, V.; Weiss, S.; et al. Cerebrospinal fluid detection of interleukin-1b in phase of remission predicts disease progression in multiple sclerosis. *J. Neuroinflammation* **2014**, *11*, 32. [CrossRef] [PubMed]

4. Stampanoni Bassi, M.; Iezzi, E.; Landi, D.; Monteleone, F.; Gilio, L.; Simonelli, I.; Musella, A.; Mandolesi, G.; De Vito, F.; Furlan, R.; et al. Delayed treatment of MS is associated with high CSF levels of IL-6 and IL-8 and worse future disease course. *J. Neurol.* **2018**, *265*, 2540–2547. [CrossRef]
5. Centonze, D.; Muzio, L.; Rossi, S.; Cavasinni, F.; De Chiara, V.; Bergami, A.; Musella, A.; D'Amelio, M.; Cavallucci, V.; Martorana, A.; et al. Inflammation triggers synaptic alteration and degeneration in experimental autoimmune encephalomyelitis. *J. Neurosci.* **2009**, *29*, 3442–3452. [CrossRef]
6. Mandolesi, G.; Musella, A.; Gentile, A.; Grasselli, G.; Haji, N.; Sepman, H.; Fresegna, D.; Bullitta, S.; De Vito, F.; Musumeci, G.; et al. Interleukin-1β alters glutamate transmission at purkinje cell synapses in a mouse model of multiple sclerosis. *J. Neurosci.* **2013**, *33*, 12105–12121. [CrossRef]
7. Stampanoni Bassi, M.; Gilio, L.; Buttari, F.; Maffei, P.; Marfia, G.A.; Restivo, D.A.; Centonze, D.; Iezzi, E. Remodeling Functional Connectivity in Multiple Sclerosis: A Challenging Therapeutic Approach. *Front. Neurosci.* **2017**, *11*, 710. [CrossRef]
8. Achard, S.; Bullmore, E. Efficiency and cost of economical brain functional networks. *PLoS Comput. Biol.* **2007**, *3*, e17. [CrossRef]
9. Hillary, F.G.; Grafman, J.H. Injured Brains and Adaptive Networks: The Benefits and Costs of Hyperconnectivity. *Trends. Cogn. Sci.* **2017**, *21*, 385–401. [CrossRef]
10. Centonze, D.; Rossi, S.; Tortiglione, A.; Picconi, B.; Prosperetti, C.; De Chiara, V.; Bernardi, G.; Calabresi, P. Synaptic plasticity during recovery from permanent occlusion of the middle cerebral artery. *Neurobiol. Dis.* **2007**, *27*, 44–53. [CrossRef]
11. Di Lazzaro, V.; Profice, P.; Pilato, F.; Capone, F.; Ranieri, F.; Pasqualetti, P.; Colosimo, C.; Pravatà, E.; Cianfoni, A.; Dileone, M. Motor cortex plasticity predicts recovery in acute stroke. *Cereb. Cortex* **2010**, *20*, 1523–1528. [CrossRef] [PubMed]
12. Watts, D.J.; Strogatz, S.H. Collective dynamics of 'small-world' networks. *Nature* **1998**, *393*, 440–442. [CrossRef] [PubMed]
13. Sporns, O.; Tononi, G.; Kötter, R. The human connectome: A structural description of the human brain. *PLoS Comput. Biol.* **2005**, *1*, e42. [CrossRef] [PubMed]
14. Stam, C.J.; Reijneveld, J.C. Graph theoretical analysis of complex networks in the brain. *Nonlinear Biomed. Phys.* **2007**, *1*, 3. [CrossRef]
15. Bullmore, E.; Sporns, O. Complex brain networks: Graph theoretical analysis of structural and functional systems. *Nat. Rev. Neurosci.* **2009**, *10*, 186–198. [CrossRef]
16. Stam, C.J. Modern network science of neurological disorders. *Nat. Rev. Neurosci.* **2014**, *15*, 683–695. [CrossRef] [PubMed]
17. Rossini, P.M.; Di Iorio, R.; Bentivoglio, M.; Bertini, G.; Ferreri, F.; Gerloff, C.; Ilmoniemi, R.J.; Miraglia, F.; Nitsche, M.A.; Pestilli, F.; et al. Methods for analysis of brain connectivity: An IFCN-sponsored review. *Clin. Neurophysiol.* **2019**, *130*, 1833–1858. [CrossRef]
18. Biswal, B.B.; Yetkin, F.Z.; Haughton, V.M.; Hyde, J.S. Functional connectivity in the motor cortex of resting human brain using echo-planar MRI. *Magn. Reson. Med.* **1995**, *34*, 537–541. [CrossRef]
19. Fox, M.D.; Raichle, M.E. Spontaneous fluctuations in brain activity observed with functional magnetic resonance imaging. *Nat. Rev. Neurosci.* **2007**, *8*, 700–711. [CrossRef]
20. Sporns, O.; Honey, C.J.; Kötter, R. Identification and classification of hubs in brain networks. *PLoS ONE* **2007**, *2*, e1049. [CrossRef]
21. Van den Heuvel, M.P.; Sporns, O. An anatomical substrate for integration among functional networks in human cortex. *J. Neurosci.* **2013**, *33*, 14489–14500. [CrossRef] [PubMed]
22. Barabasi, A.L.; Albert, R. Emergence of scaling in random networks. *Science* **1999**, *286*, 509–512. [CrossRef] [PubMed]
23. Achard, S.; Salvador, R.; Whitcher, B.; Suckling, J.; Bullmore, E. A resilient, low-frequency, small-world human brain functional network with highly connected association cortical hubs. *J. Neurosci.* **2006**, *26*, 63–72. [CrossRef] [PubMed]
24. Crossley, N.A.; Mechelli, A.; Scott, J.; Carletti, F.; Fox, P.T.; McGuire, P.; Bullmore, E.T. The hubs of the human connectome are generally implicated in the anatomy of brain disorders. *Brain* **2014**, *137*, 2382–2395. [CrossRef]
25. Turrigiano, G.G.; Nelson, S.B. Homeostatic plasticity in the developing nervous system. *Nat. Rev. Neurosci.* **2004**, *5*, 97–107. [CrossRef]

26. Citri, A.; Malenka, R.C. Synaptic plasticity: Multiple forms, functions, and mechanisms. *Neuropsychopharmacology* **2008**, *33*, 18–41. [CrossRef]
27. Bliss, T.V.; Lømo, T. Long-lasting potentiation of synaptic transmission in the dentate area of the anaesthetized rabbit following stimulation of the perforant path. *J. Physiol.* **1973**, *232*, 331–356. [CrossRef] [PubMed]
28. Mori, F.; Kusayanagi, H.; Nicoletti, C.G.; Weiss, S.; Marciani, M.G.; Centonze, D. Cortical plasticity predicts recovery from relapse in multiple sclerosis. *Mult. Scler.* **2014**, *20*, 451–457. [CrossRef]
29. Bliss, T.V.; Collingridge, G.L. A synaptic model of memory: Long-term potentiation in the hippocampus. *Nature* **1993**, *361*, 31–39. [CrossRef]
30. Harris, E.W.; Ganong, A.H.; Cotman, C.W. Long-term potentiation in the hippocampus involves activation of N-methyl-D-aspartate receptors. *Brain Res.* **1984**, *323*, 132–137. [CrossRef]
31. Malenka, R.C.; Bear, M.F. LTP and LTD: An embarrassment of riches. *Neuron* **2004**, *44*, 5–21. [CrossRef]
32. Turrigiano, G.G.; Leslie, K.R.; Desai, N.S.; Rutherford, L.C.; Nelson, S.B. Activity-dependent scaling of quantal amplitude in neocortical neurons. *Nature* **1998**, *391*, 892–896. [CrossRef]
33. Turrigiano, G.G.; Nelson, S.B. Hebb and homeostasis in neuronal plasticity. *Curr. Opin. Neurobiol.* **2000**, *10*, 358–364. [CrossRef]
34. Lissin, D.V.; Gomperts, S.N.; Carroll, R.C.; Christine, C.W.; Kalman, D.; Kitamura, M.; Hardy, S.; Nicoll, R.A.; Malenka, R.C.; von Zastrow, M. Activity differentially regulates the surface expression of synaptic AMPA and NMDA glutamate receptors. *Proc. Natl. Acad. Sci. USA* **1998**, *95*, 7097–7102. [CrossRef] [PubMed]
35. Desai, N.S.; Cudmore, R.H.; Nelson, S.B.; Turrigiano, G.G. Critical periods for experience-dependent synaptic scaling in visual cortex. *Nat. Neurosci.* **2002**, *5*, 783–789. [CrossRef] [PubMed]
36. Turrigiano, G.G. Homeostatic synaptic plasticity: Local and global mechanisms for stabilizing neuronal function. *Cold. Spring Harb. Perspect. Biol.* **2012**, *4*, a005736. [CrossRef]
37. Stampanoni Bassi, M.; Mori, F.; Buttari, F.; Marfia, G.A.; Sancesario, A.; Centonze, D.; Iezzi, E. Neurophysiology of synaptic functioning in multiple sclerosis. *Clin. Neurophysiol.* **2017**, *128*, 1148–1157. [CrossRef]
38. Rossi, S.; Studer, V.; Motta, C.; De Chiara, V.; Barbieri, F.; Bernardi, G.; Centonze, D. Inflammation inhibits GABA transmission in multiple sclerosis. *Mult. Scler.* **2012**, *18*, 1633–1635. [CrossRef]
39. Rossi, S.; Furlan, R.; De Chiara, V.; Motta, C.; Studer, V.; Mori, F.; Musella, A.; Bergami, A.; Muzio, L.; Bernardi, G.; et al. Interleukin-1 β causes synaptic hyperexcitability in multiple sclerosis. *Ann. Neurol.* **2012**, *71*, 76–83. [CrossRef]
40. Mori, F.; Kusayanagi, H.; Buttari, F.; Centini, B.; Monteleone, F.; Nicoletti, C.G.; Bernardi, G.; Verdun Di Cantogno, E.; Marciani, M.G.; Centonze, D. Early treatment with high-dose interferon beta-1a reverses cognitive and cortical plasticity deficits in multiple sclerosis. *Funct. Neurol.* **2012**, *27*, 163–168. [PubMed]
41. Mori, F.; Nisticò, R.; Mandolesi, G.; Piccinin, S.; Mango, D.; Kusayanagi, H.; Berretta, N.; Bergami, A.; Gentile, A.; Musella, A.; et al. Interleukin-1β promotes long-term potentiation in patients with multiple sclerosis. *Neuromolecular. Med.* **2014**, *16*, 38–51. [CrossRef] [PubMed]
42. Di Filippo, M.; Chiasserini, D.; Gardoni, F.; Viviani, B.; Tozzi, A.; Giampà, C.; Costa, C.; Tantucci, M.; Zianni, E.; Boraso, M.; et al. Effects of central and peripheral inflammation on hippocampal synaptic plasticity. *Neurobiol. Dis.* **2013**, *52*, 229–236. [CrossRef] [PubMed]
43. Nisticò, R.; Mango, D.; Mandolesi, G.; Piccinin, S.; Berretta, N.; Pignatelli, M.; Feligioni, M.; Musella, A.; Gentile, A.; Mori, F.; et al. Inflammation subverts hippocampal synaptic plasticity in experimental multiple sclerosis. *PLoS ONE* **2013**, *8*, e54666. [CrossRef] [PubMed]
44. Bellinger, F.P.; Madamba, S.G.; Campbell, I.L.; Siggins, G.R. Reduced long-term potentiation in the dentate gyrus of transgenic mice with cerebral overexpression of interleukin-6. *Neurosci. Lett.* **1995**, *198*, 95–98. [CrossRef]
45. Li, A.J.; Katafuchi, T.; Oda, S.; Hori, T.; Oomura, Y. Interleukin-6 inhibits long-term potentiation in rat hippocampal slices. *Brain. Res.* **1997**, *748*, 30–38. [CrossRef]
46. Schneider, H.; Pitossi, F.; Balschun, D.; Wagner, A.; del Rey, A.; Besedovsky, H.O. A neuromodulatory role of interleukin- 1β in the hippocampus. *Proc. Natl. Acad. Sci. USA* **1998**, *95*, 7778–7783. [CrossRef]
47. Coogan, A.N.; O'Neill, L.A.; O'Connor, J.J. The P38 mitogen-activated protein kinase inhibitor SB203580 antagonizes the inhibitory effects of interleukin-1β on long-term potentiation in the rat dentate gyrus in vitro. *Neuroscience* **1999**, *93*, 57–69. [CrossRef]

48. Avital, A.; Goshen, I.; Kamsler, A.; Segal, M.; Iverfeldt, K.; Richter-Levin, G.; Yirmiya, R. Impaired interleukin-1 signaling is associated with deficits in hippocampal memory processes and neural plasticity. *Hippocampus* **2003**, *13*, 826–834. [CrossRef]
49. Stampanoni Bassi, M.; Iezzi, E.; Mori, F.; Simonelli, I.; Gilio, L.; Buttari, F.; Sica, F.; De Paolis, N.; Mandolesi, G.; Musella, A.; et al. Interleukin-6 Disrupts Synaptic Plasticity and Impairs Tissue Damage Compensation in Multiple Sclerosis. *Neurorehab. Neural. Rep.* **2019**, *33*, 825–835. [CrossRef]
50. Mori, F.; Rossi, S.; Sancesario, G.; Codecà, C.; Mataluni, G.; Monteleone, F.; Buttari, F.; Kusayanagi, H.; Castelli, M.; Motta, C.; et al. Cognitive and cortical plasticity deficits correlate with altered amyloidβ CSF levels in multiple sclerosis. *Neuropsychopharmacology* **2011**, *36*, 559–568. [CrossRef]
51. Stampanoni Bassi, M.; Garofalo, S.; Marfia, G.A.; Gilio, L.; Simonelli, I.; Finardi, A.; Furlan, R.; Sancesario, G.M.; Di Giandomenico, J.; Storto, M.; et al. Amyloid-β Homeostasis Bridges Inflammation, Synaptic Plasticity Deficits and Cognitive Dysfunction in Multiple Sclerosis. *Front. Mol. Neurosci.* **2017**, *10*, 390. [CrossRef] [PubMed]
52. Stellwagen, D.; Malenka, L.C. Synaptic scaling mediated by glial TNF-α. *Nature* **2006**, *440*, 1054–1059. [CrossRef] [PubMed]
53. Becker, D.; Deller, T.; Vlachos, A. Tumor necrosis factor (TNF)-receptor 1 and 2 mediate homeostatic synaptic plasticity of denervated mouse dentate granule cells. *Sci. Rep.* **2015**, *5*, 12726. [CrossRef] [PubMed]
54. Lewitus, G.M.; Pribiag, H.; Duseja, R.; St-Hilaire, M.; Stellwagen, D. An adaptive role of TNFα in the regulation of striatal synapses. *J. Neurosci.* **2014**, *34*, 6146–6155. [CrossRef]
55. Ren, W.; Lui, Y.; Zhou, L.; Li, W.; Zhong, Y.; Pang, R.; Xin, W.; Wei, X.; Wang, J.; Zhuet, H.; et al. Peripheral nerve injury leads to working memory deficits and dysfunction of the hippocampus by upregulation of TNF-α in rodents. *Neuropsychopharmacology* **2011**, *36*, 979–992. [CrossRef]
56. Haji, N.; Mandolesi, G.; Gentile, A.; Sacchetti, L.; Fresegna, D.; Rossi, S.; Musella, A.; Sepman, H.; Motta, C.; Studer, V.; et al. TNF-α-mediated anxiety in a mouse model of multiple sclerosis. *Exp. Neurol.* **2012**, *237*, 296–303. [CrossRef]
57. Rossi, S.; Motta, C.; Studer, V.; Barbieri, F.; Buttari, F.; Bergami, A.; Sancesario, G.; Bernardini, S.; De Angelis, G.; Martino, G.; et al. Tumor necrosis factor is elevated in progressive multiple sclerosis and causes excitotoxic neurodegeneration. *Mult. Scler.* **2014**, *20*, 304–312. [CrossRef]
58. Valentin-Torres, A.; Savarin, C.; Hinton, D.R.; Phares, T.W.; Bergmann, C.C.; Stohlman, S.A. Sustained TNF production by central nervous system infiltrating macrophages promotes progressive autoimmune encephalomyelitis. *J. Neuroinflamm.* **2016**, *13*, 46. [CrossRef]
59. He, Y.; Dagher, A.; Chen, Z.; Charil, A.; Zijdenbos, A.; Worsley, K.; Evans, A. Impaired small-world efficiency in structural cortical networks in multiple sclerosis associated with white matter lesion load. *Brain* **2009**, *132*, 3366–3379. [CrossRef]
60. Filippi, M.; van den Heuvel, M.P.; Fornito, A.; He, Y.; Hulshoff Pol, H.E.; Agosta, F.; Comi, G.; Rocca, M.A. Assessment of system dysfunction in the brain through MRI-based connectomics. *Lancet Neurol.* **2013**, *12*, 1189–1199. [CrossRef]
61. Schoonheim, M.M.; Geurts, J.J.; Landi, D.; Douw, L.; van der Meer, M.L.; Vrenken, H.; Polman, C.H.; Barkhof, F.; Stam, C.J. Functional connectivity changes in multiple sclerosis patients: A graph analytical study of MEG resting state data. *Hum. Brain Mapp.* **2013**, *34*, 52–61. [CrossRef] [PubMed]
62. Tewarie, P.; Schoonheim, M.M.; Stam, C.J.; van der Meer, M.L.; van Dijk, B.W.; Barkhof, F.; Polman, C.H.; Hillebrand, A. Cognitive and clinical dysfunction, altered MEG resting-state networks and thalamic atrophy in multiple sclerosis. *PLoS ONE* **2014**, *8*, e69318. [CrossRef] [PubMed]
63. Rocca, M.A.; Valsasina, P.; Meani, A.; Falini, A.; Comi, G.; Filippi, M. Impaired functional integration in multiple sclerosis: A graph theory study. *Brain Struct. Funct.* **2016**, *221*, 115–131. [CrossRef] [PubMed]
64. Liu, Y.; Dai, Z.; Duan, Y.; Huang, J.; Ren, Z.; Liu, Z.; Dong, H.; Shu, N.; Vrenken, H.; Wattjes, M.P.; et al. Whole brain functional connectivity in clinically isolated syndrome without conventional brain MRI lesions. *Eur. Radiol.* **2016**, *26*, 2982–2991. [CrossRef]
65. Shu, N.; Duan, Y.; Xia, M.; Schoonheim, M.M.; Huang, J.; Ren, Z.; Sun, Z.; Ye, J.; Dong, H.; Shi, F.; et al. Disrupted topological organization of structural and functional brain connectomes in clinically isolated syndrome and multiple sclerosis. *Sci. Rep.* **2016**, *6*, 29383. [CrossRef]

66. Fleischer, V.; Radetz, A.; Ciolac, D.; Muthuraman, M.; Gonzalez-Escamilla, G.; Zipp, F.; Groppa, S. Graph Theoretical Framework of Brain Networks in Multiple Sclerosis: A Review of Concepts. *Neuroscience* **2019**, *403*, 35–53. [CrossRef]
67. Roosendaal, S.D.; Schoonheim, M.M.; Hulst, H.E.; Sanz-Arigita, E.J.; Smith, S.M.; Geurts, J.J.; Barkhof, G. Resting state networks change in clinically isolated syndrome. *Brain* **2010**, *133*, 1612–1621. [CrossRef]
68. Faivre, A.; Rico, A.; Zaaraoui, W.; Crespy, L.; Reuter, F.; Wybrecht, D.; Soulier, E.; Malikova, I.; Confort-Gouny, S.; Cozzone, P.J.; et al. Assessing brain connectivity at rest is clinically relevant in early multiple sclerosis. *Mult. Scler.* **2012**, *18*, 1251–1258. [CrossRef]
69. Liu, Y.; Wang, H.; Duan, Y.; Huang, J.; Ren, Z.; Ye, J.; Dong, H.; Shi, F.; Li, K.; Wang, J. Functional brain network alterations in clinically isolated syndrome and multiple sclerosis: A graph-based connectome study. *Radiology* **2017**, *282*, 534–541. [CrossRef]
70. Hawellek, D.J.; Hipp, J.F.; Lewis, C.M.; Corbetta, M.; Engel, A.K. Increased functional connectivity indicates the severity of cognitive impairment in multiple sclerosis. *Proc. Natl. Acad. Sci. USA* **2011**, *108*, 19066–19071. [CrossRef]
71. Liu, Y.; Duan, Y.; Dong, H.; Barkhof, F.; Li, K.; Shu, N. Disrupted Module Efficiency of Structural and Functional Brain Connectomes in Clinically Isolated Syndrome and Multiple Sclerosis. *Front. Hum. Neurosci.* **2018**, *12*, 138. [CrossRef] [PubMed]
72. Ciccarelli, O.; Barkhof, F.; Bodini, B.; De Stefano, N.; Golay, X.; Nicolay, K.; Pelletier, D.; Pouwels, P.J.; Smith, S.A.; Wheeler-Kingshott, C.A.; et al. Pathogenesis of multiple sclerosis: Insights from molecular and metabolic imaging. *Lancet Neurol.* **2014**, *13*, 807–822. [CrossRef]
73. Rocca, M.A.; Valsasina, P.; Martinelli, V.; Misci, P.; Falini, A.; Comi, G.; Filippi, M. Large-scale neuronal network dysfunction in relapsing-remitting multiple sclerosis. *Neurology* **2012**, *79*, 1449–1457. [CrossRef] [PubMed]
74. Rocca, M.A.; Valsasina, P.; Leavitt, V.M.; Rodegher, M.; Radaelli, M.; Riccitelli, G.C.; Martinelli, V.; Martinelli-Boneschi, F.; Falini, A.; Comi, G.; et al. Functional network connectivity abnormalities in multiple sclerosis: Correlations with disability and cognitive impairment. *Mult. Scler.* **2018**, *24*, 459–471. [CrossRef] [PubMed]
75. Cruz-Gómez, Á.J.; Ventura-Campos, N.; Belenguer, A.; Ávila, C.; Forn, C. The link between resting-state functional connectivity and cognition in MS patients. *Mult. Scler.* **2014**, *20*, 338–348. [CrossRef] [PubMed]
76. Zhou, F.; Zhuang, Y.; Gong, H.; Wang, B.; Wang, X.; Chen, Q.; Wu, L.; Wan, H. Altered inter-subregion connectivity of the default mode network in relapsing remitting multiple sclerosis: A functional and structural connectivity study. *PLoS ONE* **2014**, *9*, e101198. [CrossRef]
77. Filippi, M.; Cercignani, M.; Inglese, M.; Horsfield, M.A.; Comi, G. Diffusion tensor magnetic resonance imaging in multiple sclerosis. *Neurology* **2001**, *56*, 304–311. [CrossRef]
78. Kocevar, G.; Stamile, C.; Hannoun, S.; Cotton, F.; Vukusic, S.; Durand-Dubief, F.; Sappey-Marinier, D. Graph theory-based brain connectivity for automatic classification of multiple sclerosis clinical courses. *Front. Neurosci.* **2016**, *10*, 478. [CrossRef]
79. Wu, G.F.; Matthew, R.; Parks, C.A.L.; Ances, B.M.; Van Stavern, G.P. An eye on brain integrity: Acute optic neuritis affects resting state functional connectivity impact of acute on on rs-fcMRI. *Investig. Ophthalmol. Vis. Sci.* **2015**, *56*, 2541–2546. [CrossRef]
80. Hannestad, J.; Gallezot, J.D.; Schafbauer, T.; Lim, K.; Kloczynski, T.; Morris, E.D.; Carson, R.E.; Ding, Y.; Cosgrove, K.P. Cosgrove Endotoxin-induced systemic inflammation activates microglia:[^{11}C]PBR28 positron emission tomography in nonhuman primates. *Neuroimage* **2012**, *63*, 232–239. [CrossRef]
81. Dipasquale, O.; Cooper, E.A.; Tibble, J.; Voon, V.; Baglio, F.; Baselli, G.; Cercignani, M.; Harrison, N.A. Interferon-α acutely impairs whole-brain functional connectivity network architecture–A preliminary study. *Brain Behav. Immun.* **2016**, *58*, 31–39. [CrossRef] [PubMed]
82. Labrenz, F.; Wrede, K.; Forsting, M.; Engler, H.; Schedlowski, M.; Elsenbruch, S.; Benson, S. Alterations in functional connectivity of resting state networks during experimental endotoxemia–an exploratory study in healthy men. *Brain Behav. Immun.* **2016**, *54*, 17–26. [CrossRef] [PubMed]
83. Lekander, M.; Karshikoff, B.; Johansson, E.; Soop, A.; Fransson, P.; Lundström, J.N.; Andreasson, A.; Ingvar, M.; Petrovic, P.; Axelsson, J.; et al. Intrinsic functional connectivity of insular cortex and symptoms of sickness during acute experimental inflammation. *Brain Behav. Immun.* **2016**, *56*, 34–41. [CrossRef] [PubMed]

84. Capuron, L.; Pagnoni, G.; Demetrashvili, M.; Woolwine, B.J.; Nemeroff, C.B.; Berns, G.S.; Miller, A.H. Anterior cingulate activation and error processing during interferon-alpha treatment. *Biol. Psychiatry* **2005**, *58*, 190–196. [CrossRef] [PubMed]
85. Harrison, N.A.; Brydon, L.; Walker, C.; Gray, M.A.; Steptoe, A.; Dolan, R.J.; Critchley, H.D. Neural origins of human sickness in interoceptive responses to inflammation. *Biol. Psychiatry* **2009**, *66*, 415–422. [CrossRef] [PubMed]
86. Marsland, A.L.; Kuan, D.C.H.; Sheu, L.K.; Krajina, K.; Kraynak, T.E.; Manuck, S.B.; Gianaros, P.J. Systemic inflammation and resting state connectivity of the default mode network. *Brain Behav. Immun.* **2017**, *62*, 162–170. [CrossRef] [PubMed]
87. Tona, F.; Petsas, N.; Sbardella, E.; Prosperini, L.; Carmellini, M.; Pozzilli, C.; Pantano, P. Multiple sclerosis: Altered thalamic resting-state functional connectivity and its effect on cognitive function. *Radiology* **2014**, *271*, 814–821. [CrossRef]
88. Muthuraman, M.; Fleischer, V.; Kolber, P.; Luessi, F.; Zipp, F.; Groppa, S. Structural brain network characteristics can differentiate CIS from early RRMS. *Front. Neurosci.* **2016**, *10*, 14. [CrossRef]
89. Fleischer, V.; Koirala, N.; Droby, A.; Gracien, R.M.; Deichmann, R.; Ziemann, U.; Meuth, S.G.; Muthuraman, M.; Zipp, F.; Groppa, S. Longitudinal cortical network reorganization in early relapsing-remitting multiple sclerosis. *Ther. Adv. Neurol. Disord.* **2019**, *12*, 1–15. [CrossRef]
90. Latora, V.; Marchiori, M. Efficient behavior of small-world networks. *Phys. Rev. Lett.* **2001**, *87*, 198701. [CrossRef]
91. Kashtan, N.; Alon, U. Spontaneous evolution of modularity and network motifs. *Proc. Natl. Acad. Sci. USA* **2005**, *102*, 13773–13778. [CrossRef] [PubMed]
92. Shu, N.; Duan, Y.; Huang, J.; Ren, Z.; Liu, Z.; Dong, H.; Barkhof, F.; Li, K.; Liu, Y. Progressive brain rich-club network disruption from clinically isolated syndrome towards multiple sclerosis. *Neuroimage Clin.* **2018**, *19*, 232–239. [CrossRef] [PubMed]
93. Koubiyr, I.; Deloire, M.; Besson, P.; Coupé, P.; Dulau, C.; Pelletier, J.; Tourdias, T.; Audoin, B.; Brochet, B.; Ranjeva, J.P.; et al. Longitudinal study of functional brain network reorganization in clinically isolated syndrome. *Mult. Scler.* **2018**, *27*, 1–13. [CrossRef] [PubMed]
94. Termenon, M.; Achard, S.; Jaillard, A.; Delon-Martin, C. The "Hub Disruption Index," a Reliable Index Sensitive to the Brain Networks Reorganization. A Study of the Contralesional Hemisphere in Stroke. *Front. Comput. Neurosci.* **2016**, *10*, 84. [CrossRef] [PubMed]
95. Achard, S.; Delon-Martin, C.; Vértes, P.E.; Renard, F.; Schenck, M.; Schneider, F.; Heinrich, C.; Kremer, S.; Bullmore, E.T. Hubs of brain functional networks are radically reorganized in comatose patients. *Proc. Natl. Acad. Sci. USA* **2012**, *109*, 20608–20613. [CrossRef] [PubMed]
96. Stellmann, J.P.; Hodecker, S.; Cheng, B.; Wanke, N.; Young, K.L.; Hilgetag, C.; Gerloff, C.; Heesen, C.; Thomalla, G.; Siemonsen, S. Reduced rich-club connectivity is related to disability in primary progressive MS. *Neurol. Neuroimmunol. Neuroinflamm.* **2017**, *4*, e375. [CrossRef]
97. Mori, F.; Rossi, S.; Piccinin, S.; Motta, C.; Mango, D.; Kusayanagi, H.; Bergami, A.; Studer, V.; Nicoletti, C.G.; Buttari, F.; et al. Synaptic plasticity and PDGF signaling defects underlie clinical progression in multiple sclerosis. *J. Neurosci.* **2013**, *33*, 19112–19119. [CrossRef]
98. Nicoletti, C.G.; Monteleone, F.; Marfia, G.A.; Usiello, A.; Buttari, F.; Centonze, D.; Mori, F. Oral D-Aspartate enhances synaptic plasticity reserve in progressive multiple sclerosis. *Mult. Scler.* **2019**, *7*, 1–8. [CrossRef]
99. Stampanoni Bassi, M.; Iezzi, E.; Marfia, G.A.; Simonelli, I.; Musella, A.; Mandolesi, G.; Fresegna, D.; Pasqualetti, P.; Furlan, R.; Finardi, A.; et al. Platelet-derived growth factor predicts prolonged relapse-free period in multiple sclerosis. *J. Neuroinflamm.* **2018**, *15*, 108. [CrossRef]

International Journal of
Molecular Sciences

MDPI

Review

Synaptic Plasticity Shapes Brain Connectivity: Implications for Network Topology

Mario Stampanoni Bassi [1], Ennio Iezzi [1], Luana Gilio [1], Diego Centonze [1,2,*] and Fabio Buttari [1]

[1] Unit of Neurology & Neurorehabilitation, IRCCS Neuromed, 86077 Pozzilli, Italy; m.stampanonibassi@gmail.com (M.S.B.); ennio.iezzi@neuromed.it (E.I.); gilio.luana@gmail.com (L.G.); fabio.buttari@gmail.com (F.B.)

[2] Laboratory of Synaptic Immunopathology, Department of Systems Medicine, Tor Vergata University, 00173 Rome, Italy

* Correspondence: centonze@uniroma2.it; Tel.: +39-0865-929170

Received: 31 October 2019; Accepted: 6 December 2019; Published: 8 December 2019

Abstract: Studies of brain network connectivity improved understanding on brain changes and adaptation in response to different pathologies. Synaptic plasticity, the ability of neurons to modify their connections, is involved in brain network remodeling following different types of brain damage (e.g., vascular, neurodegenerative, inflammatory). Although synaptic plasticity mechanisms have been extensively elucidated, how neural plasticity can shape network organization is far from being completely understood. Similarities existing between synaptic plasticity and principles governing brain network organization could be helpful to define brain network properties and reorganization profiles after damage. In this review, we discuss how different forms of synaptic plasticity, including homeostatic and anti-homeostatic mechanisms, could be directly involved in generating specific brain network characteristics. We propose that long-term potentiation could represent the neurophysiological basis for the formation of highly connected nodes (hubs). Conversely, homeostatic plasticity may contribute to stabilize network activity preventing poor and excessive connectivity in the peripheral nodes. In addition, synaptic plasticity dysfunction may drive brain network disruption in neuropsychiatric conditions such as Alzheimer's disease and schizophrenia. Optimal network architecture, characterized by efficient information processing and resilience, and reorganization after damage strictly depend on the balance between these forms of plasticity.

Keywords: brain networks; connectivity; synaptic plasticity; Alzheimer's disease (AD); schizophrenia; long-term potentiation (LTP); synaptic scaling; resting state functional MRI (rs-fMRI)

1. Introduction

The functional properties of the brain are largely determined by the characteristics of its neurons and the pattern of synaptic connections between them. In the last century, how information is coded within a neuron and flows between neurons through synapses has been deeply investigated. However, little is known about how neural plasticity shapes network organization.

Comprehension of brain networks organization has been fueled by the application of procedures able to investigate brain connectivity in vivo [1,2] based on new theoretical/mathematical approaches (i.e., graph theory) to extract several measures that describe network architecture and functioning [3,4]. This approach led to the identification of specific features of brain networks, such as modularity and the presence of network hubs, that provide efficient information processing and elevated resistance to damage [5]. Furthermore, studies in patients with neurological diseases offered the chance to explore brain network reorganization following different types of damage [6–8].

Synaptic plasticity refers to the ability of neurons to modify the strength of their connections and is an important neurophysiological process involved in brain networks development and reorganization after damage [9]. Plasticity and network organization are highly intermingled, although they are generally studied as independent phenomena. Different forms of synaptic plasticity, namely, anti-homeostatic (i.e., Hebbian) and homeostatic plasticity (i.e., synaptic scaling), have been described. A fine balance between these two forms of synaptic plasticity could be crucial to maintain an optimal brain network architecture [9].

In the present review article, we will try to link together evidence from different research fields on the relationship between synaptic plasticity and principles of brain network organization. We will suggest that some key features of brain networks architecture may result from the fine tuning between different forms of synaptic plasticity. This perspective may be helpful to understand how networks adapt in response to brain damage and to explain mechanisms of network disruption in neuropsychiatric conditions such as Alzheimer's disease and schizophrenia.

2. Brain Network Organization

Classical studies explored neuronal connections using reconstructions of electron micrographs of serial sections to map neuronal local connectivity and have been used to reconstruct the structure and connectivity of simple nervous systems, such as that of *Caenorhabditis elegans* and *Drosophila* [4,10,11]. The application of new neurophysiological and imaging tools offered the opportunity to study the activity of different brain areas simultaneously, with different degrees of spatial and temporal resolution. Electroencephalography (EEG) and magnetoencephalography (MEG) can be used to analyze functional connectivity (FC) through the analysis of temporal correlations between spontaneous activity of different regions and are characterized by elevated temporal resolution. Functional MRI (fMRI) has a higher spatial resolution and represents the most used approach to explore brain network organization in vivo [12,13]. Resting state-fMRI (rs-fMRI) offers the chance to explore overall brain connectivity in healthy individuals and in different pathological conditions [14,15]. In fMRI studies, FC is calculated by the analysis of temporal correlations between spontaneous activity of blood-oxygen-level-dependent (BOLD) signals coming from different brain regions [2,16]. Conversely, structural connectivity (SC) can be calculated using fiber tractography from diffusion tensor imaging (DTI) or studying the correlations in cortical thickness between areas. Anatomically connected regions show stronger FC [17,18]; however, functional interactions may also occur in brain areas not directly connected [17–19], and therefore, the relationship between SC and FC requires further investigation.

Network-based studies use SC and FC data to create a comprehensive map of brain connections [20–22]. The graph theory represents the most useful approach to model brain networks [4,23]. According to the graph theory, a complex network can be represented as a set of nodes and edges, respectively indicating the basic elements of the network and the relationships between them. This approach can be used to describe complex networks with different spatial resolution, from microscale to large-scale networks [1,5]. In large-scale networks, nodes can represent EEG channels or regions of interest identified on MRI. The definition of edges originates from the analysis of SC and FC between nodes. The structure of a graph can be further analyzed to extract a set of quantitative measures describing specific properties of the network, including global and local efficiency, modularity, and degree distribution [5]. An important parameter is the wiring cost, expressing the energetic expenditure due to fiber tracts length and number of synaptic connections [4].

Brain networks typically show a small-world topology characterized by the prevalence of high locally connected nodes with a relatively small number of long-distance connections, optimizing efficient network communication and limiting wiring costs increase [24]. A simple measure able to provide essential information about brain network organization is node degree, that is the number of connections to a single node. The degree distribution of a graph $P(k)$ can be defined as the fraction of nodes having degree k. Brain networks show scale-free degree distribution, with a large prevalence of nodes with low degree and a small number of highly connected nodes, termed hubs, that support efficient global information processing [25]. Furthermore, brain networks are organized into modules composed of locally connected nodes, with long-distance connections between modules mainly confined to hub regions [26]. Such organization supports both functional segregation of information, which is provided by local communities of neurons highly interconnected, and integration, which measures the efficiency of global information transfer and the ability of networks to integrate distributed information [27].

Hubs are critical to provide efficient integration. Moreover, hubs show high inter-connectivity generating a subnetwork of densely interconnected hub regions, the rich club, which critically contribute to network efficiency and resilience to damage [4,26]. According to the preferential attachment theory, during the development of brain networks, the formation of hubs mostly follows a "rich get richer" principle, that is, the more connected nodes have greater chances to further increase their connectivity [28].

Considering the energetic expense of hyperconnectivity, this network architecture represents an efficient tradeoff between cost and efficiency, concentrating most connections to selected strategic nodes [24]. Accordingly, in invertebrates and mammalians, the brain shows the same organizational principles. Similar characteristics have been identified in several real-world complex networks, and this architecture provides high resilience to random damage due to the numerical prevalence of low-connected nodes [5,29]. Conversely, targeted attack to hubs may dramatically impact overall network efficiency [30], as evidenced in specific neuropsychiatric conditions such as Alzheimer's disease (AD) [31] or schizophrenia [32,33].

3. Synaptic Plasticity

Neurons can modify the efficacy of synaptic connections through different forms of synaptic plasticity, including anti-homeostatic and homeostatic mechanisms [9,34].

Long term potentiation (LTP) is one of the most studied forms of synaptic plasticity and has been associated to learning and memory processes [35,36], as well as to clinical recovery after brain damage [37]. LTP has been extensively investigated in hippocampal neurons [34,35] and consists of persisting enhancement of synaptic excitability, accompanied by structural rearrangement occurring at both the presynaptic and post-synaptic terminal [38,39]. LTP induction is associated to remodeling of dendritic spines, including increased spine volume, stability and clustering [40–42]. LTP depends on the activation of N-methyl-D-aspartate receptors (NMDARs) [43,44], and some key properties of this form of synaptic plasticity directly arise from the functional characteristics of this receptor [38,45]. LTP is cooperative, as the concomitant activation of multiple synapses favors the induction of this form of synaptic plasticity. Moreover, LTP is associative, meaning that a weak stimulus can be reinforced if associated with a strong one. Finally, LTP is input-specific, as only activated synapses in a neuron undergo potentiation. One important characteristic is that LTP is associated to increased neuronal excitability, which in turn facilitates further induction of LTP in a positive feedback loop, making LTP an anti-homeostatic phenomenon [9]. It is important to mention that activation of NMDARs could also induce a different form of anti-homeostatic plasticity, represented by long-term depression (LTD), which is associated to a lasting reduction in synaptic excitability [9,46]. Moreover, LTD induction is associated to changes in dendritic spine morphology, including marked spine shrinkage leading to the elimination of dendritic spines [47,48]. Notably, LTD-induced spine retraction could be reversed by subsequent LTP [48]. Although through opposite effects, LTP and LTD mutually interact to refine neural connections during the development and to regulate cognitive processes.

Anti-homeostatic plasticity alone can lead to uncontrolled increases or decreases of neuronal excitability (Figure 1A). The total amount of excitatory drive toward a neuron must be tightly regulated, which is difficult to obtain if synapses are independently modified; therefore, other mechanisms are required to stabilize neuronal activity. Persistent increase or decrease of neuronal excitability is associated to compensatory synaptic scaling (Figure 1B). Synaptic scaling of excitatory synapses is not regulated by NMDARs and mainly relies on the activity of α-amino-3-hydroxy-5-methyl-4-isoxazolepropionic acid receptors (AMPARs). Changes in the expression and clustering of AMPARs induce an increase (upscaling) or decrease (downscaling) of neuronal excitability in response to opposite changes in the strength of their synaptic excitatory inputs [49,50]. Trafficking of surface AMPARs is regulated by the expression of Arc/Arg gene [51,52]. Unlike LTP, synaptic scaling is a homeostatic negative feedback mechanism and represents a form of hetero-synaptic plasticity, as it lacks input-specificity and involves all synapses of a given neuron.

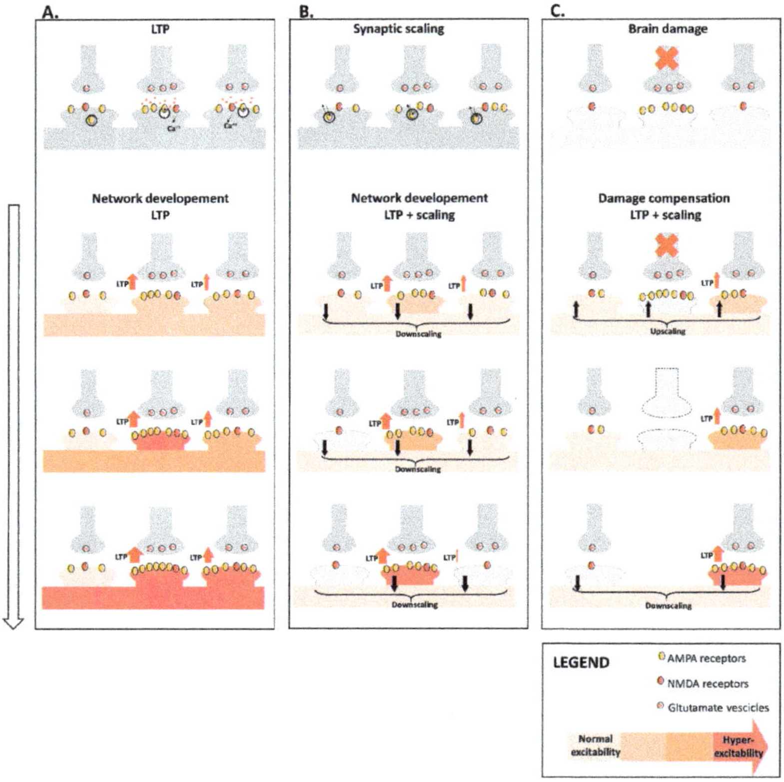

Figure 1. Different forms of synaptic plasticity cooperate to regulate neuronal activity. (**A**) LTP is input specific as it involves only active synapses. This form of NMDA-mediated plasticity implies calcium entrance in the post-synaptic terminal, which in turn induces amplified expression of AMPA receptors and increase of synaptic excitability, favoring further LTP expression. LTP is an anti-homeostatic form of plasticity and could promote uncontrolled enhancement of synaptic activity, leading to neuronal hyperexcitability and network instability during brain networks development. (**B**) Synaptic scaling is a homeostatic form of plasticity independent of NMDA receptors activation, involves all synapses and is mediated by increased (upscaling) or decreased (downscaling) expression of AMPA receptors. A balance between anti-homeostatic and homeostatic plasticity could promote optimal network organization associated to efficient information processing, with coexistence of potentiated and silent synapses (grey spines), allowing specific and segregated information processing, preventing excessive increase of overall excitability. (**C**) Brain damage induces acute disconnection depriving neurons of their synaptic inputs. Synaptic scaling and LTP may cooperate to restore neuronal excitability, promoting initial hyperexcitability (induced by compensatory upscaling) and favoring chronic reorganization properly balancing homeostatic and anti-homeostatic plasticity. Abbreviations: long-term potentiation (LTP); *N*-methyl-D-aspartate (NMDA); α-amino-3-hydroxy-5-methyl-4-isoxazolepropionic acid (AMPA).

Homeostatic and anti-homeostatic plasticity may cooperate to maintain proper neuronal activity, preventing hyper- or hypoexcitability and concur to reestablish neuronal activity after brain damage (Figure 1C). LTP induction can be influenced by overall neuronal excitability so that synaptic upscaling or changes in inhibitory balance may increase synaptic activity favoring LTP expression [53,54]. Accordingly, after brain damage, reduced GABAergic transmission in surrounding neurons could promote functional reorganization [55,56]. Furthermore, downscaling of excitatory synapses is important to prevent chronic hyperexcitability and to promote, in concert with LTP, selective increases of synaptic activity.

4. Synaptic Plasticity and Brain Network Organization

Addressing the relationship between synaptic plasticity and brain network organization is particularly difficult due to multiple reciprocal influences between brain network structure and function. Indeed, network architecture strongly influences neuronal activity [57–64], and patterns of neuronal activity may differently shape synaptic connections.

A fundamental characteristic of neuronal networks is the ability to produce rhythmic oscillations in different frequency ranges, providing integration of brain functioning in physiological conditions such as those occurring during sleep and awake [65,66]. In particular, temporal and spatial variations of frequency are useful to obtain coordinated information processing during sensory, motor and cognitive activities which are subserved by synchronous oscillations of neuronal networks [67–69]. The excitability state of neurons changes during oscillations so that firing probability is higher during the depolarizing phase, whereas during the hyperpolarizing phase, neurons show less propensity to fire in response to excitatory inputs [70,71]. Synchronous bursting of neuronal population may induce long-lasting changes in connectivity. In particular, high-frequency bursting induces LTP, whereas low-frequency activity is associated to LTD [35,72–74]. Furthermore, the temporal correlation between converging inputs to neurons can bidirectionally modulate synaptic strength according to the so-called spike timing-dependent plasticity (STDP). In particular, the firing of the presynaptic neuron can respectively induce LTP or LTD if occurring shortly before or after the firing of the postsynaptic neuron [75,76]. STDP provides an additional rule for Hebbian plasticity based on the temporal association of converging activity, bridging brain network organization and neuronal activity. This mechanism is required to form neuronal communities showing high connectivity and strongly coordinated activity during specific processing [77]. Oscillatory activity and STDP interact to shape effective coupling between anatomically connected areas. Intriguingly, the degree of synchrony of neuronal discharges and neuronal firing rate could be independently adjusted [78]. Accordingly, neurons cycling in phase with each other tend to show synchronized activity favoring synaptic LTP [79]

Previous synaptic history and state of neuronal excitability further complicate the relationship between neuronal activity and connectivity. It has been demonstrated that repeated synaptic activation may influence subsequent induction of Hebbian plasticity. In particular, considering that low and prolonged calcium entry is associated to LTD whereas high calcium influx likely mediates LTP [80], it has been proposed that changes in calcium levels into dendritic spines can modify plasticity induction [81].

Similarities existing between synaptic plasticity mechanisms and specific features of brain network organization suggest that different forms of plasticity could be directly involved in generating specific brain networks characteristics. LTP is anti-homeostatic, input-specific, activity-dependent and associative. Due to its properties, LTP could be directly implicated in generating highly connected nodes, allowing the establishment of strong and specific connections by independently acting at each synapse [82]. In particular, the preferential attachment theory of hub formation suggests the existence of an associative, positive feedback mechanisms which strongly follow the Hebbian plasticity rules.

Experimental studies in rats have elegantly shown that LTP induction in the perforant pathway induced remodeling of hippocampal long-range connections [83,84]. In particular, increased interhemispheric communication and increased connectivity has been found between the hippocampus, the prefrontal cortex and the nucleus accumbens [83,84]. These data suggest that network effects of LTP induction at hubs' level is associated to long-lasting widespread network remodeling of brain connectivity.

It has been consistently shown that the isolated effect of anti-homeostatic positive feedback plasticity leads to perturbation in the stability of neuronal networks, that must be counterbalanced by negative feedback homeostatic plasticity to maintain network activity in an optimal range [9,85]. While LTP is anti-homeostatic and represents a possible substrate for hubs generation, we propose that homeostatic plasticity, in particular synaptic scaling, may intervene to maintain low connectivity (but still some connectivity) in the peripheral nodes of the network.

The study of brain network reorganization in response to brain damage could help to shed light on the relationship between synaptic plasticity and brain network remodeling. Particular neuropsychiatric conditions, such as schizophrenia and AD, in which altered plasticity is one main neurophysiological feature [86,87], may represent useful models to investigate how synaptic plasticity alterations impact brain network architecture. Notably, the central role of plasticity alteration has also been proposed in other neurological conditions such as temporal lobe epilepsy. In epileptic models, the recurrence of seizures has been associated with imbalanced excitatory and inhibitory synaptic transmission leading to hypersynchronized neuronal activity [88,89], and epileptogenesis has been linked to altered expression of hippocampal LTP and LTD at glutamatergic and GABAergic synapses, respectively [90].

4.1. *Synaptic Plasticity Promotes Brain Network Reorganization after Damage*

Both brain network architecture and synaptic plasticity play an important role in clinical compensation of brain damage. As previously discussed, specific characteristics of brain networks organization provide high resistance to random damage [5]. Notably, anti-homeostatic and homeostatic plasticity could be both involved in promoting an efficient network reorganization after brain damage [91,92].

Experimental studies pointed out that the efficiency of synaptic plasticity, and particularly of LTP, critically influences clinical recovery (Figure 2). In animal models of brain damage (i.e., focal ischemia), symptoms compensation relies on the ability of surviving neurons to increase their excitability, as shown by a positive correlation between improvement in clinical scores and increased excitatory glutamatergic transmission in perilesional area [37]. Synaptic plasticity can be explored non-invasively in humans using transcranial magnetic stimulation (TMS). It has been demonstrated that the amount of LTP-like plasticity inducible with different TMS protocols after brain damage, the so-called LTP reserve, correlated with the degree of clinical recovery [93,94]. These results strongly suggest that LTP, specifically enhancing synaptic efficacy, is a fundamental requisite for network remodeling after brain damage.

Widespread increase in brain functional connectivity represents a common response to different types of damage, including traumatic brain injury [95–97], stroke [98], Parkinson's disease [99,100] and mild cognitive impairment (MCI) [101,102]. This early-phase adaptation could be useful to counterbalance connectivity decline and restore network functionality, delaying symptoms onset. It has been recently proposed that selective remodeling of hub connectivity could represent an efficient mechanism to restore network activity containing the wiring cost. The central role played by LTP in clinical recovery agrees with the role of hubs as the preferential site for connectivity increases, according to the positive feedback nature of both phenomena. This pattern of reorganization requires normal functioning network hubs, and a prominent involvement of hubs has been accordingly evidenced in different neurological conditions [8].

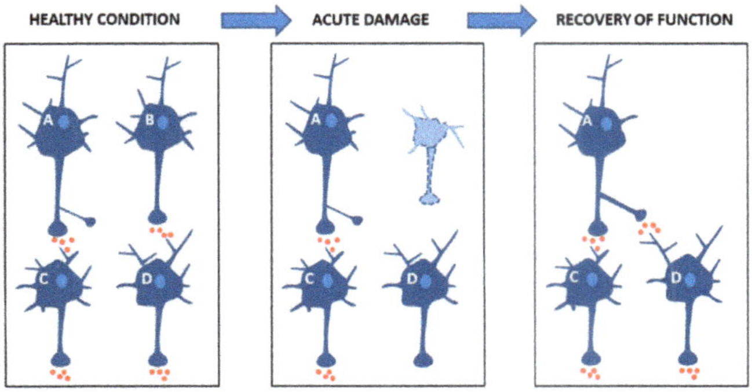

Figure 2. Synaptic plasticity promotes recovery after neural damage. Healthy condition: a schematic model representing neuronal excitatory connections. Neuron C and D receive synaptic excitatory inputs from neurons A and B respectively. Acute damage: damage to neuron B deprives neuron D of excitatory synaptic input leading to disconnection and symptoms appearance. Recovery of function: clinical recovery is associated to increased excitability of the surviving A neuron that unmasks latent synaptic connections through LTP and restores synaptic activity of neuron D.

Importantly, connectivity increases should be tightly regulated to preserve an optimal tradeoff between cost and efficiency [8]. Homeostatic plasticity cooperates with LTP to determine optimal brain network reorganization in both acute and chronic remodeling. In the early phases, synaptic upscaling could induce widespread hyperexcitability favoring network hyperconnectivity, with the aim of partially restoring network efficiency. Moreover, neuronal hyperexcitability favors LTP induction at hub level, further increasing hubs connectivity. Homeostatic plasticity changes may therefore regulate Hebbian plasticity expression. This could be particularly relevant also in the late phases of network reorganization when efficient downscaling is needed to limit excessive connectivity increase, preventing chronic diffuse hyperconnectivity and selectively shaping hub remodeling.

Another line of evidence, strongly suggesting a strict relationship between LTP induction and hub remodeling, arises from the paradigm of cognitive reserve. Accordingly, higher levels of education, cognitive abilities, occupation and physical activity have been correlated with reduced functional impact of brain structural damage as demonstrated in healthy aging subjects and in neurological patients [103–107]. In preclinical studies, environmental enrichment with physical, cognitive, and social stimuli improved the performance in different behavioral tasks exploring memory and learning [108] and enhanced LTP induction [109–113].

In humans, cognitive reserve has been linked to increased connectivity of hub regions. In healthy elder subjects, higher cognitive reserve correlated with increased metabolism and functional connectivity of the anterior cingulate cortex [114]. Similarly, higher cognitive reserve has been associated with enhanced functional connectivity in the left frontal cortex and reduced cognitive impairment in MCI and AD patients [115,116]. It has been proposed that cognitive reserve may promote brain network resilience increasing hubs connectivity, thus enhancing the resistance of hubs to damage [117,118].

4.2. Synaptic Plasticity Dysfunction May Drive Brain Network Disruption

AD and schizophrenia could represent useful models to explore the relationship between LTP expression and hubs connectivity. In particular, in AD and schizophrenia, impaired plasticity [87,119] may be responsible for reduced hubs degree and centrality, and decreased rich club connectivity [7, 31,120]. In particular, impaired synaptic plasticity alters the synchrony of both local and distributed neuronal oscillations and could promote brain network dysfunction [69,121,122].

AD is a neurodegenerative disease characterized by accumulation of amyloid-β (Aβ) and tau protein [123] associated with prominent cognitive decline [124]. In the hippocampus of AD patients synaptic alterations have been evidenced since the early phases of the disease [125,126]. In particular, it has been proposed that early synaptic plasticity impairment could represent a main cause of memory deficits in AD, even independently of neurodegeneration [86,119]. Studies in animal models of AD documented lacking hippocampal LTP induction [127–129], and pathological LTD enhancement [130]. Accordingly, it has been observed that elevated levels of soluble Aβ oligomers could reduce LTP and promote LTD expression in the hippocampus [131–134]. It has been suggested that also impaired homeostatic plasticity could contribute to the clinical manifestations and disease progression in AD. In particular, altered interaction between homeostatic and anti-homeostatic plasticity in AD could ultimately promote synaptic loss [135,136]. In line with experimental data, in early AD patients TMS studies have confirmed that LTP-like plasticity is abolished, and LTD-like plasticity induction is favored [86,119].

Reduced small world topology and altered connectivity, particularly in associative areas, have been reported in AD [7,15,137–142]. In particular, it has been shown a specific involvement of hub regions [7,30,142,143] with decreased centrality of the hippocampus and the default mode network [31]. In addition, impaired hubs connectivity has been correlated with worse cognitive performance and reduced CSF levels of Aβ 1–42 [31]. It has been evidenced that hubs show increased Aβ deposition in MCI, AD and even in older healthy subjects [120,141,144,145]. In fact, a growing body of evidence leads to the hypothesis that chronic hyperconnectivity and enhanced neuronal activity could expose hubs to Aβ deposition and neurodegeneration [8,143]. Altered synaptic plasticity expression could explain the peculiar hubs vulnerability described in AD. In particular, impaired LTP and favored LTD may specifically disrupt hubs connectivity leading to compensatory maladaptive upscaling in healthy neighboring neurons, resulting in chronic hyperexcitability [135].

Schizophrenia is a highly disabling psychiatric condition characterized by positive and negative symptoms, including hallucinations and delusions, with most of the patients showing a progressive clinical decline [146]. The pathogenesis of schizophrenia has been classically linked to neurotransmitters alteration [147], neurodevelopmental disorders [148] and disconnection [149].

Recently, disrupted synaptic plasticity has been proposed as a possible pathophysiological marker of schizophrenia [87,150,151]. Accordingly, reduced spine density has been described in the prefrontal and temporal cortices of schizophrenic patients [152–155], and altered expression and function of NMDARs and AMPARs have been reported [156–160]. NMDARs dysfunction seems to be particularly relevant, as NMDAR antagonists could produce symptoms which strongly resemble schizophrenia manifestations [161]. Impaired LTP and LTD-like plasticity has been consistently reported in patients with schizophrenia [150,151,162]. Using a TMS protocol useful to explore cortical connectivity and in particular spike timing-dependent plasticity [163,164], reduced LTP-like plasticity has been shown between posterior parietal and frontal cortices in schizophrenia compared to control subjects [87].

Previous studies exploring brain network organization in schizophrenia showed alterations of several properties [165–170]. In particular, reduced hub connectivity [171] and rich club organization have been reported in schizophrenic patients [172]. Decreased connectivity within the frontal cortex has been considered a pathophysiological hallmark and decreased centrality in cortical and subcortical frontal areas has been reported accordingly [166,173,174]. In line with the disconnection hypothesis, impaired hub connectivity and reduced rich club efficiency may alter overall brain connectivity in schizophrenia [149].

Changes in synaptic plasticity expression may explain the hubs loss seen in schizophrenia. It has been suggested that spike timing-dependent plasticity alterations may be responsible of progressive disconnection of prefrontal circuits [175]. Altered NMDA-mediated synaptic plasticity reduces temporal correlation between converging inputs to connected neurons and could lead to additional activity-dependent disconnection of prefrontal networks [175]. In fact, subverted spike timing-dependent plasticity could induce LTD instead of LTP in prefrontal networks, ultimately producing progressive disruption of anterior hubs and generating a persistent disconnection within the rich club.

5. Conclusions

Synaptic plasticity mechanisms and specific features of brain network share common principles that contribute to explain how neural plasticity influences brain network organization. Indeed, different forms of synaptic plasticity could be directly involved in generating specific brain networks' characteristics. A cooperative, associative, input-specific and anti-homeostatic Hebbian plasticity is well suited to form brain networks characterized by modules and hubs, providing segregation and integration of information. LTP could be implicated in generating highly connected nodes that are crucially involved in network remodeling after brain damage and also represent the specific target of pathophysiological processes in different neuropsychiatric conditions. Conversely, homeostatic forms of synaptic plasticity intervene to prevent excessive connectivity in the peripheral nodes, stabilizing network activity and preventing excessive cost-efficiency increase. Finally, the fine tuning between homeostatic and anti-homeostatic plasticity plays a key role in recovery after damage and may help to understand how brain networks reorganize in response to different neurological conditions. Further studies combining neurophysiological investigations and fMRI measures are required to better define the relationship between synaptic plasticity and brain network topology.

Author Contributions: Conceptualization, M.S.B, L.G., D.C. and F.B.; methodology, M.S.B, D.C. and F.B.; writing—original draft preparation, M.S.B. and E.I.; writing—review and editing, M.S.B., E.I., L.G., D.C. and F.B.; visualization, M.S.B., E.I. and L.G.; supervision, D.C.; funding acquisition, D.C.

Funding: This research was funded by the Italian Ministry of Health (Ricerca Corrente), by the 5 × 1000 grant to IRCCS Neuromed, and by the Fondazione Italiana Sclerosi Multipla (FISM) COD. 2019/S/1 to DC.

Conflicts of Interest: The author(s) declared the following potential conflicts of interest with respect to the research, authorship, and/or publication of this article: F.B. acted as Advisory Board members of Teva and Roche and received honoraria for speaking or consultation fees from Merck Serono, Teva, Biogen Idec, Sanofi, and Novartis and non-financial support from Merck Serono, Teva, Biogen Idec, and Sanofi. D.C. is an Advisory Board member of Almirall, Bayer Schering, Biogen, GW Pharmaceuticals, Merck Serono, Novartis, Roche, Sanofi-Genzyme, and Teva and received honoraria for speaking or consultation fees from Almirall, Bayer Schering, Biogen, GW Pharmaceuticals, Merck Serono, Novartis, Roche, Sanofi-Genzyme, and Teva. He is also the principal investigator in clinical trials for Bayer Schering, Biogen, Merck Serono, Mitsubishi, Novartis, Roche, Sanofi-Genzyme, and Teva. His preclinical and clinical research was supported by grants from Bayer Schering, Biogen Idec, Celgene, Merck Serono, Novartis, Roche, Sanofi-Genzyme and Teva. The funders had no role in the design of the study; in the collection, analyses, or interpretation of data; in the writing of the manuscript; or in the decision to publish the results. M.S.B.: E.I. and L.G., declare no conflict of interest.

Abbreviations

EEG	Electroencephalography
MEG	Magnetoencephalography
FC	Functional Connectivity
fMRI	Functional MRI
rs-fMRI	Resting state-fMRI
BOLD	Blood-Oxygen-Level-Dependent
SC	Structural Connectivity
DTI	Diffusion Tensor Imaging
P(k)	Degree Distribution of a Graph
AD	Alzheimer's disease
LTP	Long term potentiation
NMDARs	N-methyl-D-aspartate receptors
LTD	Long-Term Depression
AMPARs	α-amino-3-hydroxy-5-methyl-4-isoxazolepropionic acid receptors
STDP	Spike Timing-Dependent Plasticity
TMS	Transcranial Magnetic Stimulation
MCI	Mild Cognitive Impairment
Aβ	Amyloid-β

References

1. Sporns, O.; Tononi, G.; Kötter, R. The human connectome: A structural description of the human brain. *PLoS Comput. Biol.* **2005**, *1*, e42. [CrossRef] [PubMed]
2. Fox, M.D.; Raichle, M.E. Spontaneous fluctuations in brain activity observed with functional magnetic resonance imaging. *Nat. Rev. Neurosci.* **2007**, *8*, 700–711. [CrossRef] [PubMed]
3. Watts, D.J.; Strogatz, S.H. Collective dynamics of 'small-world' networks. *Nature* **1998**, *393*, 440–442. [CrossRef]
4. Bullmore, E.; Sporns, O. Complex brain networks: Graph theoretical analysis of structural and functional systems. *Nat. Rev. Neurosci.* **2009**, *10*, 186–198. [CrossRef] [PubMed]
5. Achard, S.; Bullmore, E. Efficiency and cost of economical brain functional networks. *PLoS Comput. Biol.* **2007**, *3*, e17. [CrossRef] [PubMed]
6. Rubinov, M.; Knock, S.A.; Stam, C.J.; Micheloyannis, S.; Harris, A.W.; Williams, L.M.; Breakspear, M. Small-world properties of nonlinear brain activity in schizophrenia. *Hum. Brain Mapp.* **2009**, *30*, 403–416. [CrossRef]
7. Stam, C.J.; de Haan, W.; Daffertshofer, A.; Jones, B.F.; Manshanden, I.; Van Cappellen van Walsum, A.M.; Montez, T.; Verbunt, J.P.; de Munck, J.C.; van Dijk, B.W.; et al. Graph theoretical analysis of magnetoencephalographic functional connectivity in Alzheimer's disease. *Brain* **2009**, *132*, 213–224. [CrossRef]
8. Hillary, F.G.; Grafman, J.H. Injured Brains and Adaptive Networks: The Benefits and Costs of Hyperconnectivity. *Trends. Cogn. Sci.* **2017**, *21*, 385–401. [CrossRef]
9. Turrigiano, G.G.; Nelson, S.B. Homeostatic plasticity in the developing nervous system. *Nat. Rev. Neurosci.* **2004**, *5*, 97–107. [CrossRef]
10. Varshney, L.R.; Chen, B.L.; Paniagua, E.; Hall, D.H.; Chklovskii, D.B. Structural properties of the Caenorhabditis elegans neuronal network. *PLoS Comput. Biol.* **2011**, *7*, e1001066. [CrossRef]
11. Shih, C.T.; Sporns, O.; Yuan, S.L.; Su, T.S.; Lin, Y.J.; Chuang, C.C.; Wang, T.Y.; Lo, C.C.; Greenspan, R.J.; Chiang, A.S. Connectomics-based analysis of information flow in the Drosophila brain. *Curr. Biol.* **2015**, *25*, 1249–1258. [CrossRef] [PubMed]
12. Jensen, O.; Spaak, E.; Zumer, J.M. Human brain oscillations: From physiological mechanisms to analysis and cognition. In *Magnetoencephalography: From Signals to Dynamic Cortical Networks*; Supek, S., Aine, C.J., Eds.; Springer: Berlin, Germany, 2014; pp. 359–403.
13. Rossini, P.M.; Di Iorio, R.; Bentivoglio, M.; Bertini, G.; Ferreri, F.; Gerloff, C.; Ilmoniemi, R.J.; Miraglia, F.; Nitsche, M.A.; Pestilli, F.; et al. Methods for analysis of brain connectivity: An IFCN-sponsored review. *Clin. Neurophysiol.* **2019**, *130*, 1833–1858. [CrossRef] [PubMed]
14. Bassett, D.S.; Bullmore, E.T.; Verchinski, B.A.; Mattay, V.S.; Weinberger, D.R.; Meyer-Lindenberg, A. Hierarchical organization of human cortical networks in health and schizophrenia. *J. Neurosci.* **2008**, *28*, 9239–9248. [CrossRef] [PubMed]
15. He, Y.; Chen, Z.; Evans, A. Structural insights into aberrant topological patterns of large-scale cortical networks in Alzheimer's disease. *J. Neurosci.* **2008**, *28*, 4756–4766. [CrossRef]
16. Biswal, B.B.; Yetkin, F.Z.; Haughton, V.M.; Hyde, J.S. Functional connectivity in the motor cortex of resting human brain using echo-planar MRI. *Magn. Reson. Med.* **1995**, *34*, 537–541. [CrossRef]
17. Honey, C.J.; Kötter, R.; Breakspear, M.; Sporns, O. Network structure of cerebral cortex shapes functional connectivity on multiple time scales. *Proc. Natl. Acad. Sci. USA* **2007**, *104*, 10240–10245. [CrossRef]
18. Rubinov, M.; Sporns, O.; van Leeuwen, C.; Breakspear, M. Symbiotic relationship between brain structure and dynamics. *BMC Neurosci.* **2009**, *10*, 55. [CrossRef]
19. Honey, C.J.; Sporns, O.; Cammoun, L.; Gigandet, X.; Thiran, J.P.; Meuli, R.; Hagmann, P. Predicting human resting-state functional connectivity from structural connectivity. *Proc. Natl. Acad. Sci. USA* **2009**, *106*, 2035–2040. [CrossRef]
20. Rubinov, M.; Sporns, O. Complex network measures of brain connectivity: Uses and interpretations. *Neuroimage* **2010**, *52*, 1059–1069. [CrossRef]
21. Smith, S.M.; Miller, K.L.; Salimi-Khorshidi, G.; Webster, M.; Beckmann, C.F.; Nichols, T.E.; Ramsey, J.D.; Woolrich, M.W. Network modeling methods for FMRI. *Neuroimage* **2011**, *54*, 875–891. [CrossRef]

22. Friston, K.J. Functional and effective connectivity: A review. *Brain Connect.* **2011**, *1*, 13–36. [CrossRef] [PubMed]
23. Stam, C.J.; Reijneveld, J.C. Graph theoretical analysis of complex networks in the brain. *Nonlinear Biomed. Phys.* **2007**, *1*, 3. [CrossRef] [PubMed]
24. Bullmore, E.; Sporns, O. The economy of brain network organization. *Nat. Rev. Neurosci.* **2012**, *13*, 336–349. [CrossRef] [PubMed]
25. Sporns, O.; Honey, C.J.; Kötter, R. Identification and classification of hubs in brain networks. *PLoS ONE* **2007**, *2*, e1049. [CrossRef] [PubMed]
26. Van den Heuvel, M.P.; Sporns, O. An anatomical substrate for integration among functional networks in human cortex. *J. Neurosci.* **2013**, *33*, 14489–14500. [CrossRef] [PubMed]
27. Sporns, O. Network attributes for segregation and integration in the human brain. *Curr. Opin. Neurobiol.* **2013**, *23*, 162–171. [CrossRef] [PubMed]
28. Barabasi, A.L.; Albert, R. Emergence of scaling in random networks. *Science* **1999**, *286*, 509–512. [CrossRef]
29. Achard, S.; Salvador, R.; Whitcher, B.; Suckling, J.; Bullmore, E. A resilient, low-frequency, small-world human brain functional network with highly connected association cortical hubs. *J. Neurosci.* **2006**, *26*, 63–72. [CrossRef]
30. Crossley, N.A.; Mechelli, A.; Scott, J.; Carletti, F.; Fox, P.T.; McGuire, P.; Bullmore, E.T. The hubs of the human connectome are generally implicated in the anatomy of brain disorders. *Brain* **2014**, *137*, 2382–2395. [CrossRef]
31. Yu, M.; Engels, M.M.A.; Hillebrand, A.; van Straaten, E.C.W.; Gouw, A.A.; Teunissen, C.; van der Flier, W.M.; Scheltens, P.; Stam, C.J. Selective impairment of hippocampus and posterior hub areas in Alzheimer's disease: An MEG-based multiplex network study. *Brain* **2017**, *140*, 1466–1485. [CrossRef]
32. Rubinov, M.; Bullmore, E. Fledgling pathoconnectomics of psychiatric disorders. *Trends Cogn. Sci.* **2013**, *17*, 641–647. [CrossRef] [PubMed]
33. Klauser, P.; Baker, S.T.; Cropley, V.L.; Bousman, C.; Fornito, A.; Cocchi, L.; Fullerton, J.M.; Rasser, P.; Schall, U.; Henskens, F.; et al. White matter disruptions in schizophrenia are spatially widespread and topologically converge on brain network hubs. *Schizophr. Bull.* **2016**, *43*, 425–435. [CrossRef] [PubMed]
34. Citri, A.; Malenka, R.C. Synaptic plasticity: Multiple forms, functions, and mechanisms. *Neuropsychopharmacology* **2008**, *33*, 18–41. [CrossRef] [PubMed]
35. Bliss, T.V.; Lømo, T. Long-lasting potentiation of synaptic transmission in the dentate area of the anaesthetized rabbit following stimulation of the perforant path. *J. Physiol.* **1973**, *232*, 331–356. [CrossRef]
36. Morris, R.G.; Anderson, E.; Lynch, G.S.; Baudry, M. Selective impairment of learning and blockade of long-term potentiation by an N-methyl-D-aspartate receptor antagonist, AP5. *Nature* **1986**, *319*, 774–776 [CrossRef]
37. Centonze, D.; Rossi, S.; Tortiglione, A.; Picconi, B.; Prosperetti, C.; De Chiara, V.; Bernardi, G.; Calabresi, P. Synaptic plasticity during recovery from permanent occlusion of the middle cerebral artery. *Neurobiol. Dis.* **2007**, *27*, 44–53. [CrossRef]
38. Collingridge, G.L.; Kehl, S.J.; McLennan, H. Excitatory amino acids in synaptic transmission in the Schaffer collateral-commissural pathway of the rat hippocampus. *J. Physiol.* **1983**, *334*, 33–46. [CrossRef]
39. Malenka, R.C.; Kauer, J.A.; Zucker, R.S.; Nicoll, R.A. Postsynaptic calcium is sufficient for potentiation of hippocampal synaptic transmission. *Science* **1988**, *242*, 81–84. [CrossRef]
40. Engert, F.; Bonhoeffer, T. Dendritic spine changes associated with hippocampal long-term synaptic plasticity. *Nature* **1999**, *399*, 66–70. [CrossRef]
41. De Roo, M.; Klauser, P.; Mendez, P.; Poglia, L.; Muller, D. Activity-dependent PSD formation and stabilization of newly formed spines in hippocampal slice cultures. *Cereb. Cortex* **2008**, *18*, 151–161. [CrossRef]
42. Hill, T.C.; Zito, K. LTP-induced long-term stabilization of individual nascent dendritic spines. *J. Neurosci.* **2013**, *33*, 678–686. [CrossRef] [PubMed]
43. Bliss, T.V.; Collingridge, G.L. A synaptic model of memory: Long-term potentiation in the hippocampus. *Nature* **1993**, *361*, 31–39. [CrossRef] [PubMed]
44. Harris, E.W.; Ganong, A.H.; Cotman, C.W. Long-term potentiation in the hippocampus involves activation of N-methyl-D-aspartate receptors. *Brain Res.* **1984**, *323*, 132–137. [CrossRef]
45. Malinow, R.; Schulman, H.; Tsien, R.W. Inhibition of postsynaptic PKC or CaMKII blocks induction but not expression of LTP. *Science* **1989**, *245*, 862–866. [CrossRef]

46. Malenka, R.C.; Bear, M.F. LTP and LTD: An embarrassment of riches. *Neuron* **2004**, *44*, 5–21. [CrossRef]
47. Tada, T.; Sheng, M. Molecular mechanisms of dendritic spine morphogenesis. *Curr. Opin. Neurobiol.* **2006**, *16*, 95–101. [CrossRef]
48. Zhou, Q.; Homma, K.J.; Poo, M.M. Shrinkage of dendritic spines associated with long-term depression of hippocampal synapses. *Neuron* **2004**, *44*, 749–757. [CrossRef]
49. Turrigiano, G.G.; Leslie, K.R.; Desai, N.S.; Rutherford, L.C.; Nelson, S.B. Activity-dependent scaling of quantal amplitude in neocortical neurons. *Nature* **1998**, *391*, 892–896. [CrossRef]
50. Lissin, D.V.; Gomperts, S.N.; Carroll, R.C.; Christine, C.W.; Kalman, D.; Kitamura, M.; Hardy, S.; Nicoll, R.A.; Malenka, R.C.; von Zastrow, M. Activity differentially regulates the surface expression of synaptic AMPA and NMDA glutamate receptors. *Proc. Natl. Acad. Sci. USA* **1998**, *95*, 7097–7102. [CrossRef]
51. Tzingounis, V.; Nicoll, R.A. Arc/Arg3.1: Linking gene expression to synaptic plasticity and memory. *Neuron* **2006**, *52*, 403–407. [CrossRef]
52. Shepherd, J.D.; Rumbaugh, G.; Wu, J.; Chowdhury, S.; Plath, N.; Kuhl, D.; Huganir, R.L.; Worley, P.F. Arc/Arg3.1 mediates homeostatic synaptic scaling of AMPA receptors. *Neuron* **2006**, *52*, 475–484. [CrossRef] [PubMed]
53. Jacobs, K.M.; Donoghue, J.P. Reshaping the cortical motor map by unmasking latent intracortical connections. *Science* **1991**, *251*, 944–947. [CrossRef] [PubMed]
54. Ziemann, U.; Rothwell, J.C.; Ridding, M.C. Interaction between intracortical inhibition and facilitation in human motor cortex. *J. Physiol.* **1996**, *496*, 873–881. [CrossRef] [PubMed]
55. Baroncelli, L.; Braschi, C.; Spolidoro, M.; Begenisic, T.; Maffei, L.; Sale, A. Brain plasticity and disease: A matter of inhibition. *Neural. Plast.* **2011**, *2011*. [CrossRef]
56. Imbrosci, B.; Mittmann, T. Functional consequences of the disturbances in the GABA-mediated inhibition induced by injuries in the cerebral cortex. *Neural. Plast.* **2011**, *2011*. [CrossRef]
57. Zhao, L.; Beverlin, B.I.; Netoff, T.; Nykamp, D.Q. Synchronization from second order network connectivity statistics. *Front. Comput. Neurosci.* **2011**, *5*, 28. [CrossRef]
58. Roxin, A. The role of degree distribution in shaping the dynamics in networks of sparsely connected spiking neurons. *Front. Comput. Neurosci.* **2011**, *5*, 8. [CrossRef]
59. Pernice, V.; Staude, B.; Cardanobile, S.; Rotter, S. How structure determines correlations in neuronal networks. *PLoS Comput. Biol.* **2011**, *7*, e1002059. [CrossRef]
60. Pernice, V.; Deger, M.; Cardanobile, S.; Rotter, S. The relevance of network micro-structure for neural dynamics. *Front. Comput. Neurosci.* **2013**, *7*, 72. [CrossRef]
61. Trousdale, J.; Hu, Y.; Shea-Brown, E.; Josić, K. Impact of network structure and cellular response on spike time correlations. *PLoS Comput. Biol.* **2012**, *8*, e1002408. [CrossRef]
62. Hu, Y.; Trousdale, J.; Josić, K.; Shea-Brown, E. Motif statistics and spike correlations in neuronal networks. *BMC Neurosci.* **2013**, *2013*, P03012. [CrossRef]
63. Hu, Y.; Trousdale, J.; Josić, K.; Shea-Brown, E. Local paths to global coherence: Cutting networks down to size. *Phys. Rev. E Stat. Nonlin. Soft Matter Phys.* **2014**, *89*, 032802. [CrossRef]
64. Helias, M.; Tetzlaff, T.; Diesmann, M. The correlation structure of local neuronal networks intrinsically results from recurrent dynamics. *PLoS Comput. Biol.* **2014**, *10*, e1003428. [CrossRef]
65. Buzsáki, G. Two-stage model of memory trace formation: A role for "noisy" brain states. *Neuroscience* **1989**, *31*, 551–570. [CrossRef]
66. Wilson, M.A.; McNaughton, B.L. Reactivation of hippocampal ensemble memories during sleep. *Science* **1994**, *29*, 676–679. [CrossRef]
67. Babiloni, C.; Vecchio, F.; Lizio, R.; Ferri, R.; Rodriguez, G.; Marzano, N.; Frisoni, G.B.; Rossini, P.M. Resting state cortical rhythms in mild cognitive impairment and Alzheimer's disease: Electroencephalographic evidence. *J. Alzheimers Dis.* **2011**, *26*, 201–214. [CrossRef]
68. Buzsáki, G.; Watson, B.O. Brain rhythms and neural syntax: Implications for efficient coding of cognitive content and neuropsychiatric disease. *Dialogues Clin. Neurosci.* **2012**, *14*, 345–367. [PubMed]
69. Uhlhaas, P.J.; Singer, W. Abnormal neural oscillations and synchrony in schizophrenia. *Nat. Rev. Neurosci.* **2010**, *11*, 100–113. [CrossRef]
70. Wang, X.J.; Buzsáki, G. Gamma oscillation by synaptic inhibition in a hippocampal interneuronal network model. *J. Neurosci.* **1996**, *16*, 6402–6413. [CrossRef]

71. Whittington, M.A.; Traub, R.D.; Jefferys, J.G. Synchronized oscillations in interneuron networks driven by metabotropic glutamate receptor activation. *Nature* **1995**, *373*, 612–615. [CrossRef]
72. Lynch, G.; Larson, J.; Kelso, S.; Barrionuevo, G.; Schottler, F. Intracellular injections of EGTA block induction of hippocampal long-term potentiation. *Nature* **1983**, *305*, 719–721. [CrossRef]
73. Stanton, P.K.; Sejnowski, T.J. Associative long-term depression in the hippocampus induced by Hebbian covariance. *Nature* **1989**, *339*, 215–218. [CrossRef]
74. Dudek, S.M.; Bear, M.F. Homosynaptic long-term depression in area CA1 of hippocampus and effects of N-methyl-D-aspartate receptor blockade. *Proc. Natl. Acad. Sci. USA* **1992**, *89*, 4363–4367. [CrossRef]
75. Markram, H.; Lübke, J.; Frotscher, M.; Sakmann, B. Regulation of synaptic efficacy by coincidence of postsynaptic APs and EPSPs. *Science* **1997**, *275*, 213–215. [CrossRef]
76. Markram, H.; Gerstner, W.; Sjostrom, P.J. A history of spike-timing-dependent plasticity. *Front. Synaptic Neurosci.* **2011**, *3*, 4. [CrossRef]
77. Harris, K.D. Neural signatures of cell assembly organization. *Nat. Rev. Neurosci.* **2005**, *6*, 399–407. [CrossRef]
78. Singer, W. Neuronal synchrony: A versatile code for the definition of relations? *Neuron* **1999**, *24*, 49–65. [CrossRef]
79. Zanos, S.; Rembado, I.; Chen, D.; Fetz, E.E. Phase-Locked Stimulation during Cortical Beta Oscillations Produces Bidirectional Synaptic Plasticity in Awake Monkeys. *Curr. Biol.* **2018**, *28*, R879–R882. [CrossRef]
80. Nevian, T.; Sakmann, B. Spine Ca^{2+} signaling in spike-timing-dependent plasticity. *J. Neurosci.* **2006**, *26*, 11001–11013. [CrossRef]
81. Egelman, D.M.; Montague, P.R. Calcium dynamics in the extracellular space of mammalian neural tissue. *Biophys. J.* **1999**, *76*, 1856–1867. [CrossRef]
82. Abbott, L.F.; Nelson, S.B. Synaptic plasticity: Taming the beast. *Nat. Neurosci.* **2000**, *3*, 1178–1183. [CrossRef]
83. Canals, S.; Beyerlein, M.; Merkle, H.; Logothetis, N.K. Functional MRI evidence for LTP-induced neural network reorganization. *Curr. Biol.* **2009**, *19*, 398–403. [CrossRef]
84. Alvarez-Salvado, E.; Pallarés, V.; Moreno, A.; Canals, S. Functional MRI of long-term potentiation: Imaging network plasticity. *Philos. Trans. R. Soc. B Biol. Sci.* **2013**, *369*, 20130152. [CrossRef]
85. Miller, K.D. Synaptic economics: Competition and cooperation in synaptic plasticity. *Neuron* **1996**, *17*, 371–374. [CrossRef]
86. Koch, G.; Di Lorenzo, F.; Bonnì, S.; Ponzo, V.; Caltagirone, C.; Martorana, A. Impaired LTP- but not LTD-like cortical plasticity in Alzheimer's disease patients. *J. Alzheimers Dis.* **2012**, *31*, 593–599. [CrossRef]
87. Ribolsi, M.; Lisi, G.; Ponzo, V.; Siracusano, A.; Caltagirone, C.; Niolu, C.; Koch, G. Left hemispheric breakdown of LTP-like cortico-cortical plasticity in schizophrenic patients. *Clin. Neurophysiol.* **2017**, *128*, 2037–2042. [CrossRef]
88. Guerriero, R.M.; Giza, C.C.; Rotenberg, A. Glutamate and GABA imbalance following traumatic brain injury. *Curr. Neurol. Neurosci. Rep.* **2015**, *15*, 27. [CrossRef]
89. Khazipov, R.; Valeeva, G.; Khalilov, I. Depolarizing GABA and developmental epilepsies. *CNS Neurosci. Ther.* **2015**, *21*, 83–91. [CrossRef]
90. Bonansco, C.; Fuenzalida, M. Plasticity of hippocampal excitatory-inhibitory balance: Missing the synaptic control in the epileptic brain. *Neural Plast.* **2016**, *2016*, 8607038. [CrossRef]
91. Desai, N.S.; Cudmore, R.H.; Nelson, S.B.; Turrigiano, G.G. Critical periods for experience-dependent synaptic scaling in visual cortex. *Nat. Neurosci.* **2002**, *5*, 783–789. [CrossRef]
92. Turrigiano, G. Homeostatic synaptic plasticity: Local and global mechanisms for stabilizing neuronal function. *Cold. Spring Harb. Perspect. Biol.* **2012**, *4*, a005736. [CrossRef]
93. Di Lazzaro, V.; Profice, P.; Pilato, F.; Capone, F.; Ranieri, F.; Pasqualetti, P.; Colosimo, C.; Pravatà, E.; Cianfoni, A.; Dileone, M. Motor cortex plasticity predicts recovery in acute stroke. *Cereb. Cortex* **2010**, *20*, 1523–1528. [CrossRef]
94. Mori, F.; Kusayanagi, H.; Nicoletti, C.G.; Weiss, S.; Marciani, M.G.; Centonze, D. Cortical plasticity predicts recovery from relapse in multiple sclerosis. *Mult. Scler.* **2014**, *20*, 451–457. [CrossRef] [PubMed]
95. Hillary, F.G.; Rajtmajer, S.M.; Roman, C.A.; Medaglia, J.D.; Slocomb-Dluzen, J.E.; Calhoun, V.D.; Good, D.C.; Wylie, G.R. The rich get richer: Brain injury elicits hyperconnectivity in core subnetworks. *PLoS ONE* **2014**, *9*, e113545. [CrossRef] [PubMed]

96. Iraji, A.; Benson, R.R.; Welch, R.D.; O'Neil, B.J.; Woodard, J.L.; Ayaz, S.I.; Kulek, A.; Mika, V.; Medado, P.; Soltanian-Zadeh, H.; et al. Resting state functional connectivity in mild traumatic brain injury at the acute stage: Independent component and seed-based analyses. *J. Neurotrauma* **2016**, *32*, 1031–1045. [CrossRef]
97. Bharath, R.D.; Munivenkatappa, A.; Gohel, S.; Panda, R.; Saini, J.; Rajeswaran, J.; Shukla, D.; Bhagavatula, I.D.; Biswal, B.B. Recovery of resting brain connectivity ensuing mild traumatic brain injury. *Front. Hum. Neurosci.* **2015**, *9*, 513. [CrossRef] [PubMed]
98. Nicolo, P.; Rizk, S.; Magnin, C.; Pietro, M.D.; Schnider, A.; Guggisberg, A.G. Coherent neural oscillations predict future motor and language improvement after stroke. *Brain* **2015**, *138*, 3048–3060. [CrossRef]
99. Gorges, M.; Müller, H.P.; Lulé, D.; LANDSCAPE Consortium; Pinkhardt, E.H.; Ludolph, A.C.; Kassubek, J. To rise and to fall: Functional connectivity in cognitively normal and cognitively impaired patients with Parkinson's disease. *Neurobiol. Aging* **2014**, *36*, 1727–1735. [CrossRef]
100. Fernández-Seara, M.A.; Mengual, E.; Vidorreta, M.; Castellanos, G.; Irigoyen, J.; Erro, E.; Pastor, M.A. Resting state functional connectivity of the subthalamic nucleus in Parkinson's disease assessed using arterial spin-labeled perfusion fMRI. *Hum. Brain Mapp.* **2015**, *36*, 1937–1950. [CrossRef]
101. Cohen, A.D.; Price, J.C.; Weissfeld, L.A.; James, J.; Rosario, B.L.; Bi, W.; Nebes, R.D.; Saxton, J.A.; Snitz, B.E.; Aizenstein, H.A.; et al. Basal Cerebral Metabolism May Modulate the Cognitive Effects of Aβ in Mild Cognitive Impairment: An Example of Brain Reserve. *J. Neurosci.* **2009**, *29*, 14770–14778. [CrossRef]
102. Gour, N.; Ranjeva, J.P.; Ceccaldi, M.; Confort-Gouny, S.; Barbeau, E.; Soulier, E.; Guye, M.; Didic, M.; Felician, O. Basal functional connectivity within the anterior temporal network is associated with performance on declarative memory tasks. *Neuroimage* **2011**, *58*, 687–697. [CrossRef] [PubMed]
103. Stern, Y. Cognitive reserve in ageing and Alzheimer's disease. *Lancet Neurol.* **2012**, *11*, 1006–1012. [CrossRef]
104. Valenzuela, M.J.; Sachdev, P. Brain reserve and dementia: A systematic review. *Psychol. Med.* **2006**, *36*, 441–454. [CrossRef]
105. Okonkwo, O.C.; Schultz, S.A.; Oh, J.M.; Larson, J.; Edwards, D.; Cook, D.; Koscik, R.; Gallagher, C.L.; Dowling, N.M.; Carlsson, C.M.; et al. Physical activity attenuates age-related biomarker alterations in preclinical AD. *Neurology* **2014**, *83*, 1753–1760. [CrossRef] [PubMed]
106. Tolppanen, A.M.; Solomon, A.; Kulmala, J.; Kåreholt, I.; Ngandu, T.; Rusanen, M.; Laatikainen, T.; Soininen, H.; Kivipelto, M. Leisure-time physical activity from mid- to late life, body mass index, and risk of dementia. *Alzheimers Dement.* **2015**, *11*, 434–443. [CrossRef] [PubMed]
107. Duzel, E.; van Praag, H.; Sendtner, M. Can physical exercise in old age improve memory and hippocampal function? *Brain* **2016**, *139*, 662–673. [CrossRef] [PubMed]
108. Frick, K.M.; Stearns, N.A.; Pan, J.Y.; Berger-Sweeney, J. Effects of environmental enrichment on spatial memory and neurochemistry in middle-aged mice. *Learn. Mem.* **2003**, *10*, 187–198. [CrossRef]
109. Leggio, M.G.; Mandolesi, L.; Federico, F.; Spirito, F.; Ricci, B.; Gelfo, F.; Petrosini, L. Environmental enrichment promotes improved spatial abilities and enhanced dendritic growth in the rat. *Behav. Brain Res.* **2005**, *163*, 78–90. [CrossRef]
110. Malik, R.; Chattarji, S. Enhanced intrinsic excitability and EPSP-spike coupling accompany enriched environment-induced facilitation of LTP in hippocampal CA1 pyramidal neurons. *J. Neurophysiol.* **2012**, *107*, 1366–1378. [CrossRef]
111. Hullinger, R.; O'Riordan, K.; Burger, C. Environmental enrichment improves learning and memory and long-term potentiation in young adult rats through a mechanism requiring mGluR5 signaling and sustained activation of p70s6k. *Neurobiol. Learn. Mem.* **2015**, *125*, 126–134. [CrossRef]
112. Stein, L.R.; O'Dell, K.A.; Funatsu, M.; Zorumski, C.F.; Izumi, Y. Short-term environmental enrichment enhances synaptic plasticity in hippocampal slices from aged rats. *Neuroscience* **2016**, *329*, 294–305. [CrossRef] [PubMed]
113. Cortese, G.P.; Olin, A.; O'Riordan, K.; Hullinger, R.; Burger, C. Environmental enrichment improves hippocampal function in aged rats by enhancing learning and memory, LTP, and mGluR5-Homer1c activity. *Neurobiol. Aging* **2018**, *63*, 1–11. [CrossRef] [PubMed]
114. Arenaza-Urquijo, E.M.; Landeau, B.; La Joie, R.; Mevel, K.; Mézenge, F.; Perrotin, A.; Desgranges, B.; Bartrés-Faz, D.; Eustache, F.; Chételat, G. Relationships between years of education and gray matter volume, metabolism and functional connectivity in healthy elders. *Neuroimage* **2013**, *83*, 450–457. [CrossRef] [PubMed]

115. Franzmeier, N.; Duering, M.; Weiner, M.; Dichgans, M.; Ewers, M.; Alzheimer's Disease Neuroimaging Initiative (ADNI). Left frontal cortex connectivity underlies cognitive reserve in prodromal Alzheimer disease. *Neurology* **2017**, *88*, 1054–1061. [CrossRef]
116. Franzmeier, N.; Düzel, E.; Jessen, F.; Buerger, K.; Levin, J.; Duering, M.; Dichgans, M.; Haass, C.; Suárez-Calvet, M.; Fagan, A.M.; et al. Left frontal hub connectivity delays cognitive impairment in autosomal-dominant and sporadic Alzheimer's disease. *Brain* **2018**, *141*, 1186–1200. [CrossRef]
117. Sumowski, J.F.; Wylie, G.R.; Deluca, J.; Chiaravalloti, N. Intellectual enrichment is linked to cerebral efficiency in multiple sclerosis: Functional magnetic resonance imaging evidence for cognitive reserve. *Brain* **2010**, *133*, 362–374. [CrossRef]
118. Santarnecchi, E.; Rossi, S.; Rossi, A. The smarter, the stronger: Intelligence level correlates with brain resilience to systematic insults. *Cortex* **2015**, 293–309. [CrossRef]
119. Di Lorenzo, F.; Ponzo, V.; Bonnì, S.; Motta, C.; Negrão Serra, P.C.; Bozzali, M.; Caltagirone, C.; Martorana, A.; Koch, G. Long-term potentiation-like cortical plasticity is disrupted in Alzheimer's disease patients independently from age of onset. *Ann. Neurol.* **2016**, *80*, 202–210. [CrossRef]
120. Buckner, R.L.; Sepulcre, J.; Talukdar, T.; Krienen, F.M.; Liu, H.; Hedden, T.; Andrews-Hanna, J.R.; Sperling, R.A.; Johnson, K.A. Cortical hubs revealed by intrinsic functional connectivity: Mapping, assessment of stability, and relation to Alzheimer's disease. *J. Neurosci.* **2009**, *29*, 1860–1873. [CrossRef]
121. Uhlhaas, P.J.; Singer, W. Neural synchrony in brain disorders: Relevance for cognitive dysfunctions and pathophysiology. *Neuron* **2006**, *52*, 155–168. [CrossRef]
122. Yener, G.G.; Başar, E. Brain oscillations as biomarkers in neuropsychiatric disorders: Following an interactive panel discussion and synopsis. *Suppl. Clin. Neurophysiol.* **2013**, *62*, 343–363. [CrossRef] [PubMed]
123. Jack, C.R., Jr.; Holtzman, D.M. Biomarker modeling of Alzheimer's disease. *Neuron* **2013**, *80*, 1347–1358. [CrossRef] [PubMed]
124. Weintraub, S.; Wicklund, A.H.; Salmon, D.P. The neuropsychological profile of Alzheimer disease. *Cold. Spring Harb. Perspect. Med.* **2012**, *2*, a006171. [CrossRef] [PubMed]
125. Masliah, E.; Mallory, M.; Alford, M.; DeTeresa, R.; Hansen, L.A.; McKeel, D.W., Jr.; Morris, J.C. Altered expression of synaptic proteins occurs early during progression of Alzheimer's disease. *Neurology* **2001**, *56*, 127–129. [CrossRef]
126. Scheff, S.W.; Price, D.A.; Schmitt, F.A.; Dekosky, S.T.; Mufson, E.J. Synaptic alterations in CA1 in mild Alzheimer disease and mild cognitive impairment. *Neurology* **2007**, *68*, 1501–1508. [CrossRef]
127. Chapman, P.F.; White, G.L.; Jones, M.W.; Cooper-Blacketer, D.; Marshall, V.J.; Irizarry, M.; Younkin, L.; Good, M.A.; Bliss, T.V.; Hyman, B.T.; et al. Impaired synaptic plasticity and learning in aged amyloid precursor protein transgenic mice. *Nat. Neurosci.* **1999**, *2*, 271–276. [CrossRef]
128. Hsia, A.Y.; Masliah, E.; McConlogue, L.; Yu, G.Q.; Tatsuno, G.; Hu, K.; Kholodenko, D.; Malenka, R.C.; Nicoll, R.A.; Mucke, L. Plaque-independent disruption of neural circuits in Alzheimer's disease mouse models. *Proc. Natl. Acad. Sci. USA* **1999**, *96*, 3228–3233. [CrossRef]
129. Jacobsen, J.S.; Wu, C.C.; Redwine, J.M.; Comery, T.A.; Arias, R.; Bowlby, M.; Martone, R.; Morrison, J.H.; Pangalos, M.N.; Reinhart, P.H.; et al. Early-onset behavioral and synaptic deficits in a mouse model of Alzheimer's disease. *Proc. Natl. Acad. Sci. USA* **2006**, *103*, 5161–5166. [CrossRef]
130. Li, S.; Hong, S.; Shepardson, N.E.; Walsh, D.M.; Shankar, G.M.; Selkoe, D. Soluble oligomers of amyloid Beta protein facilitate hippocampal long-term depression by disrupting neuronal glutamate uptake. *Neuron* **2009**, *62*, 788–801. [CrossRef]
131. Hsieh, H.; Boehm, J.; Sato, C.; Iwatsubo, T.; Tomita, T.; Sisodia, S.; Malinow, R. AMPAR removal underlies Aβ-induced synaptic depression and dendritic spine loss. *Neuron* **2006**, *52*, 831–843. [CrossRef]
132. Shankar, G.M.; Bloodgood, B.L.; Townsend, M.; Walsh, D.M.; Selkoe, D.J.; Sabatini, B.L. Natural Oligomers of the Alzheimer Amyloid-β Protein Induce Reversible Synapse Loss by Modulating an NMDA-Type Glutamate Receptor-Dependent Signaling Pathway. *J. Neurosci.* **2007**, *27*, 2866–2875. [CrossRef]
133. Shankar, G.M.; Walsh, D.M. Alzheimer's disease: Synaptic dysfunction and Aβ. *Mol. Neurodegen.* **2009**, *4*. [CrossRef] [PubMed]
134. Parihar, M.S.; Brewer, G.J. Amyloid-β as a modulator of synaptic plasticity. *J. Alzheimers Dis.* **2010**, *22*, 741–763. [CrossRef] [PubMed]
135. Small, D.H. Network dysfunction in Alzheimer's disease: Does synaptic scaling drive disease progression? *Trends Mol. Med.* **2008**, *14*, 103–108. [CrossRef]

136. Jang, S.S.; Chung, H.J. Emerging Link between Alzheimer's Disease and Homeostatic Synaptic Plasticity. *Neural. Plast.* **2016**, *2016*. [CrossRef]
137. Stam, C.J.; Jones, B.F.; Nolte, G.; Breakspear, M.; Scheltens, P. Small-world networks and functional connectivity in Alzheimer's disease. *Cereb. Cortex* **2007**, *17*, 92–99. [CrossRef]
138. De Haan, W.; Pijnenburg, Y.A.; Strijers, R.L.; Van der Made, Y.; Van der Flier, W.M.; Scheltens, P.; Stam, C.J. Functional neural network analysis in frontotemporal dementia and Alzheimer's disease using EEG and graph theory. *BMC Neurosci.* **2009**, *10*, 101. [CrossRef]
139. De Haan, W.; van der Flier, W.M.; Wang, H.; Van Mieghem, P.F.; Scheltens, P.; Stam, C.J. Disruption of functional brain networks in Alzheimer's disease: What can we learn from graph spectral analysis of resting-state magnetoencephalography? *Brain Connect.* **2012**, *2*, 45–55. [CrossRef]
140. Vecchio, F.; Miraglia, F.; Marra, C.; Quaranta, D.; Vita, M.G.; Bramanti, P.; Rossini, P.M. Human brain networks in cognitive decline: A graph theoretical analysis of cortical connectivity from EEG data. *J. Alzheimers Dis.* **2014**, *41*, 113–127. [CrossRef]
141. Canuet, L.; Pusil, S.; Lopez, M.E.; Bajo, R.; Pineda-Pardo, J.A.; Cuesta, P.; Gálvez, G.; Gaztelu, J.M.; Lourido, D.; García-Ribas, G.; et al. Network disruption and cerebrospinal fluid amyloid-beta and phospho-tau levels in mild cognitive impairment. *J. Neurosci.* **2015**, *35*, 10325–10330. [CrossRef]
142. Dai, Z.; Yan, C.; Li, K.; Wang, Z.; Wang, J.; Cao, M.; Lin, Q.; Shu, N.; Xia, M.; Bi, Y.; et al. Identifying and mapping connectivity patterns of brain network hubs in Alzheimer's disease. *Cereb. Cortex* **2015**, *25*, 3723–3742. [CrossRef] [PubMed]
143. De Haan, W.; Mott, K.; van Straaten, E.C.; Scheltens, P.; Stam, C.J. Activity dependent degeneration explains hub vulnerability in Alzheimer's disease. *PLoS Comput. Biol.* **2012**, *8*, e1002582. [CrossRef] [PubMed]
144. Sperling, R.A.; Laviolette, P.S.; O'Keefe, K.; O'Brien, J.; Rentz, D.M.; Pihlajamaki, M.; Marshall, G.; Hyman, B.T.; Selkoe, D.J.; Hedden, T.; et al. Amyloid deposition is associated with impaired default network function in older persons without dementia. *Neuron* **2009**, *63*, 178–188. [CrossRef] [PubMed]
145. Drzezga, A.; Becker, J.A.; Van Dijk, K.R.; Sreenivasan, A.; Talukdar, T.; Sullivan, C.; Schultz, A.P.; Sepulcre, J.; Putcha, D.; Greve, D.; et al. Neuronal dysfunction and disconnection of cortical hubs in non-demented subjects with elevated amyloid burden. *Brain* **2011**, *134*, 1635–1646. [CrossRef]
146. Buchanan, R.W.; Carpenter, W.T. Schizophrenia: Introduction and overview. In *Comprehensive Textbook of Psychiatry*; Sadock, B.J., Sadock, V.A., Eds.; Lippincott, Williams, and Wilkins: Philadelphia, PA, USA, 2000; Volume 1, pp. 1096–1110.
147. Coyle, J.T. The glutamatergic dysfunction hypothesis for schizophrenia. *Harv. Rev. Psychiat.* **1996**, *3*, 241–253. [CrossRef] [PubMed]
148. Lewis, D.A.; Levitt, P. Schizophrenia as a disorder of neurodevelopment. *Annu. Rev. Neurosci.* **2002**, *25*, 409–432. [CrossRef]
149. Rubinov, M.; Bullmore, E.T. Schizophrenia and abnormal brain network hubs. *Dialogues Clin. Neurosci.* **2013**, *15*, 339–349. [PubMed]
150. Hasan, A.; Nitsche, M.A.; Rein, B.; Schneider-Axmann, T.; Guse, B.; Gruber, O.; Falkai, P.; Wobrock, T. Dysfunctional long term potentiation-like plasticity in schizophrenia revealed by transcranial direct current stimulation. *Behav. Brain Res.* **2011**, *224*, 15–22. [CrossRef]
151. Hasan, A.; Nitsche, M.A.; Herrmann, M.; Schneider-Axmann, T.; Marshall, L.; Gruber, O. Impaired long-term depression in schizophrenia: A cathodal tDCS pilot study. *Brain Stim.* **2012**, *5*, 475–483. [CrossRef]
152. Glantz, L.A.; Lewis, D.A. Decreased dendritic spine density on prefrontal cortical pyramidal neurons in schizophrenia. *Arch. Gen. Psychiatry* **2000**, *57*, 65–73. [CrossRef]
153. Hill, J.J.; Hashimoto, T.; Lewis, D.A. Molecular mechanisms contributing to dendritic spine alterations in the prefrontal cortex of subjects with schizophrenia. *Mol. Psychiatry* **2006**, *11*, 557–566. [CrossRef] [PubMed]
154. Sweet, R.A.; Henteleff, R.A.; Zhang, W.; Sampson, A.R.; Lewis, D.A. Reduced dendritic spine density in auditory cortex of subjects with schizophrenia. *Neuropsychopharmacology* **2009**, *34*, 374–389. [CrossRef] [PubMed]
155. Glausier, J.R.; Lewis, D.A. Dendritic spine pathology in schizophrenia. *Neuroscience* **2013**, *251*, 90–107. [CrossRef] [PubMed]
156. MacDonald, M.L.; Ding, Y.; Newman, J.; Hemby, S.; Penzes, P.; Lewis, D.A.; Yates, N.A.; Sweet, R.A. Altered glutamate protein coexpression network topology linked to spine loss in the auditory cortex of schizophrenia. *Biol. Psychiatry* **2015**, *77*, 959–968. [CrossRef] [PubMed]

157. Akbarian, S.; Sucher, N.J.; Bradley, D.; Tafazzoli, A.; Trinh, D.; Hetrick, W.P.; Potkin, S.G.; Sandman, C.A.; Bunney, W.E., Jr.; Jones, E.G. Selective alterations in gene expression for NMDA receptor subunits in prefrontal cortex of schizophrenics. *J. Neurosci.* **1996**, *16*, 19–30. [CrossRef] [PubMed]
158. Beneyto, M.; Meador-Woodruff, J.H. Lamina-specific abnormalities of AMPA receptor trafficking and signaling molecule transcripts in the prefrontal cortex in schizophrenia. *Synapse* **2006**, *60*, 585–598. [CrossRef]
159. Emamian, E.S.; Karayiorgou, M.; Gogos, J.A. Decreased phosphorylation of NMDA receptor type 1 at serine 897 in brains of patients with Schizophrenia. *J. Neurosci.* **2004**, *24*, 1561–1564. [CrossRef]
160. Funk, A.J.; Rumbaugh, G.; Harotunian, V.; McCullumsmith, R.E.; Meador-Woodruff, J.H. Decreased expression of NMDA receptor-associated proteins in frontal cortex of elderly patients with schizophrenia. *Neuroreport* **2009**, *20*, 1019–1022. [CrossRef]
161. Javitt, D.C. Glutamate and schizophrenia: Phencyclidine, N-methyl-D-aspartate receptors, and dopamine-glutamate interactions. *Int. Rev. Neurobiol.* **2007**, *78*, 69–108. [CrossRef]
162. Frantseva, M.V.; Fitzgerald, P.B.; Chen, R.; Möller, B.; Daigle, M.; Daskalakis, Z.J. Evidence for impaired long-term potentiation in schizophrenia and its relationship to motor skill learning. *Cereb. Cortex* **2008**, *18*, 990–996. [CrossRef]
163. Koch, G.; Ponzo, V.; Di Lorenzo, F.; Caltagirone, C.; Veniero, D. Hebbian and anti-Hebbian spike-timing-dependent plasticity of human cortico-cortical connections. *J Neurosci.* **2013**, *33*, 9725–9733. [CrossRef] [PubMed]
164. Veniero, D.; Ponzo, V.; Koch, G. Paired associative stimulation enforces the communication between interconnected areas. *J. Neurosci.* **2013**, *33*, 13773–13783. [CrossRef] [PubMed]
165. Liu, Y.; Liang, M.; Zhou, Y.; He, Y.; Hao, Y.; Song, M.; Yu, C.; Liu, H.; Liu, Z.; Jiang, T. Disrupted small-world networks in schizophrenia. *Brain* **2008**, *131*, 945–961. [CrossRef] [PubMed]
166. Lynall, M.E.; Bassett, D.S.; Kerwin, R.; McKenna, P.J.; Kitzbichler, M.; Muller, U.; Bullmore, E. Functional connectivity and brain networks in schizophrenia. *J Neurosci.* **2010**, *30*, 9477–9487. [CrossRef]
167. Rubinov, M.; Bassett, D.S. Emerging evidence of connectomic abnormalities in schizophrenia. *J. Neurosci.* **2011**, *31*, 6263–6265. [CrossRef]
168. Skudlarski, P.; Jagannathan, K.; Anderson, K.; Stevens, M.C.; Calhoun, V.D.; Skudlarska, B.A.; Pearlson, G. Brain connectivity is not only lower but different in schizophrenia: A combined anatomical and functional approach. *Biol. Psychiatry* **2010**, *68*, 61–69. [CrossRef]
169. Zalesky, A.; Fornito, A.; Seal, M.L.; Cocchi, L.; Westin, C.F.; Bullmore, E.T.; Egan, G.F.; Pantelis, C. Disrupted axonal fiber connectivity in schizophrenia. *Biol. Psychiatry* **2011**, *69*, 80–90. [CrossRef]
170. Kambeitz, J.; Kambeitz-Ilankovic, L.; Cabral, C.; Dwyer, D.B.; Calhoun, V.D.; van den Heuvel, M.P.; Falkai, P.; Koutsouleris, N.; Malchow, B. Aberrant Functional Whole-Brain Network Architecture in Patients with Schizophrenia: A Meta-analysis. *Schizophr. Bull.* **2016**, *42*, 13–21. [CrossRef]
171. Van den Heuvel, M.P.; Sporns, O.; Collin, G.; Scheewe, T.; Mandl, R.C.; Cahn, W.; Goñi, J.; Hulshoff Pol, H.E.; Kahn, R.S. Abnormal rich club organization and functional brain dynamics in schizophrenia. *JAMA Psychiat.* **2013**, *70*, 783–792. [CrossRef]
172. Zhao, X.; Tian, L.; Yan, J.; Yue, W.; Yan, H.; Zhang, D. Abnormal Rich-Club Organization Associated with Compromised Cognitive Function in Patients with Schizophrenia and Their Unaffected Parents. *Neurosci. Bull.* **2017**, *33*, 445–454. [CrossRef]
173. Van den Heuvel, M.P.; Mandl, R.C.W.; Stam, C.J.; Kahn, R.S.; Hulshoff Pol, H.E. Aberrant frontal and temperal network structure in schizophrenia: A graph theoretical analysis. *J. Neurosci.* **2010**, *30*, 15915–15926. [CrossRef] [PubMed]
174. Gollo, L.L.; Roberts, J.A.; Cropley, V.L.; Di Biase, M.A.; Pantelis, C.; Zalesky, A.; Breakspear, M. Fragility and volatility of structural hubs in the human connectome. *Nat. Neurosci.* **2018**, *21*, 1107–1116. [CrossRef] [PubMed]
175. Zick, J.L.; Blackman, R.K.; Crowe, D.A.; Amirikian, B.; DeNicola, A.L.; Netoff, T.I.; Chafee, M.V. Blocking NMDAR Disrupts Spike Timing and Decouples Monkey Prefrontal Circuits: Implications for Activity-Dependent Disconnection in Schizophrenia. *Neuron* **2018**, *98*, 1243–1255. [CrossRef] [PubMed]

International Journal of
Molecular Sciences

MDPI

Article

Loss of Non-Apoptotic Role of Caspase-3 in the PINK1 Mouse Model of Parkinson's Disease

Paola Imbriani [1,2,†], Annalisa Tassone [1,2,†], Maria Meringolo [1,2], Giulia Ponterio [1,2], Graziella Madeo [2], Antonio Pisani [1,2,*], Paola Bonsi [1,‡] and Giuseppina Martella [1,2,‡]

1 Laboratory of Neurophysiology and Plasticity, IRCCS Fondazione Santa Lucia, 00143 Rome, Italy
2 Department of Systems Medicine, University of Rome "Tor Vergata", 00133 Rome, Italy
* Correspondence: pisani@uniroma2.it
† They are Co-first Authors.
‡ They are Co-senior Authors.

Received: 24 June 2019; Accepted: 9 July 2019; Published: 11 July 2019

Abstract: Caspases are a family of conserved cysteine proteases that play key roles in multiple cellular processes, including programmed cell death and inflammation. Recent evidence shows that caspases are also involved in crucial non-apoptotic functions, such as dendrite development, axon pruning, and synaptic plasticity mechanisms underlying learning and memory processes. The activated form of caspase-3, which is known to trigger widespread damage and degeneration, can also modulate synaptic function in the adult brain. Thus, in the present study, we tested the hypothesis that caspase-3 modulates synaptic plasticity at corticostriatal synapses in the phosphatase and tensin homolog (PTEN) induced kinase 1 (PINK1) mouse model of Parkinson's disease (PD). Loss of PINK1 has been previously associated with an impairment of corticostriatal long-term depression (LTD), rescued by amphetamine-induced dopamine release. Here, we show that caspase-3 activity, measured after LTD induction, is significantly decreased in the PINK1 knockout model compared with wild-type mice. Accordingly, pretreatment of striatal slices with the caspase-3 activator α-(Trichloromethyl)-4-pyridineethanol (PETCM) rescues a physiological LTD in PINK1 knockout mice. Furthermore, the inhibition of caspase-3 prevents the amphetamine-induced rescue of LTD in the same model. Our data support a hormesis-based double role of caspase-3; when massively activated, it induces apoptosis, while at lower level of activation, it modulates physiological phenomena, like the expression of corticostriatal LTD. Exploring the non-apoptotic activation of caspase-3 may contribute to clarify the mechanisms involved in synaptic failure in PD, as well as in view of new potential pharmacological targets.

Keywords: Parkinson's disease; PINK1; caspase-3; striatum; synaptic plasticity; long-term depression

---

1. Introduction

Caspases are a highly conserved family of cysteine proteases playing a central role in the execution phase of apoptotic cell death [1,2]. Caspases are synthesized as catalytically inactive proenzymes and can be categorized in initiators and effectors. Initiator caspases (caspases-8, -9, -10, and -2) require autocatalytic cleavage for maturation. Effector caspases (caspases-3, -6, and -7) are activated by initiator caspases through a proteolytic cleavage and are responsible for degradation of structural proteins, signaling molecules, and DNA repair enzymes [3]. Active caspases have also been detected in non-apoptotic cells, including neurons, where they have been found to be active in dendrites, synapses, and growth cones [4–7], not involving death-related processes [8]. A member of this family, caspase-3, classically known as a key mediator of neuronal apoptosis, also exerts non-apoptotic functions [9,10]. In immature central nervous system, caspases-3 modulates structural functions, including developmental pruning of axons, dendrites, and synaptic connections, and is involved in axon degeneration induced by trophic factor withdrawal [11–13]. Mice lacking caspase-3 displays severe anomalies in the cerebral cortex and forebrain, owing to defects in nervous system development [14]. In the healthy adult brain, caspase-3 activity seems essential for physiological synaptic function and neurogenesis [15]. Both in vivo and in vitro findings support the involvement of caspase-3 in the molecular mechanisms underlying learning and memory processes [10,16]. Several caspase-3 substrates are key determinants of the molecular machinery of synaptic plasticity, including α-amino-3-hydroxy-5-methyl-4-isoxazolepropionic acid (AMPA) receptor subunit GluR1 [8,17–19]. Electrophysiological experiments on the terrestrial snail (Helix lucorum) showed that blockade of caspase-3 activity, obtained using the cell-permeable inhibitor Z-Devd-fmk, prevented the development of the long-term synaptic sensitization, indicating that caspase-3 is essential for long-term plasticity in invertebrate neurons [20]. A further recent electrophysiology study performed on rat hippocampal slices demonstrated that caspase-9-caspase-3 cascade, activated by pro-apoptotic molecules released from mitochondria, is required for the internalization of glutamatergic AMPA receptors from synapses and for the induction of long-term depression (LTD) [10]. Moreover, pretreatment of hippocampal slices with caspase inhibitors Z-Devd-fmk and Lehd-fmk prevents the induction of LTD at Schaffer collateral-CA1 synapses [10]. To our knowledge, the studies available so far have focused exclusively on the role of caspase-3 on hippocampal synaptic plasticity mechanisms. Here, we investigated whether caspase-3 plays a role in modulating synaptic plasticity expression at corticostriatal synapses in physiological and pathological conditions. To this aim, we used a well-characterized mouse model of autosomal recessive early-onset Parkinson's disease (PD) [21], the phosphatase and tensin homolog (PTEN) induced kinase 1 (PINK1) knockout mouse model [22]. As previously demonstrated, homozygous PINK1 knockout (PINK1$^{-/-}$) mice show an impaired expression of corticostriatal LTD [22], whereas LTD is preserved in heterozygous PINK1 (PINK1$^{+/-}$) mice [23]. Thus, we explored whether caspase-3 is involved in the synaptic plasticity machinery underlying the expression of corticostriatal LTD in the PINK1 mouse model.

2. Results

2.1. Normal Intrinsic Membrane Properties of Striatal Medium Spiny Neurons (MSNs) after Pharmacological Modulation of Caspase-3

First, we explored whether caspase-3 modulation perturbs the intrinsic membrane properties of striatal MSNs (Table 1). MSNs' membrane properties were analyzed from corticostriatal slices obtained from three different PINK1 genotypes, namely PINK1$^{+/+}$, PINK1$^{+/-}$, and PINK1$^{-/-}$, as previously described [22,24–26]. Specifically, we measured evoked firing activity (Figure 1A,B), action potential amplitude (Figure 1C), delay to spike threshold (Figure 1D), resting membrane potential (RMP) (Figure 1E), and rheobase (Figure 1F).

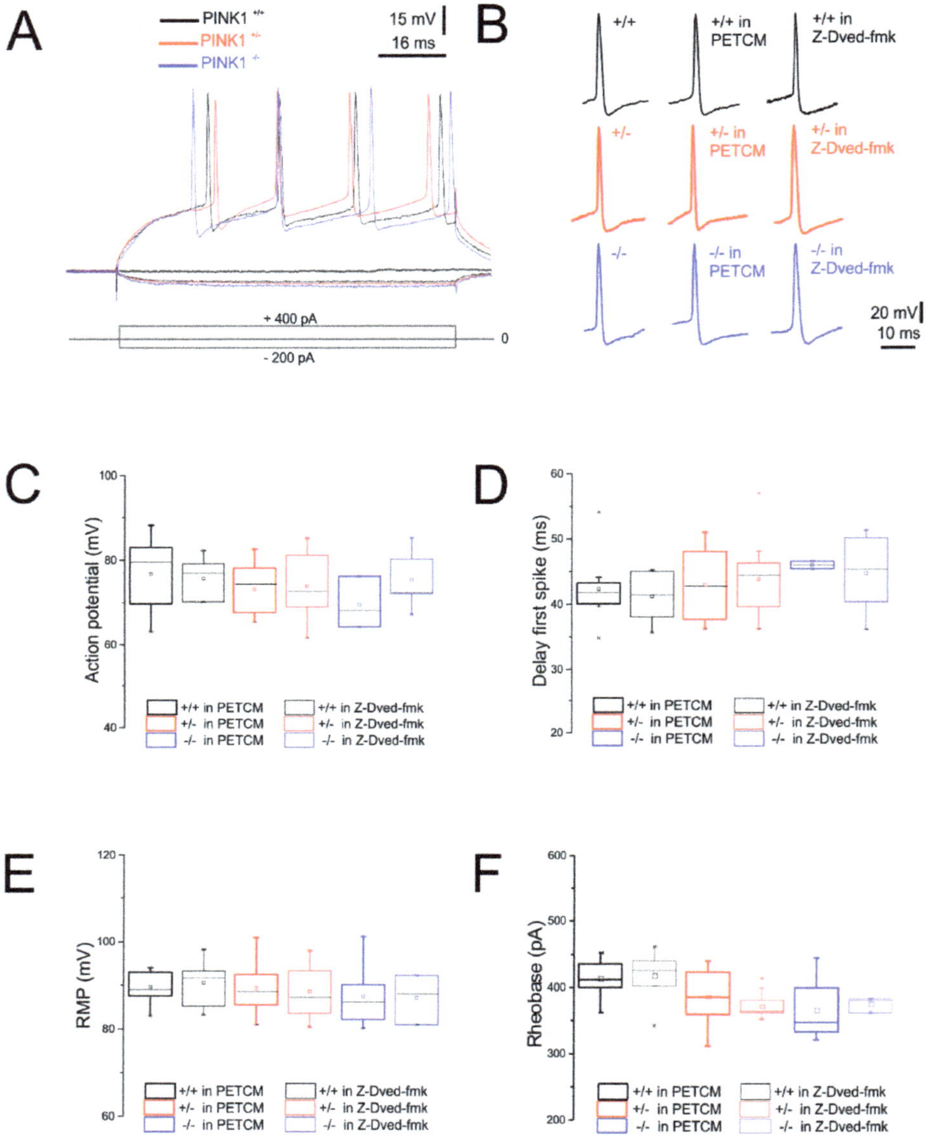

Figure 1. Intrinsic membrane properties of medium spiny neurons (MSNs) in phosphatase and tensin homolog (PTEN) induced kinase 1 (PINK1) mice after pretreatment with caspase-3 inhibitor and caspase-3 activator. (**A**) Superimposed traces showing voltage responses to current steps in both hyperpolarizing (−200 pA, 80 ms) and depolarizing (+400 pA, 80 ms) direction from PINK1$^{+/+}$ (black trace), PINK1$^{+/-}$ (red trace), and PINK1$^{-/-}$ (blue trace) MSNs. No significant differences were observed between the three genotypes. (**B**) Representative single action potentials recorded from MSNs both in basal condition and after treatment with the caspase-3 inhibitor Z-Devd-fmk (5 μM) or with the caspase-3 activator α-(Trichloromethyl)-4-pyridineethanol (PETCM 30 μM) in all PINK1 genotypes. We did not observe significant differences between groups, in terms of upward spike, firing threshold, and half amplitude duration. (**C**–**F**) Whisker plots showing (**C**) action potential amplitude (PINK1$^{+/+}$: 77.00 ± 3.8 mV; PINK1$^{+/-}$: 78.4 ± 3.6 mV; PINK1$^{-/-}$: 79.4 ± 4.02 mV; $n = 9$ for each group;

one-way analysis of variance (ANOVA) $p > 0.05$), (**D**) delay to first spike (PINK1$^{+/+}$: 40.65 ± 1.99 ms; PINK1$^{+/-}$: 42.13 ± 1.6 ms; PINK1$^{-/-}$: 44.23 ± 2.01 ms; $n = 9$ for each group; one-way ANOVA $p > 0.05$), (**E**) resting membrane potential (RMP) (PINK1$^{+/+}$: −85 ± 3.8 mV; PINK1$^{+/-}$: −87.21 ± 3.6 mV; PINK1$^{-/-}$: −79.4 ± 4.02 mV; $n = 9$ for each group; one-way ANOVA; RMP Z-Dved-fmk on PINK1$^{+/+}$: −90.53 ± 1.59 mV; Z-Dved-fmk on PINK1$^{+/-}$: −88.51 ±1.60 mV; Z-Dved-fmk on PINK1$^{-/-}$: −87.10 ± 3.1 mV; $n = 9$ for each group; two-way ANOVA $p > 0.05$), and (**F**) Rheobase (PINK1$^{+/+}$: 4201 ± 12.1 ms; PINK1$^{+/-}$: 400 ± 18.1 ms; PINK1$^{-/-}$: 380 ± 18 ms; $n = 9$ for each group; one-way ANOVA $p > 0.05$) values, without significant differences between genotypes, regardless of treatment with caspase-3 inhibitor/activator.

Table 1. Electrophysiological properties of PINK1 MSNs, in the presence of caspase-3 modulators.

	PINK1$^{+/+}$			PINK1$^{+/-}$			PINK1$^{-/-}$		
	Saline	**Z-Dedvd-fmk**	**PETCM**	**Saline**	**Z-Dedvd-fmk**	**PETCM**	**Saline**	**Z-Dedvd-fmk**	**PETCM**
RMP (mV)	−85.2 ± 3.8	−90.5 ± 1.6	−89.5 ± 1.3	−87.2 ± 3.6	−88.5 ±1.6	−89.3 ± 1.7	−79.4 ± 4.0	−87.10 ± 3.1	−87.4 ± 2.2
AP amplitude (mV)	77.0 ± 3.8	75.8 ± 1.88	76.8 ± 3.03	78.4 ± 3.6	75.0 ± 2.1	73.1 ± 2.9	79.4 ± 4.0	69.5 ± 4.3	75.3 ± 2.5
Membrane resistance (MΩ)	35.2 ± 2.4	36.8 ± 1.7	38.2 ± 1.8	38.0 ± 2.0	38.9 ± 0.7	38.7 ± 1.3	35.2 ± 3.3	34.1 ± 1.2	35.6 ± 1.1
Delay to first spike (ms)	40.5 ± 1.99	41.2 ± 1.3	42.3 ± 1.6	42.1 ± 1.6	43.8 ± 1.6	42.9 ± 1.5	44.2 ± 2.0	44.7 ± 1.8	46.0 ± 0.3
Rheobase (pA)	421 ± 12	418 ± 9	444 ± 11	400 ± 18	373 ± 6	387 ± 12	380 ± 18	376 ± 6	366 ± 15

Corticostriatal slices from PINK1$^{+/+}$, PINK1$^{+/-}$, and PINK1$^{-/-}$ mice were randomly assigned to saline solution (vehicle), or to saline solution plus either the caspase-3 inhibitor Z-Devd-fmk (5 μM) or the caspase-3 activator PETCM (30 μM). After 1 h pretreatment at room temperature, each brain slice was transferred into the recording chamber and continuously perfused with the same drug, during the entire experimental period. Pharmacological treatment with vehicle, Z-Dved-fmk, or PETCM did not alter MSNs' intrinsic excitability properties, as well as the RMP (Table 1; Figure 1A–F; one-way analysis of variance (ANOVA) $p > 0.05$).

2.2. Unaltered Glutamatergic Short-Term Synaptic Plasticity after Pharmacological Modulation of Caspase-3

Excitatory postsynaptic potentials (EPSPs) were evoked by 0.1 Hz stimulation of corticostriatal fibers (Figure 2A). Slice perfusion with either PETCM (Figure 2B,D) or Z-Devd-fmk (Figure 2C,D) did not modify amplitude and slope of the EPSPs recorded from MSNs of all genotypes (Figure 2B–D). The input–output relationship was obtained by increasing stimulus intensity and by measuring the peak amplitude and slope of the evoked EPSP. No significant differences among groups were observed (Figure 2A–D; $n \geq 6$ for each condition, two-way ANOVA $p > 0.05$). Glutamatergic short-term synaptic plasticity was assessed with a paired-pulse protocol, consisting of paired stimulation (50 ms interstimulus interval, ISI) delivered at 0.1 Hz in picrotoxin (PTX, 50 μM). Paired stimulation produced a similar response in all genotypes [22,23,27]. A comparable facilitation of synaptic transmission (PPF) was induced in PINK1$^{-/-}$ corticostriatal slices (Figure 2E) after incubation in saline solution (Figure 2Fa), PETCM (Figure 2Fb), or Z-Devd-fmk (Figure 2Fc). Similarly, PPR measured in PINK1$^{+/+}$ and PINK1$^{+/-}$ MSNs in the different experimental conditions did not show any significant differences (Figure 2E).

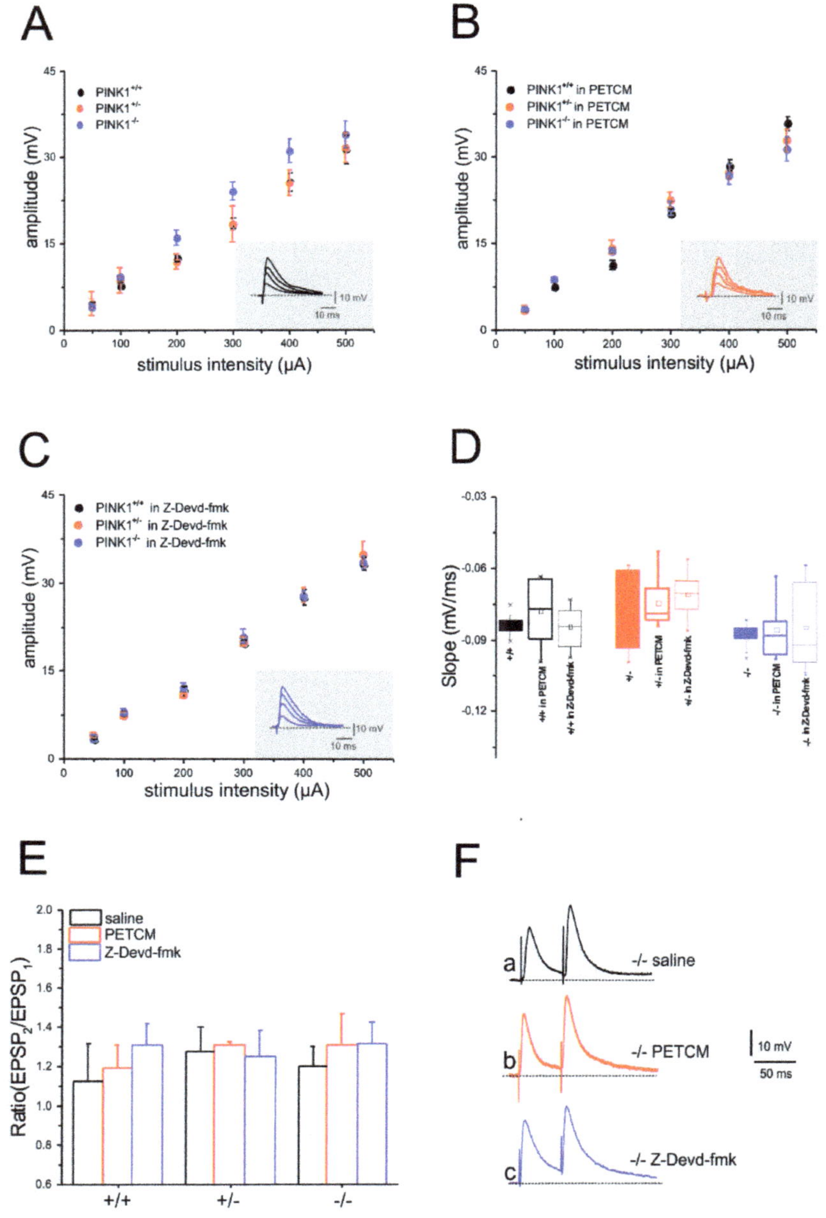

Figure 2. Pretreatment with caspase-3 inhibitor and activator does not affect evoked synaptic responses in PINK1 MSNs. Stimulation of corticostriatal fibers in the presence of Picrotoxin (PTX, 50 μM) produced glutamatergic excitatory postsynaptic potentials (EPSPs). Input–output (I/O) curves were built by measuring the amplitude of EPSPs evoked by increasing intensities of stimulation. (**A**) The I/O relationships were not significantly different between the three PINK1 genotypes. The inset shows representative superimposed traces of a single recording. (**B**) Treatment with PETCM (30 μM) did not induce significant differences neither among the three genotypes nor as compared with non-treated MSNs. (**C**) Similarly, treatment with Z-Devd-fmk (5 μM) did not affect the I/O curves in any experimental condition. (**D**) Plots showing EPSP slope values recorded from PINK1$^{+/+}$ (black), PINK1$^{+/-}$

(red), and PINK1$^{-/-}$ (blue) MSNs treated with vehicle, PETCM, or Z-Devd-fmk. No significant differences were observed among groups (50 μA stimulus: PINK1$^{+/+}$: 3.84 ± 0.27 mV/ms, n = 11; PINK1$^{+/-}$: 3.5 ± 0.66 mV/ms, n = 10; PINK1$^{-/-}$: 3.6 ± 0.3 mV/ms, n = 10; PINK1$^{+/+}$ in Z-Devd-fmk: 4.21 ± 0.3 mV/ms, n = 10; PINK1$^{+/-}$ in Z-Devd-fmk: 4.8 ± 0.61 mV/ms, n = 10; PINK1$^{-/-}$ in Z-Devd-fmk: 4.01 ± 0.72 mV/ms, n = 8; PINK1$^{+/+}$ in PETCM: 4.4 ± 0.6 mV/ms, n = 8; PINK1$^{+/-}$ in PETCM: 4.78 ± 1.02 mV/ms, n = 7; PINK1$^{-/-}$ in PETCM: 4.21 ± 0.53 mV/ms, n = 8; two-way ANOVA $p > 0.05$). **(E,F)** Paired-pulse facilitation (50 ms ISI) was similar in all PINK1 genotypes, both in saline solution and after pharmacological treatment. **(E)** The plot shows paired-pulse ratio (PPR) values, defined as EPSP2/EPSP1 ratio (PPR: PINK1$^{+/+}$: 1.2 ± 0.16, n = 6; PINK1$^{+/-}$: 1.23 ± 0.14, n = 6; PINK1$^{-/-}$: 1.28 ± 0.12, n = 6; PINK1$^{+/+}$ in Z-Devd-fmk: 1.28 ± 0.12, n = 9; PINK1$^{+/+}$ in PETCM: 1.2 ± 0.10, n = 9; PINK1$^{+/-}$ in Z-Devd-fmk: 1.32 ± 0.02, n = 9; PINK1$^{+/-}$ in PETCM: 1.30 ± 0.16, n = 9; two-way ANOVA $p >$ 0.05). **(F)** Representative paired recordings of EPSPs from PINK1$^{-/-}$ MSNs after treatment with saline solution (a, black traces), PETCM (b, red traces), and Z-Devd-fmk (c, blue traces) (PPR: PINK1$^{-/-}$ in saline: 1.32 ± 0.10, n = 9; PINK1$^{-/-}$ in Z-Devd-fmk: 1.25 ± 0.14, n = 9; PINK1$^{-/-}$ in PETCM: 1.32 ± 0.10, n = 9; one-way ANOVA $p > 0.05$).

2.3. Corticostriatal LTD Expression Requires Caspase-3 Activation

We previously reported the loss of both corticostriatal LTD and LTP in homozygous PINK1$^{-/-}$ mice [2] and sparing of LTD in heterozygous PINK1$^{+/-}$ mice [23]. Considering previous studies investigating the role of caspase-3 in the expression of hippocampal LTD [10], we aimed at assessing whether modulation of caspase-3 could influence the expression of corticostriatal LTD in the PINK1 mouse model.

In a first set of experiments, we assessed whether the inhibition of caspase-3 by Z-Devd-fmk could disrupt the physiological LTD observed both in PINK1$^{+/+}$ and PINK1$^{+/-}$ mice (Figure 3A–C). Incubation in Z-Devd-fmk (5 μM/1 h) of PINK1$^{+/+}$ (Figure 3A,D) and PINK1$^{+/-}$ (Figure 3B,D) corticostriatal slices prevented LTD expression. In light of these results, supporting caspase-3 requirement for corticostriatal LTD expression, we tested whether the activation of caspase-3 could be sufficient to rescue LTD defect in PINK1$^{-/-}$ neurons. Therefore, PINK1$^{-/-}$ corticostriatal slices were treated with the caspase-3 activator PETCM. Indeed, in this condition, we obtained a complete rescue of LTD expression in PINK1$^{-/-}$ MSNs (Figure 4A,C). Our previous results showed that amphetamine was able to rescue LTD in PINK1$^{-/-}$ corticostriatal slices. Thus, we performed further experiments in the presence of the caspase-3 inhibitor Z-Devd-fmk plus amphetamine, to verify if the activation of caspase-3 is involved in the amphetamine-mediated rescue of LTD. These experiments showed that when caspase-3 activation was blocked, amphetamine was no longer able to rescue LTD in PINK1$^{-/-}$ MSNs (Figure 4B,D).

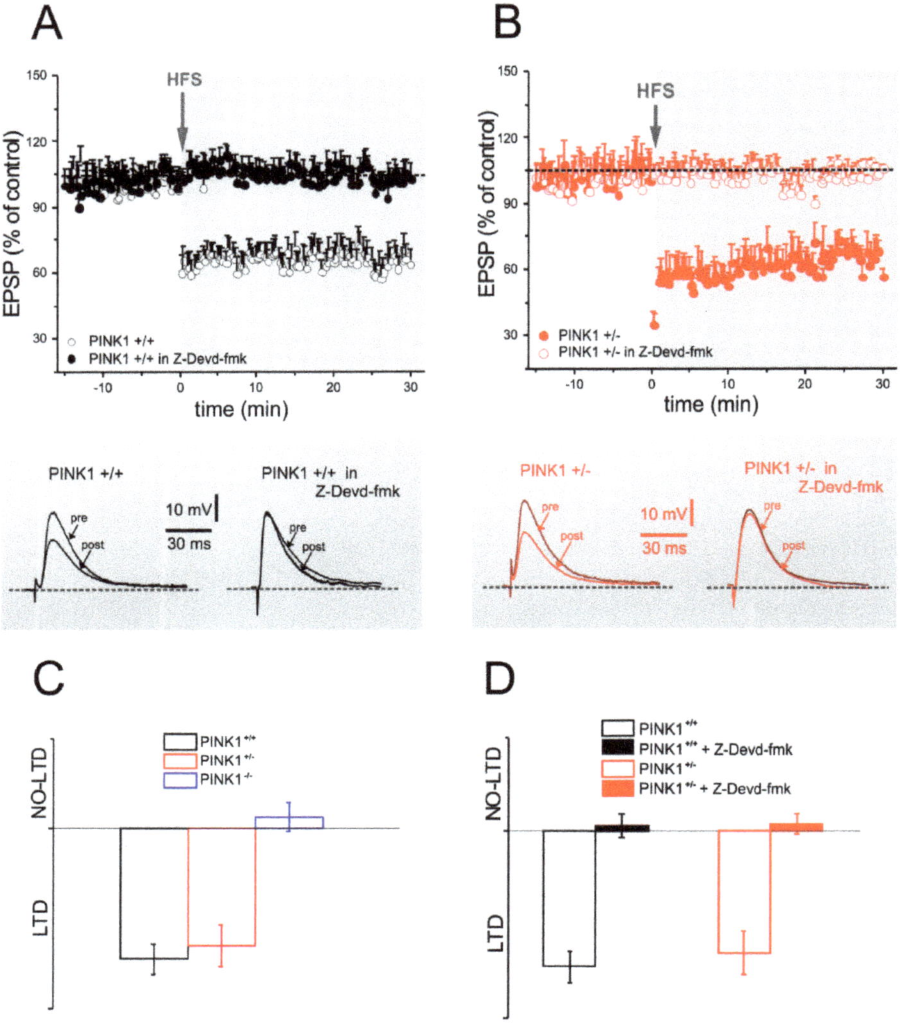

Figure 3. Long-term depression (LTD) in PINK1$^{+/+}$ and PINK1$^{+/-}$ mice is suppressed by caspase-3 inhibition. (**A**,**B**) Top. Time-course of LTD in PINK1$^{+/+}$ and PINK1$^{+/-}$ mice. Stimulus intensity was raised to reach threshold level for high-frequency stimulation (HFS). The amplitude of EPSPs was plotted over-time as percentage of the pre-HFS control EPSP. (**A**) HFS of corticostriatal glutamatergic afferents elicited a robust LTD in MSNs recorded from PINK1$^{+/+}$ mice (black dots), but not after pretreatment with the caspase-3 inhibitor Z-Dved-fmk (5 μM) (white dots) (PINK1$^{+/+}$ in Z-Devd-fmk: 98.25% ± 2.12% of control, $n = 8$; Student's t-test $p > 0.05$; PINK1$^{+/+}$ in saline: 63.70% ± 3.09% of control, $n = 6$; Student's t-test $p < 0.05$). (**B**) Similarly, Z-Dved-fmk treatment suppressed LTD expression in PINK1$^{+/-}$ mice (PINK1$^{+/-}$ in Z-Devd-fmk: 105.69% ± 3.0% of control, $n = 8$, Student's t-test $p > 0.05$; PINK1$^{+/-}$ in saline: 61.44% ± 1.28% of control, $n = 10$; Student's t-test $p < 0.05$). Each data point represents the mean ± SEM of eight independent observations for each group. Bottom. Sample traces of representative EPSPs recorded before (pre) and 30 min after (post) HFS in PINK1$^{+/+}$ and PINK1$^{+/-}$ mice, both in control condition and after treatment with Z-Dved-fmk. (**C**) Plot summarizing corticostriatal LTD expression in PINK1 mice. (**D**) The plot summarizes the loss of corticostriatal LTD after Z-Dved-fmk treatment in PINK1$^{+/+}$ and PINK1$^{+/-}$ genotypes.

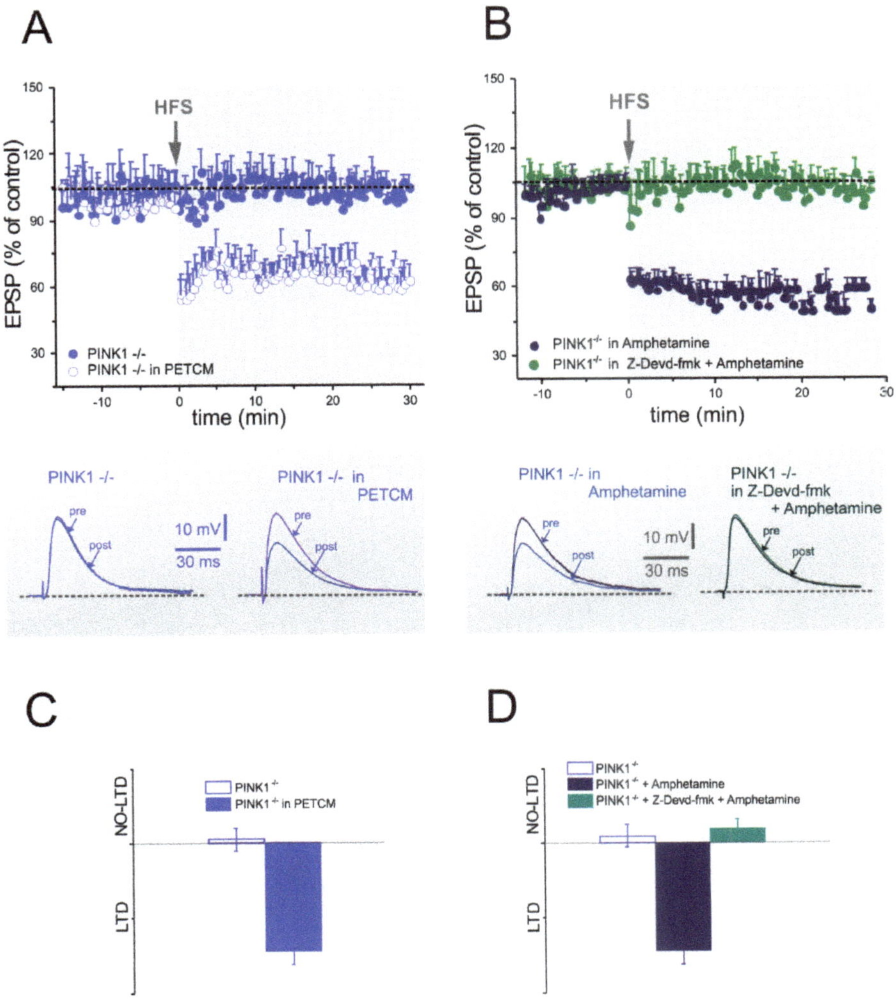

Figure 4. Long-term depression impairment in PINK1$^{-/-}$ MSNs is rescued by caspase-3 activation. (**A**,**B**) Top. Time-course of LTD in PINK1$^{-/-}$ mice. (**A**) HFS-protocol did not elicit LTD in MSNs recorded from PINK1$^{-/-}$ mice (blue dots), but induced the expression of physiological LTD after pretreatment with the caspase-3 activator PETCM (30 μM) (empty dots) (PINK1$^{-/-}$ in PETCM: 61.63% ± 1.40% of control, $n = 8$; Student's t-test $p < 0.05$; PINK1$^{-/-}$ in saline: 98.55% ± 2.7.% of control, $n = 6$; Student's t-test $p > 0.05$). (**B**) Treatment whit amphetamine (100 μM) was also able to rescue LTD in PINK1$^{-/-}$ mice (dark blue dots). However, combined treatment with amphetamine plus Z-Dved-fmk failed to rescue LTD in the same PINK1 genotype (green dots) (PINK1$^{-/-}$ in amphetamine: 60.15% ± 3.67% of control, $n = 6$; Student's t-test $p < 0.05$; PINK1$^{-/-}$ in Z-Devd-fmk + amphetamine: 101% ± 3.08% of control, $n = 6$; Student's t-test $p > 0.05$). Bottom. Sample traces of representative EPSPs recorded before (pre) and 30 min after (post) HFS in PINK1$^{-/-}$ mice, in all the described experimental conditions. (**C**,**D**) The plots summarize the rescue of corticostriatal LTD after either PETCM or amphetamine treatment and the loss of LTD after combined treatment with amphetamine plus Z-Dved-fmk in PINK1$^{-/-}$ mice.

2.4. Caspase-3 Activity is Lower in PINK1$^{-/-}$ Mice after High Frequency Stimulation

In addition to electrophysiological experiments, we performed a colorimetric assay in order to measure caspase-3 activity in all PINK1 genotypes. First, we analyzed striatal slices from wild type mice, incubated with high doses of PETCM (100 μM, 1 h, 37 °C), and compared them with naïve samples (untreated slices). As expected, PETCM was able to induce a drastic increase in caspase-3 activity (Figure S1; naïve absorbance = 0.0285 ± 0.0077, $n = 3$; PETCM absorbance = 0.14550 ± 0.01768; $n = 3$; Student's t-test $p < 0.05$), supporting the reliability of the assay. The levels of caspase-3 activation measured after PETCM treatment were in accordance with caspase-3 assay performed on cell lysates and suggestive of apoptosis [24].

Next, we repeated the colorimetric assay in striatal slices obtained from the three genotypes in different experimental conditions: non-treated slices (CTRL), post-HFS slices, and post-HFS slices preincubated with low doses of PETCM (30 μM, 1 h, 32 °C) (Figure 5). We obtained similar low levels of caspase-3 activation in the CTL condition in all genotypes. Interestingly, after HFS protocol, we observed a significant decrease of caspase-3 activation only in PINK1$^{-/-}$ mice, compared with the other genotypes. In light of these results, we treated PINK1$^{-/-}$ slices with PETCM (30 μM, 1 h, 32 °C) before HFS stimulation. In this condition, we observed rescue of PINK1$^{-/-}$ post-HFS caspase-3 activity to levels similar to PINK1$^{+/+}$ and PINK1$^{+/-}$ mice.

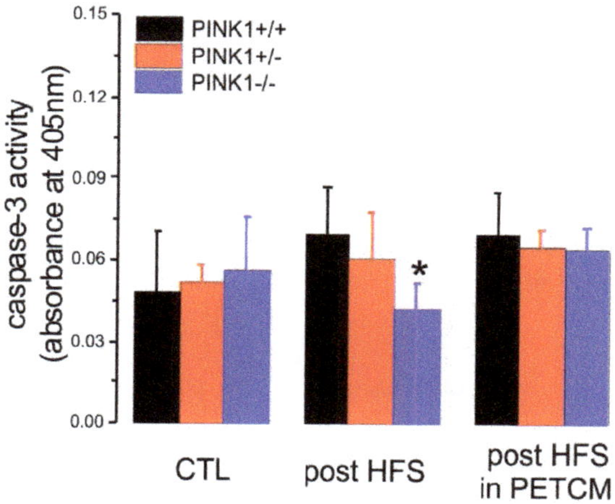

Figure 5. Caspase-3 biochemical assay showed low levels of caspase-3 in PINK1$^{-/-}$ mice after HFS. Non-treated (CTL) brain slices obtained from the three genotypes did not show significant differences in caspase-3 activity levels (PINK1$^{+/+}$: absorbance = 0.048 ± 0.023, $n = 6$; black column; PINK1$^{+/-}$: absorbance = 0.052 ± 0.007, $n = 3$; red column; PINK1$^{-/-}$: absorbance = 0.057 ± 0.020, $n = 6$; blue column; one-way ANOVA and Tukey post hoc test, $p > 0.05$). Caspase-3 activity measured post-HFS was significantly reduced in PINK1$^{-/-}$ slices compared with other genotypes (PINK1$^{+/+}$: absorbance = 0.070 ± 0.017, $n = 6$; black column; PINK1$^{+/-}$: absorbance = 0.061 ± 0.017, $n = 3$; red column; PINK1$^{-/-}$: absorbance = 0.042 ± 0.010, $n = 5$; blue column; one-way ANOVA and Tukey post hoc test, * $p < 0.05$). If slices were pretreated with the activator of caspase-3, PETCM, the same HFS protocol induced similar levels of caspase-3 activity (PINK1$^{+/+}$: absorbance = 0.070 ± 0.015, $n = 6$; black column; PINK1$^{+/-}$: absorbance = 0.066 ± 0.006, $n = 3$; red column; PINK1$^{-/-}$: absorbance = 0.064 ± 0.008, $n = 6$; blue column; one-way ANOVA and Tukey post hoc test, $p > 0.05$). Data are represented as mean ± SD.

2.5. Analysis of Glutamate Vesicular Release in the Different PINK1 Genotypes

In different animal models of monogenic PD, such as LRRK2, α-synuclein, Parkin, and DJ-1, an impairment of membrane trafficking and synaptic release has been shown [22,28–31]. This altered neurotransmitter release has been interpreted as a dysregulation of synaptic vesicle mobilization and/or trafficking. In physiological conditions, the exocytosis process ensures the correct synaptic activity, while in these forms of monogenic PD, a reduced vesicle mobilization or trafficking impairs transmitter release, thereby altering synaptic function [23,24]. To this regard, we performed additional experiments to evaluate whether PINK1 knockout could influence vesicular release. To this aim, a 30 s 30 Hz stimulation protocol [32] in PTX (50 μM) was applied to trigger multiple EPSCs. As expected, this protocol induced a synaptic depression in all PINK1 genotypes (Figure 6A–C). However, in PINK1$^{-/-}$ slices, the decline of EPSCs amplitude showed a slightly different profile, compared with PINK1$^{+/+}$ and PINK1$^{+/-}$ (Figure 6D). This different synaptic depression rate can be explained by a decrease in vesicular release probability [32,33].

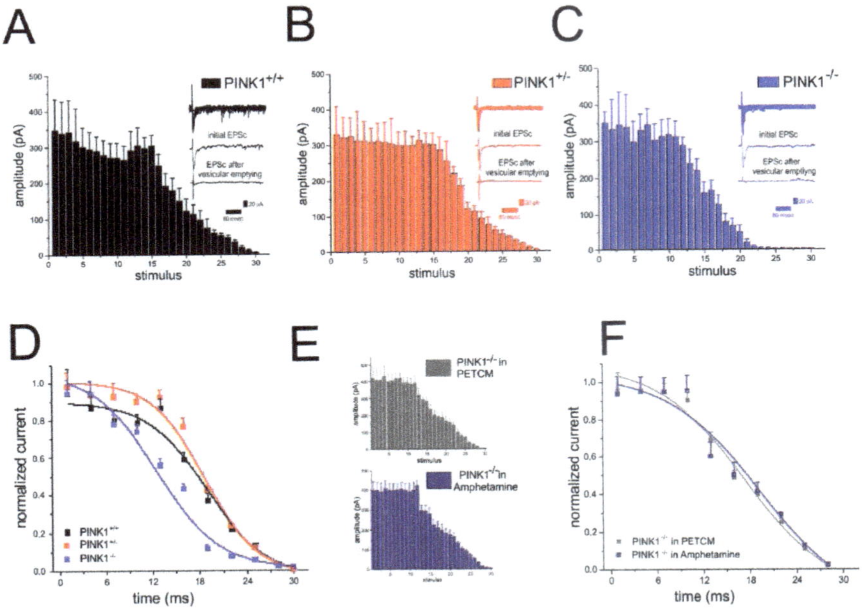

Figure 6. Effect of caspase-3 activation on synaptic responses to sustained electrical stimulation in PINK1$^{-/-}$ MSNs. (**A**) Time-course of synaptic responses to sustained stimulation (30 Hz, 30 s) in PINK1$^{+/+}$ MSNs. Current peak amplitudes are reported as mean ± SEM. In the inset, the first and last EPSC from a representative experiment are shown. (**B**) Time course of synaptic responses in PINK1$^{+/-}$ MSNs shows a similar response to sustained stimulation. (**C**) In PINK$^{-/-}$ MSNs, we observed a faster depression kinetics after the 12th stimulus. (**D**) In PINK1$^{+/+}$ MSNs, the normalized current curve shows an initial fast depression, followed by a slow fall of response. The fit was used to measure the time constant of synaptic depression, which appeared slightly different in PINK1$^{-/-}$ mice compared with PINK1$^{+/+}$ and PINK1$^{+/-}$ mice. Decay time constants: PINK1$^{+/+}$: 18.88 ± 0.8 ms, $n = 4$; PINK1$^{+/-}$: 18.68 ± 0.56 ms, $n = 5$; PINK1$^{-/-}$: 12,43 ± 2.43 ms, $n = 6$; ANOVA and Tukey post hoc, $p < 0.05$). (**E**) Time-course of synaptic responses to sustained stimulation in the presence of amphetamine (100 μM) (blue) and PETCM (30 μM) (gray). No significant differences were observed among the three genotypes when PINK1$^{-/-}$ was treated by either amphetamine. (**F**) No differences were detected between normalized curves obtained from PINK1$^{-/-}$ MSNs treated with either amphetamine or PETCM. Time constants: PINK1$^{-/-}$ plus amphetamine: 19.28 ± 0.93 ms, $n = 6$; $p > 0.05$ vs. PINK1$^{-/-}$; PINK1$^{-/-}$ plus PETCM: 17.22 ± 0.86 ms, $n = 4$; $p > 0.05$ vs. PINK1$^{-/-}$.

We previously demonstrated a consistent decrease in evoked dopamine release in PINK1$^{-/-}$ striatal slices by amperometric recordings [22,23]. This release impairment was rescued by the exogenous application of amphetamine. Recently, Avelar and colleagues demonstrated that amphetamine can increase vesicular release in the dorsal striatum, by stimulating dopamine synthesis and inhibiting dopamine degradation [34]. Moreover, amphetamine induces adaptations in glutamatergic signalling, regulating the glutamate release via glutamate transporters [35–37]. Therefore, we tested whether amphetamine could restore neurotransmitter release in PINK1$^{-/-}$ striatum. As expected, in 100 μM amphetamine, PINK1$^{-/-}$ MSNs showed a vesicular release profile similar to the other genotypes, with comparable time constants (Figure 6E,F). Given the role of caspase-3 in neuronal synaptic processes, we repeated the experimental recordings in the presence of the caspase-3 activator PETCM (30 μM, 1 h). Surprisingly, in PINK1$^{-/-}$ MSNs, PETCM treatment was able to rescue vesicular release, as witnessed by the change in time constant (Figure 6E,F). These results suggest a possible involvement of caspase-3 in striatal synaptic release.

3. Discussion

Several studies support the role of caspase-3 in non-apoptotic cell-functions, including synaptic structural remodeling and synaptic plasticity mechanisms. Here, we provide evidence that caspase-3 plays a modulatory role in the expression of LTD at corticostriatal synapses, both in physiological and pathological conditions. We demonstrate that in PINK1 wild type mice, which express a physiological corticostriatal LTD [22], inhibition of caspase-3 by Z-Devd-fmk prevents the induction of this form of plasticity. Moreover, the impairment of striatal LTD in PINK1 knockout mice [22] can be restored by activation of caspase-3 mediated by α-(Trichloromethyl)-4-pyridineethanol (PETCM). We also demonstrate that caspase-3 modulators per se were not able to induce any change in intrinsic excitability properties in MSNs, as well as in paired pulse modulation. A number of studies support an additional role of caspase-3, because, besides its well-known involvement in apoptotic mechanisms, it acts as a key modulator of many other neuronal processes, including regulation of synaptic plasticity [9,10]. In this respect, Li and colleagues explored the function of caspase-3 in synaptic plasticity in hippocampal neurons in a caspase-3 knockout mouse model, demonstrating that this protease, activated via the mitochondrial pathway, is necessary for LTD expression, acting through regulation of the inducible internalization of AMPA receptors [10]. Although the exact mechanism by which caspase-3 promotes synaptic processes remains to be elucidated, it has been hypothesized that a fine modulation of caspase-3 activity exists at various levels, including the synaptic terminal, where it ensures physiological cellular activities unrelated to cell death. Thus, whether the cell is directed towards apoptosis or plasticity depends on the intensity and duration of caspase-3 activation [15]. Accordingly, the signaling mechanisms involved in LTD are closely linked to those implicated in cell death and survival, and it is likely that caspase-3 acts on substrates that are involved in synaptic plasticity, as has been shown for the protein kinase Akt1 [10]. In fact, proteolysis of Akt1 by caspase-3 is not only required for cell death program, but may also represent a key molecular step for LTD induction [38].

A growing body of evidence suggests the importance of the mitochondrial apoptotic pathway in the induction of LTD [1,4,9,38]. The mitochondrial intrinsic apoptotic pathway is engaged by BAD activation, a pro-apoptotic protein, which belongs to the Bcl-2 family [39]. BAD activates BAX, which induces the mitochondrial outer membrane permeabilization (MOMP) [40], which, in turn, promotes the cytosolic release of cytochrome c, with consequent activation of caspase-9 and caspase-3. During the expression of LTD, the BAD-induced caspase activation is necessary to promote AMPA receptor endocytosis and the expression of long-term synaptic plasticity [39]. This is confirmed by experiments on BAD knockout mice, which show deficient caspase-3 activation and AMPA receptor internalization, and the absence of LTD [39,41]. BCL-XL, another member of the Bcl-2 family located on the outer mitochondrial membrane, plays an anti-apoptotic role by inhibiting BAX [38]; it has been demonstrated that also overexpression of BCL-XL inhibits LTD [10]. In this scenario, PINK1 protein can be considered a key actor too. PINK1 and BCL-XL colocalize at the outer mitochondrial membrane, and Arena and colleagues showed that PINK1 can protect cell against apoptosis by phosphorylating BCL-XL, preventing its pro-apoptotic cleavage [41]. The interaction between these two proteins would take place only on depolarized mitochondria, suggesting a specific role for this molecular mechanism in mitochondria physiopathology [41]. Moreover, data support a crucial function of PINK1 in mitochondria-dependent apoptosis [42]. PINK1 acts as a neuroprotective protein, protecting cells from damage-mediated mitochondrial dysfunction [43]. PINK1 mutations impair mitochondrial complex I activity, causing mitochondrial membrane depolarization and increased vulnerability to stress [44], and also affecting crucial cellular processes including neurotransmitter release from presynaptic terminals, which unavoidably impairs long-term synaptic plasticity [22,26,44]. Compelling evidence suggests a complex interaction among mitochondrial intrinsic apoptotic pathway, PINK1, and the long-term synaptic plasticity machinery, where caspase-3 may act as a sensor whose activation degree may determine either the cell fate or the expression of long-term synaptic plasticity.

We hypothesize that a fine modulation of caspase cascade exits at corticostriatal synapses, and that PINK1 deficiency, impairing mitochondrial functional integrity, may also perturb these downstream mechanisms, leading to synaptic dysfunction.

As previously demonstrated, amphetamine acts as an indirect agonist of dopamine receptor, facilitating vesicular dopamine [45] and glutamate release in the striatum [46–48], which can explain its capacity to restore synaptic plasticity deficit in the PINK1 model [22]. Besides the well-known ability of amphetamine to increase dopamine transmission, it may also trigger activation of caspase-3 [49]. This evidence could in part explain the failed activation of caspase-3 in our knockout PINK1 model, given the link between the mitochondrial BCL-2/Beclin complex and the absence of PINK1. In this regard, our results suggest that the rescue of LTD exerted by amphetamine in the PINK1 homozygous mouse can be ascribed to caspase-3 activation. This hypothesis is further supported by the finding that caspase-3 inhibition by Z-Devd-fmk completely blocked LTD amphetamine-mediated rescue. Further studies are required to establish the precise mechanism of this involvement. However, we hypothesize that caspase 3 is involved in striatal synaptic plasticity processes in our transgenic PINK1 mouse model.

4. Material and Methods

4.1. Animals

The Animal Care and Use Committees of University of Rome "Tor Vergata" and IRCCS Fondazione Santa Lucia, and the Italian Ministry of Health approved the experiments (Aut. Nr. 517/2016-PR). All the experiments were carried out in accordance to the Directive EU 2010/63 on the protection of animals used for scientific purposes, and the implementation of the directive by Italian legislature: DLS/26 04/03/2014. Mice were generated and characterized as previously reported [22]. Homozygous ($PINK1^{-/-}$) and heterozygous ($PINK1^{+/-}$) mice and their wild-type littermates ($PINK1^{+/+}$) were bred at our animal house. All experiments were performed blindly. Body weight and animal welfare were monitored throughout the duration of experiments. For genotyping, DNA was isolated from mouse-tail using the Extract-N-Amp Tissue PCR Kit (Sigma-Aldrich, Milan, Italy) and processed as previously described [26].

4.2. Electrophysiology

Mice (2–3 months of age) were sacrificed by cervical dislocation and the brain was immediately removed from the skull and cut with a vibratome (Leica Microsystems, Buccinasco, Milan, Italy) in Krebs' solution (126 NaCl, 2.5 KCl, 1.3 $MgCl_2$, 1.2 NaH_2PO_4, 2.4 $CaCl_2$, 10 glucose, 18 $NaHCO_3$, all expressed in mM), bubbled with 95% O_2 and 5% CO_2, as previously described [25,26]. Sagittal corticostriatal slices (300 μm thick) were maintained in oxygenated Krebs' solution for 60 min to recover, then transferred into a recording chamber, continuously superfused with oxygenated Krebs' solution at 32–33 °C. Sharp-electrode recordings of striatal medium spiny neurons (MSNs) were performed in the current-clamp configuration, by using intracellular borosilicate electrodes filled with 2M KCl (30–60 MΩ). Signal acquisition and off-line analysis were performed with Axoclamp 2B amplifiers and pClamp 10.2 (Axon instruments, Molecular Devices, LLC. San Jose, CA, USA).

A bipolar electrode was placed in the corpus callosum to evoke glutamatergic corticostriatal excitatory postsynaptic potentials (EPSPs), in the presence of picrotoxin (PTX, 50 µM), a gamma-aminobutyric acid-A (GABA-A) receptors antagonist. The current–voltage (I–V) relationship was assessed by applying 10 mV steps (0.3 s duration) ranging from −140 to −40 mV. A negative step of −10 mV was utilized to measure membrane resistance. Paired-pulse facilitation (PPF) was evaluated by delivering two stimuli at 50 ms interstimulus interval (ISI), in PTX (50 µM) and measuring the EPSP2/EPSP1 ratio (PPR). For vesicular release experiments, multiple EPSPs were evoked by electric stimulation (30 Hz stimulation frequency, 30 s duration) to induce glutamate release [50,51]. High-frequency supra-threshold stimulation (HFS, three trains 100 Hz, 3 s, 20 s apart) was delivered to induce long-term depression (LTD). For each cell, the amplitude of EPSPs was averaged and plotted over-time as percentage of the mean amplitude of pre-HFS control EPSP. A further set of recordings was performed after preincubation of corticostriatal slices with caspase-3 modulators, the inhibitor Z-Devd-fmk (5 µM) and the activator PETCM (30 µM), respectively, applied for 1 h, at room temperature [10].

Synaptic responses were also recorded in the whole-cell configuration of the patch-clamp technique. Neurons were visualized using infrared differential interference contrast (IR-DIC) video microscopy in the dorsal striatum, as described [52]. Recordings were made with an Axopatch 200 amplifier (Axon instruments, Molecular Devices, LLC. San Jose, CA, USA), using borosilicate glass pipettes (1.5 mm outer diameter, 0.86 inner diameter) pulled on a P-97 Puller (Sutter Instruments, Novato, CA 94949, USA). Pipette resistances ranged from 2.5 to 5 MΩ. Membrane currents were continuously monitored, and access resistance measured in voltage-clamp was in the range of 5–30 MΩ prior to electronic compensation (60%–80% routinely used). Data were acquired with Clampex 10.6 software ((Axon instruments- Molecular Devices, LLC. San Jose, CA, USA). The pipette internal solution included (in mM): K+-gluconate (125), NaCl (10), $CaCl_2$ (1.0), $MgCl_2$ (2.0), 1,2-bis (2-aminophenoxy) ethane-N,N,N,N-tetra-acetic acid (BAPTA) (1), Hepes (10), guanosine 5′-triphosphate (GTP) (0.3) Mg-adenosine triphosphate (ATP) (2.0); pH adjusted to 7.3 with potassium hydroxide KOH. Pirotoxin (50 µM) was added to block GABA-A currents.

4.3. Caspase-3 Biochemical Assay

Corticostriatal slices (300 µm thick) from PINK1$^{+/+}$, PINK1$^{+/-}$, and PINK1$^{-/-}$ mice were weighed and homogenized with a motor-driven pestle for two twenty-strokes cycles, with freezing and thawing of samples. After centrifugation at 10,000× *g* for 10 min at 4 °C, the supernatant was assessed by Bradford assay. A total of 150 µg total proteins for each genotype was processed according to manufacturer's protocol (Abcam AB39401). The 96-well plate was incubated at 37 °C for 60 min and the output was measured (OD 405 nm) on a microplate reader, Thermo Fisher Scientific Multiskan GO.4.3.

4.4. Statistical Analysis

For electrophysiological experiments, we recorded one single MSN for each brain slice, and at least six MSNs for each experimental condition. The number of animals used was at least three for each condition. Power analysis was used to determine the sample size required for different sets of experiments, in accordance to the principle of scientific reduction (Russell and Burch 1959). All data are presented as mean ± standard error mean (SEM). Data were analyzed off-line using Clampfit 10.7 and Adalta Origin 2016. Student's *t*-test was used to evaluate statistical differences between two groups. One-way analysis of variance (ANOVA) followed by post-hoc Tukey was used to compare groups. Two-way ANOVA was used to analyze the effect of multiple variables on all genotypes. An analysis of variance with Tukey's post-hoc test was performed among groups. Alpha was set at 0.05; the null hypothesis was rejected for $p \leq$ alpha and the experimental hypothesis was accepted. Vesicular release was calculated using the cumulative amplitude analysis [52,53] after protocol induction of 30 Hz/30 sec [33]. The time course of synaptic responses to sustained stimulation was plotted in function of cumulative amplitude. Data plotted represent the means of 4/6 independent recordings ± SEM. Data plotted were fitted in function of time, and the time constant, relative to rapid synaptic depression, was calculated by Boltzmann fit. All differences by genotypes were analyzed by ANOVA followed by post hoc test.

For biochemical experiments, we obtained eight different slices from each animal. Data were analyzed using Adalta Origin 2016 software. One-way ANOVA followed by post-hoc Tukey was used to compare groups. *t*-test was used for all other conditions. Statistical significant was fixed at $p < 0.05$. All the data are shown as mean ± standard deviation (SD).

4.5. Drug Sources

Picrotoxin, (+)–MK801, and CNQX were purchased from Tocris Cookson, Bristol, UK. The genotyping kit was provided by Sigma Aldrich, Milan, Italy. Caspase inhibitors and activators were purchased from Tocris Cookson, Bristol, UK. All other drugs were purchased from Panreac Química, Castellar del Vallès (Barcelona, Spain). Caspase-3 Assay Kit (Colorimetric) was provided by Abcam, Cambridge, UK.

5. Conclusions

Our findings support the idea that the mitochondrial-dependent caspase pathway plays a critical role in synaptic depression and that activity of caspase-3 is crucial for the expression of LTD in the PINK1 mouse model of PD. Indeed, in our experiments, a low level of caspase-3, completely insufficient to drive the cells toward apoptosis, could contribute to activating the molecular mechanisms involved in synaptic plasticity and could restore the impairment of LTD in our animal model. Exploring the non-apoptotic activation and modulation of caspase-3 may contribute to a better comprehension of caspases' anti-apoptotic involvement at corticostriatal level, as well as help to clarify the mechanisms involved in synaptic failure in the pathophysiology of PD [54,55], also in view of rationale therapeutic approaches for counteracting PD progression. Furthermore, in a broader view, low-non-lethal levels of caspase-3 may be involved in cellular mechanisms relevant also for other neurodegenerative diseases [56]. The evidence of a biphasic response of caspase-3 activation gives us a new prospective to investigate the pathophysiology of neurodegenerative disorders.

Supplementary Materials: Supplementary materials can be found at http://www.mdpi.com/1422-0067/20/14/3407/s1.

Author Contributions: Conceptualization, A.T. and G.M. (Giuseppina Martella); methodology, A.T., M.M., G.P. and G.M. (Giuseppina Martella); validation, P.I., A.T., M.M. and G.M. (Giuseppina Martella); formal analysis, P.I., A.T., M.M., G.P., G.M. (Graziella Madeo) and G.M. (Giuseppina Martella); investigation, G.M. (Giuseppina Martella); resources, A.P. and G.M. (Giuseppina Martella); data curation, A.T., M.M., G.P., G.M. (Graziella Madeo), P.B. and G.M. (Giuseppina Martella); writing—original draft preparation, P.I. and G.M. (Graziella Madeo); writing—review and editing, P.I. and P.B.; visualization, M.M. and G.P.; supervision, P.I., A.P. and P.B.; project administration, A.P. and G.M. (Giuseppina Martella); funding acquisition, A.P.

Funding: This paper was supported by the following grants: Fondazione Baroni 2016 to A.P.; Fondazione Cariplo to G.S.; and PRIN 2015 (2015FNWP34, Italian Ministry of University and Research) to A.P.

Acknowledgments: All the authors wish to thank Massimo Tolu for his excellent technical assistance and experimental support, and EM Valente for her fruitful.

Conflicts of Interest: The authors declare no conflict of interest.

References

1. Shalini, S.; Dorstyn, L.; Dawar, S.; Kumar, S. Old, new and emerging functions of caspases. *Cell Death Differ.* **2015**, *22*, 526–539. [CrossRef] [PubMed]
2. McIlwain, D.R.; Berger, T.; Mak, T.W. Caspase functions in cell death and disease. *Cold Spring Harb. Perspect. Biol.* **2013**, *5*, a008656. [CrossRef] [PubMed]
3. Hengartner, M.O. The biochemistry of apoptosis. *Nature* **2000**, *407*, 770–776. [CrossRef] [PubMed]
4. Gilman, C.P.; Mattson, M.P. Do apoptotic mechanisms regulate synaptic plasticity and growth-cone motility? *Neuromol. Med.* **2002**, *2*, 197–214. [CrossRef]

5. Campbell, D.S.; Holt, C.E. Apoptotic pathway and MAPKs differentially regulate chemotropic responses of retinal growth cones. *Neuron* **2003**, *37*, 939–952. [CrossRef]
6. Kuo, C.T.; Zhu, S.; Younger, S.; Jan, L.Y.; Jan, Y.N. Identification of E2/E3 ubiquitinating enzymes and caspase activity regulating Drosophila sensory neuron dendrite pruning. *Neuron* **2006**, *51*, 283–290. [CrossRef] [PubMed]
7. Williams, D.W.; Kondo, S.; Krzyzanowska, A.; Hiromi, Y.; Truman, J.W. Local caspase activity directs engulfment of dendrites during pruning. *Nat. Neurosci.* **2006**, *9*, 1234–1236. [CrossRef]
8. Chan, S.L.; Mattson, M.P. Caspase and calpain substrates: Roles in synaptic plasticity and cell death. *J. Neurosci. Res.* **1999**, *58*, 167–190. [CrossRef]
9. D'Amelio, M.; Cavallucci, V.; Cecconi, F. Neuronal caspase-3 signaling: Not only cell death. *Cell Death Differ.* **2010**, *17*, 1104–1114. [CrossRef]
10. Li, Z.; Jo, J.; Jia, J.M.; Lo, S.C.; Whitcomb, D.J.; Jiao, S.; Cho, K.; Sheng, M. Caspase-3 activation via mitochondria is required for long-term depression and AMPA receptor internalization. *Cell* **2010**, *141*, 859–871. [CrossRef]
11. Truman, J.W. Metamorphosis of the central nervous system of Drosophila. *J. Neurobiol.* **1990**, *21*, 1072–1084. [CrossRef] [PubMed]
12. Luo, L.; O'Leary, D.D.M. Axon retraction and degeneration in development and disease. *Annu. Rev. Neurosci.* **2005**, *28*, 127–156. [CrossRef] [PubMed]
13. Shen, C.; Xiong, W.C.; Mei, L. Caspase-3, shears for synapse pruning. *Dev. Cell* **2014**, *28*, 604–606. [CrossRef] [PubMed]
14. Kuida, K.; Zheng, T.S.; Na, S.; Kuan, C.; Yang, D.; Karasuyama, H.; Rakic, P.; Flavell, R.A. Decreased apoptosis in the brain and premature lethality in CPP32-deficient mice. *Nature* **1996**, *384*, 368–372. [CrossRef] [PubMed]
15. Snigdha, S.; Smith, E.D.; Prieto, G.A.; Cotman, C.W. Caspase-3 activation as a bifurcation point between plasticity and cell death. *Neurosci. Bull.* **2012**, *28*, 14–24. [CrossRef] [PubMed]
16. Dash, P.K.; Blum, S.; Moore, A.N. Caspase activity plays an essential role in long-term memory. *Neuroreport* **2000**, *11*, 2811–2816. [CrossRef] [PubMed]
17. Glazner, G.W.; Chan, S.L.; Lu, C.; Mattson, M.P. Caspase-mediated degradation of AMPA receptor subunits: A mechanism for preventing excitotoxic necrosis and ensuring apoptosis. *J. Neurosci.* **2000**, *20*, 3641–3649. [CrossRef] [PubMed]
18. Lu, C.; Fu, W.; Salvesen, G.S.; Mattson, M.P. Direct cleavage of AMPA receptor subunit GluR1 and suppression of AMPA currents by caspase-3: Implications for synaptic plasticity and excitotoxic neuronal death. *Neuromol. Med.* **2002**, *1*, 69–79.
19. Ye, B.; Sugo, N.; Hurn, P.D.; Huganir, R.L. Physiological and pathological caspase cleavage of the neuronal RasGEF GRASP-1 as detected using a cleavage site-specific antibody. *Neuroscience* **2002**, *114*, 217–227. [CrossRef]
20. Bravarenko, N.I.; Onufriev, M.V.; Stepanichev, M.Y.; Ierusalimsky, V.N.; Balaban, P.M.; Gulyaeva, N.V. Caspase-like activity is essential for long-term synaptic plasticity in the terrestrial snail Helix. *Eur. J. Neurosci.* **2006**, *23*, 129–140. [CrossRef]
21. Valente, E.M.; Abou-Sleiman, P.M.; Caputo, V.; Muqit, M.M.K.; Harvey, K.; Gispert, S.; Ali, Z.; Del Turco, D.; Bentivoglio, A.R.; Healy, D.G.; et al. Hereditary early-onset Parkinson's disease caused by mutations in PINK1. *Science* **2004**, *304*, 1158–1160. [CrossRef] [PubMed]
22. Kitada, T.; Pisani, A.; Porter, D.R.; Yamaguchi, H.; Tscherter, A.; Martella, G.; Bonsi, P.; Zhang, C.; Pothos, E.N.; Shen, J. Impaired dopamine release and synaptic plasticity in the striatum of PINK1-deficient mice. *Proc. Natl. Acad. Sci. USA* **2007**, *104*, 11441–11446. [CrossRef] [PubMed]
23. Madeo, G.; Schirinzi, T.; Martella, G.; Latagliata, E.C.; Puglisi, F.; Shen, J.; Valente, E.M.; Federici, M.; Mercuri, N.B.; Puglisi-Allegra, S.; et al. PINK1 heterozygous mutations induce subtle alterations in dopamine-dependent synaptic plasticity. *Mov. Disord.* **2014**, *29*, 41–53. [CrossRef] [PubMed]
24. Madeo, G.; Martella, G.; Schirinzi, T.; Ponterio, G.; Shen, J.; Bonsi, P.; Pisani, A. Aberrant striatal synaptic plasticity in monogenic parkinsonisms. *Neuroscience* **2012**, *211*, 126–135. [CrossRef] [PubMed]
25. Martella, G.; Maltese, M.; Nisticò, R.; Schirinzi, T.; Madeo, G.; Sciamanna, G.; Ponterio, G.; Tassone, A.; Mandolesi, G.; Vanni, V.; et al. Regional specificity of synaptic plasticity deficits in a knock-in mouse model of DYT1 dystonia. *Neurobiol. Dis.* **2014**, *65*, 124–132. [CrossRef] [PubMed]

26. Martella, G.; Madeo, G.; Maltese, M.; Vanni, V.; Puglisi, F.; Ferraro, E.; Schirinzi, T.; Valente, E.M.; Bonanni, L.; Shen, J.; et al. Exposure to low-dose rotenone precipitates synaptic plasticity alterations in PINK1 heterozygous knockout mice. *Neurobiol. Dis.* **2016**, *91*, 21–36. [CrossRef] [PubMed]
27. Martella, G.; Tassone, A.; Sciamanna, G.; Platania, P.; Cuomo, D.; Viscomi, M.T.; Bonsi, P.; Cacci, E.; Biagioni, S.; Usiello, A.; et al. Impairment of bidirectional synaptic plasticity in the striatum of a mouse model of DYT1 dystonia: Role of endogenous acetylcholine. *Brain* **2009**, *132*, 2336–2349. [CrossRef]
28. Goldberg, M.S.; Fleming, S.M.; Palacino, J.J.; Cepeda, C.; Lam, H.A.; Bhatnagar, A.; Meloni, E.G.; Wu, N.; Ackerson, L.C.; Klapstein, G.J.; et al. Parkin-deficient mice exhibit nigrostriatal deficits but not loss of dopaminergic neurons. *J. Biol. Chem.* **2003**, *278*, 43628–43635. [CrossRef]
29. Goldberg, M.S.; Pisani, A.; Haburcak, M.; Vortherms, T.A.; Kitada, T.; Costa, C.; Tong, Y.; Martella, G.; Tscherter, A.; Martins, A.; et al. Nigrostriatal dopaminergic deficits and hypokinesia caused by inactivation of the familial Parkinsonism-linked gene DJ-1. *Neuron* **2005**, *45*, 489–496. [CrossRef]
30. Kitada, T.; Pisani, A.; Karouani, M.; Haburcak, M.; Martella, G.; Tscherter, A.; Platania, P.; Wu, B.; Pothos, E.N.; Shen, J. Impaired dopamine release and synaptic plasticity in the striatum of parkin-/- mice. *J. Neurochem.* **2009**, *110*, 613–621. [CrossRef]
31. Ghiglieri, V.; Calabrese, V.; Calabresi, P. Alpha-Synuclein: From Early Synaptic Dysfunction to Neurodegeneration. *Front. Neurol.* **2018**, *9*, 295. [CrossRef] [PubMed]
32. Sara, Y.; Mozhayeva, M.G.; Liu, X.; Kavalali, E.T. Fast vesicle recycling supports neurotransmission during sustained stimulation at hippocampal synapses. *J. Neurosci.* **2002**, *22*, 1608–1617. [CrossRef] [PubMed]
33. Choi, S.; Lovinger, D.M. Decreased probability of neurotransmitter release underlies striatal long-term depression and postnatal development of corticostriatal synapses. *Proc. Natl. Acad. Sci. USA* **1997**, *94*, 2665–2670. [CrossRef] [PubMed]
34. Avelar, A.J.; Juliano, S.A.; Garris, P.A. Amphetamine augments vesicular dopamine release in the dorsal and ventral striatum through different mechanisms. *J. Neurochem.* **2013**, *125*, 373–385. [CrossRef] [PubMed]
35. Underhill, S.M.; Wheeler, D.S.; Li, M.; Watts, S.D.; Ingram, S.L.; Amara, S.G. Amphetamine modulates excitatory neurotransmission through endocytosis of the glutamate transporter EAAT3 in dopamine neurons. *Neuron* **2014**, *83*, 404–416. [CrossRef] [PubMed]
36. Reid, M.S.; Ho, L.B.; Tolliver, B.K.; Wolkowitz, O.M.; Berger, S.P. Partial reversal of stress-induced behavioral sensitization to amphetamine following metyraponetreatment. *Brain Res.* **1998**, *783*, 133–142. [CrossRef]
37. Kalivas, P.W. Cocaine and amphetamine-like psychostimulants: Neurocircuitry and glutamate neuroplasticity. *Dialogues Clin. Neurosci.* **2007**, *9*, 389–397. [PubMed]
38. Li, Z.; Sheng, M. Caspases in synaptic plasticity. *Mol. Brain* **2012**, *5*, 15. [CrossRef] [PubMed]
39. Jiao, S.; Li, Z. Nonapoptotic function of BAD and BAX in long-term depression of synaptic transmission. *Neuron* **2011**, *70*, 758–772. [CrossRef] [PubMed]
40. Danial, N.N. BAD: Undertaker by night, candyman by day. *Oncogene* **2008**, *27*, S53–S70. [CrossRef]
41. Arena, G.; Gelmetti, V.; Torosantucci, L.; Vignone, D.; Lamorte, G.; De Rosa, P.; Cilia, E.; Jonas, E.A.; Valente, E.M. PINK1 protects against cell death induced by mitochondrial depolarization, by phosphorylating Bcl-xL and impairing its pro-apoptotic cleavage. *Cell Death Differ.* **2013**, *20*, 920–930. [CrossRef] [PubMed]
42. Zhu, Y.; Hoell, P.; Ahlemeyer, B.; Krieglstein, J. PTEN: A crucial mediator of mitochondria-dependent apoptosis. *Apoptosis* **2006**, *11*, 197–207. [CrossRef]
43. Matsuda, S.; Kitagishi, Y.; Kobayashi, M. Function and characteristics of PINK1 in mitochondria. *Oxid. Med. Cell Longev.* **2013**, *2013*, 601587. [CrossRef] [PubMed]
44. Morais, V.A.; Verstreken, P.; Roethig, A.; Smet, J.; Snellinx, A.; Vanbrabant, M.; Haddad, D.; Frezza, C.; Mandemakers, W.; Vogt-Weisenhorn, D.; et al. Parkinson's disease mutations in PINK1 result in decreased Complex I activity and deficient synaptic function. *EMBO Mol. Med.* **2009**, *1*, 99–111. [CrossRef] [PubMed]
45. Calipari, E.S.; Ferris, M.J. Amphetamine mechanisms and actions at the dopamine terminal revisited. *J. Neurosci.* **2013**, *33*, 8923–8925. [CrossRef] [PubMed]
46. Nash, J.F.; Yamamoto, B.K. Effect of D-amphetamine on the extracellular concentrations of glutamate and dopamine in iprindole-treated rats. *Brain Res.* **1993**, *627*, 1–8. [CrossRef]
47. Del Arco, A.; González-Mora, J.L.; Armas, V.R.; Mora, F. Amphetamine increases the extracellular concentration of glutamate in striatum of the awake rat: Involvement of high affinity transporter mechanisms. *Neuropharmacology* **1999**, *38*, 943–954. [CrossRef]

48. Rawls, S.M.; McGinty, J.F. Delta opioid receptors regulate calcium-dependent, amphetamine-evoked glutamate levels in the rat striatum: An in vivo microdialysis study. *Brain Res.* **2000**, *861*, 296–304. [CrossRef]
49. Cunha-Oliveira, T.; Rego, A.C.; Cardoso, S.M.; Borges, F.; Swerdlow, R.H.; Macedo, T.; de Oliveira, C.R. Mitochondrial dysfunction and caspase activation in rat cortical neurons treated with cocaine or amphetamine. *Brain Res.* **2006**, *1089*, 44–54. [CrossRef]
50. Zhang, X.; Zhou, Z.; Winterer, J.; Müller, W.; Stanton, P.K. NMDA-dependent, but not group I metabotropic glutamate receptor-dependent, long-term depression at Schaffer collateral-CA1 synapses is associated with long-term reduction of release from the rapidly recycling presynaptic vesicle pool. *J. Neurosci.* **2006**, *26*, 10270–10280. [CrossRef]
51. Sciamanna, G.; Tassone, A.; Mandolesi, G.; Puglisi, F.; Ponterio, G.; Martella, G.; Madeo, G.; Bernardi, G.; Standaert, D.G.; Bonsi, P.; et al. Cholinergic dysfunction alters synaptic integration between thalamostriatal and corticostriatal inputs in DYT1 dystonia. *J. Neurosci.* **2012**, *32*, 11991–12004. [CrossRef] [PubMed]
52. Baldelli, P.; Fassio, A.; Valtorta, F.; Benfenati, F. Lack of synapsin I reduces the readily releasable pool of synaptic vesicles at central inhibitory synapses. *J. Neurosci.* **2007**, *27*, 13520–13531. [CrossRef] [PubMed]
53. Stevens, C.F.; Williams, J.H. Discharge of the readily releasable pool with action potentials at hippocampal synapses. *J. Neurophysiol.* **2007**, *98*, 3221–3229. [CrossRef] [PubMed]
54. Zhou, H.; Li, X.M.; Meinkoth, J.; Pittmanm, R.N. Akt regulates cell survival and apoptosis at a postmitochondrial level. *J. Cell Biol.* **2000**, *151*, 483–494. [CrossRef]
55. Luyet, C.; Schulze, K.; Sayar, B.S.; Howald, D.; Müller, E.J.; Galichet, A. Preclinical studies identify non-apoptotic low-level caspase-3 as therapeutic target in pemphigus vulgaris. *PLoS ONE* **2015**, *10*, e0119809. [CrossRef]
56. Parrish, A.B.; Freel, C.D.; Kornbluth, S. Cellular mechanisms controlling caspase activation and function. *Cold Spring Harb. Perspect. Biol.* **2013**, *5*, a008672. [CrossRef]

International Journal of
Molecular Sciences

MDPI

Article

Corticosterone Upregulates Gene and Protein Expression of Catecholamine Markers in Organotypic Brainstem Cultures

Carla L. Busceti [1], Rosangela Ferese [1], Domenico Bucci [1], Larisa Ryskalin [2], Stefano Gambardella [1], Michele Madonna [1], Ferdinando Nicoletti [1,3] and Francesco Fornai [1,2,*]

1 I.R.C.C.S. Neuromed, 86077 Pozzilli, Italy; carla.busceti@neuromed.it (C.L.B.); rosangela.ferese@neuromed.it (R.F.); domenico.bucci@neuromed.it (D.B.); stefano.gambardella@neuromed.it (S.G.); stabulario@neuromed.it (M.M.); nicoletti@neuromed.it (F.N.)

2 Department of Translational Research and New Technologies in Medicine and Surgery, University of Pisa, 56126 Pisa, Italy; larisa.ryskalin@unipi.it

3 Department of Physiology and Pharmacology, University Sapienza, 00185 Roma, Italy

* Correspondence: francesco.fornai@neuromed.it or francesco.fornai@med.unipi.it; Tel.: +39-050-221-8611

Received: 9 April 2019; Accepted: 12 June 2019; Published: 14 June 2019

Abstract: Glucocorticoids are produced by the adrenal cortex and regulate cell metabolism in a variety of organs. This occurs either directly, by acting on specific receptors in a variety of cells, or by stimulating catecholamine expression within neighbor cells of the adrenal medulla. In this way, the whole adrenal gland may support specific metabolic requirements to cope with stressful conditions from external environment or internal organs. In addition, glucocorticoid levels may increase significantly in the presence of inappropriate secretion from adrenal cortex or may be administered at high doses to treat inflammatory disorders. In these conditions, metabolic alterations and increased blood pressure may occur, although altered sleep-waking cycle, anxiety, and mood disorders are frequent. These latter symptoms remain unexplained at the molecular level, although they overlap remarkably with disorders affecting catecholamine nuclei of the brainstem reticular formation. In fact, the present study indicates that various doses of glucocorticoids alter the expression of genes and proteins, which are specific for reticular catecholamine neurons. In detail, corticosterone administration to organotypic mouse brainstem cultures significantly increases Tyrosine hydroxylase (TH) and Dopamine transporter (DAT), while Phenylethanolamine N-methyltransferase (PNMT) is not affected. On the other hand, Dopamine Beta-Hydroxylase (DBH) increases only after very high doses of corticosterone.

Keywords: glucocorticoids; noradrenaline; dopamine; tyrosine hydroxylase; reticular formation; dopamine transporter

1. Introduction

Glucocorticoids are produced from the adrenal cortex and may be administered as a drug. These compounds spread to the whole extracellular space to regulate a variety of functions. The amount of glucocorticoids present is key to regulating inflammation [1,2], cell proliferation [3], pain [4–7] and cell metabolism [8,9] in a variety of organs. In this way, glucocorticoids produce a variety of effects, which are seminal for homeostasis. The anatomical connections between the adrenal cortex and adrenal medulla, due to a specific capillary network, determine high glucocorticoids concentrations focally at the level of the adrenal medulla, where they promote catecholamine synthesis and shift noradrenaline into adrenaline cells through epigenetic effects [10–14]. High levels of glucocorticoids may occur during prolonged stressful conditions [15–17] or as a consequence of inappropriate production primarily within the adrenal cortex [18,19] or they may increase following abnormal pituitary ACTH production [20–22]. Similarly, exogenous administration of glucocorticoids is a common way to treat a variety of disorders, including chronic inflammatory diseases. In all these circumstances, glucocorticoids produce multiple effects, which may lead to Cushing syndrome [23]. This syndrome features altered plasma glucose level [24], abnormally high pressure [25], altered distribution of lipids in the body [26], thinning of the skin [27], and muscle atrophy, mostly in the lower limb [28]. Neurological alterations are common, featuring mood disorders [29], altered sleep-waking cycle [30], autonomic dysfunction [31], anxiety [32], and movement disorders [33]. Neuropsychiatric symptoms occurring in Cushing syndrome are still lacking a clear molecular explanation and even the brain regions being affected remain under debate. Most behavioral symptoms persist for a considerable amount of time, even when normal glucocorticoid levels are re-established back to control values. This suggests that exposure to high levels of corticosteroid produces plastic changes of neuronal phenotype, which persist over time. This is similar to the phenotypic shift which normally takes place within adrenal medulla. Thus, we wondered whether prolonged high corticosteroids levels may produce a phenotypic shift within catecholamine containing cells of central nervous system (CNS). All catecholamine nuclei are placed in a small region within the caudal part of the CNS known as the brainstem. These nuclei are highly interconnected within the brainstem reticular formation, which contains noradrenaline (NA), dopamine (DA) and adrenaline (A) cell groups. Despite these nuclei being placed in a small brain area, their axons spread to the entire CNS. This may help to understand why small nuclei, from a restricted brain region, regulate a variety of functions which are key for survival. These functions are the sleep-waking cycle, mood regulation, central control of blood pressure, and anxiety. A number of functions of these reticular nuclei overlap with those affected during exposure to high levels of glucocorticoids. In the present study we challenged the hypothesis that glucocorticoids may shift the nuclei of the reticular formation towards a catecholamine phenotype. To test this experimental issue, we selected "ad hoc" experimental settings. Since altered blood levels of corticosteroids may alter brainstem catecholamine nuclei indirectly, due to blood pressure and metabolic changes, we set up an experimental setting allowing us to establish the "pure" direct effects of glucocorticoids on brainstem reticular nuclei. Therefore, instead of an intact CNS with blood vessels and peripheral nerve connections, we used organotypic cultures of mouse brainstem to work within quite preserved short neural networks without risking the bias of non-specific systemic influences drove by blood supply or nerve afferents from the whole body. In this way we could rule out the effects of systemic glucocorticoids. Thus, a potential shift in specific gene and protein expression could be reliably attributed to a focal influence of glucocorticoids on these nuclei. This was further validated by occluding the very same effects when applying focally a glucocorticoid receptor antagonist. In order to consider the potential influence of endogenous corticosteroids in the cell growth serum (which is mandatory to sustain cell survival) we measured corticosteroid levels in the cell medium, which turned out to be way below the lowest dose administered exogenously (0.0067 μM compared with 0.1 μM, respectively). The lack of any influence in the present data of serum glucocorticoids was further demonstrated by administering the receptor antagonist, mifepristone, which did not vary the phenotype of controls, while it prevented the effects produced by exogenously administered glucocorticoids.

2. Results

2.1. TH Increases within Organotypic Cultures Following Incubation with Corticosterone

ELISA analysis showed that the medium culture contains a negligible quantity of corticosterone (0.0067 μM) compared to those selected for the study: 0.1, 0.5, 1, and 200 μM.

These doses were selected in order to reproduce a moderate (0.1–1 μM) to massive, frankly toxicant (200 μM) corticosterone stimulation on the basis of previous studies [34–37]. In order to assess the effect of glucocorticoids on the brainstem, we exposed organotypic cultures following a subdivision of the brainstem in two blocks: the anterior part (also defined as rostral brainstem, from Bregma= −4.3 to Bregma= −6.3) and the posterior part (also defined as the caudal brainstem, from Bregma= −6.3 to Bregma= −7.8) (Figure 1A–C).

Corticosterone concentrations ranging from 0.1 to 1 μM did not induce apoptosis, as shown by unchanged value of Bax/Bcl2 mRNA ratio (Figure 1D,E). However, real time PCR analysis showed a marked upregulation of TH mRNA level in the caudal part which was absent in the rostral part (Figure 2A), while the effect was significant in the caudal part (Figure 2B). Such a rostro-caudal difference is consistent with a remarkable amount of glucocorticoid receptors mRNA levels in the caudal compared with negligible amount in the rostral part of the mouse brainstem (Figure 2C). The glucocorticoid receptor–dependency of such an effect was confirmed by administering the selective glucocorticoid receptor antagonist mifepristone (10 μM), which suppresses the increase of TH mRNA levels in the caudal part of the brainstem (Figure 2D). Remarkably, mifepristone did not modify TH mRNA levels measured in controls, which indicates a lack of effective stimulation of glucocorticoid receptors potentially induced by glucocorticoids detectable in trace amounts within horse serum or even in the medium culture. The increase in TH mRNA levels was consistent with glucocorticoid-induced increase in the TH protein as roughly measured by Western blot analysis (Figure 2E), in very same part of the brainstem for the very same corticosterone doses (0.1, 0.5 or 1 μM, at 24 h). Consistently, immunohistochemistry provided evidence for an increase in TH-immune-staining affecting the caudal part of the brainstem, mostly at the level of A1/C1 nuclei (Figure 2F).

In a separate set of experiments, we assessed the effect of a neurotoxic concentration of corticosterone on the catecholamine system of the mouse brainstem. To this aim, organotypic cultures from the mouse brainstem were treated for 24 h with 200 μM of corticosterone [38–40]. The neurotoxic effect was detected as a significant increase of the Bax/Bcl2 mRNA ratio in cultures of whole mouse brainstem at 24 h of incubation with the highest dose of corticosterone (Figure 3A). Despite a pro-apoptotic evidence, real time PCR and western blot analyses following the highest dose of corticosterone indicate a robust increase in the TH mRNA (Figure 3B) and protein (Figure 3C) when measured from the whole mouse brainstem. However, no effect was observed in the rostral part (Figure 3D), thus the increase of TH mRNA was selectively due to a remarkable effect within the caudal part (Figure 3E). This recapitulates what observed for low doses of corticosterone. This was confirmed by the increased density of TH immune-staining in A1/C1 and A2/C2 regions of the caudal brainstem (Figure 3F).

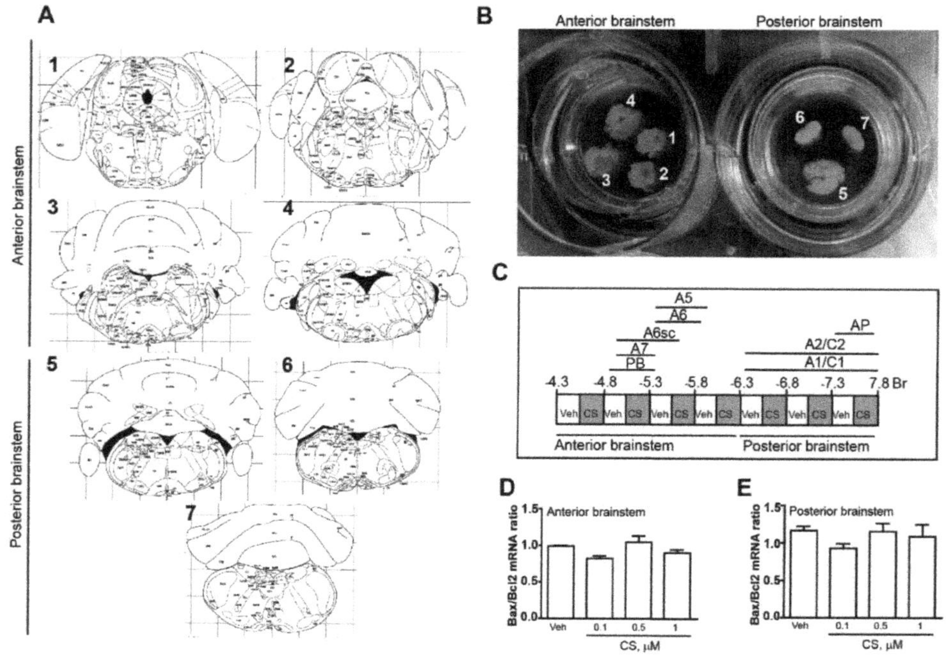

Figure 1. *Organotypic cultures of the mouse brainstem.* (**A**) Plates of brain atlas corresponding to the level chosen for rostral (1–4) and caudal (5–7) slices to set up organotypic cultures. (**B**) Authentic organotypic slices cultured from mouse brainstem. (**C**) Map of catecholamine nuclei of the mouse brainstem from pons to medulla oblongata from Bregma= −4.3 to Bregma= −7.8. The diagram shows the anatomical localization of each brainstem catecholamine nuclei. From each brainstem, 16 coronal sections (250 μm thickness) were sampled with a 250 μm interval. Eight sections were treated with vehicle (Veh), while 8 were administered various corticosterone (CS) doses. The diagram shows the rostro-caudal extension of the rostral (from Bregma= −4.3 to Bregma= −6.3) and the caudal part (from Bregma= −6.3 to Bregma= −7.8) of the mouse brainstem. (**D**,**E**) Real time PCR analysis of the Bax/Bcl2 mRNA ratio in organotypic cultures of the rostral or caudal part of the mouse brainstem at 24 h vehicle (Veh, ethanol 0.01%) or CS (0.1, 0.5 or 1 μM in ethanol 0.01%) incubation. The value reported for the vehicle in Figure 1D was standardized based on vehicle of Figure 1C. This means that in baseline conditions, the BAX/Bcl2 ratio is higher in the caudal compared with rostral brainstem.

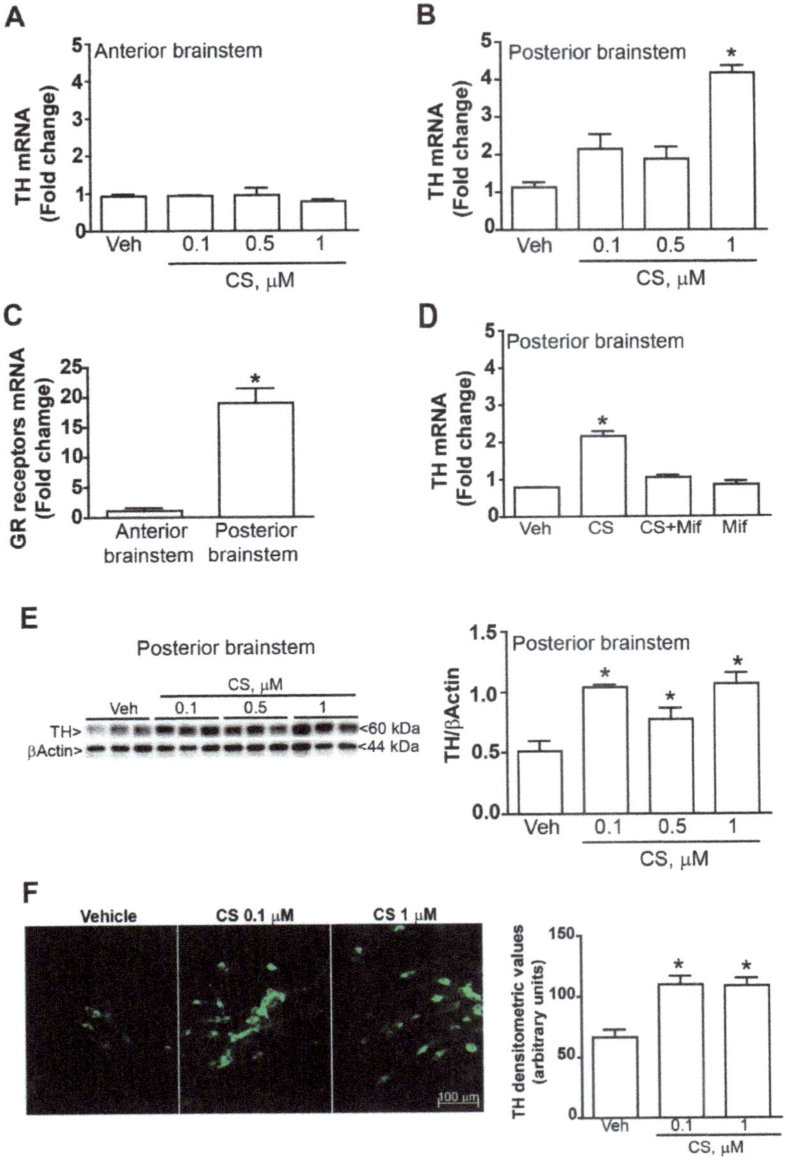

Figure 2. *Corticosterone increases TH expression in the mouse brainstem.* (**A**,**B**) Real time PCR analysis of TH mRNA in cultures from the rostral or caudal part from the mouse brainstem at 24 h vehicle (Veh, ethanol 0.01%) or CS (0.1, 0.5 or 1 μM in ethanol 0.01%) incubation. (**C**) Real time PCR analysis of glucocorticoid receptors (GR) mRNA. (**D**) Real time PCR analysis of TH mRNA in cultures of the caudal part at 24 h incubation with either vehicle (Veh, ethanol 0.01%) or CS (0.1 μM in ethanol 0.01%) in the presence or absence of the selective GR receptors antagonist mifepristone (Mif, 10 μM). (**E**) Western blot analysis of TH in organotypic cultures of the caudal part of the brainstem at 24 h incubation with either vehicle (Veh, ethanol 0.01%) or CS (0.1, 0.5 or 1 μM). (**F**) Immunofluorescence of TH in the A1/C1 catecholamine area of the caudal part of the brainstem at 24 h incubation with either vehicle (Veh, ethanol 0.01%) or CS (0.1, or 1 μM). Densitometry of immunofluorescence was expressed in arbitrary units. All values are expressed as the means±SEM. * $p < 0.05$ compared with vehicle.

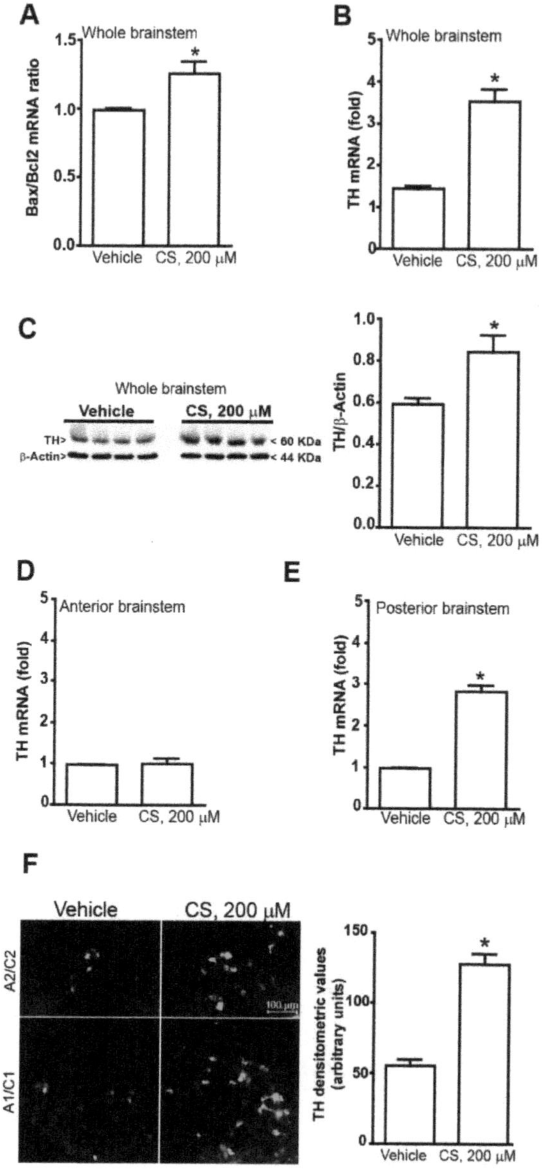

Figure 3. A high dose of corticosterone increases TH expression in the mouse brainstem. (**A**) Real time PCR analysis of the Bax/Bcl2 mRNA ratio in the whole mouse brainstem at 24 h incubation with either vehicle (Veh, ethanol 0.66%) or CS (200 μM in ethanol 0.66%). (**B**) Real time PCR analysis of TH mRNA level and (**C**) western blot analysis of TH protein within the whole mouse brainstem at 24 h incubation with either vehicle (Veh, ethanol 0.66%) or CS (200 μM in ethanol 0.66%). (**D,E**) Real time PCR analysis of TH mRNA within the rostral or caudal and part of the brainstem at 24 h incubation with either vehicle (Veh, ethanol 0.66%) or CS (200 μM in ethanol 0.66%). (**F**) Immunofluorescence of TH in the A1/C1 and A2/C2 catecholamine areas of the caudal brainstem at 24 h incubation with either vehicle (Veh, ethanol 0.66%) or CS (200 μM in ethanol 0.66%). Densitometry of immunofluorescence is expressed in arbitrary units. All values are expressed as the means ± SEM. * $p < 0.05$ compared with the vehicle.

2.2. Low Doses of Corticosterone Increase DAT mRNA and Protein without Altering PNMT and DBH

We examined whether TH increased expression was associated with changes in the expression of the noradrenergic marker DBH and/or the adrenergic marker PNMT. Our data did not show any differences in the expression of DBH and PNMT mRNA levels in organotypic cultures of the caudal part or the whole brainstem in response to treatment with low corticosterone doses (0.1, 0.5 or 1 μM) (Figure 4A,B). Similarly, Western blot analysis did not show any difference in DBH expression level in the caudal part of the brainstem in response to low concentrations of corticosterone (0.1, 0.5 or 1 μM) (Figure 4C). On the other hand, mRNA expression level of the dopaminergic marker DAT was significantly increased in cultures from the caudal part of the brainstem (Figure 4D). This effect depends on the stimulation of corticosterone receptor since it was occluded by the receptor antagonist mifepristone (Figure 4E). It is remarkable that the very same dose of mifepristone did not modify the DAT mRNA expression level in vehicle-treated cultures, thus ruling out an effective role for the negligible amount of corticosteroid we measured in the culture medium.

Similarly to mRNA levels, Western blotting of the DAT indicates an increased protein level in organotypic cultures from the caudal part of the brainstem (Figure 4F).

When the highest doses of corticosterone was administered, PNMT and DBH mRNA levels were still similar to the vehicle when measured in the whole brainstem (Figure 5A,B). In contrast, the highest dose of corticosterone increased mRNA DAT levels even when measured in the whole brainstem (Figure 5C). Unexpectedly, when the DBH protein was immunoblotted in the caudal brainstem, the highest dose of corticosterone produced a significant increase (Figure 5D), which was not documented following the low doses. This suggests that despite the DA phenotype being triggered by low doses of corticosterone, a noradrenergic phenotype may be induced by corticosterone on the mouse brainstem during extreme corticosterone stimulation.

Incidentally, when the DAT levels were tested with Ponceau-stained dot blotting the highest dose of corticosterone replicated the increase in DAT protein documented in the brainstem for the whole range of low doses at western blotting.

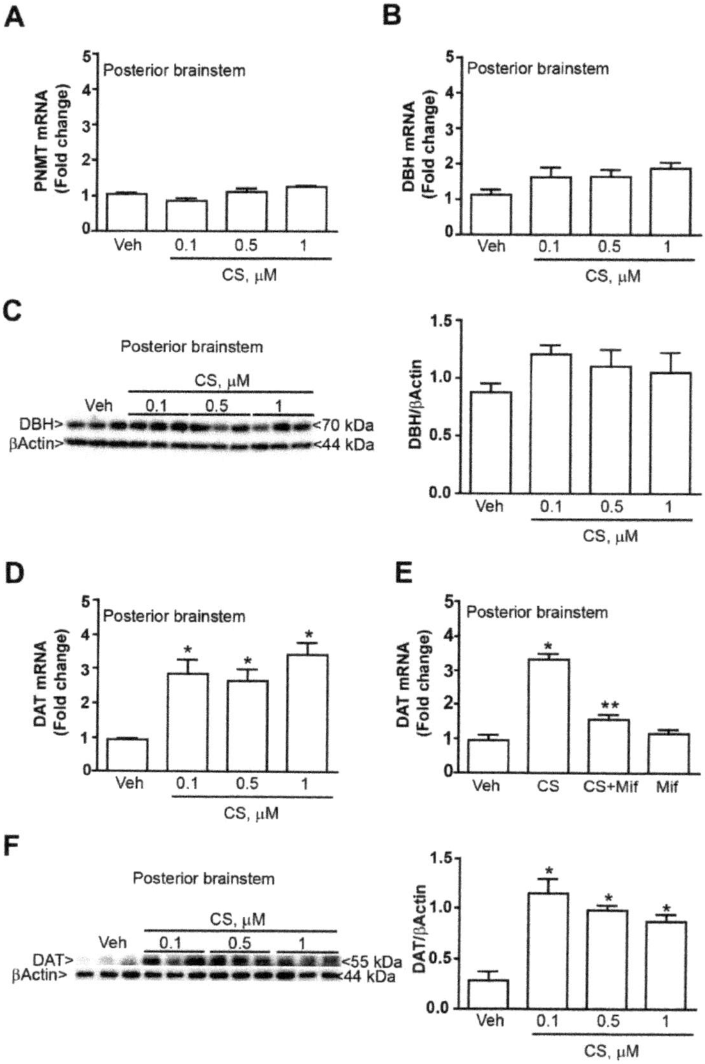

Figure 4. *Corticosterone increases DAT expression without modifying DBH and PNMT expression in the mouse brainstem.* Real time PCR of (**A**) the adrenergic marker PNMT and (**B**) the noradrenergic marker DBH within the caudal part of the brainstem at 24 h incubation with either vehicle (Veh, ethanol 0.01%) or corticosterone (CS, 0.1, 0.5 or 1 μM). (**C**) Western blot of DBH within the caudal part of the brainstem at 24 h incubation with either vehicle (Veh, ethanol 0.01%) or corticosterone (CS, 0.1, 0.5 or 1 μM). (**D**) Real time PCR of the dopaminergic marker (DAT) within the caudal part of the brainstem at 24 h incubation with either vehicle (Veh, ethanol 0.01%) or corticosterone (CS, 0.1, 0.5 or 1 μM). (**E**) Real time PCR of the DAT within the caudal part of the brainstem at 24 h incubation with either vehicle or CS (0.1 μM), in the presence or absence of the selective glucocorticoid receptors antagonist mifepristone (Mif, 10 μM). (**F**) Western blot of DAT within the caudal part of the brainstem at 24 h incubation with either vehicle (Veh, ethanol 0.01%) or corticosterone (CS, 0.1, 0.5 or 1 μM). The lines of the βActin are the same showed in the Figure 2E as the western blot for DAT was performed by incubating with the anti-DAT primary antibody the very same membrane used for TH immunoblotting. All values are expressed as the means±SEM. * $p < 0.05$ compared with the vehicle. ** $p < 0.05$ compared with the vehicle and CS.

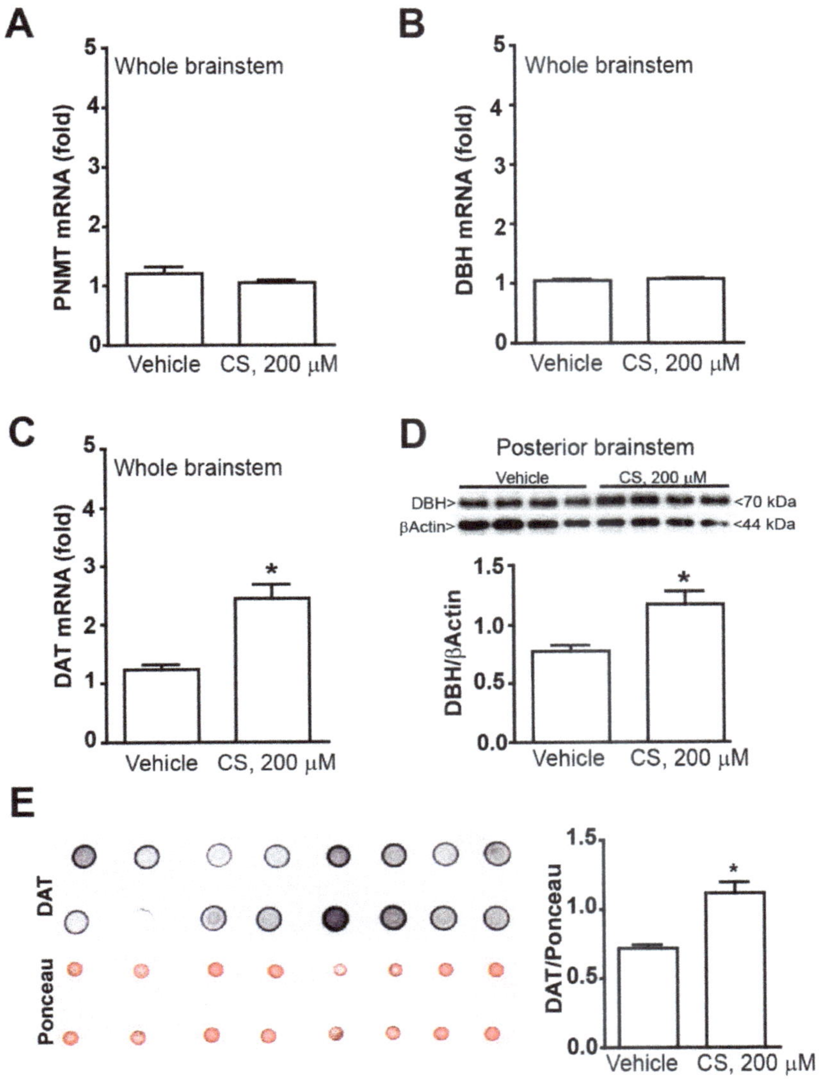

Figure 5. *The highest dose of corticosterone increases DBH expression without modifying PNMT expression in the mouse brainstem.* Real time PCR of (**A**) the adrenergic marker PNMT, (**B**) the noradrenergic marker DBH and (**C**) the dopaminergic marker (DAT) in the whole mouse brainstem at 24 h incubation with either vehicle (Veh, ethanol 0.66%) or CS (200 μM in ethanol 0.66%). (**D**) Western blot of DBH within the caudal part of the brainstem at 24 h incubation with either vehicle (Veh, ethanol 0.66%) or CS (200 μM in ethanol 0.66%). (**E**) Dot blots of DAT in the whole mouse brainstem at 24 h incubation with either vehicle (Veh, ethanol 0.66%) or CS (200 μM in ethanol 0.66%) support data shown at Western blotting. Values are expressed as the means±SEM. * $p < 0.05$ compared with the vehicle.

3. Discussion

In the present manuscript, an increased expression of specific catecholamine-related genes and proteins within mouse caudal brainstem reticular nuclei is documented. These effects occur differently following different doses of corticosterone, ranging from the effects occurring during stressful conditions, or during Cushing syndrome, up to a high toxic concentration.

These data were obtained using organotypic cultures from the mouse brainstem. The cultures were obtained by dissecting thick slices sampled along the whole rostro-caudal extent of the pons and medulla oblongata, ruling out the midbrain.

This experimental setting allowed us to assess the pure effect of corticosterone exposure on the catecholamine system within brainstem local networks, while ruling out the potential bias due to systemic effects induced by whole body corticosterone administration.

These organotypic cultures allow visualizing neuronal aggregates in their coronal extent with a fair preservation of local circuitries. Of course, the effects of distant neural connections are lost in these experimental conditions, although the authentic effects of corticosterone on catecholamine markers can be inferred more reliably. At caudal brainstem level, a marked increase in the catecholamine enzyme TH is documented. This is a key finding since TH is the rate-limiting step in catecholamine biosynthesis. The increase of TH documented here consisted both in mRNA and protein level. The tissue immunofluorescence confirmed increased density of TH immune-staining at the level of A1/C1 area in the caudal brainstem. Remarkably, the *Th* gene owns highly conserved glucocorticoid-responsive elements [41,42]. In the present study, the susceptibility of the caudal compared with rostral brainstem to corticosterone-induced TH expression was found to be consistently related to the impressive expression of glucocorticoid receptors in the caudal brainstem compared with the rostral brainstem. The site-specific regulation of TH expression according to a regional gradient of glucocorticoids receptors may underlie phenotype development of catecholamine neurons under glucocorticoids stimulation, which appear at the birthdate in rodent pups [43].

High density of glucocorticoid receptors also occurs within mesencephalic DA nuclei where, based on their nuclear translocation they are supposed to alter cell phenotype to influence the meso-limbic and meso-striatal DA system [44]. Within this context, it is noteworthy that the amount of meso-striatal DA axons is increased following administration of prednisolone in rodents following damage with the DA neurotoxin 6-hydroxydopamine (6-OHDA) [45]. Remarkably, DA uptake was already reported to be increased as a consequence of increased DAT expression in mice following restraint stress [46]. Nonetheless, so far no study was able to assess an epigenetic mechanism underlying these effects. Further studies analyzing the innumerous connections among the cascade which regulate gene expression will eventually elucidate the molecular mechanism underlying the glucocorticoid regulation of the DAT gene. Such an effort is expected to produce some results during future years since so far, no evidence in available genomic database is present to point to a significant molecular hypothesis. A similar conundrum exists for the *Dbh* gene which in our extreme experimental conditions (the highest dose of corticosterone) was increased as well.

In line with previous studies in the midbrain, the catecholamine nuclei being investigated here also possess glucocorticoid receptors [47–51].

This is in line with data obtained in the present study showing an increase in TH mRNA and protein levels in the brainstem, which correspond to a higher TH immunoreactivity as shown in representative figures of A1/C1 as well as A2/C2. These data suggest that glucocorticoid administration produces a phenotypic shift towards a catecholamine phenotype of hind brainstem reticular neurons. The present data do not contradict the finding of Makino et al. [52] and Zhang et al. [53] who published that stress was also inducing TH expression in a glucocorticoid-independent manner.

In fact, the experimental setting provided by organotypic cultures allowed us to dissect only the direct effects of glucocorticoids on these nuclei independently of systemic effects (increased blood pressure, altered blood glucose and others) which may be produced by systemic glucocorticoids levels, which may in fact increase TH expression in a glucocorticoid-independent manner.

A specific point of the manuscript consists in the measurement of glucocorticoid effects in the caudal brainstem catecholamine nuclei. In order to explain the different response to in vitro treatment with corticosterone between the anterior and the posterior portion of the mouse brainstem, we performed a real time PCR quantification of glucocorticoid receptor mRNA. This study allowed for demonstrating higher level of receptor expression in the posterior compared with the anterior mouse brainstem. This is in line with the data provided here and it confirms the high expression of glucocorticoid receptors in the catecholamine cell groups of the posterior brainstem described by Harfstrand et al. [48].

Remarkably, there may be opposite changes in TH expression depending on which catecholamine nucleus in the reticular formation is considered. Thus, LC which possess a low amount of glucocorticoid receptor appear to undergo a reduced TH expression under the effects of glucocorticoids [49], which confirms a reduced number of glucocorticoid receptors we measured here at rostral level as well as the lack of response of these nuclei to glucocorticoids. In contrast, the nucleus of the solitary tract (which roughly correspond to ala cinerea and the A2/C2 area in its catecholamine containing cells) was reported to increase TH content following exposure to glucocorticoids [54]. Again, this confirms the impressive amount of glucocorticoid receptor at caudal level of the brainstem measured here (Figure 2C). Apart from glucocorticoids modulation of TH expression, to our knowledge, the present study was the first to investigate the expression of mRNA and protein levels of DAT, which is specific DA marker. Despite no alterations were observed neither for PNMT nor for DBH at low doses, glucocorticoids induce an increase in mRNA and protein levels for DAT in the caudal brainstem reticular formation. Remarkably, glucocorticoids may modulate DAT expression in the lower brainstem, which is critical for a variety of functions. So far, increased DAT expression was only investigated in the midbrain by Virdee et al. [55]. These authors demonstrated that antenatal glucocorticoid exposure increases in the adulthood the number of midbrain DA neurons concomitantly with striatal DA fibers. just like what was previously shown in the adult substantia nigra and VTA [56,57]. Consistently, Niwa et al. [58] demonstrated that glucocorticoids mediate the stress-induced increase in the expression of TH within midbrain DA neurons in the adolescent brain. Altogether, these findings demonstrated that midbrain DA neurons are sensitive to glucocorticoids, which increase their catecholamine phenotype. The findings provided here extend these effects to catecholamine neurons of the caudal brainstem showing that a potential phenotypic shift takes place within reticular nuclei due to an increase in DA markers. This is expected to play a key role in the effects produced by the caudal catecholamine cell groups such as blood pressure, breathing, autonomic functions, alertness and anxiety. Remarkably, these correspond to some items, which vary during the Cushing syndrome.

4. Materials and Methods

4.1. Organotypic Brainstem Cultures

For organotypic brainstem, cultures were used postnatal day 18 C57Bl6/J derived from breading in house C57Bl6/J mice (Charles River, Calco, LC, Italy). The animals were rapidly sacrificed, brains dissected and coronally cut with the vibratome Leica VT1200S (Leica Biosystems, Buccinasco, MI, Italy). Slices of 250 μm of thickness were cut along the whole rostro-caudal extension of the pons and medulla oblongata (from Bregma= −4.3 to Bregma= −7.8) without including the midbrain (Figure 1A). We obtained 16 sections/brainstem for an extension of 4 mm. All sections were collected in sterile artificial cerebrospinal fluid (ACSF, 1 mM calcium chloride, 10 mM D-glucose, 4 mM potassium chloride, 5 mM magnesium chloride, 26 mM sodium bicarbonate, 246 mM sucrose) and placed onto a sterile 0.4 μm pore membrane (Merck Millipore, Billerica, MA, USA, Cat# PICM03050, lot n°: R6DA39745) within a 6-well plate. Sections were cultured in 6-well plates at 37 °C and 5% CO2 with 1 mL/well of the following culture medium: 50% MEM/HEPES, 25% Heat Inactivated Horse Serum, 25% Hank's solution, 2 mM NaHCO3, 6.5 mg/mL glucose, 2 mM glutamine, 100 IU/mL penicillin and 100 mg streptomycin, pH 7.2. Sections were incubated for five days and the medium was changed every two days. All animal experiments were approved by Italian Ministry of Health (N° 1065/2016-PR, 7 November 2016).

4.2. Corticosterone ELISA

The corticosterone levels in the culture medium was measured with a corticosterone enzyme-linked immunosorbent assay kit (ELISA; Abcam, Minneapolis, MN, code: ab108821; lot number: GR321740-9), according to the manufacturer's instructions.

4.3. Experimental Planning

For cultures of whole mouse brainstem, from each brainstem we obtained 16 sections (250 μm thickness) for an extension of 4 mm (from Bregma= −4.3 to Bregma= −7.8). Of these, 8 (sampled every 500 μm) were treated with vehicle (0.01% ethanol as control for CS 0.1, 0.5 or 1 μM and 0.66% ethanol as control for CS 200 μM) and 8 (sampled every 500 μm) with different concentrations of corticosterone (Sigma Aldrich, Milan, Italy, Cat# C2505; lot n°: SLBJ5337V). For cultures of the anterior portion of the mouse brainstem, sections sampled from Bregma -4.3 to Bregma -6.3 were used. For cultures of the posterior portion of the mouse brainstem, sections sampled from Bregma -6.3 to Bregma −7.8 (Figure 1B) were used. The concentration of corticosterone used for the experiments were selected based on previous studies in order to expose the cultures to stress-like/Cushing-like concentrations (CS: 0.1; 0.5 or 1 μM; [34–37]) or to a neurotoxic paradigm (CS: 200 μM; [38–40]).

The corticosterone was dissolved in 0.01% ethanol for CS 0.1, 0.5 and 1 μM or in 0.66% ethanol for CS 200 μM. The groups of control were incubated with vehicle (0.01% ethanol or 0.66% ethanol).

Following 24 h of incubation, the organotypic brainstem cultures were pooled (from the whole brainstem: from Bregma= −4.3 to Bregma= −7.8; or from the anterior brainstem: from Bregma= −4.3 to Bregma= −6.3 and the posterior brainstem from Bregma= −6.3 to Bregma= −7.8) and used for Real-Time PCR analyses of Bax, Bcl2, *Th*, *Dbh*, *Pnmt*, *Dat* and *GR* (*Glucocorticoid receptor*) mRNA levels. Separate cultures were subjected to the same experimental condition and used for western blot analysis of TH, dot blot analysis of DBH and DAT or immunohistochemical analysis of TH.

In additional set of experiments, organotypic cultures of the posterior portion of the mouse brainstem were treated for 24 h with CS 0.1 μM in the presence of absence of the selective glucocorticoid receptor antagonist mifepristone (10 μM, Sigma Aldrich, Cat# M8046-100MG; lot n°: WXBC5203V). At the end of incubation, the slices were pooled and used for Real-Time PCR analyses of *Th* and *Dat* mRNA levels.

4.4. Western Blot

Tissues were homogenized at 4°C in ice-cold lysis buffer with phosphatase and protease inhibitor. Twenty-five μg of proteins were incubated for 1 h with a mouse monoclonal anti-TH (1:10000, Sigma Aldrich, Cat# T1299 RRID:AB_477560; lot n°: 015M4759V) or a mouse monoclonal anti β-Actin (1:50000, Sigma Aldrich, Cat# A5441 RRID:AB_476744, lot n°: 028K4826) antibodies and then with secondary peroxidase coupled anti-mouse antibody (1:3000; Merck Millipore, Calbiochem®, Cat# 401215-2ML RRID:AB_437766; lot n°: 2782376). Immunostaining was revealed by enhanced chemiluminescence luminosity (GE Healthcare, Milan, Italy). Densitometric analysis was performed with ImageJ software.

4.5. Dot Blot

For dot blot analyses 1.2 μg of proteins were spotted onto the nitrocellulose membranes. The membranes were blocked for 2 h with 5% non-fat dry milk and then incubated overnight with primary monoclonal rat anti-DAT (1:1000, Merck Millipore, Cat# MAB369 RRID:AB_2190413; lot n°: 2664466) or monoclonal mouse anti-DBH (1:1000, Merck Millipore, Cat# MAB308 RRID:AB_2245740; lot n°: 2688638) antibodies. Filters were washed 3 times with TTBS buffer and then incubated for 1 h with secondary peroxidase-coupled anti-rat (1:3000, Merck Millipore, Calbiochem®, Cat# 401416 RRID:AB_437801; lot n°: D00170203) or anti-mouse (1:3000, Merck Millipore, Calbiochem®, Cat# 401215-2ML RRID:AB_437766; lot n°: 2782376) antibodies. Immunostaining was revealed by enhanced chemiluminescence luminosity (GE Healthcare). For the normalization process, membranes were stained with Ponceau S (Sigma Aldrich, Cat# P3504, lot n°: MKBF2200V, 1 mg/mL in Acetic acid/H2O 1/20) for 10 s. Densitometric analysis was performed with ImageJ software.

4.6. Real Time PCR Analysis

Total RNA was isolated using Trizol Reagent (Thermo Fisher Scientific, Invitrogen™, Waltham, MA, USA, Cat# 10296010, lot n°: 69083302) according to the manufacturer's instructions. The concentration and purity of RNA samples were determined using Nanodrop 2000 (Thermo Fisher Scientific, Life Technologies). Total RNA (100 ng) was reverse transcribed (RT) with SuperScript® VILOTM (Thermo Fisher Scientific, Invitrogen™, Cat# 100011931, lot n°: 1807301) with Oligo dT primers.

Amplification and detection were performed on a CFX Connect™ Real Time System (Bio-Rad, Hercules, CA, USA). PCR mix including 10 µL SYBR Green PCR Master (Applied Biosystems, Foster City, CA, USA, Cat# 4367659, lot n°: 1610538), 0,5 µM of each primer and 1 µL of RT reaction mix, was amplified as follows: 95°C for 10 min followed by 40 cycles of 95°C for 1 min and 30 s, 54°C for 1 min. The following primers have been designed using GenBank (http://www.ncbi.nlm.nih.gov/): *Bax* (NM_007527): 5′-aagctgagcgagtgtctc- 3′, 5′-agttgaagttgccatcagc-3′; *Bcl2* (NM_009741): 5′-ggtggtggaggaactctt-3′, 5′-ggctgagcagggtcttca-3′; *Th* (NM_009377): 5′-acctggagtactttgtgcg-3′, 5′-ctttgtgtattccacgtgtg-3′; *Dbh* (NM_138942): 5′-agccaagactaccagctgc-3′, 5′-gtagctcagtgatatagca-3′; *Pnmt* (NM_008890): 5′-acatcaccatgacagactt-3′, 5′-ctggaagctagtaagatct-3′; *Dat* (NM_010020): 5′-cacagaggactcgagatc-3′, 5′-atgcaggcctgaggtgtt-3′, *GR* 5′-aatgagaccagatgtgagttc-3′, 5′-tagcggcatgctggaca-3′ and *Beta-Globin* (NM_008219): 5′-ctaaggtgaaggctcatg- 3′; 5′-gataggcagcctgcact-3′.

Predesigned TaqMan Gene Expression assays for *Th* was obtained from Applied Biosystems. *Gapdh* was used as the endogenous control to confirm the correct work to SYBR assay.

Positive controls (DNA), negative control (distilled water), and RT-negative controls (total RNA sample) were included in each run.

The relative quantification was calculated using comparative Ct method (also known as the ΔΔCT method) [59,60] and *Beta-Globin* and *Gapdh* were selected as internal references. Ct values correspond to mean values of each PCR performed in triplicate. Gene expression was confirmed in two independent experiments.

4.7. Immunohistochemistry

Following incubation with corticosterone (200 µM, 24 h), cultures were fixed in 2.5% PFA/4% sucrose for 1.5 h and incubated overnight with a mouse monoclonal anti-TH (1:100; Sigma Aldrich, Cat# T1299 RRID:AB_477560; lot n°: 015M4759V) and then for 1 h with a secondary fluorescein anti-mouse antibody (1:50; Vector Laboratories, Burlingame, CA, Cat# BA-1000 RRID:AB_2313606; lot n°: ZA0324).

4.8. Statistical Analysis

All analyses were performed using the Student's t test (Figures 2C, 3 and 5) or One-way Anova + Fisher post hoc (Figure 1, Figure 2A,B,D–F and Figure 4). Differences at $p < 0.05$ were considered statistically significant.

Author Contributions: C.L.B. performed experiments on organotypic brainstem cultures and wrote the manuscript, M.C. performed western blot and dot blot analysis, D.B. and P.D. performed immunohistochemical analysis and revised manuscript, R.F. performed RTPCR analysis, S.G. performed RTPCR analysis and revised manuscript. M.M. breed the animals, L.R. revised manuscript, F.N. and F.F. supervised research and revised manuscript.

Funding: This research was supported by the Italian Ministry of Health (Ricerca Corrente to IRCCS Neuromed)

Conflicts of Interest: The authors declare no conflict of interest.

Abbreviations

6-OHDA	6-hydroxydopamine
A	Adrenaline
CNS	Central nervous system
CS	Corticosterone
DA	Dopamine
DAT	Dopamine transporter
DBH	Beta-Hydroxylase
GR	Glucocorticoid receptors
LC	Locus coeruleus
NA	Noradrenaline
PNMT	Phenylethanolamine N-methyltransferase
TH	Tyrosine hydroxylase
VTA	Ventral tegmental area

References

1. Cain, D.W.; Cidlowski, J.A. Immune regulation by glucocorticoids. *Nat. Rev. Immunol.* **2017**, *17*, 233–247. [CrossRef] [PubMed]
2. Straub, R.H.; Cutolo, M. Glucocorticoids and chronic inflammation. *Rheumatology* **2016**, *55*, ii6–ii14. [CrossRef] [PubMed]
3. Dickmeis, T.; Foulkes, N.S. Glucocorticoids and circadian clock control of cell proliferation: At the interface between three dynamic systems. *Mol. Cell. Endocrinol.* **2011**, *331*, 11–22. [CrossRef] [PubMed]
4. Wang, S.; Lim, G.; Zeng, Q.; Sung, B.; Ai, Y.; Guo, G.; Yang, L.; Mao, J. Expression of central glucocorticoid receptors after peripheral nerve injury contributes to neuropathic pain behaviors in rats. *J. Neurosci.* **2004**, *24*, 8595–8605. [CrossRef] [PubMed]
5. Wang, S.; Lim, G.; Zeng, Q.; Sung, B.; Yang, L.; Mao, J. Central glucocorticoid receptors modulate the expression and function of spinal NMDA receptors after peripheral nerve injury. *J. Neurosci.* **2005**, *25*, 488–495. [CrossRef] [PubMed]
6. Takasaki, I.; Kurihara, T.; Saegusa, H.; Zong, S.; Tanabe, T. Effects of glucocorticoid receptor antagonists on allodynia and hyperalgesia in mouse model of neuropathic pain. *Eur. J. Pharmacol.* **2005**, *524*, 80–83. [CrossRef] [PubMed]
7. Alexander, J.K.; DeVries, A.C.; Kigerl, K.A.; Dahlman, J.M.; Popovich, P.G. Stress exacerbates neuropathic pain via glucocorticoid and NMDA receptor activation. *Brain Behav. Immun.* **2009**, *23*, 851–860. [CrossRef]
8. De Guia, R.M.; Rose, A.J.; Herzig, S. Glucocorticoid hormones and energy homeostasis. *Horm. Mol. Biol. Clin. Investig.* **2014**, *19*, 117–128. [CrossRef]
9. Rafacho, A.; Ortsäter, H.; Nadal, A.; Quesada, I. Glucocorticoid treatment and endocrine pancreas function: Implications for glucose homeostasis, insulin resistance and diabetes. *J. Endocrinol.* **2014**, *223*, R49–R62. [CrossRef]
10. Wong, D.L.; Hayashi, R.J.; Ciaranello, R.D. Regulation of biogenic amine methyltransferases by glucocorticoids via S-adenosylmethionine and its metabolizing enzymes, methionine adenosyltransferase and S-adenosylhomocysteine hydrolase. *Brain Res.* **1985**, *330*, 209–216. [CrossRef]
11. Wong, D.L.; Siddall, B.; Wang, W. Hormonal control of rat adrenal phenylethanolamine N-methyltransferase. Enzyme activity, the final critical pathway. *Neuropsychopharmacology* **1995**, *13*, 223–234. [CrossRef]
12. Tai, T.C.; Wong, D.L. Phenylethanolamine N-methyltransferase gene regulation by cAMP-dependent protein kinase A and protein kinase C signaling pathways. *Ann. N. Y. Acad. Sci.* **2002**, *971*, 83–85. [CrossRef] [PubMed]
13. Tai, T.C.; Claycomb, R.; Siddall, B.J.; Bell, R.A.; Kvetnansky, R.; Wong, D.L. Stress-induced changes in epinephrine expression in the adrenal medulla in vivo. *J. Neurochem.* **2007**, *101*, 1108–1118. [CrossRef] [PubMed]
14. Sharara-Chami, R.I.; Joachim, M.; Pacak, K.; Majzoub, J.A. Glucocorticoid treatment-effect on adrenal medullary catecholamine production. *Shock* **2010**, *33*, 213–217. [CrossRef] [PubMed]

15. Hasan, K.M.; Rahman, M.S.; Arif, K.M.; Sobhani, M.E. Psychological stress and aging: Role of glucocorticoids (GCs). *Age* **2012**, *34*, 1421–1433. [CrossRef] [PubMed]
16. Vyas, S.; Rodrigues, A.J.; Silva, J.M.; Tronche, F.; Almeida, O.F.; Sousa, N.; Sotiropoulos, I. Chronic Stress and Glucocorticoids: From Neuronal Plasticity to Neurodegeneration. *Neural Plast.* **2016**, *2016*. [CrossRef] [PubMed]
17. De Quervain, D.; Schwabe, L.; Roozendaal, B. Stress, glucocorticoids and memory: Implications for treating fear-related disorders. *Nat. Rev. Neurosci.* **2017**, *18*, 7–19. [CrossRef] [PubMed]
18. Stratakis, C.A. Cushing syndrome caused by adrenocortical tumors and hyperplasias (corticotropin-independent Cushing syndrome). *Endocr. Dev.* **2008**, *13*, 117–132. [CrossRef] [PubMed]
19. Lacroix, A. ACTH-independent macronodular adrenal hyperplasia. *Best Pract. Res. Clin. Endocrinol. Metab.* **2009**, *23*, 245–259. [CrossRef]
20. Arnaldi, G.; Angeli, A.; Atkinson, A.B.; Bertagna, X.; Cavagnini, F.; Chrousos, G.P.; Fava, G.A.; Findling, J.W.; Gaillard, R.C.; Grossman, A.B.; et al. Diagnosis and complications of Cushing's syndrome: A consensus statement. *J. Clin. Endocrinol. Metab.* **2003**, *88*, 5593–5602. [CrossRef]
21. Newell-Price, J.; Bertagna, X.; Grossman, A.B.; Nieman, L.K. Cushing's syndrome. *Lancet* **2006**, *367*, 1605–1617. [CrossRef]
22. Biller, B.M.; Grossman, A.B.; Stewart, P.M.; Melmed, S.; Bertagna, X.; Bertherat, J.; Buchfelder, M.; Colao, A.; Hermus, A.R.; Hofland, L.J.; et al. Treatment of adrenocorticotropin-dependent Cushing's syndrome: A consensus statement. *J. Clin. Endocrinol. Metab.* **2008**, *93*, 2454–2462. [CrossRef] [PubMed]
23. Nieman, L.K. Cushing's syndrome: Update on signs, symptoms and biochemical screening. *Eur. J. Endocrinol.* **2015**, *173*, M33–M38. [CrossRef]
24. Pivonello, R.; De Leo, M.; Vitale, P.; Cozzolino, A.; Simeoli, C.; De Martino, M.C.; Lombardi, G.; Colao, A. Pathophysiology of diabetes mellitus in Cushing's syndrome. *Neuroendocrinology* **2010**, *92*, 77–81. [CrossRef] [PubMed]
25. Isidori, A.M.; Graziadio, C.; Paragliola, R.M.; Cozzolino, A.; Ambrogio, A.G.; Colao, A.; Corsello, S.M.; Pivonello, R. The hypertension of Cushing's syndrome: Controversies in the pathophysiology and focus on cardiovascular complications. *J. Hypertens.* **2015**, *33*, 44–60. [CrossRef] [PubMed]
26. Ferraù, F.; Korbonits, M. Metabolic comorbidities in Cushing's syndrome. *Eur. J. Endocrinol.* **2015**, *173*, M133–M157. [CrossRef] [PubMed]
27. Stratakis, C.A. Skin manifestations of Cushing's syndrome. *Rev. Endocr. Metab. Disord.* **2016**, *17*, 283–286. [CrossRef]
28. Cohen, S.; Nathan, J.A.; Goldberg, A.L. Muscle wasting in disease: Molecular mechanisms and promising therapies. *Nat. Rev. Drug Discov.* **2015**, *14*, 58–74. [CrossRef]
29. Bratek, A.; Koźmin-Burzyńska, A.; Górniak, E.; Krysta, K. Psychiatric disorders associated with Cushing's syndrome. *Psychiatr. Danub.* **2015**, *27*, S339–S343.
30. Shipley, J.E.; Schteingart, D.E.; Tandon, R.; Starkman, M.N. Sleep architecture and sleep apnea in patients with Cushing's disease. *Sleep* **1992**, *15*, 514–518. [CrossRef]
31. Chandran, D.S.; Ali, N.; Jaryal, A.K.; Jyotsna, V.P.; Deepak, K.K. Decreased autonomic modulation of heart rate and altered cardiac sympathovagal balance in patients with Cushing's syndrome: Role of endogenous hypercortisolism. *Neuroendocrinology* **2013**, *97*, 309–317. [CrossRef] [PubMed]
32. Pivonello, R.; Simeoli, C.; De Martino, M.C.; Cozzolino, A.; De Leo, M.; Iacuaniello, D.; Pivonello, C.; Negri, M.; Pellecchia, M.T.; Iasevoli, F.; et al. Neuropsychiatric disorders in Cushing's syndrome. *Front. Neurosci.* **2015**, *9*, 129. [CrossRef] [PubMed]
33. Brüne, M.; Schröder, S.G. Neuroleptic-induced akathisia and early onset tardive dyskinesia in affective disorder due to Cushing's syndrome. *Gen. Hosp. Psychiatry* **1997**, *19*, 445–447. [CrossRef]
34. Stenzel-Poore, M.P.; Cameron, V.A.; Vaughan, J.; Sawchenko, P.E.; Vale, W. Development of Cushing's syndrome in corticotropin-releasing factor transgenic mice. *Endocrinology* **1992**, *130*, 3378–3386. [CrossRef] [PubMed]
35. Westphal, C.H.; Muller, L.; Zhou, A.; Zhu, X.; Bonner-Weir, S.; Schambelan, M.; Steiner, D.F.; Lindberg, I.; Leder, P. The neuroendocrine protein 7B2 is required for peptide hormone processing in vivo and provides a novel mechanism for pituitary Cushing's disease. *Cell* **1999**, *96*, 689–700. [CrossRef]
36. Sun, Z.; Fan, Y.; Zha, Q.; Zhu, M.Y. Corticosterone Up-regulates Expression and Function of Norepinephrine Transporter in SK-N-BE(2)C Cells. *J. Neurochem.* **2010**, *113*, 105–116. [CrossRef] [PubMed]

37. Berry, J.N.; Meredith, A.; Saunders, M.A.; Sharrett-Field, L.J.; Reynolds, A.R.; Bardo, M.T.; James, R.; Pauly, J.R.; Prendergast, M.A. Corticosterone enhances *N*-methyl-D-aspartate receptor signaling to promote isolated ventral tegmental area activity in a reconstituted mesolimbic dopamine pathway. *Brain Res. Bull.* **2016**, *120*, 159–165. [CrossRef] [PubMed]
38. Gao, M.; Zhou, H.; Li, X. Curcumin Protects PC12 Cells from Corticosterone-Induced Cytotoxicity: Possible Involvement of the ERK1/2 Pathway. *Basic Clin. Pharmacol. Toxicol.* **2009**, *104*, 236–240. [CrossRef] [PubMed]
39. Mao, Q.Q.; Zhong, X.M.; Qiu, F.M.; Li, Z.Y.; Huang, Z. Protective effects of paeoniflorin against corticosterone-induced neurotoxicity in PC12 cells. *Phytother. Res.* **2012**, *26*, 969–973. [CrossRef] [PubMed]
40. Wu, F.; Li, H.; Zhao, L.; Li, X.; You, J.; Jiang, Q.; Li, S.; Jin, L.; Xu, Y. Protective effects of aqueous extract from Acanthopanax senticosus against corticosterone-induced neurotoxicity in PC12 cells. *J. Ethnopharmacol.* **2013**, *148*, 861–868. [CrossRef] [PubMed]
41. Polman, J.A.; Welten, J.E.; Bosch, D.S.; de Jonge, R.T.; Balog, J.; van der Maarel, S.M.; de Kloet, E.R.; Datson, N.A. A genome-wide signature of glucocorticoid receptor binding in neuronal PC12 cells. *BMC Neurosci.* **2012**, *13*, 118. [CrossRef] [PubMed]
42. Sheela Rani, C.S.; Soto-Pina, A.; Iacovitti, L.; Strong, R. Evolutionary conservation of an atypical glucocorticoid-responsive element in the human tyrosine hydroxylase gene. *J. Neurochem.* **2013**, *126*, 19–28. [CrossRef]
43. Kalinina, T.S.; Shishkina, G.T.; Dygalo, N.N. Induction of tyrosine hydroxylase gene expression by glucocorticoids in the perinatal rat brain is age-dependent. *Neurochem. Res.* **2012**, *37*, 811–818. [CrossRef] [PubMed]
44. Hensleigh, E.; Pritchard, L.M. Glucocorticoid receptor expression and sub-cellular localization in dopamine neurons of the rat midbrain. *Neurosci. Lett.* **2013**, *556*, 191–195. [CrossRef] [PubMed]
45. Rodríguez, S.; Uchida, K.; Nakayama, H. Striatal TH-immunopositive fibers recover after an intrastriatal injection of 6-hydroxydopamine in golden hamsters treated with prednisolone: Roles of tumor necrosis factor-α and inducible nitric oxidesynthase in neurodegeneration. *Neurosci. Res.* **2013**, *76*, 83–92. [CrossRef] [PubMed]
46. Copeland, B.J.; Neff, N.H.; Hadjiconstantinou, M. Enhanced dopamine uptake in the striatum following repeated restraint stress. *Synapse* **2005**, *57*, 167–174. [CrossRef] [PubMed]
47. Czyrak, A.; Chocyk, A. Search for the presence of glucocorticoid receptors in dopaminergic neurons of rat ventral tegmental area and substantia nigra. *Pol. J. Pharmacol.* **2001**, *53*, 681–684.
48. Harfstrand, A.; Fuxe, K.; Cintra, A.; Agnati, L.F.; Zini, I.; Wikstrom, A.C.; Okret, S.; Yu, Z.Y.; Goldstein, M.; Steinbusch, H.; et al. Glucocorticoid receptor immunoreactivity in monoaminergic neurons of rat brain. *Proc. Natl. Acad. Sci. USA* **1986**, *83*, 9779–9783. [CrossRef]
49. Heydendael, W.; Jacobson, L. Glucocorticoid status affects antidepressant regulation of locus coeruleus tyrosine hydroxylase and dorsal raphe tryptophan hydroxylase gene expression. *Brain Res.* **2009**, *1288*, 69–78. [CrossRef]
50. Li, M.; Han, F.; Shi, Y. Expression of locus coeruleus mineralocorticoid receptor and glucocorticoid receptor in rats under single-prolonged stress. *Neurol. Sci.* **2011**, *32*, 625–631. [CrossRef]
51. Nakagaki, T.; Hirooka, Y.; Matsukawa, R.; Nishihara, M.; Nakano, M.; Ito, K.; Hoka, S.; Sunagawa, K. Activation of mineralocorticoid receptors in the rostral ventrolateral medulla is involved in hypertensive mechanisms in stroke-prone spontaneously hypertensive rats. *Hypertens. Res.* **2012**, *35*, 470–476. [CrossRef] [PubMed]
52. Makino, S.; Smith, M.A.; Gold, P.W. Regulatory role of glucocorticoids and glucocorticoid receptor mRNA levels on tyrosine hydroxylase gene expression in the locus coeruleus during repeated immobilization stress. *Brain Res.* **2002**, *943*, 216–223. [CrossRef]
53. Zhang, R.; Jankord, R.; Flak, J.N.; Solomon, M.B.; D'Alessio, D.A.; Herman, J.P. Role of glucocorticoids in tuning hindbrain stress integration. *J. Neurosci.* **2010**, *30*, 14907–14914. [CrossRef] [PubMed]
54. Núñez, C.; Földes, A.; Pérez-Flores, D.; García-Borrón, J.C.; Laorden, M.L.; Kovács, K.J.; Milanés, M.V. Elevated glucocorticoid levels are responsible for induction of tyrosine hydroxylase mRNA expression, phosphorylation, and enzyme activity in the nucleus of the solitary tract during morphine withdrawal. *Endocrinology* **2009**, *150*, 3118–3127. [CrossRef] [PubMed]

55. Virdee, K.; McArthur, S.; Brischoux, F.; Caprioli, D.; Ungless, M.A.; Robbins, T.W.; Dalley, J.W.; Gillies, G.E. Antenatal glucocorticoid treatment induces adaptations in adult midbrain dopamine neurons, which underpin sexually dimorphic behavioral resilience. *Neuropsychopharmacology* **2014**, *39*, 339–350. [CrossRef] [PubMed]
56. McArthur, S.; McHale, E.; Dalley, J.W.; Buckingham, J.C.; Gillies, G.E. Altered mesencephalic dopaminergic populations in adulthood as a consequence of brief perinatal glucocorticoid exposure. *J. Neuroendocrinol.* **2005**, *17*, 475–482. [CrossRef]
57. McArthur, S.; McHale, E.; Gillies, G.E. The size and distribution of midbrain dopaminergic populations are permanently altered by perinatal glucocorticoid exposure in a sex- region- and time-specific manner. *Neuropsychopharmacology* **2007**, *32*, 1462–1476. [CrossRef]
58. Niwa, M.; Jaaro-Peled, H.; Tankou, S.; Seshadri, S.; Hikida, T.; Matsumoto, Y.; Cascella, N.G.; Kano, S.; Ozaki, N.; Nabeshima, T.; et al. Adolescent stress-induced epigenetic control of dopaminergic neurons via glucocorticoids. *Science* **2013**, *339*, 335–339. [CrossRef]
59. Schefe, J.H.; Lehmann, K.E.; Buschmann, I.R.; Unger, T.; Funke-Kaiser, H. Quantitative real-time RT- PCR data analysis: Current concepts and the novel "gene expression's CT difference" formula. *J. Mol. Med.* **2006**, *84*, 901–910. [CrossRef]
60. Yuan, J.S.; Reed, A.; Chen, F.; Stewart, C.N., Jr. Statistical analysis of real-time PCR data. *BMC Bioinform.* **2006**, *7*, 85. [CrossRef]

International Journal of *Molecular Sciences*

Review

Cell Clearing Systems Bridging Neuro-Immunity and Synaptic Plasticity

Fiona Limanaqi [1,†], Francesca Biagioni [2,†], Carla Letizia Busceti [2], Larisa Ryskalin [1], Paola Soldani [1], Alessandro Frati [2] and Francesco Fornai [1,2,*]

1 Human Anatomy, Department of Translational Research and New Technologies in Medicine and Surgery, University of Pisa, Via Roma 55, 56126 Pisa (PI), Italy; f.limanaqi@studenti.unipi.it (F.L.); larisa.ryskalin@unipi.it (L.R.); paola.soldani@dmu.unipi.it (P.S.)

2 I.R.C.C.S Neuromed, Via Atinense, 86077 Pozzilli (IS), Italy; francesca.biagioni@neuromed.it (F.B.); carla.busceti@neuromed.it (C.L.B.); alessandro.frati@uniroma1.it (A.F.)

* Correspondence: francesco.fornai@neuromed.it or francesco.fornai@med.unipi.it; Tel.: +39-050-2218611

† These authors equally contributed to the present work.

Received: 9 April 2019; Accepted: 30 April 2019; Published: 4 May 2019

Abstract: In recent years, functional interconnections emerged between synaptic transmission, inflammatory/immune mediators, and central nervous system (CNS) (patho)-physiology. Such interconnections rose up to a level that involves synaptic plasticity, both concerning its molecular mechanisms and the clinical outcomes related to its behavioral abnormalities. Within this context, synaptic plasticity, apart from being modulated by classic CNS molecules, is strongly affected by the immune system, and vice versa. This is not surprising, given the common molecular pathways that operate at the cross-road between the CNS and immune system. When searching for a common pathway bridging neuro-immune and synaptic dysregulations, the two major cell-clearing cell clearing systems, namely the ubiquitin proteasome system (UPS) and autophagy, take center stage. In fact, just like is happening for the turnover of key proteins involved in neurotransmitter release, antigen processing within both peripheral and CNS-resident antigen presenting cells is carried out by UPS and autophagy. Recent evidence unravelling the functional cross-talk between the cell-clearing pathways challenged the traditional concept of autophagy and UPS as independent systems. In fact, autophagy and UPS are simultaneously affected in a variety of CNS disorders where synaptic and inflammatory/immune alterations concur. In this review, we discuss the role of autophagy and UPS in bridging synaptic plasticity with neuro-immunity, while posing a special emphasis on their interactions, which may be key to defining the role of immunity in synaptic plasticity in health and disease.

Keywords: autophagy; proteasome; immunoproteasome; mTOR; T-cells; glia; dopamine; glutamate; neuro-inflammation

1. Introduction

In recent years, unexpected connections have emerged between synaptic transmission, inflammatory/immune mediators, and brain (patho)-physiology [1–3]. In fact, the prevailing dogma that portrayed the nervous and immune system as two independent entities has been progressively replaced by new levels of functional connections and commonalities [4–6]. This interconnection rose up to a level that involves synaptic plasticity concerning both its molecular mechanisms and the clinical outcomes related to behavioral abnormalities [7,8]. Synaptic plasticity refers to those activity-dependent changes in the strength or efficacy of synaptic transmission, which occur continuously upon exposure to either positive or negative stimuli, such as learning, exercise, stress, or substance abuse, as well as the subsequent mood conditions [8]. Modifications of the neural circuits entail a variety of cellular and molecular events, encompassing neurotransmitter release; ionic activity; and metabolic, epigenetic, and transcriptional changes, which converge to shape the neuronal proteome and phenotype in an attempt to restore homeostasis [9–11]. The ability to re-establish and/or sustain baseline brain functions depends on a plethora of synchronized activities, which indeed involve both neuronal- and immune-related mechanisms. In this scenario, neurotransmitters and immune-related molecules adopt a common language to fine-tune brain functions [12–15]. In fact, classic immune molecules, including cytokines, major histocompatibility complex (MHC) molecules, and T-cells, are deeply involved in central nervous system (CNS) plasticity, while CNS factors, mostly neurotransmitters encompassing dopamine (DA) and glutamate (GLUT), actively participate in shaping immune functions [14]. Neuro-immune surveillance is a critical component for brain function, as circulating T-cells that recognize CNS antigens (Ags) are key in supporting the brain's plasticity, both in health and disease [8]. The functional anatomy from which the molecular interplay between the immune system and brain matter stems, was recently identified at the level of lymphatic pathways operating in the perivascular (also known as "glymphatic") and dural meningeal spaces [16–18]. Lymphatic flows foster the drainage of the brain interstitial fluid into the cerebrospinal fluid, and then back again into the bloodstream, or even directly into the secondary lymphoid organs. Functionally, this translates into a clearance of potentially threatening interstitial solutes and the drainage of CNS-derived Ag peptides to the deep cervical lymph-nodes to be captured and processed by antigen presenting cells (APCs) [19,20]. Within this context, synaptic plasticity, apart from being modulated by classic CNS molecules, is strongly affected by the immune system. This is not surprising, given the common molecular pathways that operate at the cross-road between the nervous- and immune-system. In fact, just like what is happening for the key proteins involved in neurotransmitter release [21,22], Ag processing within APCs is carried out by the two major cell-clearing machineries, ubiquitin proteasome (UPS) and autophagy [23–25]. In detail, UPS and autophagy operate both in the CNS and immune system, to ensure protein turnover and homeostasis. In the CNS, UPS- and autophagy-dependent protein degradation is seminal to protect neurons from potentially harmful proteins, and to modulate neurotransmitter release and synaptic plasticity [21,26–28]. Similarly, in the immune system, UPS and autophagy cleave endogenously- and exogenously-derived proteins to produce Ag peptides, which bind to MHC molecules class I and II [23–25,29]. Indeed, these pathways converge when the CNS components are cleared by immunocompetent mechanisms [24,29]. Thus, CNS-derived Ags bound to MHC-I and –II may be exposed on the plasma membrane of APCs, for presentation to CD8+ and CD4+ T-lymphocytes, respectively [29,30]. The associative binding of MHC molecules with T-cells receptors (TCR), coupled with co-stimulatory signals and the presentation of CNS-derived Ags, fosters the activation of naïve T-cells in the periphery, while mounting CNS-directed adaptive immune responses, which may produce either beneficial or detrimental effects already pertaining to the field of CNS plasticity [2,14,31–33]. Still, at anatomical level, the sympathetic innervation of both primary and secondary lymphoid organs provides a means of functional connection between the immune- and nervous-system [34]. In fact, catecholamine, and mostly DA released from sympathetic nerve terminals, is an active regulator of the metabolism, fate, and activity of naïve CD4+ and CD8+ T-cells [35,36]. This is achieved through the binding of DA to its cognate receptors and transporters, which are abundantly expressed on lymphoid cells. Likewise, the GLUT released in the bloodstream or within the CNS modulates T-cells activity, through binding to its cognate receptors, which are expressed on T-cells [2,37]. In this way, neurotransmitters and CNS-derived Ag presentation synergize to define the pool of immunocompetent cells that travel back and forth between the brain and periphery, to guarantee neuro–immune surveillance and synaptic plasticity. Antigen presentation and immune responses may also occur directly in the brain upon interactions between CNS circulating T-cells and glia, or even neurons [38–42]. Unexpectedly, recent studies showed that naïve T-cells are able to cross CNS barriers and infiltrate the brain parenchyma [38–47]. This is magnified during pro-inflammatory conditions when the glia and even neurons operate as competent APCs, as they become able to process and present Ags via MHC molecules [39,47,48]. At the same time, T-cells possess all of the machinery that is necessary for releasing and responding to neurotransmitters, just like neurons and glia do [35,37]. The existence of such a bi-directional dialogue between nerve and immune cells has now challenged the classical dichotomy between inflammatory and degenerative disorders of the CNS. In fact, defective or inappropriate communication between the immune and nervous system gives rise to a chain of events, where inflammatory/immune and synaptic alterations intermingle to produce CNS disorders, encompassing neuro-developmental, neurodegenerative, and auto-immune diseases [2,12,13,20,49]. When searching for a common pathway bridging neuro–immune and synaptic dysregulations, UPS and autophagy machineries take center stage. The dysregulations of both UPS and autophagy characterize a plethora of CNS disorders, where synaptic and neuro-inflammatory/immune alterations co-exist, such as Parkinson's, Alzheimer's, and Huntingtin's diseases (PD, AD, and HD); epilepsy; ischemia; brain tumors; multiple sclerosis (MS), and psychiatric and substance-abuse disorders [6,21,50–77]. The reason for such a common dysregulation of UPS and autophagy in etiologically different CNS disorders is rooted in their pleiotropic catalytic functions, which are seminal for both synaptic plasticity and neuro-immunity [30,78–87]. Despite being traditionally considered as independent systems, recent evidence has unraveled a functional cross-talk between UPS and autophagy, which occurs at both biochemical and morphological levels [73,88–90]. Thus, it is not surprising that autophagy and UPS share most of their substrates and functions, and they operate dynamically and coordinately in both nerve and immune cells so as to modulate neurotransmission, oxidative/inflammatory stress response, and immunity [91–94]. This is accomplished through the degradation and turnover of proteins, including those involved in endocytic and secretory pathways, transcription factors, and oxidized and/or immunogenic proteins. The present review aims to analyze those molecular interactions that are related to both UPS and autophagy, and that enable neurons and immune cells to surveil synaptic and neuro–immune activity. Apart from being well known triggers of synaptic plasticity, environmental agents such as pathogens, inflammatory cytokines, free radicals, and abnormal neurotransmitter release can profoundly affect cell-clearing systems [51,52,94–100]. As a proof of concept, when a dysregulation of cell-clearing systems occurs, the altered communication between the nervous and immune cells translates into maladaptive plasticity, which may underlie behavioral alterations. Given the variety of specific regulatory signals and molecules involved in the interplay between UPS and autophagy, a better understanding of their interactions is key in order to define the role of immunity in synaptic plasticity in health and disease.

2. Cell Clearing Systems: Tracing the Path of the Interplay between Proteasome and Autophagy

Autophagy and UPS ensure eukaryotic cell proteostasis by clearing unfolded, misfolded, oxidized, or disordered proteins, so as to prevent their accumulation, aggregation, and spreading [60,101–106]. Besides being seminal in extreme cell conditions when cell survival is jeopardized, autophagy and UPS activities operate in baseline conditions in order to keep the turnover of proteins that naturally occur within a living cell steady. In fact, as actors of protein degradation, autophagy and UPS regulate most cell functions encompassing cell cycle and division, cell differentiation and development, endo- and exo-cytosis, and, specifically, synaptic strength and Ag processing [6,21,22,25–27,107–111]. Autophagy initiates with the formation of double-layered membrane vacuoles, named phagophores. The maturation and sealing of the phagophore leads to the formation of the autophagosome, which stains for autophagy markers such as beclin-1 (the orthologue of yeast Atg6) and LC3 (Atg8) [112,113]. The autophagosome shuttles a variety of substrates, including ubiquitinated proteins and whole organelles (e.g., mitochondria, endoplasmic reticula, ribosomes, and synaptic vesicles) to the lysosomal compartment, which is gifted with a rich enzymatic activity. The merging of the autophagosome with endosomes and lysosomes generates the catalytic organelle autophagolysosome, where the degradation and recycling of "in bulk" sequestered cytosolic cargoes occurs [112,113]. Protein tagging with ubiquitin chains, which is carried out by the UPS system, represents a sorting signal for either UPS- or autophagy-dependent protein degradation [114]. Protein ubiquitination is an ATP-dependent process that is accomplished by three enzymes, namely ubiquitin-activating (E1), ubiquitin-conjugating (E2), and ubiquitin-ligase (E3). Several proteins operate at the cross-road between UPS and autophagy, to regulate the sorting and shuttling of ubiquitinated substrates towards either system. Among these proteins, which indeed constitute a much longer list, three are worth mentioning, namely (i) Parkin, (ii) histone deacetylase 6 (HDAC6), and (iii) Sequestosome-1 (SQSTM1)/p62.

(i) Parkin is an ubiquitin-ligase enzyme (E3 ligase), which mediates protein polyubiquitination and serves as a signal for targeting misfolded proteins to the aggresome, where autophagy is recruited [115]. Parkin-dependent ubiquitination triggers the removal of the pro-apoptotic proteins BAX and BCL-2 by either UPS or autophagy, and it is seminal to induce mitophagy, that is, mitochondria-specific autophagy [116]. After ubiquitin linkage, Parkin also induces the coupling of target proteins with dynein motor complexes via the adaptor protein HDAC6 in order to facilitate their transport to the aggresome, where autophagy is recruited;

(ii) HDAC6 is a microtubule-associated histone deacetylase, which shuttles polyubiquitinated substrates along the microtubules for autophagosomal engulfment, while fostering lysosomes transport to the site of autophagy occurrence. In detail, HDAC6 binds the polyubiquitin chains [117,118] or even C-terminal regions of free ubiquitin [119] via a C-terminal zinc finger-containing domain (called BUZ domain). Then, HDAC6 binds to the microtubule-associated dynein motors to shuttle the polyubiquinated proteins to the aggresomes, while fostering the recruitment of autophagy to the aggresomes [120,121]. Again, HDAC6 participates in the fusion of autophagosomes with lysosomes for final autophagy degradation [122,123]. Remarkably, HDAC6 activity is essential for autophagy, to compensate for protein degradation and rescue cell survival when UPS is impaired [124], thus providing a functional link between autophagy and UPS.

(iii) SQSTM1/p62 is a ubiquitin-binding scaffold protein that links ubiquitinated proteins to autophagy machinery in order to enable their degradation [125]. This occurs through a direct interaction between SQSTM1/p62 and ubiquitinated proteins via a C-terminal UBA domain, and their subsequent binding to autophagy proteins such as LC3 and GABARAP family proteins. As p62 is itself degraded by autophagy, it is widely used as a marker of autophagy flux [126].

Once tagged with ubiquitin, proteins are recognized by autophagy and/or the proteasome 26S (P26S) multimeric complex, which is formed by a catalytic core (P20S) and two regulatory subunits (P19S, also known as PA700) capping the ends of P20S [127]. P19S binds the poly-ubiquitin chain and cleaves it from the substrate. In this way, the unfolded substrate enters the P20S to be degraded by the β1, β2, and β5 catalytic subunits of the P20S, which own chymotrypsin-like, trypsin-like, and caspase-like activity, respectively. Despite being traditionally considered as cytosolic catalytic machinery, UPS also associates with vesicular organelles, including precursor synaptic vesicles (SVs), Golgi-derived vesicles, mitochondria, and lysosomes [128]. Far from being a mere phenomenon of morphological co-localization, the association of UPS with vesicular structures probably underlies a sophisticated functional cooperation. In fact, vacuolar organelles may serve as a ferryboat to shuttle UPS in different cell-compartments, while the UPS handles the turnover of vesicle-associated proteins. This is in line with the recent studies characterizing a novel organelle named "autophagoproteasome", where the autophagy and UPS markers co-localize [73,88] (Figure 1). The formation of this specific vacuolar compartment is hindered by the administration of the neurotoxic abused drug methamphetamine (meth), while its rescue via the inhibition of the mammalian target of rapamycin (mTOR) correlates with cell protection and survival [73]. This is line with studies indicating the mTOR pathway as a common modulator of both UPS- and autophagy-dependent protein degradation [89]. These findings configure mTOR inhibition as a potential strategy to synergistically enhance autophagy and UPS-dependent protein degradation. mTOR is a ubiquitously expressed serine-threonine kinase, which senses and integrates several environmental and intracellular cues to orchestrate major processes, such as cell growth and metabolism [55,75,129]. mTOR has been widely implicated in synaptic plasticity, inflammation, and immunity, although this was merely related to the role in protein synthesis. In the last decades, mTOR has been posed at the center stage on a variety of cell functions, mostly related to autophagy and UPS. The emerging mechanisms linking mTOR with autophagy and UPS unravel a close interdependency between the cell-clearing systems. In detail, the duration and amplitude of the autophagy response depends on the stability of the serine/threonine kinase ULK1/Atg1, which, in turn, is coordinately regulated by UPS and mTOR [130]. ULK1 acts at multiple steps of autophagy initiation and response, in part by phosphorylating autophagy proteins, including Atg13, Beclin 1, and Atg9 [131]. mTOR activation inhibits ULK1 kinase activity (and thus autophagy initiation) via phosphorylation, and also coordinates ULK1 de novo protein synthesis [130,132]. In this context, UPS behaves as a sentinel in sensing and regulating mTOR/ULK1-dependent autophagy. In fact, during the early stages of autophagy, UPS mediates the K63-linked polyubiquitination of ULK1 via the AMBRA1–TRAF6 (E3 ligase) complex to maintain its stability, self-association, and kinase activity [133]. Conversely, during prolonged nutrient starvation, UPS targets ULK1 for degradation, following Cullin/KLHL20-dependent K48-linked polyubiquitination, thus providing a feedback control of the autophagy response [134]. In turn, autophagy may control UPS efficacy and activity through the degradation of inactive UPS subunits, which are shuttled to autophagosomes, a phenomenon known as "proteophagy" [135–137]. This may explain the intriguing effects that are observed on autophagy upon UPS inhibition, and vice versa, while remarking on the importance of autophagy-UPS cross-talk in cell homeostasis. In fact, the inhibition of either autophagy or UPS alone may produce detrimental effects to cell survival [138–143], which are bound to impaired protein turnover by both cell-clearing systems. For instance, autophagy inhibition leads to the accumulation of ubiquitinated substrates by affecting UPS either upstream, or at the level of its catalytic activity [144,145]. Conversely, UPS inhibition may induce an enhancement of autophagy as an early compensatory response to cope with protein overload and grant cell-survival [146–148]. Such an effect turns out to be only transitory, as UPS dysfunction at later stages impedes mitophagy and decreases the levels of essential autophagy proteins, such as Atg9 and LC3B [93]. This is not surprising, as UPS is essential for endo–lysosome membrane fusion [149,150], which, in turn, is involved in the late steps of autophagy. In fact, UPS modulates the activity of Rab GTPases (GTP-bound Ras proteins in the brain), which are involved in all cell-trafficking mechanisms, including autophagy-dependent endocytosis and autophagy membrane fusion [150–153]. These findings indicate that the synergistic and compensatory functional interplay between autophagy and UPS needs to be taken into account in experimental approaches modulating either systems alone. On the one hand, this may lead to confounding outcomes when assessing the effects of autophagy and UPS alone; on the other, it calls for investigating the potential strategies that can simultaneously rescue the defects of autophagy and UPS. In keeping with this, it is worth of mentioning that UPS exists as two alternative isoforms, the standard 26S proteasome and the immuno-proteasome (SP and IP, respectively). It is remarkable that the mTOR pathway also modulates the switch between these alternative subtypes of UPS, which evolution has preserved in order to optimize different tasks according to specific cell demands [154–157]. In fact, SP is ubiquitously expressed in all eukaryotic cells, and it is generally enhanced by mTOR inhibition, while IP is an alternative, cytokine-inducible form that is downregulated by mTOR inhibition. Despite overlapping in structure and functions, these alternative UPS isoforms differ in catalytic subunits and substrate specificity [83–85,101–103]. In fact, the IP operates constitutively in all immune-related cells, including professional APCs (e.g., dendritic cells—DC) and lymphocytes, and thus, it is mostly involved in potentiating innate and adaptive immunity [158].

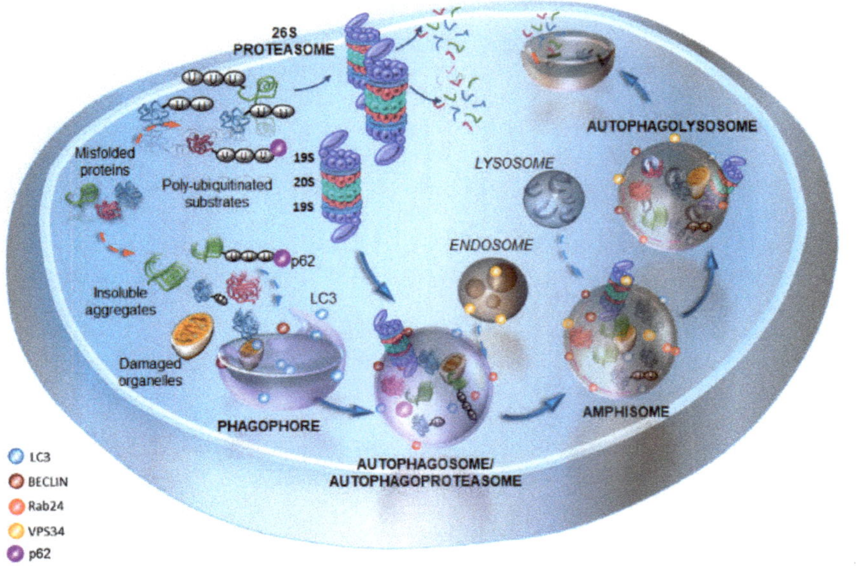

Figure 1. A schematic overview of the cell-clearing systems, with a focus on the interplay between autophagy and proteasome. The cartoon offers a rough schematization of the autophagy pathway, starting from the phagophore biogenesis staining for LC3 and Beclin1, which engulfs cytoplasmic portions containing insoluble aggregates, ubiquitinated substrates, and damaged organelles. The phagophore membrane seals to from the autophagosome, which then fuses with the late endosomes to generate the amphisome. This latter fuses with lysosomes to generate the autophagosome (staining for LC3, Beclin1, Rab24, VPS34, and p62) where cargo degradation occurs. At the same time, ubiquitinated substrates can be degraded by the 26S ubiquitin proteasome (UPS), which is formed by the regulatory subunits 19s and the catalytic subunits 20s. Signals such as p62 contribute to sort ubiquitinated proteins for either UPS or autophagy degradation. At the same time, p62 may serve as a signal to promote the merging of UPS and autophagy into a single organelle, "autophagoproteasome", where potentiated cell-clearance may take place. Alternatively, the autophagy-dependent degradation of inactive UPS subunits may occur within this compartment, a phenomenon that is named "proteophagy". Dotted arrows indicate the formation of insoluble aggregates from misfolded proteins and their ubiquitination (red), the shuttling of substrates to the phagophore (blue), and the fusion of endosomes and lysosomes with autophagy vacuoles (blue). Solid blue arrows indicate the progression of the autophagy machinery, the shuttling of ubiquitinated substrates to the UPS, and the shuttling of the UPS within autophagosomes.

Remarkably, persistent oxidative/inflammatory stress may concomitantly affect autophagy flux and IP–SP switch, either in the immune periphery or within the CNS. This is expected to alter the clearing capacity and/or substrate specificity of the cell-clearing systems, while triggering a cascade of molecular events that synergize to produce synaptic dysfunctions/toxicity, along with a loss of auto-immune tolerance up to the development of CNS-directed inflammatory and auto-immune reactions. In the present manuscript, we discuss the role of autophagy, SP, and IP at the level of classic neuronal and immunological synapses, while posing a special emphasis on their effects at the level of hybrid junctions, which establish "neuro-immunological synapses" between immune and nerve cells. This is critical to comprehend those autophagy and UPS-dependent mechanisms that finely tune T-cells populations that migrate to the CNS. Again, the role of autophagy and UPS is seminal to disclose those molecular events, which induce neurons and glia to behave as competent APCs, and, as such, become possible targets for auto-immune damage.

3. Autophagy and Proteasome Tune Synaptic Plasticity by Modulating Neurotransmission and Immunity

Autophagy and SP are constitutively expressed in neurons either in the cell body, nucleus, or synapses, where they modulate synaptic plasticity by surveilling oxidative stress, gene transcription, and neurotransmitter release [22,159–171]. Autophagy and SP operate at various sub-cellular levels in both pre- and post-synaptic sites, and in detail, they are the following:

(i) intersect with secretory pathways to modulate SV trafficking, as well as the size and number of SV pools [21,128,161,172,173];

(ii) degrade protein isoforms and presynaptic chaperone proteins such as alpha-synuclein, beta amyloid, and tau [62,65,108,174,175], which, when altered in either amount or conformation, can drive synaptic dysfunctions [176–179];

(iii) modulate the rate and duration of neurotransmitter release (including DA and GLUT) by degrading whole SVs (in the case of autophagy), as well as soluble Nsf attachment protein receptor (SNARE) and accessory proteins, which are involved in SV exo-/endo-cytosis [21,22,152,161,162,166–168,180–183];

(iv) foster the internalization and degradation of DA and GLUT receptors, which are coupled with downstream intracellular cascades driving metabolic, transcriptional, and epigenetic changes within neurons [11,98,184–187]. As such, autophagy and SP are deeply involved in those mechanisms driving synaptic plasticity, such as long term-potentiation and -depression, which are directly related to neuronal and behavioral phenotypes. In fact, the inhibition of either SP or autophagy in experimental models produces profound alterations in neurotransmitter release and the expression of neurotransmitter receptors [31,161,173,181–183]. Reiterated stimuli that alter neurotransmitter activity are seminal to induce maladaptive changes in synaptic strength and connectivity, which translate into long-lasting psychomotor changes. In the last decades, experimental evidence has accumulated, suggesting that early synaptic alterations may represent a major event fostering neuronal degeneration [2,188,189]. This is best exemplified by the mechanisms of action of abused drugs such as meth, which produces psychiatric alterations including addiction and psychoses, and even neurotoxicity affecting the DA terminals, DA cell bodies, and post-synaptic neurons of the DA circuitry within the striatum, iso-cortex, and limbic brain areas [11,190–192]. This occurs through the joined contribution of epigenetic events and protein alterations (oxidation, aggregation, and spreading) arising from abnormal DA release, the abnormal pulsatile stimulation of DA receptors, and also the increased responsiveness of neurons to GLUT and GLUT exitotoxicity. In fact, abnormal levels of DA and abnormal stimulation of DA receptors play a key role in GLUT excitotoxicity, which stands for the over-activation of specific types of GLUT receptors, resulting in neuronal death, tissue damage, and loss of brain function, as it occurs both during meth toxicity and in various neurological diseases [2,193].

Both autophagy and UPS are severely affected by meth administration [70–73], while the mTOR inhibitor rapamycin prevents both the behavioral and neurotoxic effects of meth by rescuing autophagy and UPS [73,194]. This is in line with several studies showing that the genetic or pharmacological occlusion of autophagy and UPS leads to the accumulation of ubiquitinated protein-aggregates and recapitulates neurodegeneration [138–143,195]. As a support to these findings, SP and autophagy dysfunctions occur in human brain disorders characterized by early synaptic dysfunctions, which precede protein aggregation [6,21,75,196–198]. On the other hand, mTOR inhibition, which is supposed to restore both autophagy and UPS activity, ameliorates early psychomotor and cognitive behavioral alterations by recuing neurotransmission defects and by restoring proteostasis in a variety of CNS disorders, both in humans and experimental models [75,175,194,199–202].

3.1. *Autophagy- and Proteasome-Dependent Neurotransmission Linking Immune-Cells' Activity and Synaptic Plasticity*

As modulators of neurotransmitter release, autophagy and SP also modulate CNS-directed immune responses by operating at the level of the neuro-immunological synapse, which may be established between the sympathetic nerve terminals and T-cells within lymphoid organs [12]. Remarkably, both SP and autophagy modulate the release of DA [161,162,181–183], which besides being crucial for brain functions such as movement, cognition, attention, memory, and reward [203], also orchestrates the differentiation, maturation, selection, trafficking, and migration of T-lymphocytes [34–36,204–206]. In fact, T-cells express G-coupled D1-like (D1 and D5) and D2-like (D2, D3, and D4) DA receptors, and just like it occurs for the neurons, the magnitude and duration of the DA release is key to trigger the specific metabolic and intracellular cascades, switching T-cells phenotype and function [35,36,206]. As thoroughly revised elsewhere, depending on the DA concentration and the pattern of stimulation of specific DA receptors, naïve T-cells may be induced to differentiate into either memory, regulatory, or effector cells, including CD4+ T helper (Th) 1, 2, or 17, and CD8+ cytotoxic T-lymphocyte (CTL) phenotype [34–36,206]. In this context, the autophagy- and SP-dependent surveillance of the DA release at the level of the neuro–immunological synapse is expected to guarantee the physiological stimulation of the DA-receptors placed on the T-cells, and control the neuro–immune activity (Figure 2). The circulation of T-lymphocytes in the brain occurs physiologically, since the early development, and persists during adulthood, to guarantee synaptic plasticity [8,207,208]. For instance, both CD4+ and CD8+ T-cells are essential for spinogenesis and GLUT synaptic function in the hippocampus [209]. In addition, CD8+ T cells regulate the hippocampal volume by promoting neurogenesis [210]. Intriguingly, peripheral and brain-infiltrating T-cells, besides regulating GLUT synaptic transmission and plasticity, are regulated themselves by GLUT [2,37]. Both autophagy and UPS modulate, and are in turn modulated by GLUT transmission [95,211–213]. GLUT is a major excitatory neurotransmitter, which besides being critical for the brain's development and function, participates in tuning the T-cells activity. In fact, ionotropic and metabotropic GLUT receptors are differently expressed among resting and activated T-cells, as well as in different T-cells subtypes [2,37]. At low physiological concentrations, GLUT promotes T-cell adhesion, migration, proliferation, and protection of activated T-cells from Ag-induced apoptotic cell death. Yet, depending on the abnormalities concerning either the GLUT concentration, stimulation of specific GLUT receptors, or the presence of other converging stimuli (such as inflammatory cytokines or other neurotransmitters), GLUT may profoundly affect T-cells activity, thus playing an active role in immune diseases [2,37]. Remarkably, brain infiltrating T-cells were shown to respond to GLUT by activating a neuroprotective pathway, thus providing a potential feedback regulatory mechanism to limit GLUT excitotoxic damage in the CNS [214]. A loss of GLUT-mediated responsiveness of T-cells has been described in MS [215]. Furthermore, various alterations in CNS-circulating T-lymphocyte populations are described in both classic and autoimmune degenerative disorders, such as PD, AD, and MS [216]. Emerging evidence also indicates an association between early inflammatory mechanisms underlying neurodegeneration, and synaptic alterations involving abnormal levels of DA and/or GLUT, as well as the deregulation of their receptors on T-cells [2,37,217–219]. In line with this, specific modulators of DA and/or GLUT activity, may have beneficial effects, not only in classic neurodegenerative diseases, but also in auto-immune CNS disorders such as MS [2,35,217].

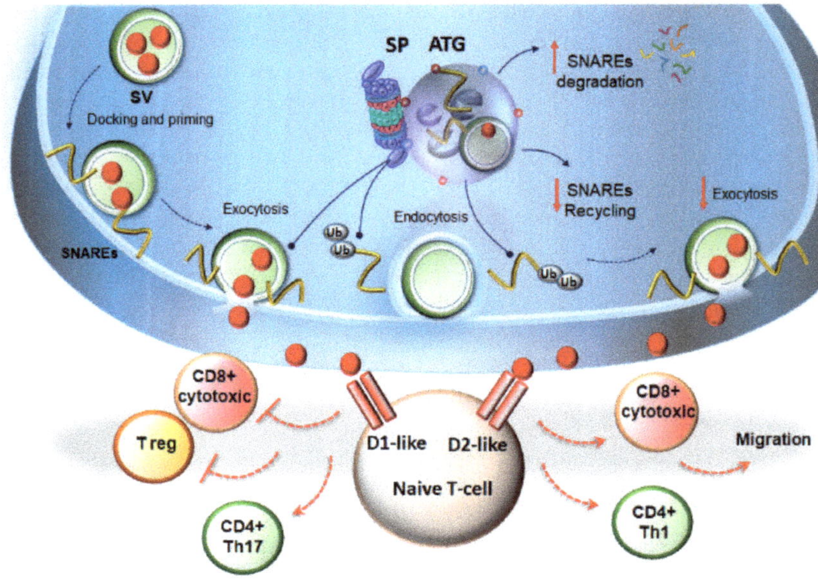

Figure 2. Autophagy and proteasome modulate immune activity by surveilling dopamine (DA) release. The standard proteasome (SP) and autophagy blunt DA release by degrading entire synaptic vesicles (SVs), as well as soluble Nsf attachment protein receptor (SNARE)- and SNARE associated-proteins, which foster synaptic vesicle exocytosis. In fact, they both prevent the rapid recycling of SV proteins back to the plasma membrane, which would otherwise lead to a further round of exocytosis. In this way, SP- and autophagy-dependent amount and duration of DA released at the level of the neuro-immunological synapse surveils the stimulation of DA receptors expressed on T-cells. This is seminal to modulate the differentiation of T-cells toward cytotoxic-, regulatory-, or helper-T-cells, as well as T-cell migration in periphery. For instance, the abnormal stimulation of D1-like receptors increases cyclic adenosine monophosphate (cAMP) levels to inhibit cytotoxic CD8+ cells; it impairs the differentiation and activity of T-regulatory cells, while inducing polarization of naive CD4+ cells toward the Th17 phenotype. On the other hand, the stimulation of D2-like receptors induces the differentiation of CD8+ cells into cytotoxic T-lymphocytes, induces the polarization of naive CD4+ cells toward the Th1 phenotype, and controls T-cell migration and adhesion. Dotted blue arrows indicate the progression of the SV cycle. Solid blue arrows indicate the targeting and shuttling of SNARE proteins to the UPS and autophagy. Solid red arrows indicate the increase/decrease in SNAREs degradation/recycling and exocytosis rate. Dotted red lines indicate the induction (arrows) or inhibition (lines) of naïve T-cells differentiation towards various phenotypes following abnormal stimulation of DA receptors.

Besides the effects in T-cells within the lymphoid organs, DA and GLUT release may also modulate the activity of immune cells, including glia and T-cells, directly in the CNS. In line with this, a number of studies pointed to the unexpected ability of naïve CD4+ and CD8+ T-cells to infiltrate the brain parenchyma [38–47]. This occurs mostly during pro-inflammatory conditions, which enhance naïve T-cell recruitment in the CNS, while fostering T-cells activation and phenotypic commitment once they encounter activated glial cells exposing MHC-bound Ags. Thus, the effects of DA and GLUT on brain-infiltrating naïve T-cells may synergize with local Ag presentation in order to dictate the activation or suppression of T-cells directly in the CNS. However, as a general consensus view, only peripherally activated T-cells can migrate into the brain. If they encounter a CNS-resident APCs exposing the cognate Ag, T-cells become re-activated and recruit their effector machineries to produce cytotoxicity or cytokine release. In this scenario, the effects of DA and GLUT focus mostly on the glial cells, which behave as CNS-resident immune cells. Inflammatory cytokines (such as those released by brain infiltrating T-cells) synergize with neurotransmitters to induce the activation of the glial cells, which encompass the morphological changes, increased proliferation rate, and ability to operate as APCs [9,220,221]. Nonetheless, because of an intimate functional association with the synapses, the glia is deeply involved in synaptic plasticity [9]. Experimental studies suggest that activated microglia are responsible for the synaptic alterations observed in a variety of neurological disorders [2,222]. Both microglia and astrocytes express many different neurotransmitter receptors (including DA and GLUT receptors), which, when stimulated, foster the release of soluble factors acting, in turn, on neurons to alter neurotransmitter release, neurotransmitter receptor activation, and synaptic efficacy [2,9]. Besides neurotransmitters such as DA and GLUT, these include mediators of constitutive immunity such as tumor necrosis factor alpha (TNFα), interferon gamma (IFNγ), and interleukin 1 beta (IL-1β); pentraxins; and growth factors such as brain-derived neurotrophic factor (BDNF), which altogether influence synaptic activity, mostly by enhancing the long-term potentiation of excitatory transmission [2,9,77,81,223]. Once again, in this context, autophagy and UPS configure as actors in the communication between T-cells, glia, and neurons, as they (i) surveil neurotransmitter release, which is important for glial, lymphocytic, and neuronal activity; (ii) determine the metabolism and activity of the glia and lymphocytes and the subsequent production and release of soluble factors (paragraph 3.2); and (iii) process Ag peptides, which are presented to T-cells by DCs, glial cells, or neurons (paragraph 3.3).

3.2. *Cell Clearing System in the Metabolism and Fate of Immune Cells*

Similar to what occurs in the neurons, the activity of immune cells largely depends on UPS and autophagy. The multitude of metabolic changes that occur upon glial and T-cells activation is tightly intermingled with autophagy and UPS activity. As detailed in the previous paragraph, both UPS and autophagy regulate DA and GLUT release, and are also involved in the turnover of DA and GLUT receptors, thus influencing the metabolic cascades, which participate in T-cells and glia activation. Moreover, both autophagy and UPS modulate the turnover of inflammatory-related transcription factors such as nuclear factor k beta (NF-κB), which, in turn, fosters the production of cytokines by glia or T-cells [25,224–226]. Within T-cells, autophagy and UPS directly govern the metabolic cascades, which dictate T-cells differentiation, function, and activity [6,158,226]. Again, such an overlapping task may be bound to mTOR activity, which is deeply involved in T-cell metabolism [227]. Moreover, the co-existence of UPS and autophagy degradation pathways enables APCs (including peripheral DCs and glia) to present either endogenous or exogenous Ag peptides, which is key for determining the T-cells state [30]. While autophagy operates constitutively in all immune cells, the UPS exists mostly as IP, which is an alternative, cytokine-inducible isoform of the SP possessing enhanced chymotrypsin-like activity and peculiar structural features, compared with SP [228–232]. In fact, within IP, β1, β2, and β5 subunits of the SP-20S catalytic core are replaced with β1i, β2i, and β5i. Among these, β1i possesses a chymotrypsin-like activity contrarily to the standard counterpart possessing a caspase-like activity. Thus, compared with SP, IP produces, more efficiently, Ag peptides with C-terminal hydrophobic amino acids, which are suitable for binding the groove of MHC class I molecules [228–232]. The major task of IP is to process either endogenous or exogenous proteins, and generate Ag peptides, which are first complexed to MHC-I in the endoplasmic reticulum, and then exposed on the plasma membrane of APCs for either direct or cross-presentation to CD8+ T-lymphocytes. Thus, the generation of defined T-cell epitopes and the expression of MHC-I molecules largely depend on the IP activity [6,35,228–232]. Similar to IP, autophagy is key in adaptive immunity, though it is mostly implicated in the MHC-II restricted presentation of exogenously-derived Ag to CD4+ T cells [23,30]. Nonetheless, a few reports demonstrated that autophagy can also process and load endogenous (viral) peptides to MHC-I [233]. Remarkably, autophagy is implicated in MHC class I molecules internalization and degradation, thus influencing MHC-I stability at the plasma membrane of APCs, and subsequent CD8+ T-cell responses [234,235]. In fact, autophagy inactivation within APCs occludes the surface internalization of MHC-I molecules, leading to an increased Ag presentation and enhanced CD8+ T cell responses against viral peptides, both in vitro and in vivo [233]. Thus, autophagy fosters MHC II-restricted Ag presentation, while controlling MHC-I expression [233–235]. Autophagy also provides an alternative pathway to the direct IP- and MHC-I-dependent Ag presentation pathway [236]. In this case, Ags normally targeted to autophagy and exposed by MHC-II can also be loaded to MHC-I in recycling endosomes, which is seminal to trigger adaptive immune response upon viral infections. In the immune periphery, autophagy- and UPS-dependent Ag processing is also seminal for T-cells thymic selection. In the thymus, specialized forms of IP operate together with SP and autophagy to finely-tune T-cell proliferation, along with positive and negative T-cell selection [228,237]. In this way, UPS (SP and IP) and autophagy coordinately guarantee immune-tolerance and define the pool of immunocompetent T-cells, which are released in the bloodstream to reach secondary lymphoid organs, and subsequently, the brain.

4. Autophagy and Proteasome Linking Altered Immunity and Synaptic Plasticity with Neurodegeneration

In neurons and glial cells, autophagy and SP operate constitutively, while the IP is generally induced by the pro-inflammatory cytokines IFNγ and TNFα, and by oxidative stress [85,158,229]. These challenging conditions contribute to disassemble SP for the sake of IP induction, which is likely to cope with the protein overload, as it is endowed with an enhanced catalytic activity [238,239]. Remarkably, the IP cleaves both microbial- and oxidized/aggregated-proteins to produce immunogenic peptides, which are exposed on glial and neuronal MHC-I for presentation to CD8+ T cells. In fact, IP is able to degrade aggregation-prone proteins such as alpha-synuclein and beta amyloid, which are conventionally degraded by SP and autophagy, although some debate still exists concerning the degradation rate and efficacy of IP compared with SP [238,240,241]. In any case, the IP-dependent degradation of aggregation-prone proteins produces Ag peptides, which activate adaptive immunity [6,238]. This provides an oxidation-linked explanation for the baseline activity of UPS in neuro-immune surveillance [85,238]. IP recruitment may serve as a compensatory pro-survival mechanism, allowing cells to quickly expand the peptides repertoire and aid immune defense in a challenged organism. This is supported by the fact that IP also operates in baseline conditions in neurons and glia, which indeed express low amounts of IP and MHC-I, even in the absence of cytokine stimulation [238,242,243]. In line with this, MHC-I-selective expression within the neurons and glia throughout the brain and spinal cord extends well beyond a classic antigen-presenting role. In fact, the MHC-I neuronal expression is key in early neuronal development, axonal regeneration, synaptic plasticity, reward, and memory [243–246]. Nonetheless, IP induction is a tightly regulated and transient response, as cells must rapidly switch back to SP once the IP function is no longer required [247]. Abnormal IP expression and the subsequent MHC-I-dependent Ag presentation enhances the APC-like behavior of neurons, and, as such, increases their susceptibility to CD8+ auto-immune attack. In fact, a dramatic increase in the amount of IP is bound to an abnormal auto-immune response in a variety of CNS disorders [6]. As recently reviewed, IP is significantly and constantly up-regulated in the glia and neurons, both in patients and experimental models of classic and auto-immune neurodegenerative disorders [6]. Nonetheless, the functional role of IP induction differs between auto-immune compared with classic neurodegenerative disorders. In neurodegenerative disorders such as PD, AD, and HD, the upregulation of IP occurs as a compensatory response to cope with inflammatory conditions that develop during proteinopathy, when SP is downregulated [6,238,248,249]. In fact, general UPS inhibitors targeting both SP and IP produce a detrimental effect, which recapitulates neurodegeneration, while selective IP inhibitors have only limited beneficial effects in the models of neurodegenerative disorders [6]. On the other hand, in auto-immune disorders, including MS and experimental autoimmune encephalomyelitis (EAE), IP inhibitors significantly ameliorate neurological and inflammatory disease scores [6,250]. There is also evidence indicating that a combination of autophagy and IP inhibitors may be an effective strategy against EAE [251]. In keeping with this, a number of studies reported that exposure to cytokines, such as IFN-γ, also up-regulates autophagy to promote the activation of innate and adaptive auto-immunity [252]. In this context, autophagy has been suggested to represent a tolerance-avoidance mechanism, being strongly recruited during CD4+ T-cells activation [253]. Instead, autophagy inhibition induces a long-lasting state of hypo-responsiveness within T-cells [253]. In vivo, autophagy inhibition during Ag priming induces T-cell energy, and decreases the severity of disease in EAE [253]. On the other hand, studies in humans showed that autophagy activity is not increased in neither the peripheral nor brain-circulating CD4+ T cells of MS patients compared with controls, despite having increased Atg5 gene and protein levels [254]. Other studies centered on the role of microglia-related inflammation suggest that autophagy induction via mTOR inhibition contributes to reducing both demyelination and inflammation in EAE [255]. As far as it concerns neurodegenerative disorders, autophagy induction seems to play a beneficial effect in counteracting acute and chronic inflammation [256]. For instance, in an in vitro model of PD, TNF-α was shown to impair autophagy flux in microglia, while fostering microglia polarization towards the pro-inflammatory phenotype M1 [257]. The inhibition of autophagy consistently aggravates M1 polarization induced by TNF-α, and remarkably, autophagy inhibition alone is sufficient to trigger microglia activation toward M1 status, along with producing neurotoxicity [257]. Conversely, the upregulation of autophagy via serum deprivation or pharmacologic activators (rapamycin and resveratrol) promotes microglia polarization toward the M2 phenotype, thus fostering inflammation resolution and preventing neurotoxicity [257]. Again, enhancing autophagy in the microglia in an in vitro model of AD promotes the degradation of the phagocytosed fibrils of amyloid beta, along with restraining the inflammasome activation and pro-inflammatory cytokine release [258]. This is in line with findings indicating that impaired autophagy in microglia associates with synaptic defects, and with the subsequent psychiatric alterations observed in experimental models [76,77]. Again, the disruption of autophagy within neurons occurs following infection-induced microglial activation, which results in neurodegeneration [52]. This is in line with the plethora of studies pointing at autophagy dysfunction in neurodegenerative disorders, such as AD, PD, and HD. In these disorders, a progressive dysfunction of autophagy within the CNS is reminiscent of that reported for SP. In a scenario where the autophagy–UPS interplay appears critical, it is worth of considering some overlapping molecular mechanisms that may operate in various CNS disorders to foster neuro-inflammation and maladaptive synaptic plasticity through IP induction and concomitant SP-autophagy downregulation. Remarkably, autophagy and UPS activities are influenced by the same intracellular cascades that are triggered by the DA receptors expressed on the neurons and glia. In fact, signaling pathways placed downstream to plasma membrane D1-like and D2-like DA receptors converge on the mTORC1 pathway [96], which, in turn, may either suppress or enhance the baseline SP/IP and autophagy activities, depending on the pattern of stimulation of the specific DA receptors. Thus, a feedback loop is established between DA signaling and mTOR-dependent cell-clearing systems in neurons, glia, or even in T-cells. The intrinsic oxidative potential of DA, along with the abnormal stimulation of DA receptors, are primary candidates fostering protein oxidation, inflammation, impairment of autophagy flux, SP disassembly, and the subsequent IP upregulation [48,71,96,97,99,259,260] (Figure 3). This is supported by the effects of exogenously administered DA precursors in enhancing neuronal Ag presentation via MHC-I, and the subsequent activation of CTLs [48], which, in fact, is a major task of the IP.

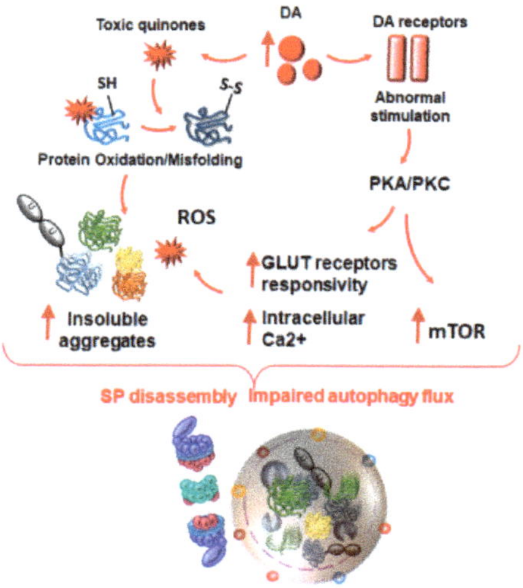

Figure 3. The effects of abnormal DA release and stimulation of DA receptors on proteasome and autophagy. An abnormal amount of intracellular DA may foster the loss of compartmentalized physiological oxidative deamination of DA, which readily undergoes auto-oxidation to produce toxic quinones and highly reactive chemical species such as reactive oxygen species (ROS). In turn, these react with sulfhydryl groups and promote the structural modifications of proteins, lipids, and nucleic acids within the DA axon terminals and surrounding compartments. Structural modifications of proteins translate into the formation of insoluble aggregates overwhelming both the SP and autophagy degradative potential. At the same, the abnormal release of DA produces an abnormal stimulation of post-synaptic DA receptors, which are coupled with intracellular cascades such as protein kinase A and C (PKA/PKC). The non-canonical activation of these cascades promotes the hyper-phosphorylation and activation of glutamate (GLUT) receptors and ion channels, which foster GLUT hyper-responsivity and Ca2+ uptake converging in the increase of oxidative stress. Again, the intracellular cascades placed downstream of the DA receptors (mostly D1-like) converge on activating the mammalian target of rapamycin (mTOR) pathway, thus promoting SP and autophagy downregulation. Red arrows in bold indicate increased levels. Plain red arrows indicate the formation of DA-derived toxic quinones and oxidized/misfolded proteins up to insoluble aggregates, and the progression of the various metabolic cascades that arise from abnormal stimulation of DA receptors.

As a consequence of the autophagy and SP dysregulation, indigested misfolded or oxidized substrates may perpetuate inflammation through the release of danger-associated molecular pattern molecules (DAMPs) (Figure 4). In fact, DAMPs activate NF-κB and the inflammasome to release cytokines such as IFNγ, along with spreading misfolded proteins, advanced glycation end-products (AGEs) and free radicals, which all converge to induce the upregulation of IP within neighboring cells via autocrine or paracrine mechanisms. DAMPs may also stimulate toll-like 4 receptor (TLR4) to impair both SP and autophagy [75]. In this scenario, impaired SP and autophagy can neither digest potentially harmful DAMPs, nor restrain the release of DA and GLUT, which may add on glia activation and the release of pro-inflammatory signals recruiting T-cells within the CNS. In this way, IP upregulation leads to an overproduction of neuronal and glial antigens co-expressed with MHC-I molecules to prime cytotoxic CD8+ T cell response (Figure 4). At the same time, autophagy cannot efficiently provide for the internalization of MHC-I molecules or the degradation of damaged proteins and organelles, which fuels inflammation and immune activation. Thus, alterations of autophagy and UPS may explain why a variety of CNS disorders feature concomitant alterations in neurotransmitter activity, oxidative-inflammatory stress, and inappropriate immune response, which synergize to alter synaptic plasticity and damage neurons.

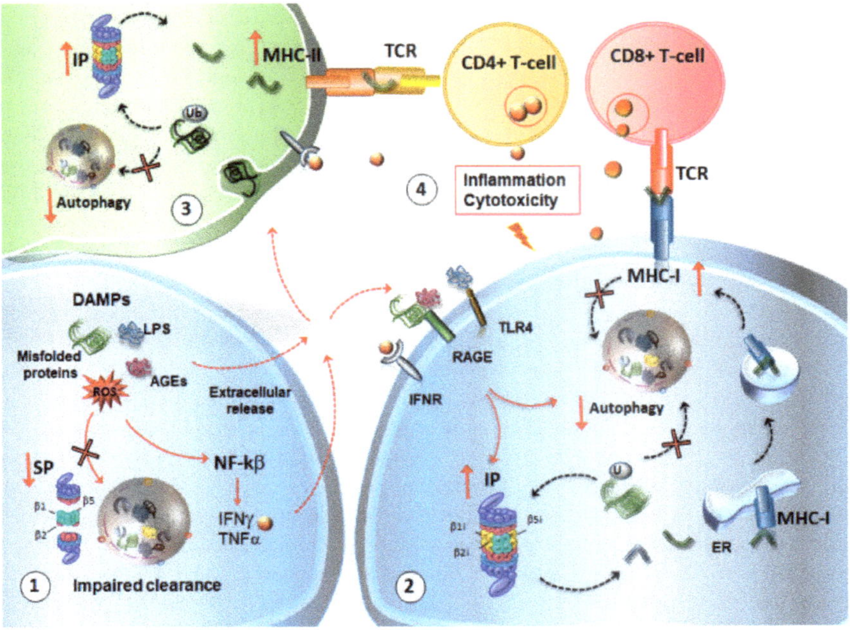

Figure 4. Molecular mechanisms bridging neuro-inflammation, immunoproteasome (IP) induction and autophagy-SP dysfunction in the central nervous system (CNS). (1) As a consequence of autophagy and SP dysregulation, indigested misfolded or oxidized substrates may perpetuate inflammation through the release of danger-associated molecular pattern molecules (DAMPs), such as advanced glycation end-products (AGEs), lipopolysaccharides (LPS), and ROS. While DAMPs activate NF-κB and the inflammasome to release cytokines such as interferon gamma (IFNγ), the spreading of indigested misfolded proteins, AGEs, and free radicals in the extracellular space occurs. (2) All of these factors converge to induce an upregulation of IP within the neighboring cells via autocrine or paracrine mechanisms. DAMPs may also stimulate AGE receptors, IFN receptors, and toll-like 4 receptor (TLR4), to converge on molecular pathways such as mTOR, which, in turn, induce IP upregulation and SP-autophagy downregulation. In this scenario, IP upregulation leads to an overproduction of CNS self-antigens co-expressed with major histocompatibility complex (MHC)-I molecules to prime cytotoxic CD8+ T cell response. At the same time, autophagy cannot efficiently provide for the internalization of MHC-I molecules or the degradation of damaged proteins and organelles. (3) Within glial cells, the same DAMPs and cytokines that foster glial activation may contribute to the impairment of autophagy and SP, while up-regulating IP, which is able to process and cross-present phagocytosed proteins via MHC-II for the activation of CD4+ T cells. (4) In this way, IP upregulation, in an attempt to compensate for SP-autophagy downregulation, may fuel inflammation and auto-immune activation to promote altered synaptic plasticity and neuronal damage. Red arrows in bold indicate decreased/increased levels/activity of SP, IP, autophagy, MHC-I, MHC-II. Plain red arrows indicate intra-cellular signaling cascades. Dotted red arrows indicate the extracellular release of DAMPs, cytokines, and their binding to cognate receptors in neighbor cells or phagocytosis by glial cells. Dotted black arrows indicate the shuttling of substrates towards UPS or autophagy, the formation of Ag peptides deriving from UPS cleavage, and the progression of MHC-I-Ag complex from the ER and endosomes to the plasma membrane. The red frame indicates the final effect produced by CD4+ and CD8+ T-cells activation on neurons and glia.

5. Conclusions and Future Directions

The evidence reviewed here suggests that autophagy and UPS are key mediators of synaptic plasticity, being placed at the cross-road between neurotransmission and immune activity. Nonetheless, we are just scratching the surface of the intricate molecular mechanisms that translate autophagy and UPS alterations into specific CNS disorders. Further experimental studies are needed to dissect the correlation between autophagy and UPS status, disease-specificity, and disease-stage. In fact, different effects on autophagy and UPS may occur in auto-immune compared with classic neurodegenerative disorders, because of the different etiologies between these CNS diseases. Moreover, the interdependency between autophagy and UPS may lead to confounding outcomes when assessing the effects of specific compounds, which indeed modulate both systems rather than autophagy or UPS individually. In addition, there are several factors that may contribute to yielding controversial results on autophagy and UPS. One of these may be the biased interpretation of the autophagy or UPS status. In fact, most studies assess autophagy by measuring the amount of LC3B or the number of autophagosomes, which does not necessarily reflect an increased autophagy capacity. Rather, it may reflect a progressive downregulation of autophagy flux due to the impaired fusion of LC3-positive autophagosomes with lysosomes. Again, most studies detect UPS status through antibodies that recognize alpha subunits, which do not allow for distinguishing between the SP/IP ratio. Likewise, measuring the overall UPS catalytic activity does not allow for dissecting whether the contribution derives from SP or IP. In any case, dysregulations of both autophagy and UPS appear as a common signature in a variety of CNS disorders, where synaptic alterations synergize with inflammatory/immune reactions. This calls for further studies aimed at investigating the effects of additional compounds, which can synergistically modulate both UPS (IP and SP) and autophagy, in an attempt to find preventive and/or therapeutic strategies against early synaptic alterations. In keeping with this, mTOR modulators remain, to date, the best candidates for acting on autophagy, IP, and SP. Beyond the gold-standard mTOR inhibitor rapamycin, several phytochemicals, which recently gained increasing interest in CNS disorders due to their adaptogenic, anti-oxidant, and anti-inflammatory effects, act indeed as mTOR modulators. Testing the effects of these compounds on autophagy and SP–IP may disclose a potential correlation between their beneficial effects and cell-clearing pathways.

Author Contributions: Original draft preparation, writing, review and art work, F.L. and F.B.; review, editing, and art work, C.L.B., and L.R.; conceptualization, P.S. and A.F.; supervision, F.F., F.L. and F.B equally contributed to the present work.

Funding: This work was funded by Ministero Della Salute (Ricerca Corrente 2019).

Conflicts of Interest: The authors declare no conflict of interest.

Abbreviations

AD	Alzheimer's disease
Ag	Antigen
APC	Antigen presenting cell
BDNF	Brain derived neurotrophic factor
CNS	Central nervous system
CTL	Cytotoxic T-lymphocyte
DA	Dopamine
DAMPs	Danger-associated molecular pattern molecules
DC	Dendritic cell
EAE	Experimental autoimmune encephalomyelitis
GLUT	Glutamate
HD	Huntingtin's disease
HDAC6	Histone deacetylase 6
IFNγ	Interferon gamma
IL-1β	Interleukin 1 beta
IP	Immunoproteasome
Meth	Methamphetamine
MHC	Major histocompatibility complex
MS	Multiple sclerosis

mTOR	Mammalian target of rapamycin
NF-κB	Nuclear factor K beta
PD	Parkinson's disease
Rab GTPase	Gtp bound ras proteins in brain
SNARE	Soluble Nsf attachment protein receptor
SP	Standard proteasome
SQSTM1	Sequestosome-1
SV	Synaptic vesicle
TCR	T-cells receptor
TLR4	Toll-like receptor 4
TNFα	Tumor necrosis factor alpha
UPS	Ubiquitin proteasome

References

1. Zipp, F.; Aktas, O. The brain as a target of inflammation: common pathways link inflammatory and neurodegenerative diseases. *Trends Neurosci.* **2006**, *29*, 518–527. [CrossRef] [PubMed]
2. Centonze, D.; Muzio, L.; Rossi, S.; Furlan, R.; Bernardi, G.; Martino, G. The link between inflammation, synaptic transmission and neurodegeneration in multiple sclerosis. *Cell Death Differ.* **2010**, *17*, 1083–1091. [CrossRef] [PubMed]
3. Yirmiya, R.; Goshen, I. Immune modulation of learning, memory, neural plasticity and neurogenesis. *Brain Behav. Immun.* **2011**, *25*, 181–213. [CrossRef] [PubMed]
4. Kerschensteiner, M.; Meinl, E.; Hohlfeld, R. Neuro-immune crosstalk in CNS diseases. *Neuroscience* **2009**, *158*, 1122–1132. [CrossRef] [PubMed]
5. Kioussis, D.; Pachnis, V. Immune and nervous systems: more than just a superficial similarity? *Immunity* **2009**, *31*, 705–710. [CrossRef] [PubMed]
6. Limanaqi, F.; Biagioni, F.; Busceti, C.L.; Gaglione, A.; Fornai, F. A Sentinel in the Crosstalk Between the Nervous and Immune System: The (Immuno)-Proteasome. *Front. Immunol.* **2019**, *10*, 628. [CrossRef]
7. Jovanova-Nesic, K.D.; Jankovic, B.D. The neuronal and immune memory systems as supervisors of neural plasticity and aging of the brain: from phenomenology to coding of information. *Ann. N. Y. Acad. Sci.* **2005**, *1057*, 279–295. [CrossRef]
8. Schwartz, M.; Shechter, R. Protective autoimmunity functions by intracranial immunosurveillance to support the mind: The missing link between health and disease. *Mol. Psychiatry.* **2010**, *15*, 342–354. [CrossRef]
9. Citri, A.; Malenka, R.C. Synaptic plasticity: multiple forms, functions, and mechanisms. *Neuropsychopharmacology* **2008**, *33*, 18–41. [CrossRef]
10. Wang, J.; Hodes, G.E.; Zhang, H.; Zhang, S.; Zhao, W.; Golden, S.A.; Bi, W.; Menard, C.; Kana, V.; Leboeuf, M.; et al. Epigenetic modulation of inflammation and synaptic plasticity promotes resilience against stress in mice. *Nat. Commun.* **2018**, *9*, 477. [CrossRef]
11. Limanaqi, F.; Gambardella, S.; Biagioni, F.; Busceti, C.L.; Fornai, F. Epigenetic Effects Induced by Methamphetamine and Methamphetamine-Dependent Oxidative Stress. *Oxid. Med. Cell Longev.* **2018**, *22*, 4982453. [CrossRef] [PubMed]
12. Tournier, J.N.; Hellmann, A.Q. Neuro-immune connections: evidence for a neuro-immunological synapse. *Trends Immunol.* **2003**, *24*, 114–115. [CrossRef]
13. Lorton, D.; Lubahn, C.L.; Estus, C.; Millar, B.A.; Carter, J.L.; Wood, C.A.; Bellinger, D.L. Bidirectional communication between the brain and the immune system: implications for physiological sleep and disorders with disrupted sleep. *Neuroimmunomodulation* **2006**, *13*, 357–374. [CrossRef] [PubMed]
14. Lucin, K.M.; Wyss-Coray, T. Immune activation in brain aging and neurodegeneration: too much or too little? *Neuron* **2009**, *64*, 110–122. [CrossRef] [PubMed]
15. Tian, L.; Rauvala, H.; Gahmberg, C.G. Neuronal regulation of immune responses in the central nervous system. *Trends Immunol.* **2009**, *30*, 91–99. [CrossRef] [PubMed]

16. Iliff, J.J.; Wang, M.; Liao, Y.; Plogg, B.A.; Peng, W.; Gundersen, G.A.; Benveniste, H.; Vates, G.E.; Deane, R.; Goldman, S.A.; et al. A paravascular pathway facilitates CSF flow through the brain parenchyma and the clearance of interstitial solutes, including amyloid beta. *Sci. Transl. Med.* **2012**, *4*, 147ra111. [CrossRef] [PubMed]
17. Plog, B.A.; Nedergaard, M. The Glymphatic System in Central Nervous System Health and Disease: Past, Present, and Future. *Annu. Rev. Pathol.* **2018**, *24*, 379–394. [CrossRef] [PubMed]
18. Louveau, A.; Smirnov, I.; Keyes, T.J.; Eccles, J.D.; Rouhani, S.J.; Peske, J.D.; Derecki, N.C.; Castle, D.; Mandell, J.W.; Lee, K.S.; et al. Structural and functional features of central nervous system lymphatic vessels. *Nature* **2016**, *523*, 337–341, Erratum in: *Nature* **2016**, *533*, 278. [CrossRef]
19. Verheggen, I.C.M.; Van Boxtel, M.P.J.; Verhey, F.R.J.; Jansen, J.F.A.; Backes, W.H. Interaction between blood-brain barrier and glymphatic system in solute clearance. *Neurosci. Biobehav. Rev.* **2018**, *90*, 26–33. [CrossRef]
20. Louveau, A.; Harris, T.H.; Kipnis, J. Revisiting the Mechanisms of CNS Immune Privilege. *Trends Immunol.* **2015**, *36*, 569–577. [CrossRef]
21. Limanaqi, F.; Biagioni, F.; Gambardella, S.; Ryskalin, L.; Fornai, F. Interdependency Between Autophagy and Synaptic Vesicle Trafficking: Implications for Dopamine Release. *Front. Mol. Neurosci.* **2018**, *11*, 299. [CrossRef]
22. Speese, S.D.; Trotta, N.; Rodesch, C.K.; Aravamudan, B.; Broadie, K. The ubiquitin proteasome system acutely regulates presynaptic protein turnover and synaptic efficacy. *Curr. Biol.* **2003**, *13*, 899–910. [CrossRef]
23. Münz, C. Autophagy proteins in antigen processing for presentation on MHC molecules. *Immunol. Rev.* **2016**, *272*, 17–27. [CrossRef] [PubMed]
24. Palmowski, M.J.; Gileadi, U.; Salio, M.; Gallimore, A.; Millrain, M.; James, E.; Addey, C.; Scott, D.; Dyson, J.; Simpson, E.; et al. Role of immunoproteasomes in cross-presentation. *J. Immunol.* **2006**, *177*, 983–990. [CrossRef] [PubMed]
25. Basler, M.; Kirk, C.J.; Groettrup, M. The immunoproteasome in antigen processing and other immunological functions. *Curr. Opin. Immunol.* **2013**, *25*, 74–80. [CrossRef]
26. Hegde, A.N. The ubiquitin-proteasome pathway and synaptic plasticity. *Learn. Mem.* **2010**, *17*, 314–327. [CrossRef] [PubMed]
27. Hegde, A.N. Proteolysis, synaptic plasticity and memory. *Neurobiol. Learn. Mem.* **2017**, *138*, 98–110. [CrossRef] [PubMed]
28. Liang, Y. Emerging Concepts and Functions of Autophagy as a Regulator of Synaptic Components and Plasticity. *Cells* **2019**, *8*, 34. [CrossRef]
29. Leung, C.S. Endogenous Antigen Presentation of MHC Class II Epitopes through Non-Autophagic Pathways. *Front. Immunol.* **2015**, *6*, 464. [CrossRef]
30. Valečka, J.; Almeida, C.R.; Su, B.; Pierre, P.; Gatti, E. Autophagy and MHC-restricted antigen presentation. *Mol. Immunol.* **2018**, *99*, 163–170. [CrossRef]
31. Shechter, R.; London, A.; Schwartz, M. Orchestrated leukocyte recruitment to immune-privileged sites: absolute barriers versus educational gates. *Nat. Rev. Immunol.* **2013**, *13*, 206–218. [CrossRef] [PubMed]
32. Schwab, N.; Schneider-Hohendorf, T.; Wiendl, H. Trafficking of lymphocytes into the CNS. *Oncotarget* **2015**, *6*, 17863–17864. [CrossRef] [PubMed]
33. Goverman, J. Autoimmune T cell responses in the central nervous system. *Nat. Rev. Immunol.* **2009**, *9*, 393–407. [CrossRef]
34. Mignini, F.; Streccioni, V.; Amenta, F. Autonomic innervation of immune organs and neuroimmune modulation. *Auton. Autacoid. Pharmacol.* **2003**, *23*, 1–25. [CrossRef]
35. Levite, M. Neurotransmitters activate T-cells and elicit crucial functions via neurotransmitter receptors. *Curr. Opin. Pharmacol.* **2008**, *8*, 460–471. [CrossRef] [PubMed]
36. Sarkar, C.; Basu, B.; Chakroborty, D.; Dasgupta, P.S.; Basu, S. The immunoregulatory role of dopamine: an update. *Brain Behav. Immun.* **2009**, *24*, 525–528. [CrossRef] [PubMed]
37. Ganor, Y.; Levite, M. The neurotransmitter glutamate and human T cells: glutamate receptors and glutamate-induced direct and potent effects on normal human T cells, cancerous human leukemia and lymphoma T cells, and autoimmune human T cells. *J. Neural Transm. (Vienna)* **2014**, *121*, 983–1006. [CrossRef]

38. Brabb, T.; von Dassow, P.; Ordonez, N.; Schnabel, B.; Duke, B.; Goverman, J. In situ tolerance within the central nervous system as a mechanism for preventing autoimmunity. *J. Exp. Med.* **2000**, *192*, 871–880. [CrossRef] [PubMed]
39. Liu, Y.; Teige, I.; Birnir, B.; Issazadeh-Navikas, S. Neuron-mediated generation of regulatory T cells from encephalitogenic T cells suppresses EAE. *Nat. Med.* **2006**, *12*, 518–525. [CrossRef]
40. Na, S.Y.; Cao, Y.; Toben, C.; Nitschke, L.; Stadelmann, C.; Gold, R.; Schimpl, A.; Hünig, T. Naive CD8 T-cells initiate spontaneous autoimmunity to a sequestered model antigen of the central nervous system. *Brain* **2008**, *13*, 2353–2365. [CrossRef] [PubMed]
41. Sosa, R.A.; Forsthuber, T.G. The critical role of antigen-presentation-induced cytokine crosstalk in the central nervous system in multiple sclerosis and experimental autoimmune encephalomyelitis. *J. Interferon Cytokine Res.* **2011**, *31*, 753–768. [CrossRef] [PubMed]
42. Jarry, U.; Jeannin, P.; Pineau, L.; Donnou, S.; Delneste, Y.; Couez, D. Efficiently stimulated adult microglia cross-prime naive CD8+ T cells injected in the brain. *Eur. J. Immunol.* **2013**, *43*, 1173–1184. [CrossRef] [PubMed]
43. Krakowski, M.L.; Owens, T. Naive T lymphocytes traffic to inflamed central nervous system, but require antigen recognition for activation. *Eur. J. Immunol.* **2000**, *30*, 1002–1009. [CrossRef]
44. McMahon, E.J.; Bailey, S.L.; Castenada, C.V.; Waldner, H.; Miller, S.D. Epitope spreading initiates in the CNS in two mouse models of multiple sclerosis. *Nat. Med.* **2005**, *11*, 335–339. [CrossRef] [PubMed]
45. Cose, S.; Brammer, C.; Khanna, K.M.; Masopust, D.; Lefrancois, L. Evidence that a significant number of naive T cells enter non-lymphoid organs as part of a normal migratory pathway. *Eur. J. Immunol.* **2006**, *36*, 1423–1433. [CrossRef]
46. Herz, J.; Paterka, M.; Niesner, R.A.; Brandt, A.U.; Siffrin, V.; Leuenberger, T.; Birkenstock, J.; Mossakowski, A.; Glumm, R.; Zipp, F.; et al. In vivo imaging of lymphocytes in the CNS reveals different behaviour of naive T cells in health and autoimmunity. *J. Neuroinflamm.* **2011**, *8*, 131. [CrossRef] [PubMed]
47. Chastain, E.M.; Duncan, D.S.; Rodgers, J.M.; Miller, S.D. The role of antigen presenting cells in multiple sclerosis. *Biochim. Biophys. Acta.* **2010**, *1812*, 265–274. [CrossRef]
48. Cebrián, C.; Zucca, F.A.; Mauri, P.; Steinbeck, J.A.; Studer, L.; Scherzer, C.R.; Kanter, E.; Budhu, S.; Mandelbaum, J.; Vonsattel, J.P.; et al. MHC-I expression renders catecholaminergic neurons susceptible to T-cell-mediated degeneration. *Nat. Commun.* **2014**, *5*, 3633. [CrossRef]
49. Schwartz, M.; Deczkowska, A. Neurological Disease as a Failure of Brain-Immune Crosstalk: The Multiple Faces of Neuroinflammation. *Trends Immunol.* **2016**, *37*, 668–679. [CrossRef]
50. Ferretti, M.T.; Bruno, M.A.; Ducatenzeiler, A.; Klein, W.L.; Cuello, A.C. Intracellular Aβ-oligomers and early inflammation in a model of Alzheimer's disease. *Neurobiol. Aging* **2012**, *33*, 1329–1342. [CrossRef]
51. Alirezaei, M.; Kemball, C.C.; Whitton, J.L. Autophagy, inflammation and neurodegenerative disease. *Eur. J. Neurosci.* **2010**, *33*, 197–204. [CrossRef]
52. Alirezaei, M.; Kiosses, W.B.; Flynn, C.T.; Brady, N.R.; Fox, H.S. Disruption of neuronal autophagy by infected microglia results in neurodegeneration. *PLoS One.* **2008**, *3*, e2906. [CrossRef]
53. Plaza-Zabala, A.; Sierra-Torre, V.; Sierra, A. Autophagy and Microglia: Novel Partners in Neurodegeneration and Aging. *Int. J. Mol. Sci.* **2017**, *18*, 598. [CrossRef] [PubMed]
54. Ferrucci, M.; Biagioni, F.; Ryskalin, L.; Limanaqi, F.; Gambardella, S.; Frati, A.; Fornai, F. Ambiguous Effects of Autophagy Activation Following Hypoperfusion/Ischemia. *Int. J. Mol. Sci.* **2018**, *19*, 2756. [CrossRef] [PubMed]
55. Ryskalin, L.; Limanaqi, F.; Biagioni, F.; Frati, A.; Esposito, V.; Calierno, M.T.; Lenzi, P.; Fornai, F. The emerging role of m-TOR up-regulation in brain Astrocytoma. *Histol. Histopathol.* **2017**, *32*, 413–431. [CrossRef]
56. Fabrizi, C.; Pompili, E.; De Vito, S.; Somma, F.; Catizone, A.; Ricci, G.; Lenzi, P.; Fornai, F.; Fumagalli, L. Impairment of the autophagic flux in astrocytes intoxicated by trimethyltin. *Neurotoxicol.* **2016**, *52*, 12–22. [CrossRef]
57. Giorgi, F.S.; Biagioni, F.; Lenzi, P.; Frati, A.; Fornai, F. The role of autophagy in epileptogenesis and in epilepsy-induced neuronal alterations. *J. Neural. Transm. (Vienna)* **2015**, *122*, 849–862. [CrossRef]
58. Lenzi, P.; Marongiu, R.; Falleni, A.; Gelmetti, V.; Busceti, C.L.; Michiorri, S.; Valente, E.M.; Fornai, F. A subcellular analysis of genetic modulation of PINK1 on mitochondrial alterations, autophagy and cell death. *Arch. Ital. Biol.* **2012**, *150*, 194–217. [CrossRef]

59. Pasquali, L.; Ruggieri, S.; Murri, L.; Paparelli, A.; Fornai, F. Does autophagy worsen or improve the survival of dopaminergic neurons? *Parkinsonism Relat. Disord.* **2009**, *15*, S24–S27. [CrossRef]
60. Ferrucci, M.; Pasquali, L.; Ruggieri, S.; Paparelli, A.; Fornai, F. Alpha-synuclein and autophagy as common steps in neurodegeneration. *Parkinsonism Relat. Disord.* **2008**, *14*, S180–S184. [CrossRef]
61. Isidoro, C.; Biagioni, F.; Giorgi, F.S.; Fulceri, F.; Paparelli, A.; Fornai, F. The role of autophagy on the survival of dopamine neurons. *Curr. Top. Med. Chem.* **2009**, *9*, 869–879.
62. Ryskalin, L.; Busceti, C.L.; Limanaqi, F.; Biagioni, F.; Gambardella, S.; Fornai, F. A Focus on the Beneficial Effects of Alpha Synuclein and a Re-Appraisal of Synucleinopathies. *Curr. Protein Pept. Sci.* **2018**, *19*, 598–611. [CrossRef]
63. Thibaudeau, T.A.; Anderson, R.T.; Smith, D.M. A common mechanism of proteasome impairment by neurodegenerative disease-associated oligomers. *Nat. Commun.* **2018**, *9*, 1097. [CrossRef]
64. Jansen, A.H.P.; Reits, E.A.J.; Hol, E.M. The ubiquitin proteasome system in glia and its role in neurodegenerative diseases. *Front. Mol. Neurosci.* **2014**, *7*, 73. [CrossRef] [PubMed]
65. Wang, J.; Wang, C.E.; Orr, A.; Tydlacka, S.; Li, S.H.; Li, X.J. Impaired ubiquitin-proteasome system activity in the synapses of Huntington's disease mice. *J. Cell Biol.* **2008**, *180*, 1177–1189. [CrossRef] [PubMed]
66. McNaught, K.S.; Belizaire, R.; Isacson, O.; Jenner, P.; Olanow, C.W. Altered proteasomal function in sporadic Parkinson's disease. *Exp. Neurol.* **2003**, *179*, 38–46. [CrossRef] [PubMed]
67. Zheng, J.; Bizzozero, O.A. Decreased activity of the 20S proteasome in the brain white matter and gray matter of patients with multiple sclerosis. *J. Neurochem.* **2011**, *117*, 143–153. [CrossRef]
68. van Scheppingen, J.; Broekaart, D.W.; Scholl, T.; Zuidberg, M.R.; Anink, J.J.; Spliet, W.G.; van Rijen, P.C.; Czech, T.; Hainfellner, J.A.; Feucht, M.; et al. Dysregulation of the (immuno)proteasome pathway in malformations of cortical development. *J. Neuroinflammation* **2016**, *13*, 202. [CrossRef]
69. Graham, S.H.; Liu, H. Life and death in the trash heap: The ubiquitin proteasome pathway and UCHL1 in brain aging, neurodegenerative disease and cerebral Ischemia. *Ageing Res. Rev.* **2017**, *34*, 30–38. [CrossRef]
70. Castino, R.; Lazzeri, G.; Lenzi, P.; Bellio, N.; Follo, C.; Ferrucci, M.; Fornai, F.; Isidoro, C. Suppression of autophagy precipitates neuronal cell death following low doses of methamphetamine. *J. Neurochem.* **2008**, *106*, 1426–1439. [CrossRef]
71. Moszczynska, A.; Yamamoto, B.K. Methamphetamine oxidatively damages parkin and decreases the activity of 26S proteasome in vivo. *J. Neurochem.* **2011**, *116*, 1005–1017. [CrossRef]
72. Lin, M.; Chandramani-Shivalingappa, P.; Jin, H.; Ghosh, A.; Anantharam, V.; Ali, S.; Kanthasamy, A.G.; Kanthasamy, A. Methamphetamine-induced neurotoxicity linked to ubiquitin-proteasome system dysfunction and autophagy-related changes that can be modulated by protein kinase C delta in dopaminergic neuronal cells. *Neuroscience* **2012**, *210*, 308–332. [CrossRef]
73. Lazzeri, G.; Biagioni, F.; Fulceri, F.; Busceti, C.L.; Scavuzzo, M.C.; Ippolito, C.; Salvetti, A.; Lenzi, P.; Fornai, F. mTOR Modulates Methamphetamine-Induced Toxicity through Cell Clearing Systems. *Oxid. Med. Cell. Longev.* **6124**. [CrossRef] [PubMed]
74. Pla, A.; Pascual, M.; Renau-Piqueras, J.; Guerri, C. TLR4 mediates the impairment of ubiquitin-proteasome and autophagy-lysosome pathways induced by ethanol treatment in brain. *Cell Death Dis.* **2014**, *5*, e1066. [CrossRef] [PubMed]
75. Ryskalin, L.; Limanaqi, F.; Frati, A.; Busceti, C.L.; Fornai, F. mTOR-Related Brain Dysfunctions in Neuropsychiatric Disorders. *Int. J. Mol. Sci.* **2018**, *19*, 2226. [CrossRef] [PubMed]
76. Kim, H.J.; Cho, M.H.; Shim, W.H.; Kim, J.K.; Jeon, E.Y.; Kim, D.H.; Yoon, S.Y. Deficient autophagy in microglia impairs synaptic pruning and causes social behavioral defects. *Mol. Psychiatry* **2017**, *22*, 1576–1584. [CrossRef] [PubMed]
77. Tan, X.; Du, X.; Jiang, Y.; Botchway, B.O.A.; Hu, Z.; Fang, M. Inhibition of Autophagy in Microglia Alters Depressive-Like Behavior via BDNF Pathway in Postpartum Depression. *Front. Psychiatry* **2018**, *9*, 434. [CrossRef] [PubMed]
78. Rubinsztein, D.C.; Bento, C.F.; Deretic, V. Therapeutic targeting of autophagy in neurodegenerative and infectious diseases. *J. Exp. Med.* **2015**, *212*, 979–990. [CrossRef]
79. Levine, B.; Mizushima, N.; Virgin, H.W. Autophagy in immunity and inflammation. *Nature* **2011**, *469*, 323–335. [CrossRef]
80. Cadwell, K. Crosstalk between autophagy and inflammatory signalling pathways: balancing defence and homeostasis. *Nat. Rev. Immunol.* **2016**, *16*, 661–675. [CrossRef]

81. Nikoletopoulou, V.; Sidiropoulou, K.; Kallergi, E.; Dalezios, Y.; Tavernarakis, N. Modulation of Autophagy by BDNF Underlies Synaptic Plasticity. *Cell Metab.* **2017**, *26*, 230–242. [CrossRef] [PubMed]
82. Shehata, M.; Inokuchi, K. Does autophagy work in synaptic plasticity and memory? *Rev. Neurosci.* **2014**, *25*, 543–557. [CrossRef]
83. Niedermann, G.; Grimm, R.; Geier, E.; Maurer, M.; Realini, C.; Gartmann, C.; Soll, J.; Omura, S.; Rechsteiner, M.C.; Baumeister, W.; et al. Potential Immunocompetence of Proteolytic Fragments Produced by Proteasomes before Evolution of the Vertebrate Immune System. *J. Exp. Med.* **1997**, *186*, 209–220. [CrossRef] [PubMed]
84. Ferrington, D.A.; Gregerson, D.S. Immunoproteasomes: structure, function, and antigen presentation. *Prog. Mol. Biol. Transl. Sci.* **2012**, *109*, 75–112. [PubMed]
85. Johnston-Carey, H.K.; Pomatto, L.C.D.; Davies, K.J.A. The Immunoproteasome in Oxidative Stress, Aging, and Disease. *Crit. Rev. Biochem. Mol. Biol.* **2015**, *51*, 268–281. [CrossRef] [PubMed]
86. Hakim, V.; Cohen, L.D.; Zuchman, R.; Ziv, T.; Ziv, N.E. The effects of proteasomal inhibition on synaptic proteostasis. *EMBO J.* **2016**, *35*, 2238–2262. [CrossRef] [PubMed]
87. Bingol, B.; Schuman, E.M. Synaptic protein degradation by the ubiquitin proteasome system. *Curr. Opin. Neurobiol.* **2005**, *15*, 536–541. [CrossRef] [PubMed]
88. Lenzi, P.; Lazzeri, G.; Biagioni, F.; Busceti, C.L.; Gambardella, S.; Salvetti, A.; Fornai, F. The autophagoproteasome a novel cell clearing organelle in baseline and stimulated conditions. *Front. Neuroanat.* **2016**, *10*, 78. [CrossRef] [PubMed]
89. Zhao, J.; Zhai, B.; Gygi, S.P.; Goldberg, A.L. mTOR inhibition activates overall protein degradation by the ubiquitin proteasome system as well as by autophagy. *Proc. Natl. Acad. Sci. USA* **2015**, *112*, 15790–15797. [CrossRef]
90. Cohen-Kaplan, V.; Ciechanover, A.; Livneh, I. p62 at the crossroad of the ubiquitin-proteasome system and autophagy. *Oncotarget* **2016**, *7*, 83833–83834. [CrossRef]
91. Ji, C.H.; Kwon, Y.T. Crosstalk and Interplay between the Ubiquitin-Proteasome System and Autophagy. *Mol. Cells* **2017**, *40*, 441–449. [CrossRef] [PubMed]
92. Korolchuk, V.I.; Menzies, F.M.; Rubinsztein, D.C. Mechanisms of cross-talk between the ubiquitin-proteasome and autophagy-lysosome systems. *FEBS Lett.* **2010**, *584*, 1393–1398. [CrossRef] [PubMed]
93. Ugun-Klusek, A.; Tatham, M.H.; Elkharaz, J.; Constantin-Teodosiu, D.; Lawler, K.; Mohamed, H.; Paine, S.M.; Anderson, G.; John Mayer, R.; Lowe, J.; et al. Continued 26S proteasome dysfunction in mouse brain cortical neurons impairs autophagy and the Keap1-Nrf2 oxidative defence pathway. *Cell Death Dis.* **2017**, *8*, e2531. [CrossRef] [PubMed]
94. Figueiredo-Pereira, M.E.; Rockwell, P.; Schmidt-Glenewinkel, T.; Serrano, P. Neuroinflammation and J2 prostaglandins: linking impairment of the ubiquitin-proteasome pathway and mitochondria to neurodegeneration. *Front. Mol. Neurosci.* **2015**, *7*, 104. [CrossRef]
95. Kulbe, J.R.; Mulcahy Levy, J.M.; Coultrap, S.J.; Thorburn, A.; Bayer, K.U. Excitotoxic glutamate insults block autophagic flux in hippocampal neurons. *Brain Res.* **2014**, *1542*, 12–19. [CrossRef] [PubMed]
96. Wang, D.; Ji, X.; Liu, J.; Li, Z.; Zhang, X. Dopamine Receptor Subtypes Differentially Regulate Autophagy. *Int. J. Mol. Sci.* **2018**, *19*, 1540. [CrossRef] [PubMed]
97. He, X.; Yuan, W.; Li, Z.; Hou, Y.; Liu, F.; Feng, J. 6-Hydroxydopamine induces autophagic flux dysfunction by impairing transcription factor EB activation and lysosomal function in dopaminergic neurons and SH-SY5Y cells. *Toxicol. Lett.* **2018**, *283*, 58–68. [CrossRef]
98. Du, D.; Hu, L.; Wu, J.; Wu, Q.; Cheng, W.; Guo, Y.; Guan, R.; Wang, Y.; Chen, X.; Yan, X.; et al. Neuroinflammation contributes to autophagy flux blockage in the neurons of rostral ventrolateral medulla in stress-induced hypertension rats. *J. Neuroinflammation* **2017**, *14*. [CrossRef]
99. Barroso-Chinea, P.; Thiolat, M.L.; Bido, S.; Martinez, A.; Doudnikoff, E.; Baufreton, J.; Bourdenx, M.; Bloch, B.; Bezard, E.; Martin-Negrier, M.L. D1 dopamine receptor stimulation impairs striatal proteasome activity in Parkinsonism through 26S proteasome disassembly. *Neurobiol. Dis.* **2015**, *78*, 77–87. [CrossRef]
100. Caldeira, M.V.; Curcio, M.; Leal, G.; Salazar, I.L.; Mele, M.; Santos, A.R.; Melo, C.V.; Pereira, P.; Canzoniero, L.M.; Duarte, C.B. Excitotoxic stimulation downregulates the ubiquitin-proteasome system through activation of NMDA receptors in cultured hippocampal neurons. *Biochim. Biophys. Acta.* **2013**, *1832*, 263–274. [CrossRef]

101. Davies, K.J. Protein modification by oxidants and the role of proteolytic enzymes. *Biochem. Soc. Trans.* **1993**, *21*, 346–353. [CrossRef] [PubMed]
102. Coux, O.; Tanaka, K.; Goldberg, A.L. Structure and functions of the 20S and 26S proteasomes. *Annu. Rev. Biochem.* **1996**, *65*, 801–847. [CrossRef] [PubMed]
103. Humbard, M.A.; Maupin-Furlow, J.A. Prokaryotic proteasomes: nanocompartments of degradation. *J. Mol. Microbiol. Biotechnol.* **2013**, *23*, 321–334. [CrossRef] [PubMed]
104. Ciechanover, A.; Kwon, Y.T. Degradation of misfolded proteins in neurodegenerative diseases: therapeutic targets and strategies. *Exp. Mol. Med.* **2015**, *47*, e147. [CrossRef] [PubMed]
105. Webb, J.L.; Ravikumar, B.; Atkins, J.; Skepper, J.N.; Rubinsztein, D.C. Alpha-Synuclein is degraded by both autophagy and the proteasome. *J. Biol. Chem.* **2003**, *278*, 25009–25013. [CrossRef] [PubMed]
106. Lee, M.J.; Lee, J.H.; Rubinsztein, D.C. Tau degradation: the ubiquitin-proteasome system versus the autophagy-lysosome system. *Prog. Neurobiol.* **2013**, *105*, 49–59. [CrossRef] [PubMed]
107. Rudolf, R.; Khan, M.M.; Wild, F.; Hashemolhosseini, S. The impact of autophagy on peripheral synapses in health and disease. *Front. Biosci. (Landmark Ed)* **2016**, *21*, 1474–1487. [CrossRef] [PubMed]
108. Akwa, Y.; Gondard, E.; Mann, A.; Capetillo-Zarate, E.; Alberdi, E.; Matute, C.; Marty, S.; Vaccari, T.; Lozano, A.M.; Baulieu, E.E.; et al. Synaptic activity protects against AD and FTD-like pathology via autophagic-lysosomal degradation. *Mol. Psychiatry* **2018**, *23*, 1530–1540. [CrossRef]
109. Shen, D.N.; Zhang, L.H.; Wei, E.Q.; Yang, Y. Autophagy in synaptic development, function, and pathology. *Neurosci. Bull.* **2015**, *31*, 416–426. [CrossRef] [PubMed]
110. Vijayan, V.; Verstreken, P. Autophagy in the presynaptic compartment in health and disease. *J. Cell Biol.* **2017**, *216*, 1895–1906. [CrossRef]
111. Münz, C. The Macroautophagy Machinery in Endo- and Exocytosis. *J. Mol. Biol.* **2017**, *429*, 473–485. [CrossRef] [PubMed]
112. Meijer, A.J.; Codogno, P. Signalling and autophagy regulation in health, aging and disease. *Mol. Asp. Med.* **2006**, *27*, 411–425. [CrossRef]
113. Klionsky, D.J.; Emr, S.D. Autophagy as a regulated pathway of cellular degradation. *Science* **2000**, *290*, 1717–1721. [CrossRef] [PubMed]
114. Zientara-Rytter, K.; Subramani, S. The Roles of Ubiquitin-Binding Protein Shuttles in the Degradative Fate of Ubiquitinated Proteins in the Ubiquitin-Proteasome System and Autophagy. *Cells* **2019**, *8*, 40. [CrossRef] [PubMed]
115. Chin, L.S.; Olzmann, J.A.; Li, L. Parkin-mediated ubiquitin signalling in aggresome formation and autophagy. *Biochem. Soc. Trans.* **2010**, *38*, 144–149. [CrossRef]
116. Cakir, Z.; Funk, K.; Lauterwasser, J.; Todt, F.; Zerbes, R.M.; Oelgeklaus, A.; Tanaka, A.; van der Laan, M.; Edlich, F. Parkin promotes proteasomal degradation of misregulated BAX. *J. Cell Sci.* **2017**, *130*, 2903–2913. [CrossRef]
117. Hook, S.S.; Orian, A.; Cowley, S.M.; Eisenman, R.N. Histone deacetylase 6 binds polyubiquitin through its zinc finger (PAZ domain) and copurifies with deubiquitinating enzymes. *Proc. Natl. Acad. Sci. USA* **2002**, *99*, 13425–13430. [CrossRef] [PubMed]
118. Olzmann, J.A.; Li, L.; Chudaev, M.V.; Chen, J.; Perez, F.A.; Palmiter, R.D.; Chin, L.S. Parkin-mediated K63-linked polyubiquitination targets misfolded DJ-1 to aggresomes via binding to HDAC6. *J. Cell Biol.* **2007**, *178*, 1025–1038. [CrossRef]
119. Ouyang, H.; Ali, Y.O.; Ravichandran, M.; Dong, A.; Qiu, W.; MacKenzie, F.; Dhe-Paganon, S.; Arrowsmith, C.H.; Zhai, R.G. Protein aggregates are recruited to aggresome by histone deacetylase 6 via unanchored ubiquitin C termini. *J. Biol. Chem.* **2012**, *287*, 2317–2327. [CrossRef] [PubMed]
120. Kawaguchi, Y.; Kovacs, J.J.; McLaurin, A.; Vance, J.M.; Ito, A.; Yao, T.P. The deacetylase HDAC6 regulates aggresome formation and cell viability in response to misfolded protein stress. *Cell.* **2003**, *115*, 727–738. [CrossRef]
121. Iwata, A.; Riley, B.E.; Johnston, J.A.; Kopito, R.R. HDAC6 and microtubules are required for autophagic degradation of aggregated huntingtin. *J. Biol. Chem.* **2005**, *280*, 40282–40292. [CrossRef]
122. Into, T.; Inomata, M.; Takayama, E.; Takigawa, T. Autophagy in regulation of Toll-like receptor signaling. *Cell Signal.* **2012**, *24*, 1150–1162. [CrossRef]

123. Lee, J.Y.; Koga, H.; Kawaguchi, Y.; Tang, W.; Wong, E.; Gao, Y.S.; Pandey, U.B.; Kaushik, S.; Tresse, E.; Lu, J.; et al. HDAC6 controls autophagosome maturation essential for ubiquitin-selective quality-control autophagy. *EMBO J.* **2010**, *29*, 969–980. [CrossRef] [PubMed]
124. Pandey, U.B.; Nie, Z.; Batlevi, Y.; McCray, B.A.; Ritson, G.P.; Nedelsky, N.B.; Schwartz, S.L.; DiProspero, N.A.; Knight, M.A.; Schuldiner, O.; et al. HDAC6 rescues neurodegeneration and provides an essential link between autophagy and the UPS. *Nature* **2007**, *447*, 859–863. [CrossRef] [PubMed]
125. Liu, W.J.; Ye, L.; Huang, W.F.; Guo, L.J.; Xu, Z.G.; Wu, H.L.; Yang, C.; Liu, H.F. p62 links the autophagy pathway and the ubiqutin-proteasome system upon ubiquitinated protein degradation. *Cell. Mol. Biol. Lett.* **2016**, *21*, 29. [CrossRef] [PubMed]
126. Bjørkøy, G.; Lamark, T.; Pankiv, S.; Øvervatn, A.; Brech, A.; Johansen, T. Monitoring autophagic degradation of p62/SQSTM1. *Methods Enzymol.* **2009**, *452*, 181–197. [CrossRef] [PubMed]
127. Livneh, I.; Cohen-Kaplan, V.; Cohen-Rosenzweig, C.; Avni, N.; Ciechanover, A. The life cycle of the 26S proteasome: from birth, through regulation and function, and onto its death. *Cell Res.* **2016**, *26*, 869–885. [CrossRef] [PubMed]
128. Otero, M.G.; Alloatti, M.; Cromberg, L.E.; Almenar-Queralt, A.; Encalada, S.E.; Pozo Devoto, V.M.; Bruno, L.; Goldstein, L.S.; Falzone, T.L. Fast axonal transport of the proteasome complex depends on membrane interaction and molecular motor function. *J. Cell. Sci.* **2014**, *127*, 1537–1549. [CrossRef] [PubMed]
129. Ryskalin, L.; Lazzeri, G.; Flaibani, M.; Biagioni, F.; Gambardella, S.; Frati, A.; Fornai, F. mTOR-Dependent Cell Proliferation in the Brain. *Biomed. Res. Int.* **2017**, *2017*, 7082696. [CrossRef] [PubMed]
130. Zhao, Y.G.; Zhang, H. ULK1 cycling: The ups and downs of the autophagy response. *J. Cell Biol.* **2016**, *215*, 757–759. [CrossRef] [PubMed]
131. Papinski, D.; Kraft, C. Regulation of autophagy by signaling through the Atg1/ULK1 complex. *J. Mol. Biol.* **2016**, *428*, 1725–1741. [CrossRef]
132. Kim, J.; Kundu, M.; Viollet, B.; Guan, K.L. AMPK and mTOR regulate autophagy through direct phosphorylation of Ulk1. *Nat. Cell Biol.* **2011**, *13*, 132–141. [CrossRef]
133. Nazio, F.; Carinci, M.; Valacca, C.; Bielli, P.; Strappazzon, F.; Antonioli, M.; Ciccosanti, F.; Rodolfo, C.; Campello, S.; Fimia, G.M.; et al. Fine-tuning of ULK1 mRNA and protein levels is required for autophagy oscillation. *J. Cell Biol.* **2016**, *215*, 841–856. [CrossRef]
134. Liu, C.C.; Lin, Y.C.; Chen, Y.H.; Chen, C.M.; Pang, L.Y.; Chen, H.A.; Wu, P.R.; Lin, M.Y.; Jiang, S.T.; Tsai, T.F.; et al. Cul3-KLHL20 ubiquitin ligase governs the turnover of ULK1 and VPS34 complexes to control autophagy termination. *Mol. Cell.* **2016**, *61*, 84–97. [CrossRef]
135. Marshall, R.S.; Vierstra, R.D. To save or degrade: balancing proteasome homeostasis to maximize cell survival. *Autophagy* **2018**, *14*, 2029–2031. [CrossRef] [PubMed]
136. Marshall, R.S.; Vierstra, R.D. Eat or be eaten: The autophagic plight of inactive 26S proteasomes. *Autophagy* **2015**, *11*, 1927–1928. [CrossRef] [PubMed]
137. Cohen-Kaplan, V.; Livneh, I.; Avni, N.; Fabre, B.; Ziv, T.; Kwon, Y.T.; Ciechanover, A. p62-and ubiquitin-dependent stress-induced autophagy of the mammalian 26S proteasome. *Proc. Natl. Acad. Sci. USA* **2016**, *113*, E7490–E7499. [CrossRef] [PubMed]
138. Hara, T.; Nakamura, K.; Matsui, M.; Yamamoto, A.; Nakahara, Y.; Suzuki-Migishima, R.; Yokoyama, M.; Mishima, K.; Saito, I.; Okano, H.; et al. Suppression of basal autophagy in neural cells causes neurodegenerative disease in mice. *Nature* **2006**, *441*, 885–889. [CrossRef]
139. Komatsu, M.; Waguri, S.; Chiba, T.; Murata, S.; Iwata, J.; Tanida, I.; Ueno, T.; Koike, M.; Uchiyama, Y.; Kominami, E.; et al. Loss of autophagy in the central nervous system causes neurodegeneration in mice. *Nature* **2006**, *441*, 880–884. [CrossRef]
140. Sato, S.; Uchihara, T.; Fukuda, T.; Noda, S.; Kondo, H.; Saiki, S.; Komatsu, M.; Uchiyama, Y.; Tanaka, K.; Hattori, N. Loss of autophagy in dopaminergic neurons causes Lewy pathology and motor dysfunction in aged mice. *Sci. Rep.* **2018**, *8*, 2813. [CrossRef]
141. Romero-Granados, R.; Fontán-Lozano, Á.; Aguilar-Montilla, F.J.; Carrión, Á.M. Postnatal proteasome inhibition induces neurodegeneration and cognitive deficiencies in adult mice: a new model of neurodevelopment syndrome. *PLoS ONE* **2011**, *6*, e28927. [CrossRef] [PubMed]
142. Fornai, F.; Lenzi, P.; Gesi, M.; Ferrucci, M.; Lazzeri, G.; Busceti, C.L.; Ruffoli, R.; Soldani, P.; Ruggieri, S.; Alessandri, M.G.; et al. Fine structure and biochemical mechanisms underlying nigrostriatal inclusions and cell death after proteasome inhibition. *J. Neurosci.* **2003**, *23*, 8955–8966. [CrossRef]

143. Fornai, F.; Lenzi, P.; Gesi, M.; Ferrucci, M.; Lazzeri, G.; Capobianco, L.; de Blasi, A.; Battaglia, G.; Nicoletti, F.; Ruggieri, S.; et al. Similarities between methamphetamine toxicity and proteasome inhibition. *Ann. N. Y. Acad. Sci.* **2004**, *1025*, 162–170. [CrossRef]
144. Korolchuk, V.I.; Mansilla, A.; Menzies, F.M.; Rubinsztein, D.C. Autophagy inhibition compromises degradation of ubiquitin-proteasome pathway substrates. *Mol. Cell* **2009**, *33*, 517–527. [CrossRef] [PubMed]
145. Yang, F.; Yang, Y.P.; Mao, C.J.; Liu, L.; Zheng, H.F.; Hu, L.F.; Liu, C.F. Crosstalk between the proteasome system and autophagy in the clearance of α-synuclein. *Acta Pharmacol. Sin.* **2013**, *34*, 674–680. [CrossRef] [PubMed]
146. Zhu, K.; Dunner, K.; McConkey, D.J. Proteasome inhibitors activate autophagy as a cytoprotective response in human prostate cancer cells. *Oncogene* **2009**, *29*, 451–462. [CrossRef] [PubMed]
147. Lan, D.; Wang, W.; Zhuang, J.; Zhao, Z. Proteasome inhibitor-induced autophagy in PC12 cells overexpressing A53T mutant α-synuclein. *Mol. Med. Rep.* **2014**, *11*, 1655–1660. [CrossRef]
148. Bao, W.; Gu, Y.; Ta, L.; Wang, K.; Xu, Z. Induction of autophagy by the MG 132 proteasome inhibitor is associated with endoplasmic reticulum stress in MCF 7 cells. *Mol. Med. Rep.* **2016**, *13*, 796–804. [CrossRef]
149. Van Kerkhof, P.; dos Santos, C.M.A.; Sachse, M.; Klumperman, J.; Bu, G.; Strous, G.J. Proteasome Inhibitors Block a Late Step in Lysosomal Transport of Selected Membrane but not Soluble Proteins. Pfeffer SR, ed. *Mol. Biol. Cell* **2001**, *12*, 2556–2566. [CrossRef] [PubMed]
150. Kleijnen, M.F.; Kirkpatrick, D.S.; Gygi, S.P. The ubiquitin–proteasome system regulates membrane fusion of yeast vacuoles. *EMBO J.* **2007**, *26*, 275–287. [CrossRef] [PubMed]
151. Villarroel-Campos, D.; Henríquez, D.R.; Bodaleo, F.J.; Oguchi, M.E.; Bronfman, F.C.; Fukuda, M.; Gonzalez-Billault, C. Rab35 Functions in Axon Elongation Are Regulated by P53-Related Protein Kinase in a Mechanism That Involves Rab35 Protein Degradation and the Microtubule-Associated Protein 1B. *J. Neurosci.* **2016**, *36*, 7298–7313. [CrossRef] [PubMed]
152. Sheehan, P.; Zhu, M.; Beskow, A.; Vollmer, C.; Waites, C.L. Activity-dependent degradation of synaptic vesicle proteins requires Rab35 and the ESCRT pathway. *J. Neurosci.* **2016**, *36*, 8668–8686. [CrossRef]
153. Shin, D.; Na, W.; Lee, J.H.; Kim, G.; Baek, J.; Park, S.H.; Choi, C.Y.; Lee, S. Site-specific monoubiquitination downregulates Rab5 by disrupting effector binding and guanine nucleotide conversion. *Elife* **2017**, *6*, e29154. [CrossRef] [PubMed]
154. Yun, Y.S.; Kim, K.H.; Tschida, B.; Chen, L.; Largaespada, D.; Kim, D-H. mTORC1 Coordinates Protein Synthesis and Immunoproteasome Formation via PRAS40 to Prevent Accumulation of Protein Stress. *Mol. Cell* **2016**, *61*, 625–639. [CrossRef]
155. Zhang, H.M.; Fu, J.; Hamilton, R.; Diaz, V.; Zhang, Y. The mammalian target of rapamycin modulates the immunoproteasome system in the heart. *J. Mol. Cell Cardiol.* **2015**, *86*, 158–167. [CrossRef] [PubMed]
156. Lelegren, M.; Liu, Y.; Ross, C.; Tardif, S.; Salmon, A.B. Pharmaceutical inhibition of mTOR in the common marmoset: effect of rapamycin on regulators of proteostasis in a non-human primate. *Pathobiol. Aging Age Relat. Dis.* **2016**, *6*, 31793. [CrossRef] [PubMed]
157. Chen, C.; Zou, L.X.; Lin, Q.Y.; Yan, X.; Bi, H.L.; Xie, X.; Wang, S.; Wang, Q.S.; Zhang, Y.L.; Li, H.H. Resveratrol as a new inhibitor of immunoproteasome prevents PTEN degradation and attenuates cardiac hypertrophy after pressure overload. *Redox Biol.* **2019**, *20*, 390–401. [CrossRef] [PubMed]
158. Krüger, E.; Kloetzel, P.M. Immunoproteasomes at the interface of innate and adaptive immune responses: two faces of one enzyme. *Curr. Opin. Immunol.* **2012**, *24*, 77–83. [CrossRef] [PubMed]
159. Dodson, M.; Redmann, M.; Rajasekaran, N.S.; Darley-Usmar, V.; Zhang, J. KEAP1-NRF2 signalling and autophagy in protection against oxidative and reductive proteotoxicity. *Biochem. J.* **2015**, *469*, 347–355. [CrossRef]
160. Simon, H.U.; Friis, R.; Tait, S.W.; Ryan, K.M. Retrograde signaling from autophagy modulates stress responses. *Sci. Signal.* **2017**, *10*, eaag2791. [CrossRef] [PubMed]
161. Hernandez, D.; Torres, C.A.; Setlik, W.; Cebrián, C.; Mosharov, E.V.; Tang, G.; Cheng, H.C.; Kholodilov, N.; Yarygina, O.; Burke, R.E.; et al. Regulation of presynaptic neurotransmission by macroautophagy. *Neuron* **2012**, *74*, 277–284. [CrossRef]
162. Torres, C.A.; Sulzer, D. Macroautophagy can press a brake on presynaptic neurotransmission. *Autophagy* **2012**, *8*, 1540–1541. [CrossRef] [PubMed]
163. Ding, M.; Shen, K. The role of the ubiquitin proteasome system in synapse remodeling and neurodegenerative diseases. *Bioessays* **2008**, *30*, 1075–1083. [CrossRef]

164. Bingol, B.; Sheng, M. Deconstruction for reconstruction: the role of proteolysis in neural plasticity and disease. *Neuron* **2011**, *69*, 22–32. [CrossRef]
165. Durairaj, G.; Kaiser, P. The 26S proteasome and initiation of gene transcription. *Biomolecules* **2014**, *4*, 827–847. [CrossRef] [PubMed]
166. Zhao, Y.; Hegde, A.N.; Martin, K.C. The ubiquitin proteasome system functions as an inhibitory constraint on synaptic strengthening. *Curr. Biol.* **2003**, *13*, 887–898. [CrossRef]
167. Tang, S.J.; Reis, G.; Kang, H.; Gingras, A.-C.; Sonenberg, N.; Schuman, E.M. A rapamycin-sensitive signaling pathway contributes to long-term synaptic plasticity in the hippocampus. *Proc. Natl. Acad. Sci. USA* **2002**, *99*, 467–472. [CrossRef]
168. Upadhya, S.C.; Smith, T.K.; Hegde, A.N. Ubiquitin-proteasome-mediated CREB repressor degradation during induction of long-term facilitation. *J. Neurochem.* **2004**, *91*, 210–219. [CrossRef] [PubMed]
169. Rinetti, G.V.; Schweizer, F.E. Ubiquitination acutely regulates presynaptic neurotransmitter release in mammalian neurons. *J. Neurosci.* **2010**, *30*, 3157–3166. [CrossRef] [PubMed]
170. Dong, C.; Bach, S.V.; Haynes, K.A.; Hegde, A.N. Proteasome Modulates Positive and Negative Translational Regulators in Long-Term Synaptic Plasticity. *J. Neurosci.* **2014**, *34*, 3171–3182. [CrossRef] [PubMed]
171. Aravamudan, B.; Broadie, K. Synaptic Drosophila UNC-13 is regulated by antagonistic G-protein pathways via a proteasome-dependent degradation mechanism. *J. Neurobiol.* **2003**, *54*, 417–438. [CrossRef] [PubMed]
172. Wentzel, C.; Delvendahl, I.; Sydlik, S.; Georgiev, O.; Müller, M. Dysbindin links presynaptic proteasome function to homeostatic recruitment of low release probability vesicles. *Nat. Commun.* **2018**, *9*, 267. [CrossRef] [PubMed]
173. Willeumier, K.; Pulst, S.M.; Schweizer, F.E. Proteasome Inhibition Triggers Activity-Dependent Increase in the Size of the Recycling Vesicle Pool in Cultured Hippocampal Neurons. *J. Neurosci.* **2006**, *26*, 11333–11341. [CrossRef]
174. Singh, A.K.; Kashyap, M.P.; Tripathi, V.K.; Singh, S.; Garg, G.; Rizvi, S.I. Neuroprotection Through Rapamycin-Induced Activation of Autophagy and PI3K/Akt1/mTOR/CREB Signaling Against Amyloid-β-Induced Oxidative Stress, Synaptic/Neurotransmission Dysfunction, and Neurodegeneration in Adult Rats. *Mol. Neurobiol.* **2017**, *54*, 5815–5828. [CrossRef] [PubMed]
175. Sunkaria, A.; Yadav, A.; Bhardwaj, S.; Sandhir, R. Postnatal Proteasome Inhibition Promotes Amyloid-β Aggregation in Hippocampus and Impairs Spatial Learning in Adult Mice. *Neuroscience* **2017**, *367*, 47–59. [CrossRef]
176. Laurén, J.; Gimbel, D.A.; Nygaard, H.B.; Gilbert, J.W.; Strittmatter, S.M. Cellular prion protein mediates impairment of synaptic plasticity by amyloid-beta oligomers. *Nature* **2009**, *457*, 1128–1132. [CrossRef] [PubMed]
177. Ghiglieri, V.; Calabrese, V.; Calabresi, P. Alpha-Synuclein: From Early Synaptic Dysfunction to Neurodegeneration. *Front. Neurol.* **2018**, *9*, 295. [CrossRef] [PubMed]
178. Phan, J.A.; Stokholm, K.; Zareba-Paslawska, J.; Jakobsen, S.; Vang, K.; Gjedde, A.; Landau, A.M.; Romero-Ramos, M. Early synaptic dysfunction induced by α-synuclein in a rat model of Parkinson's disease. *Sci. Rep.* **2017**, *7*, 6363. [CrossRef] [PubMed]
179. Li, K.; Wei, Q.; Liu, F.F.; Hu, F.; Xie, A.J.; Zhu, L.Q.; Liu, D. Synaptic Dysfunction in Alzheimer's Disease: Aβ, Tau, and Epigenetic Alterations. *Mol. Neurobiol.* **2018**, *55*, 3021–3032. [CrossRef]
180. Chin, L.S.; Vavalle, J.P.; Li, L. Staring, a novel E3 ubiquitin-protein ligase that targets syntaxin 1 for degradation. *J. Biol. Chem.* **2002**, *277*, 35071–35079. [CrossRef]
181. Subramaniam, M.; Kern, B.; Vogel, S.; Klose, V.; Schneider, G.; Roeper, J. Selective increase of in vivo firing frequencies in DA SN neurons after proteasome inhibition in the ventral midbrain. *Eur. J. Neurosci.* **2014**, *40*, 2898–2909. [CrossRef] [PubMed]
182. Konieczny, J.; Lenda, T.; Czarnecka, A. Early increase in dopamine release in the ipsilateral striatum after unilateral intranigral administration of lactacystin produces spontaneous contralateral rotations in rats. *Neuroscience* **2016**, *324*, 92–106. [CrossRef]
183. Lillethorup, T.P.; Glud, A.N.; Alstrup, A.K.O.; Mikkelsen, T.W.; Nielsen, E.H.; Zaer, H.; Doudet, D.J.; Brooks, D.J.; Sørensen, J.C.H.; Orlowski, D.; et al. Nigrostriatal proteasome inhibition impairs dopamine neurotransmission and motor function in minipigs. *Exp. Neurol.* **2018**, *303*, 142–152. [CrossRef] [PubMed]
184. Kim, O.J. A single mutation at lysine 241 alters expression and trafficking of the D2 dopamine receptor. *J. Recept. Signal Transduct. Res.* **2008**, *28*, 453–464. [CrossRef] [PubMed]

185. Rondou, P.; Haegeman, G.; Vanhoenacker, P.; Van Craenenbroeck, K. BTB protein KLHL12 targets the dopamine D4 receptor for ubiquitination by a Cul3-based E3 ligase. *J. Biol. Chem.* **2008**, *283*, 11083–11096. [CrossRef]
186. Alonso, V.; Friedman, P.A. Minireview: ubiquitination-regulated G protein-coupled receptor signaling and trafficking. *Mol. Endocrinol.* **2013**, *27*, 558–572. [CrossRef] [PubMed]
187. Peeler, J.C.; Schedin-Weiss, S.; Soula, M.; Kazmi, M.A.; Sakmar, T.P. Isopeptide and ester bond ubiquitination both regulate degradation of the human dopamine receptor 4. *J. Biol. Chem.* **2017**, *292*, 21623–21630. [CrossRef]
188. Milnerwood, A.J.; Raymond, L.A. Early synaptic pathophysiology in neurodegeneration: insights from Huntington's disease. *Trends Neurosci.* **2010**, *33*, 513–523. [CrossRef]
189. Schirinzi, T.; Madeo, G.; Martella, G.; Maltese, M.; Picconi, B.; Calabresi, P.; Pisani, A. Early synaptic dysfunction in Parkinson's disease: Insights from animal models. *Mov. Disord.* **2016**, *31*, 802–813. [CrossRef]
190. Krasnova, I.N.; Justinova, Z.; Cadet, J.L. Methamphetamine addiction: involvement of CREB and neuroinflammatory signaling pathways. *Psychopharmacology (Berl)* **2016**, *233*, 1945–1962. [CrossRef] [PubMed]
191. Krasnova, I.N.; Cadet, J.L. Methamphetamine toxicity and messengers of death. *Brain Res. Rev.* **2009**, *60*, 379–407. [CrossRef] [PubMed]
192. Moratalla, R.; Khairnar, A.; Simola, N.; Granado, N.; García-Montes, J.R.; Porceddu, P.F.; Tizabi, Y.; Costa, G.; Morelli, M. Amphetamine-related drugs neurotoxicity in humans and in experimental animals: Main mechanisms. *Prog. Neurobiol.* **2017**, *155*, 149–170. [CrossRef] [PubMed]
193. Tata, D.A.; Yamamoto, B.K. Interactions between methamphetamine and environmental stress: role of oxidative stress, glutamate and mitochondrial dysfunction. *Addiction* **2007**, *102*, 49–60. [CrossRef] [PubMed]
194. Huang, S.H.; Wu, W.R.; Lee, L.M.; Huang, P.R.; Chen, J.C. mTOR signaling in the nucleus accumbens mediates behavioral sensitization to methamphetamine. *Prog. Neuropsychopharmacol. Biol. Psychiatry* **2018**, *86*, 331–339. [CrossRef]
195. Bentea, E.; Verbruggen, L.; Massie, A. The Proteasome Inhibition Model of Parkinson's Disease. *J. Parkinsons Dis.* **2017**, *7*, 31–63. [CrossRef]
196. Keller, J.N.; Hanni, K.B.; Marksberry, W.R. Impaired proteasome function in Alzheimer's disease. *J. Neurochem.* **2000**, *75*, 436–439. [CrossRef]
197. Seo, H.; Sonntag, K.C.; Isacson, O. Generalized brain and skin proteasome inhibition in Huntington's disease. *Ann. Neurol.* **2004**, *56*, 319–328. [CrossRef]
198. Rubio, M.D.; Wood, K.; Haroutunian, V.; Meador-Woodruff, J.H. Dysfunction of the ubiquitin proteasome and ubiquitin-like systems in schizophrenia. *Neuropsychopharmacology* **2013**, *38*, 1910–1920. [CrossRef]
199. Siman, R.; Cocca, R.; Dong, Y. The mTOR Inhibitor Rapamycin Mitigates Perforant Pathway Neurodegeneration and Synapse Loss in a Mouse Model of Early-Stage Alzheimer-Type Tauopathy. *PLoS ONE* **2015**, *10*, e0142340. [CrossRef]
200. Schneider, M.; de Vries, P.J.; Schönig, K.; Rößner, V.; Waltereit, R. mTOR inhibitor reverses autistic-like social deficit behaviours in adult rats with both Tsc2 haploinsufficiency and developmental status epilepticus. *Eur. Arch. Psychiatry Clin. Neurosci.* **2017**, *267*, 455–463. [CrossRef]
201. Masini, D.; Bonito-Oliva, A.; Bertho, M.; Fisone, G. Inhibition of mTORC1 Signaling Reverts Cognitive and Affective Deficits in a Mouse Model of Parkinson's Disease. *Front. Neurol.* **2018**, *9*, 208. [CrossRef] [PubMed]
202. Kara, N.Z.; Flaisher-Grinberg, S.; Anderson, G.W.; Agam, G.; Einat, H. Mood-stabilizing effects of rapamycin and its analog temsirolimus: relevance to autophagy. *Behav. Pharmacol.* **2018**, *29*, 379–384. [CrossRef]
203. Sillitoe, R.V.; Vogel, M.W. Desire, disease, and the origins of the dopaminergic system. *Schizophr. Bull.* **2008**, *34*, 212–219. [CrossRef] [PubMed]
204. Levite, M.; Chowers, Y.; Ganor, Y.; Besser, M.; Hershkovits, R.; Cahalon, L. Dopamine interacts directly with its D3 and D2 receptors on normal human T-cells, and activates beta1 integrin function. *Eur. J. Immunol.* **2001**, *31*, 3504–3512. [CrossRef]
205. Fiserová, A.; Starec, M.; Kuldová, M.; Kovárů, H.; Páv, M.; Vannucci, L.; Pospísil, M. Effects of D2-dopamine and alpha-adrenoceptor antagonists in stress induced changes on immune responsiveness of mice. *J. Neuroimmunol.* **2002**, *130*, 55–65.

206. Buttarelli, F.R.; Fanciulli, A.; Pellicano, C.; Pontieri, F.E. The dopaminergic system in peripheral blood lymphocytes: from physiology to pharmacology and potential applications to neuropsychiatric disorders. *Curr. Neuropharmacol.* **2011**, *9*, 278–288. [PubMed]
207. Tanabe, S.; Yamashita, T. The role of immune cells in brain development and neurodevelopmental diseases. *Int. Immunol.* **2018**, *30*, 437–444. [CrossRef]
208. Smolders, J.; Heutinck, K.M.; Fransen, N.L.; Remmerswaal, E.B.M.; Hombrink, P.; Ten Berge, I.J.M.; van Lier, R.A.W.; Huitinga, I.; Hamann, J. Tissue-resident memory T-cells populate the human brain. *Nat. Commun.* **2018**, *9*, 4593. [CrossRef]
209. Zarif, H.; Hosseiny, S.; Paquet, A.; Lebrigand, K.; Arguel, M.J.; Cazareth, J.; Lazzari, A.; Heurteaux, C.; Glaichenhaus, N.; Chabry, J.; et al. CD4(+) T Cells Have a Permissive Effect on Enriched Environment-Induced Hippocampus Synaptic Plasticity. *Front. Synaptic Neurosci.* **2018**, *10*, 14. [CrossRef]
210. Zarif, H.; Nicolas, S.; Guyot, M.; Hosseiny, S.; Lazzari, A.; Canali, M.M.; Cazareth, J.; Brau, F.; Golzné, V.; Dourneau, E.; et al. CD8(+) T cells are essential for the effects of enriched environment on hippocampus-dependent behavior, hippocampal neurogenesis and synaptic plasticity. *Brain Behav. Immun.* **2018**, *69*, 235–254. [CrossRef]
211. Haas, K.F.; Miller, S.L.; Friedman, D.B.; Broadie, K. The ubiquitin-proteasome system postsynaptically regulates glutamatergic synaptic function. *Mol. Cell. Neurosci.* **2007**, *35*, 64–75. [CrossRef] [PubMed]
212. Nikoletopoulou, V.; Tavernarakis, N. Regulation and Roles of Autophagy at Synapses. *Trends Cell. Biol.* **2018**, *28*, 646–661. [CrossRef]
213. Hogins, J.; Crawford, D.C.; Jiang, X.; Mennerick, S. Presynaptic silencing is an endogenous neuroprotectant during excitotoxic insults. *Neurobiol. Dis.* **2011**, *43*, 516–525. [CrossRef] [PubMed]
214. Schori, H.; Yoles, E.; Schwartz, M. T-cell-based immunity counteracts the potential toxicity of glutamate in the central nervous system. *J. Neuroimmunol.* **2001**, *119*, 199–204. [CrossRef]
215. Lombardi, G.; Miglio, G.; Canonico, P.L.; Naldi, P.; Comi, C.; Monaco, F. Abnormal response to glutamate of T lymphocytes from multiple sclerosis patients. *Neurosci. Lett.* **2003**, *340*, 5–8. [CrossRef]
216. Sommer, A.; Winner, B.; Prots, I. The Trojan horse - neuroinflammatory impact of T-cells in neurodegenerative diseases. *Mol. Neurodegener.* **2017**, *12*, 78. [CrossRef] [PubMed]
217. Levite, M.; Marino, F.; Cosentino, M. Dopamine, T-cells and multiple sclerosis (MS). *J. Neural. Transm (Vienna)* **2017**, *124*, 525–542. [CrossRef]
218. Saunders, J.A.; Estes, K.A.; Kosloski, L.M.; Allen, H.E.; Dempsey, K.M.; Torres-Russotto, D.R.; Meza, J.L.; Santamaria, P.M.; Bertoni, J.M.; Murman, D.L.; et al. CD4+ regulatory and effector/memory T cell subsets profile motor dysfunction in Parkinson's disease. *J. Neuroimmune Pharmacol.* **2012**, *7*, 927–938. [CrossRef]
219. González, H.; Contreras, F.; Prado, C.; Elgueta, D.; Franz, D.; Bernales, S.; Pacheco, R. Dopamine receptor D3 expressed on CD4+ T-cells favors neurodegeneration of dopaminergic neurons during Parkinson's disease. *J. Immunol.* **2013**, *190*, 5048–5056. [CrossRef]
220. Talhada, D.; Rabenstein, M.; Ruscher, K. The role of dopaminergic immune cell signalling in poststroke inflammation. *Ther. Adv. Neurol. Disord.* **2018**, *11*, 1756286418774225. [CrossRef]
221. Lee, M. Neurotransmitters and microglial-mediated neuroinflammation. *Curr. Protein Pept. Sci.* **2013**, *14*, 21–32. [CrossRef] [PubMed]
222. Mastroeni, D.; Grover, A.; Leonard, B.; Joyce, J.N.; Coleman, P.D.; Kozik, B.; Bellinger, D.L.; Rogers, J. Microglial responses to dopamine in a cell culture model of Parkinson's disease. *Neurobiol. Aging* **2008**, *30*, 1805–1817. [CrossRef] [PubMed]
223. Fornai, F.; Carrizzo, A.; Ferrucci, M.; Damato, A.; Biagioni, F.; Gaglione, A.; Puca, A.A.; Vecchione, C. Brain diseases and tumorigenesis: The good and bad cops of pentraxin3. *Int. J. Biochem. Cell Biol.* **2015**, *69*, 70–74. [CrossRef] [PubMed]
224. Sun, S.C.; Chang, J.H.; Jin, J. Regulation of nuclear factor-κB in autoimmunity. *Trends Immunol.* **2013**, *34*, 282–289. [CrossRef] [PubMed]
225. Xu, H.; You, M.; Shi, H.; Hou, Y. Ubiquitin-mediated NFκB degradation pathway. *Cell. Mol. Immunol.* **2015**, *12*, 653–655. [CrossRef] [PubMed]
226. Arbogast, F.; Gros, F. Lymphocyte Autophagy in Homeostasis, Activation, and Inflammatory Diseases. *Front. Immunol.* **2018**, *9*, 1801. Erratum in: *Front. Immunol.* **2018**, *9*, 2627. [CrossRef]
227. Bronietzki, A.W.; Schuster, M.; Schmitz, I. Autophagy in T-cell development, activation and differentiation. *Immunol. Cell Biol.* **2015**, *93*, 25–34. [CrossRef] [PubMed]

228. Gaczynska, M.; Rock, K.L.; Spies, T.; Goldberg, A.L. Peptidase activities of proteasomes are differentially regulated by the major histocompatibility complex-encoded genes for LMP2 and LMP7. *Proc. Natl. Acad. Sci. USA* **1994**, *91*, 9213–9217. [CrossRef]
229. Huber, E.M.; Basler, M.; Schwab, R.; Heinemeyer, W.; Kirk, C.J.; Groettrup, M. Immuno- and constitutive proteasome crystal structures reveal differences in substrate and inhibitor specificity. *Cell* **2012**, *148*, 727–738. [CrossRef] [PubMed]
230. Driscoll, J.; Brown, M.G.; Finley, D.; Monaco, J.J. MHC-linked LMP gene products specifically alter peptidase activities of the proteasome. *Nature* **1993**, *365*, 262–264. [CrossRef]
231. Chapiro, J.; Claverol, S.; Piette, F.; Ma, W.; Stroobant, V.; Guillaume, B.; Gairin, J.E.; Morel, S.; Burlet-Schiltz, O.; Monsarrat, B.; et al. Destructive cleavage of antigenic peptides either by the immunoproteasome or by the standard proteasome results in differential antigen presentation. *J. Immunol.* **2006**, *176*, 1053–1061. [CrossRef] [PubMed]
232. Lei, B.; Abdul Hameed, M.D.; Hamza, A.; Wehenkel, M.; Muzyka, J.L.; Yao, X.J.; Kim, K.B.; Zhan, C.G. Molecular basis of the selectivity of the immunoproteasome catalytic subunit LMP2-specific inhibitor revealed by molecular modeling and dynamics simulations. *J. Phys. Chem. B.* **2010**, *114*, 12333–12339. [CrossRef]
233. Loi, M.; Müller, A.; Steinbach, K.; Niven, J.; Barreira da Silva, R.; Paul, P.; Ligeon, L.A.; Caruso, A.; Albrecht, R.A.; Becker, A.C.; et al. Macroautophagy Proteins Control MHC Class I Levels on Dendritic Cells and Shape Anti-viral CD8(+) T Cell Responses. *Cell Rep.* **2016**, *15*, 1076–1087. [CrossRef]
234. Loi, M.; Ligeon, L.A.; Münz, C. MHC Class I Internalization via Autophagy Proteins. *Methods Mol. Biol.* **2019**, *1880*, 455–477. [CrossRef] [PubMed]
235. Loi, M.; Gannagé, M.; Münz, C. ATGs help MHC class II, but inhibit MHC class I antigen presentation. *Autophagy* **2016**, *12*, 1681–1682. [CrossRef] [PubMed]
236. Tey, S.K.; Khanna, R. Autophagy mediates transporter associated with antigen processing-independent presentation of viral epitopes through MHC class I pathway. *Blood* **2012**, *120*, 994–1004. [CrossRef]
237. Nil, A.; Firat, E.; Sobek, V.; Eichmann, K.; Niedermann, G. Expression of housekeeping and immunoproteasome subunit genes is differentially regulated in positively and negatively selecting thymic stroma subsets. *Eur. J. Immunol.* **2004**, *34*, 2681–2689. [CrossRef]
238. Ugras, S.; Daniels, M.J.; Fazelinia, H.; Gould, N.S.; Yocum, A.K.; Luk, K.C.; Luna, E.; Ding, H.; McKennan, C.; Seeholzer, S.; et al. Induction of the immunoproteasome subunit Lmp7 links proteostasis and immunity in α-synuclein aggregation disorders. *EBioMedicine* **2018**, *31*, 307–319. [CrossRef]
239. Moritz, K.E.; McCormack, N.M.; Abera, M.B.; Viollet, C.; Yauger, Y.J.; Sukumar, G.; Dalgard, C.L.; Burnett, B.G. The role of the immunoproteasome in interferon-γ-mediated microglial activation. *Sci. Rep.* **2017**, *7*, 9365. [CrossRef]
240. Seifert, U.; Bialy, L.P.; Ebstein, F.; Bech-Otschir, D.; Voigt, A.; Schröter, F.; Prozorovski, T.; Lange, N.; Steffen, J.; Rieger, M.; et al. Immunoproteasomes preserve protein homeostasis upon interferon-induced oxidative stress. *Cell* **2010**, *142*, 613–624. [CrossRef]
241. Nathan, J.A.; Spinnenhirn, V.; Schmidtke, G.; Basler, M.; Groettrup, M.; Goldberg, A.L. Immuno- and constitutive proteasomes do not differ in their abilities to degrade ubiquitinated proteins. *Cell* **2013**, *152*, 1184–1194. [CrossRef]
242. Piccinini, M.; Mostert, M.; Croce, S.; Baldovino, S.; Papotti, M.; Rinaudo, M.T. Interferon-gamma-inducible subunits are incorporated in human brain 20S proteasome. *J. Neuroimmunol.* **2003**, *135*, 135–140. [CrossRef]
243. Cebrián, C.; Loike, J.D.; Sulzer, D. Neuronal MHC-I expression and its implications in synaptic function, axonal regeneration and Parkinson's and other brain diseases. *Front. Neuroanat.* **2014**, *8*. [CrossRef] [PubMed]
244. Cullheim, S.; Thams, S. Classic major histocompatibility complex class I molecules: new actors at the neuromuscular junction. *Neuroscientist.* **2010**, *16*, 600–607. [CrossRef] [PubMed]
245. Lazarczyk, M.J.; Kemmler, J.E.; Eyford, B.A.; Short, J.A.; Varghese, M.; Sowa, A.; Dickstein, D.R.; Yuk, F.J.; Puri, R.; Biron, K.E.; et al. Major Histocompatibility Complex class I proteins are critical for maintaining neuronal structural complexity in the aging brain. *Sci. Rep* **2016**, *6*, 26199. [CrossRef] [PubMed]
246. Edamura, M.; Murakami, G.; Meng, H.; Itakura, M.; Shigemoto, R.; Fukuda, A.; Nakahara, D. Functional deficiency of MHC class I enhances LTP and abolishes LTD in the nucleus accumbens of mice. *PLoS ONE* **2014**, *9*, e107099. [CrossRef]

247. Heink, S.; Ludwig, D.; Kloetzel, P.M.; Krüger, E. IFN-gamma-induced immune adaptation of the proteasome system is an accelerated and transient response. *Proc. Natl. Acad. Sci. USA* **2005**, *102*, 9241–9246. [CrossRef]
248. Mo, M.S.; Li, G.H.; Sun, C.C.; Huang, S.X.; Wei, L.; Zhang, L.M.; Zhou, M.M.; Wu, Z.H.; Guo, W.Y.; Yang, X.L.; et al. Dopaminergic neurons show increased low-molecular-mass protein 7 activity induced by 6-hydroxydopamine in vitro and in vivo. *Transl. Neurodegener.* **2018**, *7*, 19. [CrossRef]
249. Basler, M.; Mundt, S.; Muchamuel, T.; Moll, C.; Jiang, J.; Groettrup, M.; Kirk, C.J. Inhibition of the immunoproteasome ameliorates experimental autoimmune encephalomyelitis. *EMBO Mol. Med.* **2014**, *6*, 226–238. [CrossRef]
250. Fissolo, N.; Kraus, M.; Reich, M.; Ayturan, M.; Overkleeft, H.; Driessen, C.; Weissert, R. Dual inhibition of proteasomal and lysosomal proteolysis ameliorates autoimmune central nervous system inflammation. *Eur. J. Immunol.* **2008**, *38*, 2401–2411. [CrossRef]
251. Díaz-Hernández, M.; Hernández, F.; Martín-Aparicio, E.; Gómez-Ramos, P.; Morán, M.A.; Castaño, J.G.; Ferrer, I.; Avila, J.; Lucas, J.J. Neuronal induction of the immunoproteasome in Huntington's disease. *J. Neurosci.* **2003**, *23*, 11653–11661. [CrossRef] [PubMed]
252. Schmeisser, H.; Bekisz, J.; Zoon, K.C. New function of type I IFN: induction of autophagy. *J. Interferon Cytokine Res.* **2014**, *34*, 71–78. [CrossRef] [PubMed]
253. Mocholi, E.; Dowling, S.D.; Botbol, Y.; Gruber, R.C.; Ray, A.K.; Vastert, S.; Shafit-Zagardo, B.; Coffer, P.J.; Macian, F. Autophagy Is a Tolerance-Avoidance Mechanism that Modulates TCR-Mediated Signaling and Cell Metabolism to Prevent Induction of T Cell Anergy. *Cell Rep.* **2018**, *24*, 1136–1150. [CrossRef] [PubMed]
254. Paunovic, V.; Petrovic, I.V.; Milenkovic, M.; Janjetovic, K.; Pravica, V.; Dujmovic, I.; Milosevic, E.; Martinovic, V.; Mesaros, S.; Drulovic, J.; et al. Autophagy-independent increase of ATG5 expression in T cells of multiplesclerosis patients. *J. Neuroimmunol.* **2018**, *319*, 100–105. [CrossRef] [PubMed]
255. Xu, L.; Zhang, C.; Jiang, N.; He, D.; Bai, Y.; Xin, Y. Rapamycin combined with MCC950 to treat multiple sclerosis in experimental autoimmune encephalomyelitis. *J. Cell Biochem.* **2019**, *120*, 5160–5168. [CrossRef]
256. Su, P.; Zhang, J.; Wang, D.; Zhao, F.; Cao, Z.; Aschner, M.; Luo, W. The role of autophagy in modulation of neuroinflammation in microglia. *Neuroscience* **2016**, *319*, 155–167. [CrossRef] [PubMed]
257. Jin, M.M.; Wang, F.; Qi, D.; Liu, W.W.; Gu, C.; Mao, C.J.; Yang, Y.P.; Zhao, Z.; Hu, L.F.; Liu, C.F. A Critical Role of Autophagy in Regulating Microglia Polarization in Neurodegeneration. *Front. Aging Neurosci.* **2018**, *10*, 378. [CrossRef]
258. Cho, M.H.; Cho, K.; Kang, H.J.; Jeon, E.Y.; Kim, H.S.; Kwon, H.J.; Kim, H.M.; Kim, D.H.; Yoon, S.Y. Autophagy in microglia degrades extracellular β-amyloid fibrils and regulates theNLRP3 inflammasome. *Autophagy* **2014**, *10*, 1761–1775. [CrossRef]
259. Berthet, A.; Bezard, E.; Porras, G.; Fasano, S.; Barroso-Chinea, P.; Dehay, B.; Martinez, A.; Thiolat, M.L.; Nosten-Bertrand, M.; Giros, B.; et al. L-DOPA impairs proteasome activity in parkinsonism through D1 dopamine receptor. *J. Neurosci.* **2012**, *32*, 681–691. [CrossRef]
260. Lazzeri, G.; Lenzi, P.; Busceti, C.L.; Ferrucci, M.; Falleni, A.; Bruno, V.; Paparelli, A.; Fornai, F. Mechanisms involved in the formation of dopamine-induced intracellular bodies within striatal neurons. *J. Neurochem.* **2007**, *101*, 1414–1427. [CrossRef]

MDPI
St. Alban-Anlage 66
4052 Basel
Switzerland
www.mdpi.com

International Journal of Molecular Sciences Editorial Office
E-mail: ijms@mdpi.com
www.mdpi.com/journal/ijms

www.ingramcontent.com/pod-product-compliance
Lightning Source LLC
Chambersburg PA
CBHW061820210926
51850CB00006B/756
* 9 7 8 3 0 3 6 5 8 7 1 5 8 *